D0093213

TELEVISION AND AUDIO HANDBOOK

Other McGraw-Hill Reference Books of Interest

Handbooks

Avalone and Baumeister • STANDARD HANDBOOK FOR MECHANICAL ENGINEERS
Beeman • INDUSTRIAL POWER SYSTEMS HANDBOOK
Benson • AUDIO ENGINEERING HANDBOOK
Benson • TELEVISION ENGINEERING HANDBOOK
Coombs • BASIC ELECTRONIC INSTRUMENT HANDBOOK
Coombs • PRINTED CIRCUITS HANDBOOK
Croft and Summers • AMERICAN ELECTRICIANS' HANDBOOK
Di Giacomo • VLSI HANDBOOK
Fink and Beaty • STANDARD HANDBOOK FOR ELECTRICAL ENGINEERS
Fink and Christiansen • ELECTRONICS ENGINEERS' HANDBOOK
Harper • HANDBOOK OF ELECTRONIC SYSTEMS DESIGN
Harper • HANDBOOK OF THICK FILM HYBRID MICROELECTRONICS
Hicks • STANDARD HANDBOOK OF ENGINEERING CALCULATIONS
Inglis • ELECTRONIC COMMUNICATIONS HANDBOOK
Johnson and Jasik • ANTENNA ENGINEERING HANDBOOK
Juran • QUALITY CONTROL HANDBOOK
Kaufman and Seidman • HANDBOOK FOR ELECTRONICS ENGINEERING TECHNICIANS
Kaufman and Seidman • HANDBOOK OF ELECTRONICS CALCULATIONS
Kurtz • HANDBOOK OF ENGINEERING ECONOMICS
Mee and Daniel • MAGNETIC RECORDING HANDBOOK
Sherman • CD-ROM HANDBOOK
Stout • HANDBOOK OF MICROPROCESSOR DESIGN AND APPLICATIONS
Stout and Kaufman • HANDBOOK OF MICROCIRCUIT DESIGN AND APPLICATION
Stout and Kaufman • HANDBOOK OF OPERATIONAL AMPLIFIER DESIGN
Tuma • ENGINEERING MATHEMATICS HANDBOOK
Williams • DESIGNER'S HANDBOOK OF INTEGRATED CIRCUITS
Williams and Taylor • ELECTRONIC FILTER DESIGN HANDBOOK

Consumer Electronics

Luther • DIGITAL VIDEO IN THE PC ENVIRONMENT
Mee and Daniel • MAGNETIC RECORDING, VOLUMES I–III
Philips International • COMPACT DISC—INTERACTIVE

Dictionaries

DICTIONARY OF COMPUTERS
DICTIONARY OF ELECTRICAL AND ELECTRONIC ENGINEERING
DICTIONARY OF ENGINEERING
DICTIONARY OF SCIENTIFIC AND TECHNICAL TERMS
Markus • ELECTRONICS DICTIONARY

TELEVISION AND AUDIO HANDBOOK

FOR TECHNICIANS AND ENGINEERS

K. Blair Benson   Editor and Coauthor

Television Technology Consultant
Fellow and Life Member, Society of Motion Picture and Television Engineers
Senior and Life Member, Institute of Electrical and Electronic Engineers
Member, Audio Engineering Society

Jerry C. Whitaker   Coauthor

Editorial Director, Broadcast Engineering and Video Systems
Fellow and Member, Society of Broadcast Engineers
Member, Audio Engineering Society
Member, International Television Association
Member, Institute of Electrical and Electronic Engineers
Member, Society of Motion Picture and Television Engineers

McGRAW-HILL PUBLISHING COMPANY

New York   St. Louis   San Francisco   Auckland   Bogotá
Caracas   Hamburg   Lisbon   London   Madrid   Mexico
Milan   Montreal   New Delhi   Oklahoma City
Paris   San Juan   São Paulo   Singapore
Sydney   Tokyo   Toronto

Library of Congress Cataloging-in-Publication Data

Benson, K. Blair.
   Television and audio handbook: for technicians and engineers/K.
  Blair Benson, Jerry C. Whitaker.
     p.   cm.
   1. Television—Handbooks, manuals, etc. 2. Sound—Recording and
reproducing—Handbooks, manuals, etc. 3. Sound—Recording and
reproducing—Digital techniques—Handbooks, manuals, etc.
I. Whitaker, Jerry C. II. Title.
TK6642.B45   1990
791.45—dc20   89-12861
                                                  CIP

1234567890 DOC/DOC 89432109

ISBN 0-07-004787-1

The editors for this book were Daniel A. Gonneau and Marci Nugent,
and the production supervisor was Richard Ausburn. This book was set in
Times Roman. It was composed by the McGraw-Hill Publishing Company
Professional & Reference Division composition unit.

Printed and bound by R. R. Donnelley and Sons Company.

*For more information about other McGraw-Hill materials,
call 1-800-2-MCGRAW in the United States. In other
countries, call your nearest McGraw-Hill office.*

CONTENTS

Chapter 8. Television Transmission   *Jerry Whitaker and Carl Bentz*   **8.1**

Chapter 9. Television Reception   *K. Blair Benson*   **9.1**

Chapter 10. Audio Equipment: Studio and Recording   *Jerry Whitaker*
and E. Stanley Busby, Jr.   **10.1**

Chapter 11. AM and FM Transmission Systems   *Jerry Whitaker*   **11.1**

Chapter 17. Reference Data and Tables   *K. Blair Benson*   17.1

CONTRIBUTORS

K. Blair Benson, Editor-in-Chief and Coauthor   *Television Technology Consultant, Norwalk, Connecticut* (CHAPTERS *1, 2, 3, 6, 7, 9, 14, 17*)

Carl A. Bentz   *Intertec Publishing Corporation, Overland Park, Kansas* (CHAPTERS *8, 10, 11, 12, 13, 16*)

Michael Betts   *The Grass Valley Group, Grass Valley California* (CHAPTER 6)

James E. Blecksmith   *The Grass Valley Group, Grass Valley California* (CHAPTER 6)

E. Stanley Busby, Jr.   *Ampex Corporation (retired), Redwood City, California* (CHAPTERS *4, 5, 7, 10, 15*)

Dr. Richard C. Cabot, P.E.   *Audio Precision, Inc., Beaverton, Oregon*

L. H. Hoke, Jr.   *Philips Consumer Electronics Company, Knoxville, Tennessee* (CHAPTER 9)

Robert Jull   *The Grass Valley Group, Grass Valley, California* (CHAPTER 6)

Charles J. Kuca   *The Grass Valley Group, Grass Valley, California* (CHAPTER 6)

James Michener   *The Grass Valley Group, Grass Valley, California* (CHAPTER 6)

Kentaro Odaka   *Sony Corporation, Tokyo, Japan* (CHAPTER 14)

Hiroshi Ogawa   *Sony Corporation, Tokyo, Japan* (CHAPTER 14)

Robert H. Perry   *Ampex Corporation, Redwood City, California* (CHAPTER 15)

Dalton H. Pritchard   *RCA Corporation (retired)* (CHAPTER 17)

Daniel Queen   *Daniel Queen Associates, New York, New York* (CHAPTER 17)

Bruce Raynor   *The Grass Valley Group, Grass Valley, California* (CHAPTER 6)

Daniel R. von Recklinghausen   *Consultant, Hudson, New Hampshire* (CHAPTER 2)

Masanobu Yamamoto   *Sony Corporation, Tokyo, Japan* (CHAPTER 14)

Jerry C. Whitaker, Coauthor   *Intertec Publishing Corporation, Overland Park, Kansas* (CHAPTERS *8, 10, 11, 12, 13, 16*)

PREFACE

Television and audio technologies evolved, until relatively recently, as unrelated fields of engineering expertise. The first television image transmission was demonstrated in 1884 by Nipkow with his scanning disk. It was not until 53 years later, in 1937, that V. K. Zworykin invented the first electronic camera tube. This new device permitted television to emerge from the laboratory and heralded the start of television broadcasting to an audience of home viewers.

Only eight years before Nipkow's television experiment, in 1876, the electrical transmission and the reproduction of voice signals was demonstrated by Alexander Graham Bell. However, the use of audio signals for practical applications other than point-to-point voice communication was not to be realized until almost a half century later when the vacuum tube, invented earlier by Flemming in 1904, provided the means for amplification of electric signals. This led to the initiation of radio broadcasting in 1920.

The introduction of the video-cassette recorder in 1972 added another dimension to television and audio in the consumer marketplace and prompted a greater viewer awareness of the recording and reproduction requirements for both picture and sound. This has, in turn, resulted in the television viewer's demand for higher-fidelity sound reproduction. For some special programs featuring classical, jazz, and rock music, broadcasters have responded by the use of FM stations to provide auxilliary stereophonic sound channels. The demand for more improved and sophisticated sound reproduction has led virtually all North American broadcasters to provide facilities for stereophonic television-sound transmission. The closer relationship of television and audio has also carried over into the recording field with the advent of music-video disks, and the use of piggy-backed audio-on-video for satellite transmissions and on video-taped signals.

Consequently, a marriage of television and audio designs has evolved, and technicians and engineers providing the operational and maintenance support in these two areas now need to broaden the scope of their knowledge in order to be prepared to cope with problems over the diverse base-band ranges of 15 to 20,000 Hz for audio and up to 20 or 30 MHz for video, and with radio frequencies from those used for AM broadcast channels to those employed for terrestrial and satellite microwave transmissions.

The *Television and Audio Handbook for Technicians and Engineers* provides a comprehensive and practical source of information and reference data for the day-to-day maintenance and operation of television and audio equipment. Fundamental principals and the circuitry and mechanics of designs currently used by industry are covered in a descriptive manner that does not require the background of an engineering degree for full understanding or for application to every-day operating and service problems.

A large portion of the text and illustrations are adapted from two McGraw-Hill handbooks: the 1,500-page *Television Engineering Handbook,* and the 1,000-page *Audio Engineering Handbook.* The extensive theoretical dissertations and mathematical explanations from these two handbooks have been included, in simplified form, only to the extent that they are essential for an understanding of equipment operation and maintenance. The individual topics may be classified broadly under the following categories:

Electron theory (Chap. 1)

Circuit and equipment-design fundamentals (Chaps. 2 through 4)

Digital-transmission fundamentals (Chap. 5)

Television and audio equipment (Chaps. 6, 7, and 10)

Television transmission and reception (Chaps. 8 and 9)

AM and FM transmission and reception (Chaps. 11 and 12)

Sound systems (Chap. 13)

Disk and magnetic-tape recording (Chaps. 14 and 15)

Measurement, test, and maintenance (Chaps. 16 and 17)

The introductory chapter acquaints the reader with the fundamentals of electronics, including electric-current conduction, magnetic and static fields and effects, the spectrum and methods of magnetic radiation, and the concepts of analog and digital signal processing.

Practical passive and active circuit designs are described in Chap. 2 to provide a basis for an understanding of the circuit elements used for the various television and audio systems applications.

The fundamental principles of television-signal generation and reproduction are dealt with in Chap. 3 with particular emphasis on the concept of image scanning of picture elements. A primer on binary numbers and digital formats and on coding and decoding is provided in Chap. 5.

The audio-signal spectrum and the fundamentals of audio-signal transmission and processing are covered in Chap. 4 with a discussion of typical sound systems and loudspeaker designs featured in Chap. 13. Chapters 6 through 10 provide information on video-signal transmission and reception, while RF transmission and reception of audio signals by means of AM and FM carriers, including stereo coding, are described in Chaps. 11 and 12, respectively.

Chapters 6 through 9 are devoted to all types of television broadcasting and receiving equipment. Chapter 10 covers audio equipment for broadcasting and recording studios, while the salient characteristics of LPs and CDs are described in Chap. 14. Test and measurement procedures for video and audio are described in detail in Chap. 16, followed by a compilation of reference graphs and tables in Chap. 17.

The authors wish to thank E. Stanley Busby and Carl Bentz for their valuable assistance in the initial phases of organizing this handbook and for their extensive contributions to various chapters.

K. Blair Benson
Jerry C. Whitaker

CHAPTER 1
ELECTRONIC ENGINEERING FUNDAMENTALS

K. Blair Benson
Television Technology Consultant, Norwalk, Connecticut

1.1 CHARACTERISTICS OF ELECTRONS

1.1.1 Electron Theory

Structure of Matter.   The atomic theory of matter specifies that each of the many chemical elements is composed of unique and identifiable particles called *atoms*. In ancient times only 10 were known in their pure, uncombined form; these were carbon, sulfur, copper, antimony, iron, tin, gold, silver, mercury, and lead. Of the several hundred now identified, less than 50 are found in an uncombined, or chemically free, form on earth.

Each atom consists of a compact nucleus of positively (*protons*) and negatively (*electrons*) charged particles. Additional electrons travel in well-defined orbits around the nucleus. The electron orbits are grouped in regions called *shells*, and the number of electrons in each orbit increases with the increase in orbit diameter in accordance with quantum-theory laws of physics. The diameter of the outer orbiting path of electrons in an atom is in the order of one-millionth (10^{-6}) millimeter, and the nucleus, one-millionth of that. These typical figures emphasize the minute size of the atom.

Magnetic Effects.   The nucleus and the free electrons for an iron atom are shown in the schematic diagram in Fig. 1.1. Note that the electrons are spinning in dif-

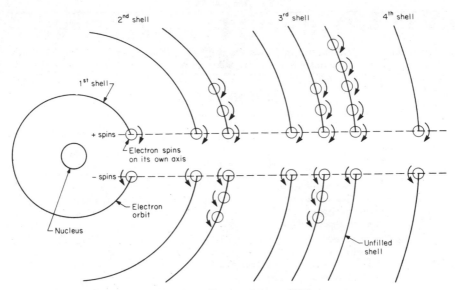

FIGURE 1.1   Schematic of iron (Fe) atom. *(Source: Benson, 1986)*

ferent directions. This rotation creates a magnetic field surrounding each electron. If the number of electrons with positive spins is equal to the number with negative spins, the net field would be zero and the atom would exhibit no magnetic field.

In the diagram, although the electrons in the first, second, and fourth shells balance each other, in the third shell five electrons have clockwise positive spins, and one a counterclockwise negative spin, which gives the iron atom in this particular electron configuration a cumulative magnetic effect.

The parallel alignment of electron spins over regions, known as *domains*, containing a large number of atoms. When a magnetic material is in a demagnetized state, the direction of magnetization in the domain is in a random order. Magnetization by an external field takes place by a change or displacement in the isolation of the domains, with the result that a large number of the atoms are aligned with their charged electrons in parallel.

Conduction in Metals.   In some elements, such as copper, the electrons in the outer shells of the atom are so weakly bound to the nucleus that they can be released by a small electrical force, or voltage. A voltage applied between two points on a length of a metallic conductor produces the flow of an electric current, and an electric field is established around the conductor. The conductivity is a constant for each metal which is unaffected by the current through or the intensity of any external electric field.

Insulators.   In some nonmetallic materials the free electrons are so tightly bound by forces in the atom that, upon the application of an external voltage, they will not separate from their atom except by an electrical force strong enough to destroy the insulating properties of the material. However, the charges will realign within the structure of their atom. This condition occurs in the insulating material (*dielectric*) of a capacitor when a voltage is applied to the two conductors encasing the dielectric.

Conduction in Semiconductors. Semiconductors are electronic conducting materials wherein the conductivity is dependent primarily upon impurities in the material. In addition to negative mobile charges of electrons, positive mobile charges are present. These positive charges are called *holes* because each exists as an absence of electrons. Holes (+) and electrons (−), because they are oppositely charged, move in opposite directions in an electric field.

The conductivity of semiconductors is highly sensitive to, and increases with, temperature.

Direct Current (DC). Direct current is defined as a unidirectional current in which there are no significant changes in the current flow. In practice the term frequently is used to identify a voltage source, in which case variations in the load can result in fluctuations in the current but not in the direction.

Direct current was used in the first systems to distribute electricity for household and industrial power. For safety reasons, and the voltage requirements of lamps and motors, distribution was at the low nominal voltage of 110. The losses in distribution circuits at this voltage seriously restricted the length of transmission lines and the size of the areas that could be covered.

Consequently, only a relatively small area could be served by a single generating plant. It was not until the development of alternating-current systems and the voltage transformer that it was feasible to transmit high levels of power at relatively low voltages over long distances for subsequent low-voltage distribution to consumers.

Alternating Current (AC). Alternating current is defined as a current that reverses direction at a periodic rate. The average value of alternating current over a period of one cycle is equal to zero. The effective value of an alternating current in the supply of energy is measured in terms of the root mean square (rms) value. The rms is the square root of the square of all the values, positive and negative, during a complete cycle, usually a sine wave. Because rms values cannot be added directly, it is necessary to perform an rms addition as shown in the equation below:

$$V_{\text{rms total}} = \sqrt{V_{\text{rms } 1}^2 + V_{\text{rms } 2}^2 + \cdots + V_{\text{rms } n}^2} \qquad (1.1)$$

As in the definition of direct current, in practice the term frequently is used to identify a voltage source.

The level of a sine-wave alternating current or voltage may be specified by two other methods of measurement in addition to rms. These are average and peak. A sine-wave signal and the rms and average levels are shown in Fig. 1.2. The levels of complex, symmetrical ac signals are specified as the peak level from the axis, as shown in Fig. 1.2.

1.1.2 Electronic Circuits

Circuit Elements. Electric circuits are composed of elements such as resistors, capacitors, inductors, and voltage and current sources, all of which may be interconnected to permit the flow of electric currents. An element is the smallest

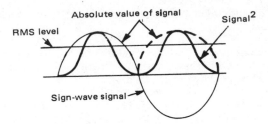

FIGURE 1.2 Root mean square (rms) measurements. The relationship of rms and average values is shown. *(Source: Benson, 1988)*

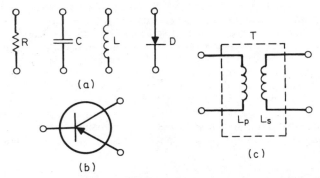

FIGURE 1.3 Schematic examples of two-, three-, and four-terminal circuit elements. (*a*) Resistor *R*, capacitor *C*, inductance *L*, and diode *D* are two-terminal elements. (*b*) *PNP* transistor three-terminal element. (*c*) Transformer *T* four-terminal element composed of primary L_p and secondary L_s windings.

component into which circuits can be subdivided. The points on a circuit element where they are connected in a circuit are called *terminals*.

Elements may have two or more terminals as shown in Fig. 1.3. The resistor, capacitor, inductor, and diode shown in Fig. 1.3*a* are two-terminal elements; the transistor in Fig. 1.3*b* is a three-terminal element; and the transformer in Fig. 1.3*c* is a four-terminal element.

Circuit elements and components also are classified as to their function in a circuit. An element is considered passive if it absorbs energy and active if it increases the level of energy in a signal. An element that receives energy from either a passive or active element is called a *load*. In addition, either passive or active elements, or components, can serve as loads.

Ohm's Law.   The basic relationship of current and voltage in a two-terminal circuit where the voltage is constant and there is only one source of voltage is given in Ohm's law. This states that the voltage *V* between the terminals of a conductor varies in accordance with the current *I*. The ratio of voltage and current, called the *resistance R*, is expressed in Ohm's law as follows:

$$E = I \times R \tag{1.2}$$

Using Ohm's law, the calculation for power in watts can be developed from $P = E \times I$ as follows:

$$P = \frac{E^2}{R} \quad \text{and} \quad P = I^2 \times R \tag{1.3}$$

Circuit Configurations. A circuit, consisting of a number of elements or components, usually amplifies or modifies a signal before delivering it to a load. The terminal to which a signal is applied is called an *input port*, or *driving port*. The pair or group of terminals that delivers a signal to a load is called the *output port*. An element or portion of a circuit between two terminals is called a *branch*. The circuit shown in Fig. 1.4 is made up of several elements and branches. R_1 is a branch; R_3 and C_1 make up a two-element branch. The secondary of transformer T, a voltage source, and R_2 also constitute a branch. The point at which three or more branches join together is called a *node*. A series connection of elements or branches, called a *path*, in which the end is connected back to the start is a *closed loop*.

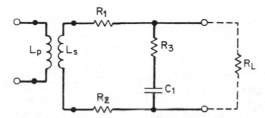

FIGURE 1.4   Circuit configuration composed of several elements and branches, and a closed loop, R_1, R_3, C_1, R_2, and L_s.

1.1.3 Circuit Analysis

Relatively complex configurations of linear circuit elements, that is, where the signal gain or loss is constant over the signal amplitude range, can be analyzed by simplification into the equivalent circuits. After the restructuring of a circuit into an equivalent form, the current and voltage characteristics at various nodes can be calculated using network-analysis theorems. These are: Kirchoff's current and voltage laws, Thevenin's theorem, and Norton's theorem.

Kirchoff's Current Law (KCL). The algebraic sum of the instantaneous currents entering a node (a common terminal of three or more branches) is zero. In other words, the currents from two branches entering a node add algebraically to the current leaving the node in a third branch.

Kirchoff's Voltage Law (KVL). The algebraic sum of instantaneous voltages around a closed loop is zero.

Thevenin's Theorem.   The behavior of a circuit at its terminals can be simulated by replacement with a voltage E from a dc source in series with an impedance Z (see Fig. 1.5*a*).

Norton's Theorem.   The behavior of a circuit at its terminals can be simulated by replacement with a dc source I in parallel with an impedance Z (see Fig. 1.5*b*).

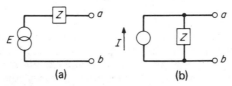

(a)   (b)

FIGURE 1.5   Equivalent circuits. (*a*) Thevenin's equivalent voltage source. (*b*) Norton's equivalent current source. *(Source: Fink, 1982) [Adapted from MG-H Electronics Engineers Handbook, Fig. 1–3 pp 1–43]*

1.2   STATIC ELECTRICITY

1.2.1   Charged Objects

Positive and Negative Charges.   The phenomenon of static electricity and related potential differences concerns configurations of conductors and insulators where no current flows and all electrical forces are unchanging; hence the term *static*. Nevertheless, static forces are present because of the number of excess electrons or protons in an object. A static charge can be induced by the application of a voltage to an object. A flow of current to or from the object can result from either a breakdown of the surrounding nonconducting material or by the connection of a conductor to the object.

Attraction and Repulsion.   Two basic laws regarding electrons and protons are:

1. Like charges exert a repelling force on each other; electrons repel other electrons and protons repel other protons.
2. Opposite charges attract each other; electrons and protons are attracted to each other.

Therefore, if two objects each contain exactly as many electrons as protons in each atom, there is no electrostatic force between the two. In this case, as described in Sec. 1.1.1, the deficiency of electrons in the nucleus, resulting in a positive charge, is neutralized by the orbiting electrons with the net result that the atoms are neutral and that externally the objects appear as neutral, or uncharged.

On the other hand, if one object is charged with an excess of protons (deficiency of electrons) and the other an excess of electrons, there will be a relatively weak attraction that diminishes rapidly with distance. An attraction also will occur between a neutral and a charged object. Examples are: the clinging of a piece of paper to a plastic or glass rod that has been rubbed on a cloth to produce a charge of excess electrons[1], or, in more practical terms, the collecting of dust by an electrostatic air-filter precipitator.

An example of the breakdown of the surrounding material, resulting in a current flow, is a bolt of lightning between oppositely charged clouds or a cloud and neutral earth.

Electrostatic Shielding. Another law, developed by Faraday, governing static electricity is that all of the charge of any conductor not carrying a current lies in the surface of the conductor. Thus, any electric fields external to a completely enclosed metal box will not penetrate beyond the surface. Conversely, fields within the box will not exert any force on objects outside the box. The box need not be a solid surface; a conduction cage or grid will suffice. This type of isolation frequently is called a *Faraday shield*.

1.3 MAGNETISM

1.3.1 Permanent and Electromagnets

Magnetic Fields and Forces. The elemental magnetic particle is the spinning electron. In magnetic materials, such as iron, cobalt, and nickel, the electrons in the third shell of the atom (see Fig. 1.1) are the source of magnetic properties. If the spins are arranged to be parallel, the atom and its associated domains or clusters of the material will exhibit a magnetic field. The magnetic field of a magnetized bar has lines of magnetic force that extend between the ends, one called the *north pole* and the other the *south pole*, as shown in Fig. 1.6a. The lines of force of a magnetic field are called *magnetic flux lines*. A pattern of the lines of force can be displayed graphically by sprinkling iron filings on a piece of paper laid over a bar magnet.

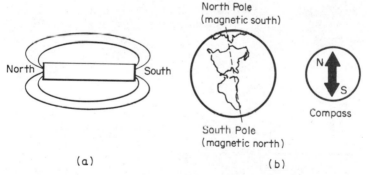

FIGURE 1.6 (a) Lines of force surrounding a bar magnet. (b) Relation of compass poles to the earth's magnetic field.

The force produced by a magnetic field can be demonstrated by placing two bar magnets side by side. It can be shown that like poles repel, and unlike north and south poles attract. This is analogous to the reactions between electrical

---

1. This was called *vis electra* when it was demonstrated in the 1550s in England using a feather and a piece of amber.

charges (see Sec. 1.2.1). The magnetic field of the earth is demonstrated by the operation of a compass, as shown in Fig. 1.6b. In actuality the magnetic north pole of the earth is a south pole because it attracts the north pole of a compass.

Electromagnetism. A current flowing in a conductor produces a magnetic field surrounding the wire as shown in Fig. 1.7a. An easy way to remember the relationship between the current and field is by the *right-hand rule*: Grasping the wire with the right hand so that the thumb is in the direction of the convention for current flow from + to −, rather than the actual electron flow from − to +, the fingers wrapped around the conductor will point in the north direction of the magnetic lines of force.

In a coil or solenoid, the direction of the magnetic field relative to the electron flow (− to +) is shown in Fig. 1.7b. The attraction and repulsion between two iron-core electromagnetic solenoids driven by direct currents is similar to that of two permanent magnets described in the preceding paragraphs.

The process of magnetizing and demagnetizing an iron-core solenoid using a current being applied to a surrounding coil can be shown graphically as a plot of the magnetizing field strength and the resultant magnetization of the material called a *hysteresis loop* (see Fig. 1.8). It will be found that the point where the field is reduced to zero, a small amount of magnetization, called *remnance*, remains.

Magnetic Shielding. In effect, the shielding of components and circuits from magnetic fields is accomplished by the introduction of a magnetic short circuit in the path between the field source and the area to be protected. The flux from a field can be redirected to flow in a partition or shield of magnetic material, rather than in the normal distribution pattern between north and south poles. The effectiveness of shielding depends primarily upon the thickness of the shield, the material, and the strength of the interfering field.

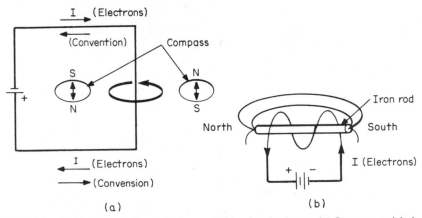

FIGURE 1.7   Magnetic field surrounding a current-carrying conductor. (*a*) Compass at right indicates the polarity and direction of a magnetic field circling a conductor carrying direct current. *I* indicates the direction of electron flow. *Note*: The convention for flow of electricity is from + to −, the reverse of the actual flow. (*b*) Direction of magnetic field for a coil or solenoid.

Some alloys are more effective than iron. However, many are less effective at high flux levels. Two or more layers of shielding, insulated to prevent circulating currents from magnetization of the shielding, are used in low-level audio components such as microphone coupling transformers.

1.3.2 Electromagnetic-Radiation Spectrum[2]

The usable spectrum of electromagnetic-radiation frequencies extends over a range from below 100 Hz for power distribution to 10^{20} for the shortest X-rays. The services using various frequency bands in the spectrum are shown in Fig. 1.9. The lower frequencies are used primarily for terrestrial broadcasting and communications. The higher frequencies include visible and near-visible infrared and ultraviolet light, and X-rays. A breakdown of the spectrum into bands, a description of the applications and services, and the propagation characteristics in those bands follow.

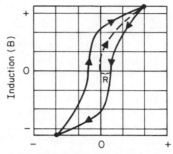

FIGURE 1.8 Graph of the magnetic hysteresis loop resulting from magnetization and demagnetization of iron. The dashed line is a plot of the induction from the initial magnetization. The solid line shows a reversal of the field and a return to the initial magnetization value. *R* is remaining magnetization (remnance) when the field is reduced to zero.

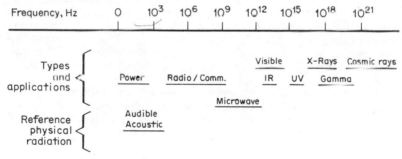

FIGURE 1.9 The electromagnetic spectrum. *(Source: Fink, 1982)*

***Low-End Spectrum Frequencies* (1 to 1000 Hz).** Electric power is transmitted by wire but not by radiation at 50 and 60 Hz, and in some limited areas, at 25 Hz. Aircraft use 400-Hz power in order to reduce the weight of iron in generators and transformers.

The restricted bandwidth that would be available for communication channels is inadequate for voice or data transmission, although some use has been made of voice communication over power distribution circuits using modulated carrier

---

2. Portions of Sec. 1.3.2 have been adapted from Donald G. Fink (editor-in-chief), *Electronics Engineers' Handbook*, McGraw-Hill, New York, 1982.

frequencies. The sound-transmission frequencies noted in Fig. 1.9 are acoustic rather than electromagnetic.

Low-End Radio Frequencies (**1000 to 100 kHz**).   These low frequencies are used for very long distance radio-telegraphic communication where extreme reliability is required and where high-power and long antennas can be erected.

Medium-Frequency Radio (**20 kHz to 2 MHz**).   The low-frequency portion of the band is used for around-the-clock communication services over moderately long distances and where adequate power is available to overcome the high level of atmospheric noise. The upper portion is used for AM radio, although the strong and quite variable sky wave occurring during the night results in substandard quality and severe fading at that time. The greatest use is for AM broadcasting, in addition to fixed and mobile service, LORAN ship and aircraft navigation, and amateur radio communication.

High-Frequency Radio (**2 to 30 MHz**).   The band provides reliable medium-range coverage during daylight and, when the transmission path is in total darkness, worldwide long-distance service, although the reliability and signal quality of the latter is dependent to a large degree upon ionosphere conditions and related long-term variations in sun-spot activity affecting sky-wave propagation. The applications are broadcasting, fixed and mobile service, and telemetering and amateur transmissions.

Very High Ultrahigh Frequencies (**30 MHz to 3 GHz**).   VHF and UHF bands, because of the greater channel bandwidth possible, can provide transmission of a large amount of information, either as television detail or data communication. Furthermore, the shorter wavelengths permit the use of highly directional parabolic or multielement antennas. Reliable long-distance communication is provided using high-power troposphere scatter techniques. The multitude of uses include, as well as television, fixed and mobile communication services, amateur radio, radio astronomy, satellite communication, telemetering, and radar.

Microwaves (**3 to 300 GHz**).   Many transmission characteristics are similar to those used for shorter optical waves, which limit the distances covered to line of sight. The uses are television, radar, and wide-band information services.

Infrared, Visible, and Ultraviolet Light.   The portion of the spectrum visible to the eye covers the gamut of transmitted colors ranging from red, through yellow, green, cyan, and blue. It is bracketed by infrared on the low-frequency side and ultraviolet (UV) on the high side. Infrared signals are used in a variety of consumer and industrial equipments for remote controls and sensor circuits in security systems. The most common use of UV waves is for excitation of phosphors to produce visible illumination.

X-Rays.   The medical and biological examination techniques and industrial and security inspection systems are the best-known applications of X-rays. X-rays in the higher-frequency range are classed as *hard* X-rays or *gamma rays*. Exposure to X-rays for long periods can result in serious irreversible damage to living cells or organisms, as well as to the human body.

1.4 ANALOG SIGNAL TRANSMISSION

1.4.1 Signal Characteristics

An analog signal represents the value of a measurable quantity as an electrical parameter. For example, in television the brightness elements in a screen are converted to a current flowing in a resistor to produce a voltage level; in audio a microphone senses sound-pressure levels to produce an electric signal of corresponding voltage levels (see Fig. 1.10). The term *analog* is used because the signal level variations over a time period are *analogous* to the variation in the original quantity.

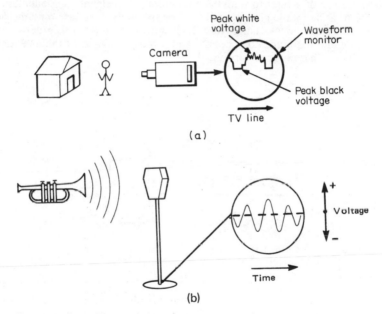

FIGURE 1.10 (*a*) Picture signal generation. (*b*) Audio signal generation. (*Source: Benson, 1988*)

Television and audio signals are generated and reproduced in analog form. A television camera scans a scene line by line in a manner similar to a person's reading a printed page, to transform the levels of a scene's brightness and color to a continuously varying electric signal. Although digital signal processing or storage, e.g., digital video-noise reduction or compact-disk audio recording, may follow analog signal generation, reproduction of the picture on a television screen or the sound by a loudspeaker is accomplished by analog signals.

Linear Circuits. In an ideal analog signal-transmission path, the amplitude levels of the signal may be continuously variable, and the input and output signals are identical in amplitude levels. This is called a *linear* transmission path. Deviations from an ideal path or circuit can be checked by transmission of a sine at the lowest frequency of the bandpass. The output waveform should be identical with the input.

1.4.2 Transmission Circuit Requirements

Audio Signals.   Nonlinearities in transmission of audio signals result in the generation of harmonic components. A pure sine-wave signal used for testing will have no higher-frequency components. If the transmission path has a nonlinear amplitude characteristic, a sine-wave signal will be distorted and the harmonics of the fundamental signal will be generated. The amount of distortion can be determined quantitatively by measurement of the amplitude of the harmonics. This requires the use of a filter to remove the fundamental signal component so that either the total harmonic distortion (THD) or the individual harmonic distortion components can be measured. Figure 1.11 shows the relationship between the fundamental and harmonic components and a block diagram of the analyzer for measurement of harmonic distortion.

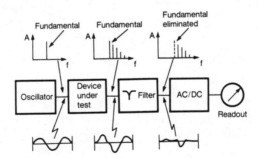

FIGURE 1.11   Basic block diagram of a harmonic-distortion analyzer. *(Source: Benson, 1988)*

In addition to the amplitude linearity, the amplitude response over the signal bandpass is an important characteristic. Normally a uniform response is required; however, for some applications in RF transmission and reception, and in recording and playback, nonuniform preemphasis and deemphasis characteristics are required. The amplitude-frequency response is measured with a sine-wave signal generator by making point-by-point measurements of gain (or loss). A more sophisticated technique uses a sweeping oscillator and a storage oscilloscope to reproduce a plot of amplitude versus frequency.

1.4.3 Television Picture and Sync-Signal Requirements

The THD technique used in audio is not suited to measurement of distortions of television video signals for two reasons:

1. Although applicable to simple amplifiers, a figure of harmonic distortion is not a direct indication of visually evident picture distortions.

2. Television transmission involves four different signal components, each with a variety of different characteristics. They are: luminance video, color video, scanning, and color-coding synchronization.

Consequently, television video components and transmission circuits require different techniques that simulate the actual picture signal for specification and measurement of transmission characteristics.

Amplitude Linearity.   The most frequently encountered picture degradation is white compression, resulting from a lower gain for highlight signals than for midgrays or blacks. The effect is a lack of fine detail in bright areas and a washed-out appearance. In extreme cases a clipping of whites may occur, either by circuit faults or by a misadjustment of clipping circuits intended to limit spurious noise and interference. The simplest approach to determine whether distortion of the video signal exists is to use a test signal composed of a sawtooth wave extending from peak black or blanking level to peak white on each line, as shown in Fig. 1.12a. The dashed line gives a qualitative indication of white compression and no information as to color signal degradations.

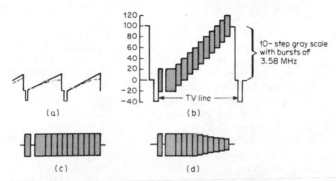

FIGURE 1.12   (a) Line-rate sawtooth test signal. Dashed line shows the effect of nonlinear transmission, in this case introducing compression of white level. (b) Test signal waveform for differential-gain and phase measurements. Bursts of 3.58 MHz (color subcarrier) are superimposed on a line-rate, 10-step gray scale. (c) Bursts of 3.58 MHz from 1.12b after removal of sync and luminance components by a high-pass filter. (d) Bursts of the 3.58-MHz signal from 1.12b after passing through a nonlinear circuit and a high-pass filter. The decrease in amplitude near the end of the line scan indicates white compression.

For a quantitative evaluation, a technique called *differential gain and phase measurement* is used wherein a low-amplitude sine-wave signal at the color-subcarrier frequency of 3.58 MHz is superimposed on a 10-level stairstep signal, the latter at horizontal-scanning rate (see Fig. 1.12b). Initial application of this technique used a sawtooth of the waveform shown in Fig. 1.12a, rather than a stairstep. The stairstep has the advantage of providing readings that can be related to specific IRE units of level measurements on a waveform oscilloscope. Color subcarrier is used for the superimposed low-amplitude signal in order to provide accurate information as to the ability of a component or system to transmit color signals without distortion.

Precise figures of differential gain are made by measuring the burst levels after removing the low-frequency sync and luminance components with a high-pass filter. The side-by-side display of the bursts, as shown in Fig. 1.12*c*, permits a comparison of burst levels and an accurate reading of the burst levels that indicates a gain (or loss) throughout the amplitude range from the midsync level of peak color-sync burst to peak white. As shown in Fig. 1.12*c*, all bursts are equal in amplitude, indicating no amplitude distortion or differential gain. In Fig. 1.12*d*, the bursts decrease in amplitude to the right, indicating an increasing compression from midgray to peak white.

Phase Linearity.   The transmission phase characteristic is a measurement of the ability of a circuit to faithfully transmit pulses. Phase distortion will produce overshoots of the video wave. Concurrently with the differential-gain measurement, the phase distortion over the 10-step range from reference black to white can be read by comparing the sine-wave bursts on an oscilloscope using a circular scan synchronized with the reference color signal.

The use of the differential gain and phase measurements described above are essential to achieving sharp pictures with accurate and believable color. Consequently, the 10-step gray scale with superimposed 10-unit subcarrier bursts has been adopted as an industry standard for differential gain and phase measurements.

Amplitude-Frequency Response.   The uniformity of video signal amplitude over the bandpass of components or of a transmission system is of an importance equal to that of amplitude linearity. High frequencies are direct measures of the ability to reproduce fine detail. In addition, since the color signal is transmitted as a modulated RF carrier at the high end of the system bandpass, the high-frequency response has a direct effect upon the fidelity of color reproduction. This characteristic can be checked with a sweep generator that produces a test signal *sweeping* over a range of frequencies covered in by a component or amplifier. However, since the sweep rate of most generators is 60 Hz, the presence of any *line clamps* that reset the sync or blanking level after each line, a signal called a *multiburst* is used. This is shown in Fig. 1.13. The bursts inserted on one

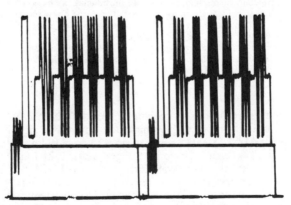

FIGURE 1.13   Multiburst signal for measurement of amplitude-frequency response over a video bandwidth of 4.2 MHz.

or more picture lines are at frequencies distributed throughout the bandpass of the video path to be checked.

Low-Frequency Amplitude and Phase Response. The lowest-frequency components of a picture video signal correspond to the 30-Hz frame rate. Therefore, response must be maintained below this frequency. In fact, because substantial phase distortions accompany a low-frequency cutoff of response, it is essential that the response in capacitively coupled circuits extend to at least 30 Hz; otherwise, smear and streaking will be present to the right of and below relatively large luminance signal areas. This type of distortion can be checked and evaluated using a *window* signal, shown in Fig. 1.14, or a pulse-and-bar signal, shown in Fig. 1.15.

(a)   (b)

FIGURE 1.14   Window signal for evaluation of low-frequency response. (*a*) Picture monitor. (*b*) Waveform monitor with horizontal sweep at twice line rate. Dashed line in *b* indicates a loss in low-frequency response.

1.5   DIGITAL SIGNAL TRANSMISSION

1.5.1   Signal Characteristics

Digital signals differ from analog in that only two steady-state levels are used for the transmission of information. The concept of requiring no more than two levels is not new, since two-level, on-off control circuits are a basic component in all electrical systems. The adaption the key-and-sounder clicks of Morse's telegraphy to continuous-wave (cw) radio transmission utilizes a two-level (on or off) transmission, but with one additional parameter: the short and long coding of dots and dashes, and these in turn are arranged in a variety of sequences to represent letters of the alphabet and numbers (alphanumeric).

A more recent application of two-level coding is the synchronizing signal in the analog transmission of television pictures. The synchronizing pulses are a digital signal of two levels consisting of blanking and sync tips. Their purpose is merely to supply information necessary for the timing of picture-scanning circuits and the on-off control of the analog picture signal. Thus, no information is transmitted in the amplitude range between blanking and sync tips; the only reasons for using the large sync-signal amplitude relative to the video signal (40 percent of the peak-white video signal) are to: (a) assure stable synchronization in the presence of any noise or other spurious signals, and (b) permit the use of relatively simple circuits to separate the sync signal from the wide-band video signal.

Digitizing an analog signal for processing or transmission requires the conversion of the signals, which are varying in level over a contiguous range of values, to a meaningful coding of only two different levels. Then for viewing or listening, these two discrete levels must be decoded into the original analog signal, again a continuously varying range of levels.

Binary Coding. Definition of the digital-transmission format requires specification of the following parameters: (1) the type of information corresponding to each of the two levels, and (2) the frequency or rate at which the information is transmitted as a bilevel signal. The digital coding of signals for television and audio systems uses a system of binary numbers in which only two digits, 0 and 1, are used. This is called a *base*, or *radix*, of 2. For comparison, our conventional decimal numbering system using 10 digits from 0 to 9 is a base (radix) of 10. It is of interest that systems of other bases are used for some more-complex mathematical applications, the principal ones being octal (8) and hexadecimal (16). See Chapter 5 for a more detailed discussion of the systems.

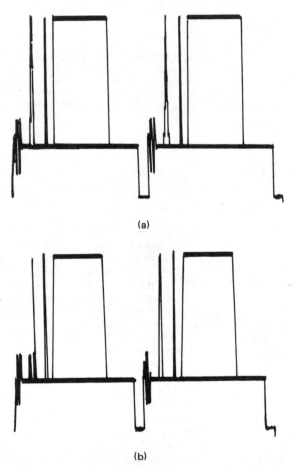

(a)

(b)

FIGURE 1.15 Pulse and bar signal for test of low-frequency response and high-frequency transient response or distortion.

Table 1.1 compares the decimal, binary, and octal digital counting systems. Note that the numbers in the decimal system are equal to the number of items counted, if used for a tabulation.

Analog-to-Digital (A/D) Conversion. Since the input and output of television and audio system are analog signals, the input must be represented as numbered

TABLE 1.1 Comparison of Counting in the Decimal, Binary, and Octal Digital Systems

Decimal	Binary	Octal
0	0	0
1	1	1
2	10	2
3	11	3
4	100	4
5	101	5
6	110	6
7	111	7
8	1000	10
9	1001	11
10	1010	12
11	1011	13
12	1100	14
13	1101	15
14	1110	16
15	1111	17

sequences corresponding to the analog levels of the signal. This is done by sampling the signal levels and assigning a binary code number to each of the samples. The rate of sampling must be substantially higher than the highest signal frequency in order to cover the bandwidth of the signal and to avoid spurious patterns (*aliasing*) generated by the interaction between the sampling signal and the higher signal frequencies. A simplified block diagram of an A/D converter (ADC) is shown in Fig. 1.16.

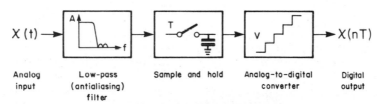

FIGURE 1.16   Analog-to-digital converter block diagram. *(Source: Benson, 1988)*

The Nyquist law for digital coding dictates that the sample rate must be at least twice the cutoff frequency of the signal to avoid these effects. This translates to a 9-MHz rate for a video bandwidth of 4.5 MHz. For an audio bandwidth of 15 kHz, the rate is 30 kHz.

Digital-to-Analog (D/A) Conversion. The digital-to-analog converter (DAC) is, in principle, quite simple. The digital stream of binary pulses is decoded into discrete, sequentially timed signals corresponding to the original sampling in the ADC. The output is an analog signal of varying levels. The time duration of each level is equal to the width of the sample taken in the A/D conversion process. The analog signal is separated from the sampling components by a low-pass filter. Figure 1.17 shows a simplified block diagram of a DAC. The deglitching sample-and-hold circuits in the center block set up the analog levels from the digital decoding and remove the unwanted high-frequency sampling components.

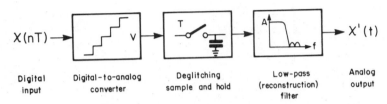

X(nT)   X'(t)

| Digital | Digital-to-analog | Deglitching | Low-pass | Analog |
| input | converter | sample and hold | (reconstruction) filter | output |

FIGURE 1.17 Digital-to-analog converter block diagram. *(Source: Benson, 1988)*

1.5.2 Why Digital

The need for at least double the analog bandwidth for equivalent digital transmission leads one to question the advantage of introducing digital paths in a television or audio system wherein the input and output signals are in the analog format. Any doubts are highlighted by the fact that both picture- and sound-monitoring equipment located throughout video and audio systems require analog signal inputs.

Technical Considerations. The most important points supporting the use of digital formats are summarized below.

- While analog signal degradations (e.g., signal-to-noise ratio, crosstalk, distortions) are accumulative and thus difficult to separate and remove from the video or audio signal, the ability to *regenerate* a digital pulse train *exactly* results in digital signals being immune to the principal circuit shortcomings affecting analog signals.

- Several different digital bit-streams can be interleaved (time-multipled). This process permits related signals to be transmitted or recorded in a single digital channel. Examples are: (a) in television, the combining of picture, sound, and second-language dialog, and (b) in audio CDs the addition of picture and data signals.

- Digital computers and digital-signal-processing techniques permit the manipulation of picture material for special effects (FX) and animation with better quality and in a fraction of the time required heretofore using motion-picture film.

- Digital noise reduction provides improved picture quality.

- Editing of sound tracks is simplified by the use of multiple generation of re-recording with little, if any, distortion or increase in noise level and the use of time compression and expansion with no change in pitch.

- Use of forward error correction (FEC) to maintain signal characteristics.

Economic Considerations. Equally important are economic factors that support the use of digital techniques. The major items are:

- Since the development of the transistor, the costs of digital solid-state devices and integrated circuits for both memory and logic are continuing to drop, and at a faster rate than the analog counterparts. In addition, the advancements in large-scale integration have cut drastically the labor costs of equipment manufacture. These production-cost reductions are reflected in lower equipment prices.
- Digital circuits do not require alignment, tuning, and other adjustments. Thus operation and maintenance of digital hardware are usually easier and more cost-effective than that of analog counterparts.
- The proliferation of computers of a variety of sizes and the increasing familiarity of technical personnel with their operation have allowed the mechanization of labor-intensive, routine-type plant operations such as maintenance and testing which, in turn have resulted in reduced labor costs and lower rates of manufacturing rejects.

Audio Transmission and Recording. The advantage most apparent to a critical listener is that the fidelity of digital sound recording and reproduction provided by moderately priced consumer products is far superior to that achieved with the most advanced professional analog equipment. The extremely low noise level permits the reproduction of an exceedingly large dynamic range. Furthermore, not only is the random-noise level so low as to be essentially nonexistent but in addition, disk playback is unaffected by imperfections in the recording medium. In analog-disk reproduction, these are apparent as pops, clicks, and random background noise. Similarly, magnetic systems are immune to random noise from the magnetic-oxide particles, and signal dropouts from irregularities in the magnetic coating do not cause an error in the digital-signal-decoding process.

Because a digital audio requires only two signal levels for transmission, amplitude distortion has practically no effect on the signal. In fact, linear amplification is not necessary. Instead, binary-coded digital signals can be regenerated over and over again without any degradation. Thus, amplification can be accomplished with very elementary solid-state circuits.

The bandwidth required, however, is considerably greater than that for the analog counterpart. In order to comply with the Nyquist criterion (see Sec. 1.5.1), the minimum sample rate for a 15-kHz audio channel is 30 kHz. For a number of reasons related to the integration of digital audio signals with television systems, worldwide, the sampling frequencies listed below have been recommended by the Audio Engineering Society:

Program material origination: 48 kHz

Consumer applications: 44.1 kHz

Another advantage of digital audio is the ability to interleave a number of signals by time-multiplexing in mixing consoles and in transmission. This feature simplifies mixing-console design and the cabling and connection requirements.

The lack of any degradation in multiple generations of duping is a major technical advantage for program production and editing. Unfortunately, it has raised a serious issue in the recording industry, which quite likely will have substantial long-term financial and artistic impact upon the recording industry. Because digital copies of audio programming can be identical to the original, unlawful duplication and sale or even home copying could have such an impact on recorded disk and tape sales that artists and producers may be reluctant to use the digital format for other than studio production or mastering.

Television Video Transmission and Recording.   The composite television video signal is composed of a number of different types of information, basically classified as scanning and color sync, luminance video, and color subcarrier. In addition, noncomposite signals are used in signal processing. Therefore, a single digital-coding system is inadequate and has resulted in the CCIR's (Consultive Committee International Radio) recommending that a family of compatible codes be adopted. One code that has gained universal support is the so-called 4:2:2, 13.5-MHz system. Unfortunately for NTSC countries, this is an international compromise that is not directly related to the NTSC color subcarrier and thus does not permit the encoding sampling to be locked to the 3.58-MHz color subcarrier. Consequently, at the time of this writing, digital video standards for the 525-line system are undecided. Nevertheless, several manufacturers of video-tape recording equipment in the United States and Japan are going their own ways.

REFERENCES

Benson, K. Blair (ed.) (1988). *Audio Engineering Handbook*. McGraw-Hill, New York.

Benson, K. Blair (editor-in-chief) (1986). *Television Engineering Handbook*. McGraw-Hill, New York.

Fink, Donald G. (editor-in-chief) (1982). *Electronics Engineers' Handbook*. McGraw-Hill, New York.

CHAPTER 2
COMPONENTS AND TYPICAL CIRCUITS

K. Blair Benson
Television Technology Consultant, Norwalk, Connecticut

2.1 PASSIVE CIRCUIT COMPONENTS

Components used in electrical circuitry may be categorized into two broad classifications as passive or active. A voltage applied to a passive component results in the flow of current and the dissipation or storage of energy. Typical passive components are resistors, coils or inductors, and capacitors. For an example, the flow of current in a resistor results in radiation of heat; from a light bulb, the radiation of light as well as heat.

On the other hand, an active component either (a) increases the level of electric energy or (b) provides available electric energy as a voltage. As an example of (a), an amplifier produces an increase in energy as a higher voltage or power level, while for (b), batteries and generators are energy sources.

2.1.1 Resistors

Resistors are components that have a nearly 0° phase shift between voltage and current over a wide range of frequencies with the average value of resistance in-

Portions of this chapter have been adapted from Daniel R. von Recklinghausen, "Amplifiers," in *Audio Engineering Handbook*, K. Blair Benson, editor, McGraw-Hill, New York, 1988.

dependent of the instantaneous value of voltage or current. Preferred values of ratings are shown in ANSI Standards Z17.1 and C83.2 or corresponding ISO or MIL standards, summarized below. Resistors are typically identified by their construction and by the resistance materials used. Fixed resistors have two or more terminals and are not adjustable. Variable resistors permit adjustment of resistance or voltage division by a control handle or with a tool.

Wire-Wound Resistors. The resistance element of most wire-wound resistors is resistance wire or ribbon wound as a single-layer helix over a ceramic or fiberglass core, which causes these resistors to have a residual series inductance that affects phase shift at high audio frequencies particularly in large-size resistors. Wire-wound resistors have low noise and are stable with temperature, with temperature coefficients normally between ±5 and 200 ppm/°C. Resistance values between 0.1 and 100,000 Ω with accuracies between 0.001 and 20 percent are available with power dissipation ratings between 1 and 250 at 70°C. The resistance element is usually covered with a vitreous enamel, which may be molded in plastic. Special construction includes such items as enclosure in an aluminum casing for heatsink mounting or special winding to reduce inductance. Resistor connections are made by self-leads or to terminals for other wires or printed circuit boards.

Metal Film Resistors. Metal film, or *cermet*, resistors have characteristics similar to wire-wound resistors except a much lower inductance. They are available as axial lead components in ⅛-, ¼-, or ½-W ratings, in chip resistor form for high-density assemblies, or as resistor networks containing multiple resistors in one package suitable for printed circuit insertion as well as in tubular form similar to high-power wire-wound resistors.

Metal film, or cermet, resistors are essentially printed circuits using a thin layer of resistance alloy on a flat or tubular ceramic or other suitable insulating substrate. The shape and thickness of the conductor pattern determine the resistance value for each metal alloy used. Resistance is trimmed by cutting into part of the conductor pattern with an abrasive or a laser. Tin oxide is also used as a resistance material.

Carbon Film Resistors. Carbon film resistors are similar in construction and characteristics to axial lead metal film resistors. Since the carbon film is a granular material, random noise may be developed because of variations in the voltage drop between granules. This noise may be of sufficient level to affect the performance of circuits providing high grain when operating at low signal levels.

Carbon Composition Resistors. Carbon composition resistors contain a cylinder of carbon-based resistive material molded into a cylinder of high-temperature plastic, which also anchors the external leads. These resistors may have noise problems similar to carbon film resistors, but their use in electronic equipment for the last 50 years has demonstrated their outstanding reliability unmatched by other components. These resistors are commonly available between 2.7 Ω and 22 Ω with tolerances of 5, 10, and 20 percent in ⅛-, ¼-, ½-, 1-, and 2-W sizes.

Control and Limiting Resistors. Resistors with a large negative temperature coefficient, or *thermistors*, are often used to measure temperature or to limit inrush current into motors or power supplies or to compensate bias circuits in transistor circuits. Resistors with a large positive temperature coefficient are used in circuits that may have to match the coefficient of copper wire. Special resistors also include those that have a low resistance when cold and become a nearly open circuit when a critical temperature or current is exceeded to protect transformers or loudspeakers.

Resistor Networks. A number of metal film or similar resistors are often packaged in a single package suitable for printed circuit mounting. They see applications in digital circuits and in DAC circuits, as well as in fixed attenuators or padding networks.

Adjustable Resistors. Cylindrical wire-wound power resistors are often made adjustable when a metal clamp can make contact to one or more turns not covered with enamel along an axial stripe.

Potentiometers are resistors with a movable arm that makes contact with a resistance element which is connected to at least two other terminals at its ends. Used as voltage dividers, these controls are most often seen as gain or equalizer controls in audio equipment. The resistance element may be circular or linear in shape, and often two or more sections are mechanically coupled or ganged for stereophonic use. Resistance materials include all those described above.

Trimmer potentiometers are similar to those above except that adjustment requires a tool.

Most potentiometers have a linear taper, which means that resistance changes linearly with control motion when measured between the movable arm and the "low," or counterclockwise, terminal. Gain controls usually have a logarithmic or audio taper so that attenuation changes linearly in decibels, a logarithmic ratio. The resistance element of a potentiometer may also contain taps that permit the connection of other components as required in an equalizer or loudness-control circuit. The matching of resistance or attenuation between sections determines channel balance in an amplifier.

Precision potentiometers as used in electronic instruments are usually constructed as 5- to 20-turn, wire-wound potentiometers.

Attenuators. Variable attenuators are adjustable resistor networks that show a calibrated increase in attenuation for each switched step. For measurement of audio, video, and RF equipment, these steps may be decades of 0.1, 1, and 10 dB per step or 3 controls for the range of 111 dB in 0.1-dB steps. Circuits for unbalanced and balanced fixed attenuators are shown in Fig. 2.1. Fixed attenuator networks may be cascaded and switched to provide step adjustment of attenuation inserted in a constant-impedance network.

Audio attenuators generally are designed for a circuit impedance of 150 Ω, although other impedances may be used for specific applications, for example, 8 Ω for feeding a loudspeaker directly. Video attenuators are designed generally to operate with unbalanced 75-Ω grounded-shield coaxial cable; RF attenuators are designed for use with 75- or 50-Ω coaxial cable.

2.1.2 Capacitors

Capacitors are passive components in which current leads voltage by nearly 90° over a wide range of frequencies. Capacitors are rated by capacitance, voltage, materials, and construction. Capacity values and voltage ratings follow ANSI Standard Z17.1, which shows preferred values.

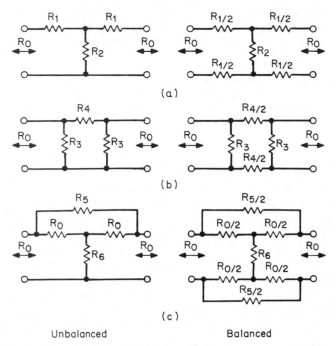

FIGURE 2.1 Unbalanced and balanced fixed-attenuator networks for equal source and load resistance. (*a*) *T* configuration. (*b*) π configuration. (*c*) Bridged-*T* configuration. *Design equations*:

$$A = \frac{\text{Input voltage}}{\text{Input voltage}} = \text{attenuation}$$

$$\frac{R_1}{R_0} = \frac{R_0}{R_3} = \frac{A - 1}{A + 1}$$

$$\frac{R_2}{R_0} = \frac{R_0}{R_4} = \frac{2}{A - (1/A)}$$

$$\frac{R_5}{R_0} = \frac{R_0}{R_6} = A - 1$$

A capacitor may have two voltage ratings. Working voltage is the normal operating voltage that should not be exceeded during operation. Use of the test or forming voltage stresses the capacitor and should occur only rarely in equipment operation. Good engineering practice is to use components only at a fraction of their ratings.

Polarized Capacitors. This type can be used in only those applications where a positive sum of all dc and peak-ac voltages is applied to the positive capacitor terminal with respect to its negative terminal. These capacitors include all tantalum and most aluminum electrolytic capacitors. They are used in power supplies or other electronic equipment where these restrictions can be met.

Losses in capacitors occur because an actual capacitor has various resistances. These losses are usually measured as the dissipation factor at a frequency of 120 Hz. Leakage resistance in parallel with the capacitor defines the time constant of discharge of a capacitor. This time constant may vary between a small fraction of a second to many hours depending on capacitor construction, materials, and other electrical leakage paths, including dirt.

The equivalent series resistance of a capacitor is largely the resistance of the conductors of the capacitor plates and the resistance of the physical and chemical system of the capacitor. When an alternating current is passed through the capacitor, the losses in the equivalent series resistance are the major causes of heat developed in a capacitor. The same resistance also determines the maximum attenuation of a filter or bypass capacitor and the loss in a coupling capacitor connected to a load, such as a tweeter loudspeaker.

Dielectric absorption of a capacitor is the residual fraction of charge remaining in a capacitor after discharge. The residual voltage appearing at the capacitor terminals after discharge is of little concern in most applications but can seriously affect the performance of ADCs that must perform precision measurements of voltage stored in a sampling capacitor.

Self-inductance of a capacitor determines the high-frequency impedance of a capacitor and its ability to bypass high-frequency currents. It is determined largely by capacitor construction and tends to be highest in capacitors with metal foil that is not wound for minimum inductance. This causes radiation of magnetic-hum fields from power supply capacitors.

Nonpolarized Capacitors.   Nonpolarized capacitors must be used in all circuits where there is no direct voltage bias across the capacitor. They are also the capacitor of choice in all uses requiring capacity tolerances of 10 percent or less.

Film Capacitors.   Plastic is a preferred dielectrical material for capacitors since it can be manufactured with minimal imperfections in thin films. A metal-foil capacitor is constructed by winding layers of metal, plastic, metal, and plastic into a cylinder and then making a connection to the two layers of metal. A metallized foil capacitor uses two layers, each of which has a very thin layer of metal evaporated on one surface, and thereby one obtains a higher capacity per volume in exchange for a higher equivalent series resistance. Metallized foil capacitors are self-repairing in the sense that the energy stored in the capacitor is often sufficient to burn away the metal layer surrounding the void in the plastic film.

Depending on dielectrical material and construction, capacitance tolerances between 1 and 20 percent are usual, as are voltage ratings from 50 to 400 V. Construction includes axial leaded capacitors with a plastic outer wrap, metal-encased units, and capacitors in a plastic box suitable for printed circuit board insertion.

Polystyrene has the lowest dielectrical absorption of 0.02 percent, a temperature coefficient of -20 to -100 ppm/°C, a temperature range to 85°C, and extremely low leakage. Capacitors between 0.001 and 2 μF can be obtained with tolerances from 0.1 to 10 percent.

Polycarbonate has an upper temperature limit of 100°C, and capacitance changes about 2 percent up to this temperature. Maximum capacitance is 27 μF.

Polypropylene has an upper temperature limit of 85°C. These capacitors are particularly suited for applications where high inrush currents occur such as in switching power supplies. Maximum value is 10 μF.

Polyester is the lowest-cost material with an upper temperature limit of 125°C and an available capacitance of up to 27 μF.

Teflon and other high-temperature materials are used mostly in aerospace applications.

Foil Capacitors. Multiple layers of silvered mica packaged in epoxy or other plastic are the construction of mica capacitors, which are available in tolerances of 1 to 20 percent in values from 10 to 10,000 pF with temperature coefficients as low as 100 ppm. Voltage ratings between 100 and 600 V are common. Mica capacitors are used mostly in high-frequency filter circuits where low loss and high stability are needed.

Paper impregnated with oil or wax is one of the oldest dielectrics used in electronic equipment and is still in use today, primarily in high-voltage power supplies and some military equipment.

Electrolytic Capacitors. Aluminum foil electrolytic capacitors can be made nonpolar by use of two cathode foils instead of anode and cathode foil in construction. With care in manufacturing, these capacitors can be obtained with capacitance tolerance as tight as 10 percent with voltage ratings of 25 to 100 V peak and values of from 1 to 1000 μF. Loudspeaker crossover circuits are one of the uses at audio frequencies. Rated up to 280 V ac, these capacitors are also in use as phaseshift capacitors for motors in tape recorders.

Ceramic Capacitors. Barium titanate and other ceramics have a high dielectrical constant and a high breakdown voltage. The exact formulation determines capacitor size, temperature range, and variation of capacitance over that range and consequently capacitor application. An alphanumeric code defines these factors, a few of which are shown here.

Ratings of Y5V capacitors range from 1000 pF to 6.8 μF at 25 to 100 V and vary +22 to −82 percent in capacitance from −30 to +85°C. Ratings of Z5U capacitors range to 1.5 μF and vary +22 to −56 percent in capacitance from +10 to +85°C. These capacitors are the smallest in size and are used mostly as bypass capacitors. X7R capacitors range from 470 pF to 1 μF and vary 15 percent in capacitance from −55 to +125°C. Nonpolarized (NPO) rated capacitors range from 10 to 47,000 pF with a temperature coefficient of 0 to +30 ppm over the same temperature range. These capacitors see service in audio circuits. High-voltage capacitors, rated up to thousands of volts, see use in television sets.

Ceramic capacitors come in various shapes, the most common being the radial-lead disk. Multilayer monolithic construction results in the smallest size, which exists both in radial-lead styles and as chip capacitors for direct surface mounting on a printed circuit board.

Ceramic capacitors use materials similar to those in ceramic microphones or phonograph pickups. Although not intentionally exposed to a polarizing potential during manufacture or use, ceramic capacitors can develop electric signals when exposed to high levels of sound or vibration.

Polarized-Capacitor Construction. Polarized capacitors have a negative terminal, the cathode, and a positive terminal, the anode, and a liquid or gel between the two layers of conductors. The actual dielectric is a thin oxide film on the cathode, which has been chemically roughened for maximum surface area. The oxide is formed with a forming voltage, higher than the normal operating voltage, applied to the capacitor during manufacture. The direct current flowing through the capacitor forms the oxide and also heats the capacitor.

Whenever an electrolytic capacitor is not used for a long period of time, some of the oxide film is degraded. It is reformed when voltage is applied again with a leakage current that decreases with time. Applying an excessive voltage to the capacitor causes a severe increase in leakage current, which may cause the electrolyte to boil. The resulting steam may escape by way of the rubber seal or may damage the capacitor. Application of a reverse voltage in excess of about 1.5 V will cause forming to begin on the unetched anode electrode. This can happen when pulse voltages superimposed on a dc voltage cause a momentary voltage reversal.

Aluminum Electrolytic Capacitors. Aluminum electrolytic capacitors use very pure aluminum foil as electrodes, which are wound into a cylinder with interlayer paper or other porous material that contains the electrolyte. Aluminum ribbon staked to the foil at the minimum inductance location is brought through the insulator to the anode terminal, while the cathode foil is similarly connected to the aluminum case and cathode terminal.

Electrolytic capacitors may have voltage ratings from 6.3 to 450 V and rated capacitances from 0.47 µF to several hundreds of microfarads at the maximum voltage to several farads at 6.3 V. Capacitance tolerance may range from ±20 to +80/−20 percent. The operating temperature range may be rated from −25 to +85°C or wider. Leakage current of an electrolytic capacitor may be rated as low as 0.002 times the capacity times the voltage rating to more than 10 times as much.

Tantalum Electrolytic Capacitors. Tantalum electrolytic capacitors are the capacitors of choice if small size, 0.33- to 100-µF range at 10 to 20 percent tolerance, low equivalent series resistance and low leakage current are needed. They are suited where the less costly aluminum electrolytic capacitors have problems of performance. These capacitors are packaged in hermetically sealed metal tubes or with axial leads in epoxy plastic.

2.1.3 Inductors and Transformers

Inductors are passive components in which voltage leads current by nearly 90° over a wide range of frequencies. Inductors are usually coils of wire wound in the form of a cylinder. The current through each turn of wire creates a magnetic field that passes through every turn of wire in the coil. When the current changes, a voltage is induced in the wire and every other wire in the changing magnetic field. The voltage induced in the same wire that carries the changing current is determined by the inductance of the coil, and voltage induced in the other wire is determined by the mutual inductance between the two coils. A transformer has at least two coils of wire closely coupled by the common magnetic core, which contains most of the magnetic field within the transformer.

Inductors and transformers may weigh less than 1 g or more than 1 ton and may have specifications ranging nearly as wide.

Losses in Inductors and Transformers. Inductors have resistive losses because of the resistance of the copper wire used to wind the coil. An additional loss occurs because the changing magnetic field causes eddy currents to flow in every conductive material in the magnetic field. Using thin magnetic laminations or powdered magnetic material reduces these currents.

Losses in inductors are measured by the Q, or quality, factor of the coil at a test frequency, typically 1000 Hz for audio frequency coils. Losses in transformers for signals are sometimes given as an insertion loss in decibels. Losses in power transformers are given as core loss in watts when there is no load connected and as a regulation in percent, measured as the relative voltage drop for each secondary winding when a rated load is connected.

Transformer loss heats the transformer and raises its temperature. For this reason, transformers are rated in watts or volt-amperes and with a temperature code designating the maximum hotspot temperature allowable for continued safe long-term operation. Class A denotes 105°C and is most common in audio equipment, with classes B and H marking higher safe temperatures. The volt-ampere rating of a power transformer must be always larger than the dc power output from the rectifier circuit connected because volt-amperes, the product of the rms currents and rms voltages in the transformer, are larger by a factor of about 1.6 than the product of the dc voltages and currents.

Winding Capacitance and Stray Coupling.   Inductors also have capacitance between the wires of the coil, which causes the coil to have a self-resonance between the winding capacitance and the self-inductance of the coil. Circuits are normally designed so that this resonance is outside of the frequency range of interest. Transformers are similarly limited. They also have capacitance to the other winding(s), which causes stray coupling. An electrostatic shield between windings reduces this problem.

Air-Core Inductors.   Air-core inductors are used primarily in radio frequency applications because of the need for values of inductance in the microhenry or lower range. The usual construction is a multilayer coil made self-supporting with adhesive-covered wire. An inner diameter of 2 times coil length and an outer diameter 2 times as large yields maximum Q, which is also proportional to coil weight.

Ferromagnetic Cores.   Ferromagnetic materials have a permeability much larger than air or vacuum and cause a proportionally higher inductance of a coil that has all its magnetic flux in this material. Ferromagnetic materials in audio and power transformers or inductors usually are made of silicon steel laminations stamped in the forms of letters E and I of such proportions that minimum metal scrap occurs in manufacture. At higher frequencies, powdered ferric oxide is used. The continued magnetization and remagnetization of silicon steel and similar materials in opposite directions does not follow the same path in both directions but encloses an area in the magnetization curve and causes a hysteresis loss at each pass or twice per ac cycle.

All ferromagnetic materials show the same behavior; only the numbers for permeability, core loss, saturation flux density, and other characteristics are different.

Shielding.   Transformers and coils can radiate magnetic fields that can induce voltages in other nearby circuits. Similarly, coils and transformers can develop voltages in their windings when subjected to magnetic fields from some other transformer, motor, or power circuit. Steel mounting frames or chassis conduct these fields offering less reluctance than air.

The simplest way to reduce the stray magnetic field from a power transformer is to wrap a copper strip as wide as the coil of wire around the transformer enclosing all three legs of the core. Shielding occurs by having a short circuit turn in the stray magnetic field outside of the core.

The next best way is to arrange wiring for minimum hum pickup such as by twisting wires or signal conductors to induce opposite polarized voltages in each twist. If that method does not reduce pickup sufficiently, magnetic shielding must be used. This method causes the magnetic flux to travel mostly in the less reluctant high-permeability material, thereby reducing the most annoying interference, hum.

This latter method is used in low-level audio transformers for microphone signals and such. These transformers may be triple-shielded with an inside can of high-permeability material, such as Mumetal, then a copper can and a second magnetic shield like the first. The high cost of these transformers restricts their use to the most critical applications.

2.1.4 Diodes and Rectifiers

A diode is a passive electronic device that has a positive anode terminal and a negative cathode terminal and has a nonlinear voltage-current characteristic. A rectifier is assembled from one or more diodes for the purpose of obtaining a direct current from an alternating current, and this term also refers to large diodes used for this purpose. Although many types of diodes exist, only those of interest in audio applications are covered.

Diode Materials. Over the years many constructions and materials have been used as diodes and rectifiers. Rectification in electrolytes with dissimilar electrodes resulted in the electrolytic rectifier. The voltage-current characteristic of conduction from a heated cathode in vacuum or low-pressure noble gases or mercury vapor is the basis of these diodes and rectifiers still in use today. Semiconductor materials such as germanium, silicon, selenium, copper-oxide, or gallium arsenide can be processed to form a *pn* junction that has a nonlinear diode characteristic. Although all these systems of rectification have seen use, the most widely used rectifier in electronic equipment today is the silicon diode. The remainder of this section deals only with these and other silicon two-terminal devices.

The pn *Junction.* When biased in a reverse direction at a voltage well below breakdown, the diode reverse current is composed of two currents. One current is caused by leakage due to contamination and is proportional to voltage. The intrinsic diode reverse current is independent of voltage but doubles for about every 10°C in temperature. The forward current of a silicon diode is approximately equal to the leakage current multiplied by e (= 2.718) raised to the power given by the ratio of forward voltage divided by 26 mV with the junction at room temperature. In practical rectifier calculations the reverse current is considered to be important in only those cases where a capacitor must hold a charge for a time and the forward voltage drop is assumed to be constant at 0.7 V, unless a wide range of currents must be considered.

Diode Capacitance and Reverse-Recovery Time. All diode junctions have a junction capacitance that is approximately inversely proportional to the square of applied reverse voltage. This capacitance rises further with applied forward voltage. When a rectifier carries current in a forward direction, the junction capacitance builds up a charge. When the voltage reverses across the junction, this charge must flow out of the junction, which now has a lower capacitance, giving rise to a current spike in the opposite direction of the forward current. After the reverse-recovery time, this spike ends, but interference may be radiated into low-level circuits. For this reason rectifier diodes are bypassed with capacitors of about 0.1 μF located close to the diodes.

Tuning diodes have a controlled reverse capacitance that varies with applied direct tuning voltage. This capacitance may vary over a 2-to-1 to as high as a 10-to-1 range and is used to change the resonant frequency of tuned RF circuits. These diodes find application in radio and television receiver circuits.

Zener Diodes and Reverse Breakdown.   When the reverse voltage on a diode is increased to a critical voltage, the reverse leakage current will increase rapidly or avalanche. This breakdown or zener voltage sets the upper voltage limit a rectifier may experience in normal operation because the peak reverse currents may become as high as the forward currents. Rectifier and other diodes have a rated peak reverse voltage, and some rectifier circuits may depend on this reverse breakdown to limit high-voltage spikes that may enter the equipment from the power line. It should also be noted that diode dissipation is very high during these periods.

The reverse breakdown voltage can be controlled in manufacture to a few percent in zener diodes used extensively in voltage-regulator circuits. It should be noted at this point that the voltage-current curve of a $p + n$ junction may go through a section where a negative resistance occurs and voltage decreases a little while current increases. This condition can give rise to noise and oscillation, which can be minimized by connecting a ceramic capacitor of about 0.02 μF and an electrolytic capacitor of perhaps 100 μF in parallel with the zener diode. Voltage-regulator diodes are available in more than 100 types covering voltages from 2.4 to 200 V with rated dissipation between ¼ and 10 W. The forward characteristics of a zener diode usually are not specified but are similar to those of a conventional diode.

Precision voltage or bandgap reference diodes make use of the difference in voltage between two diodes carrying a precise ratio of forward currents. Packaged as a two-terminal device including an operational amplifier, these devices produce stable reference voltages of 1.2, 2.5, 5, and 10 V depending on type.

Current Regulators.   A special class of diodes is the current-regulator diode used in many small signal applications where constant current is needed. These diodes are junction field-effect transistors (FETs) with the gate connected to the source and effectively operated at zero-volt bias. Only two leads are brought out. Current-regulator diodes require a minimum voltage of a few volts for good regulation. Ratings from 0.22 to 4.7 mA are available.

2.1.5 Indicators

Indicators are generally passive components that send a message to the operator of the equipment. This message is most commonly a silent visual indication that the equipment is operating in some particular phase, is ready to operate, or is not ready. Indicator lights of different colors illuminating a legend or having an adjacent legend are most commonly used. Alphanumeric codes and complete messages are often displayed on cathode ray tubes or on liquid crystal displays. These more complex displays are computer or microprocessor-controlled.

Incandescent Lamps.   Miniature light bulbs are incandescent lamps operating at low voltage between 1 and 48 V, with currents from 0.01 to 4 A and total power requirements from 0.04 to more than 20 W, resulting in light output from 0.001 to more than 20 cd. The rated life of normally 10 to 50,000 h will decrease to one-tenth of the rating if the lamp voltage is increased to 20 percent above rated value. The resistance of the filament increases with temperature, varying by as much as a factor of 16 from cold to hot.

Light-Emitting Diodes.   Solid-state lamps or light-emitting diodes (LED) are *pn*-junction lasers that generate light when diode current exceeds a critical threshold value. Visible red light is emitted from gallium arsenide phosphide junctions. Green or amber light is emitted from doped gallium phosphide junctions. The junctions have a forward voltage drop of 1.7 to 2.2 V at a normal operating current of 10 to 50 mA. No other visible colors are commercially available, although blue has been achieved in the laboratory.

The LEDs are encased singly in round or rectangular plastic cases or assembled as multiples. A linear array of LEDs is often used in an arrangement similar to a thermometer to indicate volume or transmission level in audio or video circuits. An array, typically seven segments, can form the shapes of numerals and letters by selectively applying power to some or all segments. An array of 35 lamps in a 5 by 7 matrix can be connected to power to show the shape of letters, numerals, and punctuation marks. Semiconductor integrated circuits are available to achieve these functions with groups of these digit or indicator assemblies.

Light-emitting diodes have a typical operating life of about 50,000 h but have the disadvantage of relatively high current consumption, limited colors and shapes, and poor visibility in bright light.

Fluorescent, Liquid Crystal, and Other Displays.   Electrons emitted from a heated cathode or a special cold cathode can cause molecules of low-pressure gas, such as neon, to ionize and to emit light. Neon lamps require a current-limited supply of at least 90 V to emit orange light. The most frequent use of neon lamps is to indicate the presence of power line voltage. By means of a series resistor, the current is limited to a permissible value.

Emitted electrons can also strike a target connected to the anode terminal and coated with fluorescent phosphors. By directing the electron flow to "flood" different segments, alphanumeric displays can be produced similar to LED configurations with supply voltages found in battery-operated circuits. This is a relatively simple use of fluorescence compared to the cathode ray tubes found in computer terminals and television sets.

When certain solutions of organic chemicals are exposed to an electric field, these crystal-like ions align themselves with the field and cause light of only one polarization to pass through the liquid. A second polarizer of light then causes the assembly to be a voltage-controlled attenuator of light. Liquid crystal displays come in many shapes, need low operating power, depend on external light only, and fail to function at temperatures below freezing. These displays are found in test equipment, watches, computer terminals, and more recently in miniature television sets.

2.1.6 Solid-State Control Components

Varistor.   Varistors are symmetrical nonlinear voltage-dependent resistors, behaving not unlike two zener diodes connected back to back. The current in a varistor is proportional to applied voltage raised to a power N.

Modern varistors are normally made of zinc oxide, which can be produced to have an N factor of 12 to 40. In circuits at normal operating voltages, varistors are nearly open circuits shunted by a capacitor of a few hundred to a few thousand picofarads. Upon application of a high voltage pulse, such as a lightning charge, they conduct a large current, thereby absorbing the pulse energy in the bulk of the material with only a relatively small increase in voltage, thus protecting the circuit. Varistors are available for operating voltages from 10 to 1000 V rms and can handle pulse energies from 0.1 to more than 100 J and maximum peak currents from 20 to 2000 A. Typical applications are protection of power supplies and power-switching circuits and the protection of telephone and data-communication lines.

Switches and Relays.   Solid-state components for power switching use a triac as the switching element and require a minimum load current. They have a minimum voltage drop and can cause severe distortion of an audio signal unless a direct current larger than the signal is used as a bias. In audio equipment these devices are used for power line switching only.

Solid-state switches for switching of signal voltages are typically complementary MOS field-effect transistors controlled by logic-level control signals. These switches can handle signals up to 20 V in circuits of several kilo-ohm impedance. In the off state these switches have a capacitance of a few picofarads between terminals and a leakage current measured in picoamperes. In the on state they have a resistance of tens to hundreds of ohms between terminals. These switches are available with 1 to 4 single-pole on-off switches per dual in-line package, as 4- to 8-position selector switches and as three double-throw switches. These switches operate in a fraction of a microsecond.

2.2   ACTIVE CIRCUIT COMPONENTS

Active components can generate more alternating signal power into an output load resistance than the power absorbed at the input at the same frequency. Active components are the major building blocks in system assemblies such as amplifiers and oscillators.

2.2.1   Vacuum Tubes

Vacuum tubes are the active components that enabled the amplification and control of audio, radio frequency, and other signals and helped bring about the growth of the electronics industry from a laboratory curiosity early in the twentieth century to a high state of maturity in the 1960s. Since this time the transistor and the integrated solid-state circuit have largely replaced vacuum tubes in almost all applications. The only major uses of vacuum tubes today are as displays for television sets or computers and as generators of radio frequency power in selected applications or to generate X-rays for medical and industrial use.

A heated cathode coated with rare-earth oxides in a vacuum causes a cloud of electrons to exist near the cathode. A positive anode voltage with respect to the cathode causes some of these electrons to flow as a current to the anode. A grid of wires at a location between anode and cathode and biased at a control voltage with respect to the cathode causes a greater or lesser amount of anode current to flow. Other intervening grids also control the anode current and, if biased with a positive voltage, also draw grid current from the total cathode current.

Vacuum tubes operate at lower current and higher voltage than transistors and exist only as controllers of electron flow. Consequently, vacuum-tube audio amplifiers require output transformers for best match of loudspeaker load impedance to tube output impedance.

Vacuum tubes have a limited service life because the cathode coating evaporates with time and the heater filament also fails in the same fashion. Loss of vacuum is another cause of failure. For these and other reasons transistor circuits started replacing vacuum-tube circuits in less than one decade after the invention of the transistor in 1949.

2.2.2 Bipolar Transistors

A bipolar transistor has two *pn* junctions that act exactly like the diode *pn* junctions described in Sec. 2.1.4. These junctions are the base-emitter junction and the base-collector junction. In normal use the first junction would normally have a forward bias, causing normal conduction, and the second junction would have a reverse bias. If the material of the base were very thick, the flow of electrons into the *p*-material base junction (of an *npn* transistor) would go entirely into the base junction and no current would flow in the reverse-biased collector-base junction.

If, however, the base junction were quite thin, electrons would diffuse in the semiconductor crystal lattice into the base-collector junction having been injected into the base material of the base-emitter junction. The diffusion occurs because an excess electron moving into one location would bump out an electron in the adjacent semiconductor molecule, which would bump its neighbor. Thus, a collector current would flow that would be nearly as large as the injected emitter current.

The ratio of collector to emitter current is "alpha" or the common-base current gain of the transistor, normally a value a little less than 1.000. The portion of the emitter current not flowing into the collector will flow as a base current in the same direction as the collector current. The ratio of collector current to base current, or "beta," is the conventional current gain of the transistor and may be as low as 5 in power transistors operating at maximum current levels to as high as 5000 in super-beta transistors operated in the region of maximum current gain.

It should be noted at this point that the role of emitter and collector can be interchanged by a reversal of connections to the conventional emitter and collector terminals, however, with the result of drastically changed gain and transistor overload factors. This situation can occur in amplifier circuits when energy stored in a resonant or inductive load causes voltages in excess of supply voltages. It should also be noted that this simplified transistor description omits many factors and functions that affect transistor operation.

NPN *and* **PNP** *Transistors.* Bipolar transistors are identified by the sequence of semiconductor material going from emitter to collector. *NPN* transistors operate normally with a positive voltage on the collector with respect to the emitter, with *pnp* transistors requiring a negative voltage at the collector and the flow of current being internally mostly a flow of holes or absent excess electrons in the crystal lattice at locations of flow.

Since the diffusion velocity of holes is slower than of electrons, *pnp* transistors will have more junction capacitance and slower speed than *npn* transistors of the same size. Holes and electrons in *pn* junctions are minority carriers of electric current as opposed to electrons, which are the majority carriers and which can move freely in resistors or in the conductive channel of field-effect transistors. Consequently, bipolar transistors are known as minority carrier devices.

Transistor Materials.   The most common transistor material today is silicon, which permits transistor junction temperatures as high as 200°C. The normal base-emitter voltage in use is about 0.7 V, and collector-emitter voltage ratings of up to hundreds of volts are available. At room temperature these transistors may dissipate from tens of milliwatts to hundreds of watts with proper heat removal.

In the early 1980s transistors made of gallium arsenide and similar materials became available for use in microwave and very high speed circuits taking advantage of the very high diffusion speeds and low capacitances of this material. However, the difficulty of fabrication has limited their use to only specialized applications.

Transistor Impedance and Gain.   Transistor impedances and gain are normally referred to the common-emitter connection, which also results in the highest gain. It is useful to treat the transistor parameters first as if the transistor were an ideal transistor and then examine the degradations due to nonideal behavior.

If one assumes that the transistor has a fixed current gain, then the collector current is equal to the base current multiplied by the current gain of the transistor, and the emitter current is the sum of both of these currents.

Since the collector-base junction is reverse-biased, the output impedance of the ideal transistor is very high.

Actual bipolar transistors are degradations from this ideal model. Each transistor terminal may be thought of having a resistor connected in series although these resistors are actually distributed rather than lumped components. These resistors cause the transistor to have lower gain than predicted and to have a saturation voltage in both input and output circuits. In addition, actual transistors have resistances connected between terminals that cause further reductions in available gain, particularly at low currents and with high load resistances.

In addition to resistances, actual transistors also have multiple capacitors between terminals and the ideal transistor. These capacitances are in part caused by the finite diffusion velocities in silicon and in part are physical capacitors. The effect is that the transistor current gain decreases with increasing frequency with the transistor reaching unity current gain at the transition frequency. A second effect is a feedback current from collector to base through the base-collector capacitance. The third effect is storage of energy in the output capacitor similar to energy storage in a rectifier diode. This stored energy limits the speed of turnoff of transistors, a critical factor in video-amplifier and switching-circuit design.

Transistor Connections.   In the section above transistor impedances were described using the connection most frequently used in most circuits, the common-emitter connection that results in the highest circuit gain at the lower frequencies.

Power output stages of push-pull amplifiers make use of the common-collector or emitter-follower connection. Here, the collector is directly connected to the supply voltage, and the load is connected to the emitter terminal with signal and bias voltage applied to the base terminal. The voltage gain of such a circuit is a little less than 1.000, and the load impedance at the emitter is reflected to the base circuit as if it were increased by the current gain of the transistor.

At high frequencies the base of a transistor is often grounded for high-frequency signals, which are fed to the emitter of the transistor. The input impedance of the transistor now is a low impedance, which is easily matched to radio frequency transmission lines, in part helped by the minimal capacitive feedback within the transistor.

Combinations of these transistor connections result in circuits that have their own names. Transistor analysis has dealt largely with the small signal behavior. When a transistor has to handle large signals, other limitations must be observed. When handling low-frequency signals, a transistor may be viewed as a variable-controlled resistor between the supply voltage and the load impedance. The quiescent operating point in the absence of ac signals should be chosen so that the maximum signal excursions in both positive and negative directions can be handled without limiting due to near-zero voltage across the transistor at maximum output current or near-zero current through the transistor at maximum output voltage. This is most critical in class B push-pull audio amplifiers where first one transistor stage conducts current to the load during part of one cycle and then the other stage conducts during the other part. Similar considerations also apply for distortion reduction.

Limiting conditions are also the maximum capabilities of transistors under worst-case conditions of supply voltage, load impedance, drive signal, and temperature consistent with safe operation. In no case should the maximum voltage across a transistor ever be exceeded.

Switching and Inductive-Load Ratings. When using transistors for driving relays, deflection yokes of cathode ray tubes or any other inductive or resonant load, including loudspeakers, one should be aware that current in an inductor will tend to flow in the same direction, even if interrupted by the transistor. The resultant voltage spike caused by the collapse of the magnetic field may destroy the transistor unless it is designed to handle the energy of these voltage spikes. The manufacturers of power semiconductors have special transistor types and application information relating to inductive switching circuits. In the case of audio amplifiers the use of protection diodes may be sufficient.

Transistors are often used to switch currents into a resistive load. The various junction capacitances are voltage-dependent in the same manner as the capacitance of tuning diodes that have maximum capacitance at forward voltages, becoming less at zero voltage and lowest at reverse voltages. These capacitances and the various resistances combine into the switching delay times for turn-on and turn-off. If the transistor is prevented from being saturated when turned on, shorter delay times will occur for nonsaturated switching than for saturated switching. These delay times are of importance in the design of switching amplifiers or D/A converters.

Noise. Every resistor creates noise with equal and constant energy for each hertz of bandwidth, regardless of frequency. A useful number to remember is that a 1000-Ω resistor at room temperature has an open-circuit output noise voltage of 4 nanovolts per "root-hertz." This converts to 40 nV in a 100-Hz bandwidth or 400 μV in a 10-KHz bandwidth.

Bipolar transistors also create noise in their input and output circuits, and every resistor in the circuit also contributes its own noise energy. The noise of a transistor is effectively created in its input junction, and all transistor noise ratings are referred to it.

In an ideal bipolar transistor the voltage noise at the base is created by an equivalent resistor that has a value of twice the transistor input conductance at its emitter terminal, and the current noise is created by a resistor that has the value of twice the transistor input conductance at its input terminal. This means that the current noise energy is less at the base terminal of a common-emitter stage by the current gain of the transistor when compared to the current noise at the input of a grounded-base stage.

The highest signal-to-noise ratio in an amplifier can be achieved when the source resistance of the signal source is equal to the ratio of amplifier input noise voltage and input noise current and the reactive impedances have been tuned to zero. Audio frequency amplifiers usually cannot be tuned, and minimum noise may be achieved by matching transformers or by bias current adjustment of the input transistor. With low source impedances the optimum may not be reached economically, and the equipment must then be designed to have an acceptable input noise voltage.

Practical transistors are not ideal. All transistors show a voltage and current noise energy that increases inversely with frequency. At a corner frequency this noise will then become independent of frequency. Very low noise transistors may have a corner frequency as low as a few hertz, and ordinary high-frequency devices may have a corner frequency well above the audio frequency range. Transistor noise may also be degraded by operating a transistor at more than a few percent of its maximum current rating. Poor transistor design or manufacturing techniques may result in transistors that exhibit "popcorn" noise, so named after the audible characteristics of a random low-level switching effect.

2.2.3 Field-Effect Transistors

Field-effect transistors (FETs) have a conducting channel terminated by source and drain electrodes and a gate terminal that effectively widens or narrows the channel by the electric field between the gate and each portion of the channel. No gate current is required for steady-state control.

Current flow in the channel is by majority carriers only, analogous to current flow in a resistor. Onset of conduction is not limited by diffusion speeds but by the electric field accelerating the charged electrons.

The input impedance of an FET is a capacitance. Because of this, electrostatic charges in handling may reach high voltages that are capable of breaking down gate insulation.

FET Materials. FETs for audio and video frequency applications use silicon as the semiconducting material. Most FETs in logic circuits also make use of the advanced technology in silicon processing. Until recently, only at microwave frequencies have gallium arsenide and other III-V compounds made any inroads where superior speed has outweighed the greater difficulty in processing these exotic materials.

Field-effect transistors are made both in p-channel and n-channel configurations. An n-channel FET has a positive drain voltage with respect to source voltage, and a positive increase in gate voltage causes an increase in channel current. Reverse polarities exist for p-channel devices.

n- *and* **p-***Channel Field-Effect Transistors.* An *n*-channel FET has a drain voltage that is normally positive, and a positive increase in gate-to-source voltage increases drain current and transconductance. In single-gate field-effect transistors, drain and source terminals may often be interchanged without affecting circuit performance; however, power handling and other factors may be different. Such an interchange is not possible when two FETs are interconnected internally to form a dual-gate cascode-connected FET, or matched pairs, or when channel conductance is controlled by gates on two sides of the channel as in insulated-gate FETs.

FET Impedance and Gain. The input impedance of a field-effect transistor is very high at both audio and video frequencies and is primarily capacitive. The input capacitance consists of the gate-source capacitance in parallel with the gate-drain capacitance multiplied by the stage gain + 1, assuming the FET has its source at an ac-ground potential.

The output impedance of a common-source FET is also primarily capacitive as long as the drain voltage is above a critical value, which, for a junction-gate FET, is equal to the sum of pinch-off voltage and gate-bias voltage. When the pinch-off voltage is applied between the gate and source terminals, the drain current is nearly shut off or the channel is pinched off. Actual FETs have a high drain resistance in parallel with this capacitance. At low drain voltages near zero volts, the drain impedance of an ideal FET is a resistor reciprocal in value to the transconductance of the FET in series with the residual end resistances between the source and drain terminals and the conducting FET channel. This permits an FET to be used as a variable resistor in circuits controlling analog signals.

At drain voltages between zero and the critical voltage, the drain current will increase with both increasing drain voltage and increasing gate voltage. This factor will cause increased saturation voltages in power amplifier circuits when compared to circuits with bipolar transistors.

2.3 INTEGRATED CIRCUITS

An integrated circuit (IC) is a combination of circuit elements that are interconnected with each other and are formed on and in a continuous substrate material. Usually, an integrated circuit is monolithic and formed by steps that produce semiconductor elements along with resistors and capacitors. A hybrid integrated circuit contains silicon chips along with circuit elements partially formed on the substrate.

The circuit elements formed in integrated circuits are more closely matched to each other than separately selected components, and these elements are in intimate thermal contact with each other. The circuit configurations used in integrated circuits take advantage of matching and thermal coupling.

2.3.1 Digital Integrated Circuits

The basis of digital circuits is the logic gate that produces a high (or 1) or low (or 0) logic-level output with the proper combination of logic-level inputs. A number of these gates are combined to form a digital circuit that is part of the hardware of computers or controllers of equipment or other circuits. A digital circuit may be extremely complex, containing up to more than 100,000 gates.

Bipolar and field-effect transistors are the active elements of digital integrated circuits divided into families such as transistor-transistor logic (TTL), high-speed complementary metal-oxide-gate semiconductor (HCMOS), and many others. Special families are memories, microprocessors, and interface circuits between transmission lines and logic circuits. Thousands of digital integrated circuit types in tens of families have been produced since the first circuits of 1960.

2.3.2 Linear Integrated Circuits

Linear integrated circuits are designed to process linear signals in their entirety or in part as opposed to digital circuits that process logic, or on-off, signals only. Major classes of linear integrated circuits are operational amplifiers, voltage regulators, digital-to-analog and analog-to-digital circuits, circuits for consumer electronic equipment and communications equipment, power control circuits, and others not as easily classified.

Operational Amplifiers. An operational amplifier has a pair of differential input terminals that have very high gain to the output for differential signals of opposite phase at each input and relatively low gain for common-mode signals that have the same phase at each input (see Fig. 2.2). An external feedback network between output and the minus (−) input and ground or signal sets the circuit gain, with the plus (+) input at signal or ground level. Most operational amplifiers require a positive and a negative power supply voltage. One to four operational amplifiers may be contained on one substrate mounted in a plastic, ceramic, or hermetically sealed metal-can package.

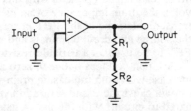

FIGURE 2.2 Operational amplifier with unbalanced input and output signals and a fixed level of feedback to set the voltage gain:

$$Voltage\ gain = \frac{1+R}{R}$$

[Source: Blair K. Benson (ed.), Audio Engineering Handbook, *McGraw-Hill, New York, 1988]*

Operational amplifiers may require external capacitors for circuit stability or may be internally compensated. Input stages may be field-effect transistors for high input impedance or may be bipolar transistors for low-offset voltage and low-voltage noise. Available types of operational amplifiers number in the hundreds.

Precision operational amplifiers generally have more tightly controlled specifications than general-purpose types.

Current and Voltage Ratings. The input-bias current of an operational amplifier is the average current drawn by each of the two inputs, + and −, from the input and feedback circuits. Any difference in dc resistance between the circuits seen by the two inputs multiplied by the input-bias current will be amplified by the circuit gain and become an output-offset voltage.

The input-offset current is the difference in bias current drawn by the two inputs, which multiplied by the sum of the total dc resistance in the input and feedback circuits and the circuit gain, becomes an additional output-offset voltage.

The input-offset voltage is the internal difference in bias voltage within the operational amplifier, which multiplied by the circuit gain, becomes an additional output-offset voltage.

If the normal input voltage is zero, the open-circuit output voltage is the sum of the three offset voltages.

2.4 TRANSISTOR AMPLIFIER DESIGN

Amplifiers are the functional building blocks of audio systems, and each of these building blocks contains several amplifier stages coupled together. An amplifier may contain its own power supply while an amplifier stage needs one or more external sources of power. The active component of each amplifier stage is usually a transistor or an FET. Other amplifying components, such as vacuum tubes, can also be used in amplifier circuits if the general principles of small and large signal voltages and current flows are followed.

2.4.1 The Single-Stage Transistor or FET Amplifier

The single-stage amplifier can best be described using a single transistor or FET connected as a common-emitter or common-source amplifier, using an *npn* transistor (Fig. 2.3*a*) or an *n*-channel FET (Fig. 2.3*b*) and treating *pnp* transistors or *p*-channel FET circuits by simply reversing the current flow and the polarity of the voltages.

DC Conditions.   At zero frequency or dc (direct current) and also at low frequencies, the transistor or FET amplifier stage requires an input voltage E_1 equal to the sum of the input voltages of the device (the transistor V_{be} or FET V_{gs}) and the voltage across the resistance R_e or R_s between the common node (ground) and the emitter or source terminal. The input current I_1 to the amplifier stage is equal to the sum of the current through the external resistor connected between ground and the base or gate and the base current I_b or gate current I_g drawn by the device. In most FET circuits the gate current may be so small that it can be neglected, while in transistor circuits the base current I_b is equal to the collector current I_c divided by the current gain beta of the transistor. The input resistance R_1 to the amplifier stage is equal to the ratio of input voltage E_1 to input current I_1.

The input voltage and the input resistance of an amplifier stage increases as the value of the emitter or source resistor becomes larger.

The output voltage E_2 of the amplifier stage, operating without any external load, is equal to the difference of supply voltage $V+$ and the product of collector or drainload resistor R_1 and collector current I_c or drain current I_d. An external load will cause the device to draw an additional current I_2, which increases the device output current.

As long as the collector-to-emitter voltage is larger than the saturation voltage of the transistor, collector current will be nearly independent of supply voltage. Similarly, the drain current of an FET will be nearly independent of drain-

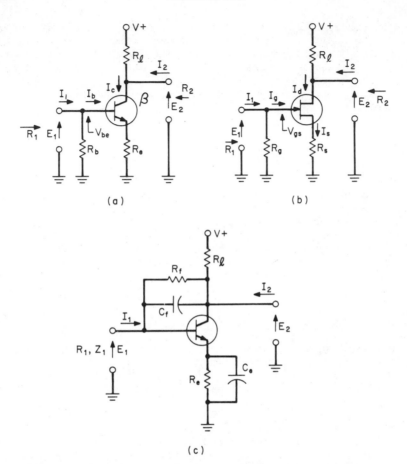

FIGURE 2.3 Single-stage amplifier circuits. (*a*) Common-emitter *npn*. (*b*) Common-source *n*-channel FET. (*c*) Single-stage with current and voltage feedback. *[Source: Blair K. Benson (ed.),* Audio Engineering Handbook, *McGraw-Hill, New York, 1988]*

to-source voltage as long as this voltage is greater than an equivalent saturation voltage. This saturation voltage is approximately equal to the difference between gate-to-source voltage and pinch-off voltage, the latter voltage being the bias voltage that causes nearly zero drain current. In some data sheets for FETs the pinch-off voltage is given under a different name as "threshold voltage." At lower supply voltages the collector or drain current will become less until it reaches zero when the drain-to-source voltage is zero or the collector-to-emitter voltage has a very small reverse value.

The output resistance R_2 of a transistor or FET amplifier stage is in effect the parallel combination of the collector or drain load resistance and the series connection of two resistors, consisting of R_e or R_s and the ratio of collector-to-emitter voltage and collector current or the equivalent drain-to-source voltage and drain current. In actual devices an additional resistor, the relatively large output resistance of the device, is connected in parallel with the output resistance of the amplifier stage.

The collector current of a single-stage transistor amplifier is equal to the base current multiplied by the current gain of the transistor. Since the current gain of a transistor may be specified as tightly as a two-to-one range at one value of collector current or may have just a minimum value, knowledge of the input current is usually not quite sufficient to specify the output current of a transistor.

Input and Output Impedance, and Voltage and Current Gain. As derived above for a common-emitter or common-source single-amplifier stage, the input impedance is the ratio of input voltage to input current, and the output impedance is the ratio of output voltage to output current. As the input current increases, the output current into the external output load resistor will increase by the current amplification factor of the stage. The output voltage will decrease because the increased current flows from the collector or drain voltage supply source into the collector or drain of the device. Therefore, the voltage amplification is a negative number having the magnitude of the ratio of output voltage change to input voltage change.

The magnitude of voltage amplification is often calculated as the product of transconductance G_m of the device and load resistance value. This can be done as long as the emitter or source resistor is zero or the resistor is bypassed with a capacitor that effectively acts as a short circuit for all signal changes of interest but allows the desired bias currents to flow through the resistor. In a bipolar transistor the transconductance is approximately equal to the emitter current multiplied by 39, which is the charge of a single electron divided by the product of Boltzmann's constant and absolute temperature in degrees Kelvin. In a field-effect transistor, this value will be less and usually proportional to input-bias voltage with reference to the pinch-off voltage.

The power gain of the device is the ratio of output power to input power, often expressed in decibels. Voltage gain or current gain may be stated in decibels but must be so marked.

AC Gain. The resistor in series with the emitter or source causes negative feedback of most of the output current, which reduces the voltage gain of the single amplifier stage and raises its input impedance (Fig. 2.3c). When this resistor R_e is bypassed with a capacitor C_e, the amplification factor will be high at high frequencies and will be reduced by approximately 3 dB at the frequency where the impedance of capacitor C_e is equal to the emitter or source input impedance of the device, which in turn is approximately equal to the inverse of the transconductance G_m of the device (Fig. 2.4a). The gain of the stage will be approximately 3 dB higher than the dc gain at the frequency where the impedance of the capacitor is equal to the emitter or source resistor. The above simplifications hold in cases where the product of transconductance and resistance values are much larger than 1.

A portion of the output voltage may also be fed back to the input, which is the base or gate terminal. This resistor R_f will lower the input impedance of the single amplifier stage, reduce current amplification, reduce output impedance of the stage, and act as a supply voltage source for the base or gate. This method is used when the source of input signals, and internal resistance R_s, is coupled with a capacitor to the base or gate and a group of devices with a spread of current gains, transconductances, or pinch-off voltages must operate with similar amplification in the same circuit. If the feedback element is also a capacitor C_f, high-

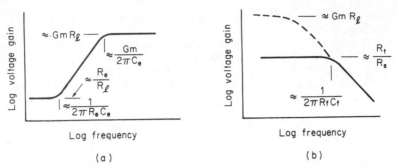

FIGURE 2.4 Feedback amplifier voltage gains. (*a*) Current feedback. (*b*) Voltage feedback. [*Source: Blair K. Benson (ed.),* Audio Engineering Handbook, *McGraw-Hill, New York, 1988]*

frequency current amplification of the stage will be reduced by approximately 3 dB when the impedance of the capacitor is equal to the feedback resistor R_f and voltage gain of the stage is high (Fig. 2.4*b*). At still higher frequencies amplification will decrease at the rate of 6 dB per octave of frequency. It should be noted at this point that the base-collector or gate-drain capacitance of the device has the same effect of limiting high-frequency amplification of the stage; however, this capacitor becomes larger as collector-base or drain-gate voltage decreases.

Feedback of the output voltage through an impedance lowers the input impedance of an amplifier stage. Voltage amplification of the stage will be affected only as this lowered input impedance loads the source of input voltage. If the source of input voltage has a finite source impedance and the amplifier stage has very high voltage amplification and reversed phase, the effective amplification for this stage will approach the ratio of feedback impedance to source impedance and also have reversed phase.

Common-Base or Common-Gate Connection. Here (Fig. 2.5*a*), the voltage amplification is the same as in the common-emitter or common-source connection; however, the input impedance is approximately the inverse of the transconductance of the device. As a benefit, the high-frequency amplification will be less affected because of the relatively lower emitter-collector or source-drain capacitance and the relatively low input impedance. This is the reason why the cascade connection (Fig. 2.5*b*) of a common-emitter amplifier stage driving a common-base amplifier stage exhibits nearly the dc amplification of a common-emitter stage with the wide bandwidth of a common-base stage. The other advantage of a common-base or common-gate amplifier stage is stable amplification at very high frequencies (VHF) and ease of matching to RF transmission-line impedances, usually 50 to 75 Ω.

Common-Collector or Common-Drain Connection. The voltage gain of a transistor or FET is slightly below 1.000. However, the input impedance of a transistor so connected will be equal to the value of the load impedance multiplied by the current gain of the device plus the inverse of the transconductance of the device (Fig. 2.5*c*). Similarly, the output impedance of the stage will be the impedance of the source of signals divided by the current gain of the transistor plus the inverse of the transconductance of the device.

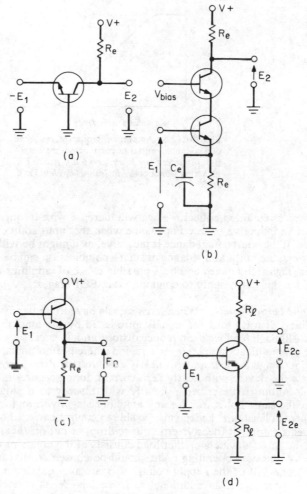

FIGURE 2.5 Transistor amplifier circuits. (*a*) Common-base *npn*. (*b*) Cascode *npn*. (*c*) Common-collector *npn*-emitter follower. (*d*) Split-load phase inverter. [*Source: Blair K. Benson (ed.),* Audio Engineering Handbook, *McGraw-Hill, New York, 1988*]

When identical resistors are connected between the collector or drain and the supply voltage and the emitter or source and ground, an increase in base or gate voltage will result in an increase of emitter or source voltage that is nearly equal to the decrease in collector or drain voltage. This type of connection is known as the *split-load phase inverter*, useful for driving push-pull amplifiers, although the output impedances at the two output terminals are unequal (Fig. 2.5*d*).

The current gain of a transistor decreases at high frequencies as the emitter-base capacitance shunts a portion of the transconductance, thereby reducing current gain until it reaches a value of 1 at the transition frequency of the transistor (Fig. 2.6). From this it can be seen that the output impedance of an emitter-

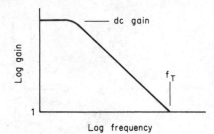

FIGURE 2.6 Amplitude-frequency response of a common-emitter or common-source amplifier. *[Source: Blair K. Benson (ed.),* Audio Engineering Handbook, *McGraw-Hill, New York, 1988]*

follower or common-collector stage will increase with frequency, having the effect of an inductive source impedance when the input source to the stage is resistive. If the source impedance is inductive, as it might be with cascaded-emitter followers, the output impedance of such a combination can be a negative value at certain high frequencies and be a possible cause of amplifier oscillation. Similar considerations also apply to common-drain FET stages.

Bias and Large Signals. When large signals have to be handled by a single-stage amplifier, distortion of the signals introduced by the amplifier must be considered. Although feedback can reduce distortion, it is necessary to ensure that each stage of amplification operates in a region where normal signals will not cause the amplifier stage to operate with nearly zero voltage drop across the device or to operate the device with nearly zero current during a portion of the cycle of the signal. Although described primarily with respect to a single-device-amplifier stage, the same holds true for any amplifier stage with multiple devices, except that here at least one device must be able to control current flow in the load without being saturated (nearly zero voltage drop) or cut off (nearly zero current).

If the single-device-amplifier load consists of the collector or drain load resistor only, the best operating point should be chosen so that in the absence of a signal, one-half of the supply voltage appears as a quiescent voltage across the load resistor R_1. If an additional resistive load R_1 is connected to the output through a coupling capacitor C_c (Fig. 2.7a), the maximum peak load current I_1 in one direction is equal to the difference between quiescent current I_1 of the stage and the current that would flow if the collector resistor and the external load resistor were connected in series across the supply voltage. In the other direction, maximum load current is limited by the quiescent voltage across the device divided by the load resistance. The quiescent current flows in the absence of an alternating signal and is caused by bias voltage or current only. Since most audio frequency signals have positive and negative peak excursions of equal probability, it is advisable to have the two peak currents be equal. This can be accomplished by increasing the quiescent current as the external load resistance decreases.

When several devices contribute current into an external load resistor (Fig. 2.7b) one useful strategy is to set bias currents so that the sum of all transconductances remains as constant as practical, which means a design for minimum distortion. This operating point for one device is near one-quarter the peak

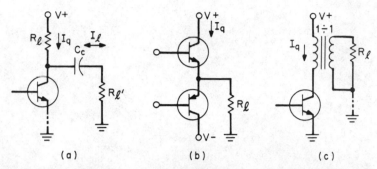

FIGURE 2.7 Output load-coupling circuits. (*a*) AC-coupled. (*b*) Series-parallel ac, push-pull half-bridge. (*c*) Single-ended transformer-coupled. *[Source: Blair K. Benson (ed.),* Audio Engineering Handbook, *McGraw-Hill, New York, 1988]*

device current for push-pull FET stages and at a lesser value for bipolar push-pull amplifiers.

When the load resistance is coupled to the single-device-amplifier stage with a transformer (Fig. 2.7c), the optimum bias current should be nearly equal to the peak current that would flow through the load impedance at the transformer with a voltage drop equal to the supply voltage.

CHAPTER 3
TELEVISION FUNDAMENTALS

K. Blair Benson
Television Technology Consultant, Norwalk, Connecticut

3.1 SYSTEM CONCEPTS

Television, meaning "seeing at a distance," may be described concisely as a system for the conversion of light rays from still or moving scenes and pictures into electric signals for transmission or storage, and subsequent reconversion into visual images on a screen. A similar function is provided in the production of motion-picture film; however, where film records the brightness variations of a complete scene on a single frame in a short exposure no longer than a fraction of a second, the brightness level of only one picture element (*pixel*) can be represented as a specific voltage level at an instant by an electric signal. Therefore, television requires that a scene be dissected into a frame composed of a mosaic of picture elements, a *picture element* being defined as the smallest area of a television picture that can be transmitted within the parameters of the system. This is accomplished by:

1. *Analyzing* an image of a scene with a photoelectric device in a sequence of horizontal scans from the top to the bottom of the image to produce an electric signal in which the brightness and color values of the individual picture elements are represented as voltage levels of a video signal

2. *Transmitting* the values of the picture elements in sequence as voltage levels of a video signal

3. *Reproducing* the image of the original scene in a video-signal display of parallel scanning lines on a viewing screen

3.1.1   History

Nipkow Mechanical-Optical Scanning System.   The first practical device for analyzing a scene to generate electric signals suitable for transmission was a scanning system proposed and built by P. Nipkow in 1884. The scanner consisted of a rotating disk, with a number of small holes or apertures arranged in a spiral, in front of a photoelectric cell. As the disk rotated, the spiral of 18 holes swept across the image of the scene from the top to bottom in a pattern of 18 parallel horizontal lines. Figure 3.1 is an outline of the 18-line Nipkow scanning system.

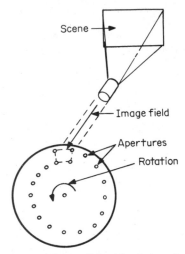

FIGURE 3.1 Nipkow disk with spiral configuration of scanning apertures.

The light from the small area covered by the apertures and the lens as the scene being televised was scanned was picked up by the light-sensitive photoelectric cell to produce an electric signal. The scanning process analyzed the scene by dissecting the scene into picture elements. The fineness of picture detail the system was capable of resolving was limited in the vertical and horizontal axes by the diameter of the area covered by the aperture in the disk. The output transmission circuit for the 18-line system introduced no bandwidth limitation to restrict the horizontal resolution.

For reproduction of the scene, a light source controlled in intensity by the electric signal was projected on a screen through a similar Nipkow disk rotated in synchronism with the pickup disk.

Despite subsequent improvements in England by J. L. Baird and in the United States by C. F. Jenkins, and in 1907 the use of Lee De Forest's vacuum-tube amplifier, the serious limitations of the mechanical approach discouraged any practical application of the Nipkow. The principal shortcomings were:

1. The inefficiency of the optical system

2. The use of rotating mechanical components

3. The lack of a high-intensity light source capable of being modulated by an electric signal at the higher frequencies required for video-signal reproduction

Nevertheless, Nipkow demonstrated a scanning process for the analysis of images by dissecting a complete scene into an orderly pattern of picture elements that could be transmitted as an electric signal and reproduced as a visual image. This approach is the basis for present-day television systems.

Electronic Imaging. The early Nipkow system scanned a scene directly with a photosensitive cell to produce an electric signal. In present-day television cameras, light rays from a scene are collected by a lens system to produce an image pattern of electrical charges on a photosensitive target surface of a vacuum-tube pickup device. The charges, corresponding in relative magnitude to the brightness levels in the scene, are converted to a video signal by *scanning* the charged photosensitive surface with an electron beam in a sequence of horizontal lines from the top to the bottom of the image frame. The video signal is developed, depending upon the type of tube, by either a modulation of the beam or by neutralization of the charges, to cause a signal current to flow in a load resistor. The use of an intermediate photoelectric transducer has two very important advantages:

1. Scanning of the image can be done by an electron beam in a vacuum or by sampling elements in a solid-state chip. No bulky mechanical components and motors are required.
2. The efficiency of the photoelectric conversion is high because of the charge-storage capability of photosensitive tubes or devices during scanning of the surface or sampling elements.

3.1.2 Image Analysis

Picture Elements. A picture element can be defined as the smallest area of a television picture capable of being outlined by an electric signal in a video transmission system. The number of picture elements (pixels) in a complete picture, and their geometric characteristics of vertical height and horizontal width, provide information as to (1) the total amount of detail that can be displayed in a scanned picture and (2) the sharpness of that detail, respectively.

The picture element (pixel) in the Nipkow mechanical scanning system was circular in shape, and the size was delineated by the lens and holes in the scanning disk. Thus, the pixel in the Nipkow system was of equal height and width. In an electronic television system a pixel is rectangular in shape, its boundries being horizontally parallel and vertically perpendicular to the scanning lines. Its center lies on the path traversed by the scanning spot (see Fig. 3.2).

The height of an *ultimate pixel* is determined by the width of the horizontal scanning line and is fixed for a given television system. The width of a pixel is dependent upon the bandwidth or cutoff frequency of the video channel and is equal to a half cycle at the channel cutoff frequency. Lowering the cutoff frequency causes the total number of pixels to decrease and the element length to become longer. In an ideal system with a mathematical-point-scanning aperture in the camera, a distortionless sharp-cutoff transmission channel, and perfectly linear transducers, the picture elements will exactly fill the raster area without overlapping.

The total number of pixels in a frame is calculated from the number of scanning lines and the bandwidth. This figure is reduced by the percentage of the frame interval used for horizontal and vertical scanning-beam retrace blanking in order to arrive at the number of active pixels. Table 3.1 lists picture-element statistics for the various television systems throughout the world.

Comparison to Film. An image in television is transmitted and reproduced as a sequential sampling of picture elements of a stored charge pattern; in contrast,

TABLE 3.1 Data on Picture Elements in Television Systems

Quantity	US/NTSC	PAL/CCIR	SECAM	HDTV (typical)
Pixels/frame	280,000	400,000	480,000	1,300,000
Pixels/raster	210,000	300,000	360,000	1,000,000
Pixels/line	440	520	620	990

film stores an image as a latent pattern of physical changes in a photographic emulsion. After chemical development, The emulsion of film yields a visual image of variable optical density, color intensity (*saturation*), and color hue. Thus, whereas television pictures are composed of a sequential and orderly reproduction of picture elements in image frames from signal voltages, images on film are a reproduction of all picture elements in random order, produced from an exposure of a complete scene in a fraction of a second.

FIGURE 3.2   Television scanning lines showing pixels of detail.

3.1.3   Image Scanning

Scanning Lines and Fields.   The image pattern of electrical charges on a camera-tube target corresponding to the brightness levels of a scene are converted to a video signal in a sequential order of picture elements in the scanning process. At the end of each horizontal-line sweep, the video signal is blanked while the beam returns rapidly to the left to start the scan of the next line until the image of the scene has been scanned from top to bottom to complete one field scan. The scan of a single field is analogous to reading a printed page where the eye scans each line from left to right, blinking at the end of each line, and progressing down to the bottom of the page.

After the completion of this first field scan, at the midpoint of the last line the beam again is blanked as it returns to the top center of the target where the process is repeated to provide a second field scan. The spot size of the beam as it impinges upon the target must be fine enough to leave unscanned areas between lines for the second scan. The pattern of scanning lines covering the area of the target, or the screen of a picture display, is called a *raster*.

Interlaced-Scanning Fields.   Because of the half-line offset for the start of the beam return to the top of the raster and for the start of the second field, the lines of the second field lie in between the lines of the first field. Thus, the lines of the two are *interlaced*. The two interlaced fields constitute a single television frame. A frame scan with the interlacing of the lines of two fields is shown in Fig. 3.3. Reproduction of the camera image on a cathode ray tube (CRT) is accomplished

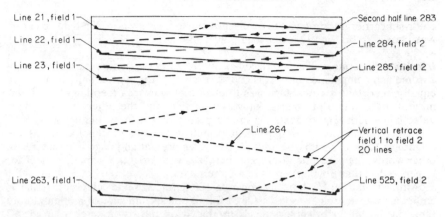

FIGURE 3.3 Interlaced-scanning pattern (raster). *(Source: Benson, 1986)*

by an identical scanning operation with the scanning beam being modulated in density by the video signal applied to an element of the electron gun to produce a variation in brightness on the phosphor screen.

Blanking of the scanning beam during the return trace is provided for in the video signal by a blacker-than-black pulse waveform. In addition, in most receivers and monitors another blanking pulse is generated from the horizontal and vertical scanning circuits and applied to the CRT electron gun to ensure a black screen during scanning retrace. The retrace lines are shown as diagonal dashed lines in Fig. 3.3. This figure illustrates an interlaced-scanning pattern (raster) of a picture display showing the horizontal and vertical scanning-beam retrace. After 21 lines for vertical blanking and sync, an electron beam progressively traces 241.5 (262.5-21) horizontal lines in sequence from the top to the bottom of the screen. At an offset of a half-line, the process is repeated with 241.5 more lines traced in the spaces between the lines of the first field.

Scanning Standards. The standards adopted by the Federal Communications Commission for monochrome television in the United States specified a system of 525 lines per frame,[1] transmitted at a frame rate of 30 Hz, with each frame composed of two interlaced fields of horizontal lines. Initially, in the development of television transmission standards, the 60-Hz power[2] line was chosen as a convenient reference for vertical scan. Furthermore, in the event of coupling of power-line hum crossing into the video-signal or scanning-deflection circuits, the visible effects were stationary and less objectionable than moving hum bars or distortion of horizontal-scanning geometry. With improvements in equipment and the introduction of compatible color programming, the power-line reference was replaced with a stable crystal oscillator.[3] For reasons relating to RF-carrier to

---

1. In the United Kingdom 625-line system, "picture" is used for NTSC's "frame."
2. 50 Hz was chosen for the 625-line systems used in Europe where the power lines operate at 50 Hz.
3. Currently, many low-cost educational and industrial cameras use the 60-Hz power line for scanning reference as a matter of convenience for interlocking two or more cameras, and for the same reason as in early monochrome broadcasting, immobilize video hum bars and scanning geometric distortions from ac power.

color-subcarrier interference, to be discussed in a later section, the field-scan rate was chosen to be slightly lower than power-line at 59.95 Hz.

Interlaced Scanning. The interlaced-scanning format, standardized for monochrome and compatible color, was chosen primarily for two partially related and equally important reasons. One was to eliminate viewer perception of the intermittent presentation of images, known as *flicker*, and the other was to reduce video bandwidth requirements for an acceptable flicker threshold level.

Perception of flicker is dependent primarily upon two conditions: (a) the brightness level of an image and (b) the relative area of an image in a picture. In other words, the perception of flicker increases with brightness and is greater for larger, rather than smaller, areas in a picture.

Although the 30-Hz transmission rate for a full 525-line television frame is comparable to the highly successful 24-frame-per-second rate of motion-picture film,[4] at the higher brightness levels produced on television screens, if all 483 lines (525 less blanking) of a television image were to be presented sequentially as single frames, viewers would observe a disturbing flicker in picture areas of high brightness. For a comparison, motion-picture theaters on the average produce a screen brightness of 10 to 25 fL (footlambert), whereas a direct-view CRT may have a highlight brightness of 50 to 80 fL.

The threshold of flicker perception for large areas is at a much lower frame rate than for higher-resolution detail. By the use of interlaced scanning, single-field images with one-half the vertical-resolution capability of the 525-line system are provided at the high flicker-perception threshold rate of 60 Hz. Higher resolution of the full 490 lines (525 less vertical blanking) of vertical detail is provided at the lower flicker-perception threshold rate of 30 Hz.

The result is a relatively flickerless picture display at a screen brightness of well over 50 to 75 fL, more than double that of motion-picture film projection. Both 60-Hz fields and 30-Hz frames have the same horizontal-resolution capability.

The second advantage of interlaced scanning, compared to progressive scanning, is a reduction in video bandwidth for an equivalent flicker threshold level. Progressive scanning of 525 lines would have to be completed in 1/60 to achieve an equivalent large-area high level of flicker perception. This would require a line scan to be completed in half the time of an interlaced scan. The bandwidth then would be double for an equivalent number of pixels per line. In other words, an equivalent of the National Television Systems Committee (NTSC) system would provide 40 rather than 80 TV lines/MHz.

The existing 525-line monochrome standards were retained for color in the recommendations of the NTSC for compatible color television in the early 1950s. The NTSC and its numerous subcommittees were composed of cognizant engineers recruited on a voluntary basis from manufacturers, broadcasters, and research laboratories throughout the United States. The NTSC system, adopted in 1953 by the Federal Communications Commission (FCC) in their *Rules and Regulations* (Part 73.699), specifies a scanning system of 525 horizontal lines per frame, with each frame consisting of two interlaced fields of 262.5 lines at a field rate of 59.94 Hz (see Fig. 3.4).

Forty-two of the 525 lines in each frame are blanked as black picture signals

---

4. Motion-picture projectors frequently use a double- or triple-segment shutter to increase the flicker rate and thus reduce the flicker perception on picture highlights. One segment of a double shutter blanks the film pulldown and the other doubles the flicker rate.

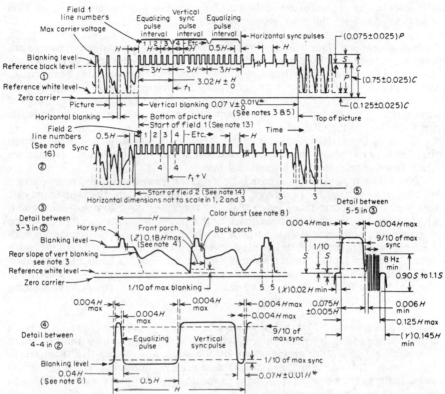

FIGURE 3.4 Television synchronizing waveform for color transmission. *Notes*: (1) H = time from start of one line to start of next line. (2) V = time from start of one field to start of next field. (3) Leading and trailing edges of vertical blanking should be complete in less than $0.1H$. (4) Leading and trailing slopes of horizontal blanking must be steep enough to preserve minimum and maximum values of $(x + y)$ and (z) under all conditions of picture content. *(5) Dimensions marked with an asterisk indicate that tolerances are permitted only for long time variations and not for successive cycles. (6) *Equalizing pulse* duration shall be between 0.45 and 0.55 of the horizontal synchronizing pulse duration. (7) Color burst follows each horizontal pulse but is omitted following the equalizing pulses and during the broad vertical pulses. (8) Color bursts to be omitted during monochrome transmission. (9) The burst frequency shall be 3.579545 MHz. The tolerance on the frequency shall be ±10 Hz with a maximum rate of change not to exceed 0.1 Hz/s. (10) The horizontal scanning frequency shall be 2/455 times the burst frequency. (11) The dimensions specified for the burst determine the times of starting and stopping the burst but not its phase. (12) Dimension P represents the peak excursion of the luminance signal from blanking level but does not include the chrominance signal. Dimension S is the synchronizing pulse amplitude above blanking level. Dimension C is the peak carrier amplitude. (13) Start of field 1 is defined by a whole line between first equalizing pulse and preceding H sync pulses. (14) Start of field 2 is defined by a half line between the first equalizing pulse and the preceding H sync pulses. (15) Field 1 line numbers start with the first equalizing pulse in field 1. (16) Field 2 line numbers start with second equalizing pulse in field 2. (17) Refer to text for further explanations and tolerances. (18) During color transmissions, the chrominance component of the picture signal may penetrate the synchronizing region and the color burst penetrates the picture region. *(Source: FCC, 1983, in Benson, 1986)*

and reserved for transmission of the vertical-scanning synchronizing signal. This results in 483 visible lines for picture information.

Camera Scanning.   The scanning operation in a present-day vidicon-type camera tube is accomplished by scanning the rear side of the photosensitive target with horizontal sweeps of an electron beam. Electrons from the scanning beam equalize the charges on the target, thus generating a video signal by causing a current to flow in an output load resistor. Alternatively, in a solid-state *charge-coupled device* (CCD) the horizontal rows of elements are sampled sequentially from the top to bottom of the CCD chip.

The sampling pattern is similar to the scanning of a pickup-tube target except that, because of the continuous vertical-scanning action in the latter, the lines are tilted down in each field by a pitch of 1 line.

The resultant electric signals are processed for transmission and reproduction on a television monitor or receiver. As an intermediate step, the signals may be recorded on magnetic tape, video disk, or motion-picture film for subsequent playback and viewing. After transmission, the pixels are reproduced in the same mosaic pattern on a direct-view cathode ray tube or other display device. In both the camera and in most picture-display devices, during the fast return of the beam after the vertical and horizontal sweeps, the picture signal is blanked by an internally generated signal to avoid a visible retrace signal. Similarly, a blacker-than-black blanking signal is incorporated with horizontal- and vertical-synchronizing pulses to form the composite video signal.

Motion Reproduction.   The illusion of continuous motion is obtained, as in motion-picture film, by presenting a series of still pictures, or "frames," each frame differing from the previous one by the degree of motion that occurred in the interval between them. The rate of frame presentation must be fast enough to avoid the appearance of jerky or uneven motion and the perception of flicker at normal viewing brightness.

Motion-picture film captures the image of an entire scene on each frame of film in a fraction of a second for projection as a series of still frames at a rate of 24/s to simulate motion. On the other hand, television pictures are made up of a series of horizontal lines, where each line captures that portion of a scene occurring in 63 μs, the duration of a single line scan. Thus, motion in television is simulated also by a series of still frames, with each frame consisting of a sequential pattern of approximately 500 horizontal lines, rather than the single exposure of motion-picture film.

3.1.4   Synchronizing and Video Signals

Signal Waveforms.   In monochrome television transmissions two basic synchronizing signals are provided to control the timing of picture-scanning deflection. These are:

1. Horizontal-sync pulses at line rate.
2. Vertical-sync pulses at field rate in the form of an interval of wide horizontal-sync pulses at field rate. Included in the interval are equalizing pulses at twice the line rate to preserve interlace in each frame between the even and odd fields that are offset by a half line.

In color transmissions, a third synchronizing signal is added during horizontal-scan blanking to provide a frequency and phase reference for color-signal-encoding circuits in cameras and decoding circuits in receivers. These synchronizing and reference signals are combined with the picture video signal to form a composite video signal.

The scanning and color-decoding circuits in receivers must follow the frequency and phase of the synchronizing signals to a high degree of precision in order to produce a stable and geometrically accurate image of the proper color hue and saturation. Any change in timing of successive vertical scans can impair the interlace of the even and odd fields in a frame.

Small errors in horizontal-scan timing of lines in a field can result in a loss of resolution in vertical-line structures. Periodic errors over several lines that may be out of the range of a receiver's horizontal-scan automatic-frequency-control (AFC) circuit will be evident as jagged vertical lines.

Industry Standards. The Federal Communications Commission (FCC) is the government body responsible for the regulation of broadcasting in the United States. The waveforms defined by the FCC for the radiated signal are shown in Fig. 3.4. The modulation extends from the synchronizing pulses at maximum carrier level (100 percent) to reference picture white at 7.5 percent. Because an increase in the amplitude of the radiated signal corresponds to a decrease, rather than an increase, in picture brightness, the polarity of modulation is termed *negative*.

Following the authorizations of the FCC *Rules and Regulations* standardizing—first monochrome and later color—transmissions, the Electronics Industry Association (EIA) released their RS170A standard for synchronizing waveforms for studio facilities. This standard defines the waveform and nominal signal levels of base-band sync and video signals, rather than the transmitted RF signal. The RS170A standard, shown in Fig. 3.5, is used as a reference in the following description of video and synchronizing signals.

Composite Video. The term *composite* is used to indicate that the signal contains:

1. Picture luminance and chrominance information
2. Timing information for synchronization of scanning and color-signal-processing circuits

Figure 3.5 shows the waveforms of the composite video signal. The negative-going portion is used to transmit information for synchronization of scanning circuits. The positive-going portion of the amplitude range is used to transmit luminance information representing brightness and, for color pictures, chrominance.

Signal Levels. At the completion of each line scan in a receiver or monitor, a horizontal-synchronizing (H-sync) pulse in the composite video signal triggers the scanning circuits to return the beam very rapidly to the left of the screen for the start of the next line scan (Fig. 3.3). During the return time, a horizontal-blanking signal at a level lower than that corresponding to the blackest portion of the scene is added to avoid the visibility of the retrace lines. In a similar manner, after completion of each field, a vertical-blanking signal blanks out the retrace signal of the scanning beam as it returns to the top of the picture to start the scan of the next field.

The small-level difference between video reference black and blanking level is

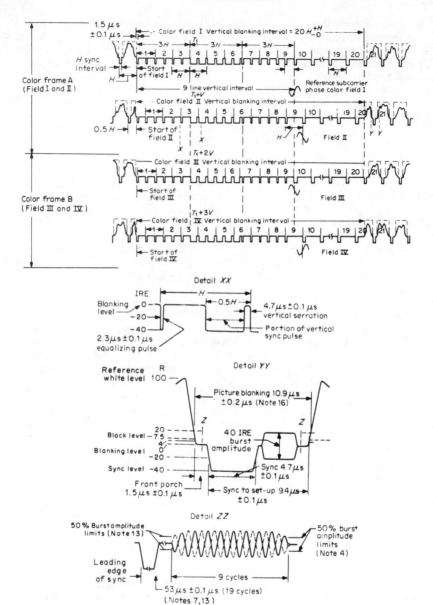

Detail XX

Detail YY

Detail ZZ

called *setup*. This is inserted as a guard interval to ensure separation of the synchronizing and video-information functions and adequate blanking of scanning-retrace lines on receivers. The need for setup initially was dictated by (a) instability or frequent drift of black-level (brightness) adjustment and the lack of retrace blanking in receivers, and (b) video-in-sync from low-frequency phase distortion in intercity common-carrier circuits causing malfunction of receiver sync circuits. In current practice, the need for setup is questionable because of the use of return-trace blanking in receivers and monitors and of improved transmission circuits. In fact, in England and many other countries zero setup is standard.

The waveform in detail YY of Fig. 3.5 shows the various reference levels of video and sync in the composite signal. The units for the levels were specified initially by the Institute of Radio Engineers (IRE) [now the Institute of Electrical and Electronic Engineers (IEEE)]. These are listed in Table 3.2.

Note in Table 3.2 that the +20 and −20 levels are specified for measurement of blanking and sync. In the measurement of sync, a negative-going transition of every period of horizontal sync is referred to as the *leading edge of H sync*. The half-amplitude point of this waveform is used as the reference for all pulse-timing measurements.

While most present-day video systems transmit only composite video, in cam-

FIGURE 3.5 EIA RS170A tentative standard, synchronizing waveforms. *Notes*: (1) Specifications apply to studio facilities. Common carrier, studio to transmitter, and transmitter characteristics are not included. (2) All tolerances and limits shown in this drawing permissible only for long time variations. (3) The burst frequency shall be 3.579545 MHz ±10 Hz. (4) The horizontal scanning frequency shall be 2/455 times the burst frequency [one scan period (H) = 63.556 μs]. (5) The vertical scanning frequency shall be 2/525 times the horizontal scanning frequency [one scan period (V) = 16.683 μs]. (6) Start of color fields I and III is defined by a whole line between the first equalizing pulse and the preceding H sync pulse. Start of color fields II and IV is defined by a half line between the first equalizing pulse and the preceding H sync pulse. Color field I: that field with positive-going zero crossings of reference subcarrier most nearly coincident with the 50% amplitude point of the leading edges of even-numbered horizontal sync pulses. Reference subcarrier is a continuous signal with the same instantaneous phase as burst. (7) The zero crossings of reference subcarrier shall be nominally coincident with the 50% point of the leading edges of all horizontal sync pulses. When the relationship between sync and subcarrier is critical for program integration, the tolerance on this coincidence is ±40° of reference subcarrier. (8) All rise times and fall times unless otherwise specified are to be 0.145 ±0.02 μs measured from 10 to 90% amplitude points. All pulse widths are measured at 50% amplitude points unless otherwise specified. (9) Tolerance on sync level, reference black level (setup), and peak-to-peak burst amplitude shall be ±2 IRE units. (10) The interval beginning with line 17 and extending through line 20 of each field may be used for test, cue, and control signals. (11) Extraneous synchronous signals during blanking intervals, including residual subcarrier, shall not exceed 1 IRE unit. Extraneous nonsynchronous signals during blanking intervals shall not exceed 0.5 IRE unit. All special-purpose signals (VITS, VIR, etc.) when added to the vertical blanking interval are excepted. Overshoot on all pulses during sync and blanking, vertical and horizontal, shall not exceed 2 IRE units. (12) Burst-envelope rise time is 0.3 + 0.2 − 0.1 μs measured between the 10 and 90% amplitude points. Burst is not present during the nine-line vertical interval. (13) The start of burst is defined by the zero crossing (positive or negative slope) preceding the first half cycle of subcarrier that is 50% or greater of the burst amplitude. Its position is nominally 19 cycles of subcarrier from the 50% amplitude point of leading edge sync (see detail ZZ). (14) The end of burst is defined by the zero crossing (positive or negative slope) following the last half cycle of subcarrier that is 50% or greater of the burst amplitude. (15) Monochrome signals shall be in accordance with this drawing except that burst is omitted and fields III and IV are identical to fields I and II respectively. (16) Occasionally measurement of picture blanking at 20 IRE units is not possible because of scene content as verified on a picture monitor. (*Source: Electronics Industry Association, in Benson, 1986*)

TABLE 3.2   Video- and Sync-Signal Levels

Signal level	Level
Reference white	100
Blanking-level width measurement	20
Color-burst sine-wave peak	+20 to −20
Reference black	7.5
Blanking	0
Sync-pulse-width measurement	−20
Sync level	−40

eras and in some portions of studio or field systems a "noncomposite" signal may be transmitted. A noncomposite signal includes the picture and blanking signal with a burst of color subcarrier for synchronization of color-encoding circuits. In these cases scanning-drive signals for camera are provided separately, and the synchronizing signals are added to the noncomposite video further on in the system.

Composite Sync.   Composite sync provides all the timing signals for synchronization of deflection circuits in cameras, picture and waveform monitors, and signal-processing equipment. It contains all the signal components of a composite video signal except video and setup. These are horizontal sync, vertical sync, equalizing pulses, and color-sync burst (see Fig. 3.6). In Fig. 3.6*b*, the vertical-scan retrace occurs approximately between points *x* and *y* in *b*. The exact timing of *x* and *y* depends upon the type of vertical-sync circuit and the vertical-hold adjustment, if one is provided.

Referring to Table 3.2, composite sync is transmitted between the IRE levels of 0 and −40 for the deflection components, and +20 and −20 for the color-synchronizing burst.

Composite sync is inserted into a television system at the following points:

1. Noncomposite video-output signal from cameras
2. Composite-sync input of processing amplifiers for VTRs
3. The output of the noncomposite section in program production switchers
4. Synchronization of monitors displaying a noncomposite video signal
5. Synchronization of VTRs, telecine cameras, special-effects generators, and other auxiliary studio equipment

A "fixed-setup" signal may be added in (1) through (3) above to permit cameras to be operated with zero setup. In this mode of operation, black level can be adjusted more precisely to the blanking level shown on the waveform oscilloscope, rather than to the setup-level indication on the waveform-monitor graticule.

Horizontal- and Vertical-Synchronizing Signals.   Horizontal drive (HD) and vertical drive (VD) are used to trigger scanning and internal blanking-generating circuits in some video-signal-generating equipment. The leading edge of horizontal drive does not correspond directly to that of horizontal sync of composite video. Instead, the leading edge is coincident with the end of active picture on a line,

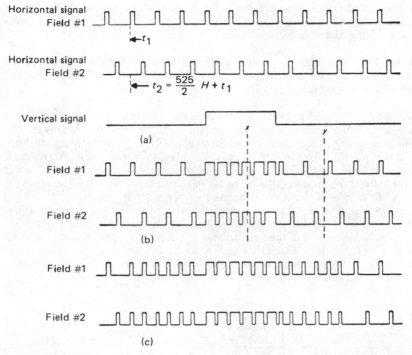

FIGURE 3.6 Composition of deflection-synchronizing signals. *(Source: Fink, 1957, in Benson, 1986)*

which is the start of horizontal blanking. The trailing edge is coincident with the trailing edge of composite sync. Horizontal drive continues without interruption during the vertical-sync interval.

The leading edge of vertical drive is coincident with the end of active picture within a field. The trailing edge is coincident with the leading edge of sync on the first line after the vertical interval. Generally, vertical drive is not used to generate a composite video signal.

Horizontal and Vertical Blanking. Blanking pulses may be provided for signal-generating equipment as a composite horizontal and vertical signal or as two separate signals. The pulse widths and rise times are suitable for transfer directly to the output video signal of a camera or other signal generator, or a processing amplifier.

Color-Frame Reference Pulse. The problems associated with matching color fields in video-tape editing have resulted in the insertion of a color-field identification pulse to facilitate production editing where the four-field repetition rate must be timed to be coincident at editing cuts between similar or identical picture material. The pulse is a 1-line-long, 9-bit code that begins on line 11 of field 1 and ends on line 12.

Burst Flag. The reference subcarrier is gated into each active picture line by the burst flag pulse. The rise and fall of the burst flag determine the shape or envelope of the burst signal in picture video.

Color Black. Color black is a composite black-video signal. It contains the proper setup with no video signal, sync and color-burst amplitudes, and the proper subcarrier-to-horizontal (ScH) phase relationship.

Vertical-Interval Reference, Test, and Data Signals. The extent and complication of color-television distibution networks has dictated the need for a means of continuous monitoring of the technical performance without programming interruption. This is possible for the sync signals, including the color burst. However, it does not provide a meaningful indication of performance in the range of levels used for video transmission. This led to the introduction by the EIA and approval by the FCC in 1974 of the vertical-interval reference (VIR) signal.

The VIR signal is intended to be representative of the program picture information (Fig. 3.7). Therefore, it has been formatted in the video-signal amplitude range a line near the end of the vertical interval (line 19) so that it can be treated like picture information. The VIR signal contains reference levels for black, 50 percent gray, and chrominance phase and amplitude, the latter being superimposed on a luminance level close to typical skin-tone luminance.

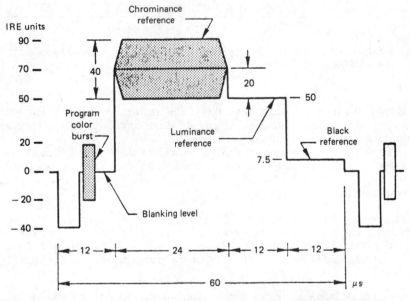

FIGURE 3.7 Vertical-interval reference (VIR) signal. *Note*: The chrominance reference and the program color burst have the same phase. *(Source: FCC, in Benson, 1986)*

The FCC *Rules and Regulations* specify several other vertical-interval test signals (VITS) for test and monitoring purposes during normal picture transmission. These are listed below.

Multiburst test signal: On line 17, field 1, sequential bursts of sine waves at frequencies of 0.5, 1.2, 2.0, 3.0, 3.58 (color subcarrier), and 4.1 MHz, preceded by a reference white-level pulse at 100 IRE units. The bursts are 40 IRE units peak to peak from 10 to 60 IRE units.

Color-bar test signal: On line 17, field 2.

Modulated stairstep, T pulses (2T and 12.5T), and white bar:: On line 18, field 2.

Scanning Control and Synchronization. The periodic timing and synchronization of horizontal and vertical scanning in cameras and picture displays requires three types of signals. These are:

1. Horizontal-sync pulses at line rate.
2. Vertical-sync pulses in the form of an interval of wide horizontal-sync pulses at field rate.
3. Equalizing pulses at twice line rate to preserve interlace between the even and odd fields in each frame. (Equalizing pulses are offset by a half line.)

In addition, color signals include bursts of a color-subcarrier sine wave to synchronize color-decoding circuits.

3.2 TELEVISION CAMERAS

3.2.1 Principles of Operation

Photosensitivity. In a television camera tube an electron beam traces a pattern of horizontal- and vertical-scanning lines over the light-sensitive surface of the photocathode. The photosensitive surface is charged to different potentials corresponding to differences in brightness of the scene focused on the surface by the camera lens and optical system. In a photoconductive surface, such as in vidicon, Plumbicon, Newvicon, or Saticon pickup tubes, as the beam passes over the rear surface of the photocathode or target, it neutralizes these charges and causes a current to flow in the signal-output lead and load resistance. In the earlier iconoscope the beam scanned the optical image side of the mosaic target of the tube causing secondary electrons to be emitted. These were collected on a signal anode to produce a signal current in a load resistor.

Storage Camera Tubes and Devices. All the tube types mentioned above have the capability of storing the electrical charge image until it is neutralized by the scanning beam. Because of the storage capability, during exposure of the photosensitive surface elements to light between the periodic discharges by the scanning beam, the charge continues to build to an increasingly higher level.

This feature is of extreme importance because of the resultant efficient use of light energy. The action may be compared to the exposure of photographic film where the longer the exposure time, the lower the light level required for the same picture contrast. Solid-state *charge-coupled devices* (CCDs) used in cameras also take advantage of the storage principle to provide high sensitivity.

Nonstorage systems are discussed in Sec. 3.2.6.

3.2.2 Storage Camera-Tube Development

Iconoscope Camera Tube. The birth of present-day television followed a development of major significance in 1923. At RCA a Russian scientist, Vladimir K. Zworykin, developed an electronic pickup device he called the *iconoscope* (from Greek for "image" and "to see"). Zworykin's television camera tube stored an image of a scene as a mosaic pattern of electrical charges that, by an electron-beam scanning process, released secondary electrons which could be read out sequentially as a video signal (see Fig. 3.8*a*).

Although the iconoscope provided good resolution, relatively high light levels, that is, studio illumination of 500 or more footcandles (fc), were necessary. In addition, picture quality was degraded by spurious flare. This was caused by photoelectrons and secondaries, resulting from the high potential of the scanning

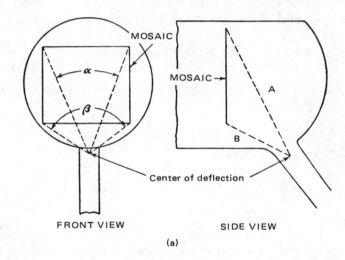

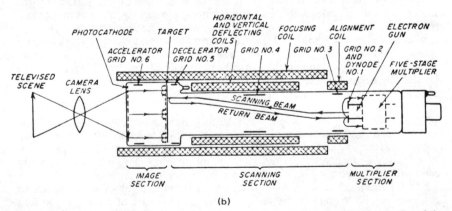

FIGURE 3.8 (*a*) Iconoscope deflection geometry. Alpha and beta are deflection angles of the scanning beam that is directed from below the axis normal to the mosaic. (*Source: Fink, 1957, in Benson, 1986*) (*b*) Schematic arrangement of the image orthicon. (*Source: Fink, 1957, in Benson, 1986*)

beam falling back at random on the storage surface. The presence of the flare signals and the lack of a reference black signal, because of the capacitive coupling through the signal plate, resulted in a gray scale that varied with scene content and necessitated virtually continuous manual correction of video-gain and blanking levels in processing the video signal.

Furthermore, since the light image was focused on the same side of the signal plate as the charge image, it was necessary to locate the electron gun and deflection coils off the optical axis in order to obstruct the light path. Consequently, since the scanning beam was directed at the signal plate at an average angle of 45°, vertical-keystone correction of the horizontal scan was needed.

Because of the unstable black level and flare signals and the imprecise gray scale, the iconoscope was not usable for three-channel color camera applications. However, long after the introduction of the image orthicon and until the advent of the vidicon, the iconoscope was used almost universally on the 525-line system for monochrome transmission of motion-picture film and slides.

With all its shortcomings, the iconoscope was the key to the introduction of the first practical all-electronic television system. Since a cathode ray picture-display tube, necessary to supplant the slowly reacting modulated light source and cumbersome rotating disk of the Nipkow display system, had been demonstrated as early as 1905, a television system composed entirely of electronic components now was feasible.

Further developments were made in iconoscope design to improve the efficiency in the image iconoscope. In the image iconoscope, a thin-film transparent photocathode was deposited on the inside of the faceplate. Electrons emitted and accelerated at a potential several hundred electronvolts from this surface were directed and focused on a target storage plate by externally applied magnetic fields. A positive-charge image was formed on the storage plate, this being the equivalent of the storage mosaic in the iconoscope. A video signal was generated by scanning the positive-charge image on the storage plate with a high-velocity beam in exactly the same manner as in the iconoscope.

Both types of iconoscopes had a light-input, video-output characteristic that compressed highlights and stretched lowlights. This less-than-unity relationship produced signals that closely matched the exponential input voltage, output-brightness characteristics of picture-display tubes, thus producing a pleasing gray scale of photographic quality.

Orthicon Camera Tube. The next development in camera tubes eliminated many of the shortcomings of the iconoscope by the use of low-velocity scanning. The *orthicon*—so named because its scanning beam landed on the target at right angles to the charge surface—used a photoemitter composed of isolated light-sensitive granuals deposited on the insulator. A similar tube, called the *CPS Emitron* for its cathode-potential stabilized-target scanning, was developed in England. The CPS Emitron target was made up of precise squares of semi-transparent photoemissive material deposited on the target insulator through a fine mesh. Both of these tubes produced high-resolution pictures with precise gray scales.

Image-Orthicon Camera Tube. The combination of three pickup-tube technologies in the image orthicon made possible studio and field operations under conditions of scene illumination from relatively comfortable studio lighting, compared to that required by the iconoscope, to a wide range of daylight and arena-stadium illumination. These were: (1) imaging the charge pattern from a photosensitive surface on an electron-storage target, (2) modulation of a scanning beam by the image charge of the target, and (3) amplification of the scanning-beam modulation signal by secondary-electron emission in a multistage multiplier.

Referring to Fig. 3.8*b*, the image of the scene being televised is focused on a transparent photocathode on the inside of the faceplate of the tube. The diameter of the photocathode on the 3-in-diameter image orthicon is 1.6 in (41 mm), the same as double-frame 35-mm film, a fortunate choice since it permits the use of already-developed conventional lenses.

Light from the scene causes a charge pattern of the image to be set up. Since the faceplate is at a negative voltage of about 450 V, electrons are emitted in proportion to the scene illumination and accelerated to the target-mesh surface, which is at nearly zero potential. The fields from the accelerator grid and the focusing coil bring the electrons into sharp focus on the target.

The target consists of an extremely fine conductive mesh closely spaced to a thin glass membrane. Most of the electrons pass through the mesh and strike the glass target, causing secondary electrons to be emitted and collected by the mesh, which is at a slightly positive voltage with respect to the target. The loss in electrons to the mesh produces a positive charge on the target glass. The secondary action results in some amplification by the secondary-emission ratio of the target causing more electrons to be emitted than strike the target. The charge pattern is stored on the glass until it is neutralized by the scanning beam from the electron gun. The supply of electrons to the target lowers its potential to that of the electron gun.

When this condition is reached, no more electrons can be deposited, and the remainder of the electrons return toward the electron gun along substantially the same path as for the scanning trip. This return beam is modulated by the loss of electrons to the target.

The modulated return beam strikes the first dynode of the multiplier section, which is also the accelerating G_2 of the electron gun. The G_2 surface has a high secondary-emission ratio. The secondaries released are attracted to the pinwheel-shaped multiplier stage assembly by the field of the second dynode where more secondaries are released. Progressively the electrons are guided through the multiplier stages to the anode, releasing more secondaries at each stage. The total current gain in the multiplier is on the order of 500 to 1000.

The light-to-signal transfer characteristic of the image orthicon is complex because of the scattered electrons and the usual practice of operating the tube over the knee of the transfer curve or, in effect, overexposing the tube to improve the signal-to-noise ratio. Consequently, conventional tubes having a wide glass-to-mesh target spacing were unsuitable for accurate color reproduction in three- or four-tube cameras. For these applications, tubes capable of close spacing, capable of providing a low noise level with a linear transfer characteristic, were used in early color-television studio broadcasts (see Fig. 3.9).

3.2.3 Photoconductive Camera Tubes

The vidicon was the first successful television camera tube to use a photoconductive surface to derive a video signal. An antimony trisulphide photoconductor target was scanned by a low-velocity electron beam to provide an output signal directly. No intermediate electron-imaging or electron-emission processes—as, for example, in the image orthicon or iconoscope—were employed. Since then,

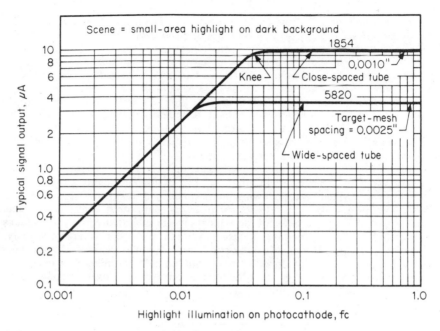

FIGURE 3.9 Transfer characteristic of wide-spaced and close-spaced image-orthicon tubes (1 fc = 10.76 lm/m^2). *(Source: Fink, 1957, in Benson, 1986)*

although a variety of tubes have been developed and identified under different trade names or the type of photoconductor used, the name *vidicon* has become the generic classification for all such photoconductive camera tubes.

The photosensitive target material of all vidicons, except the silicon-diode tube with its array of diodes on a silicon wafer, is a continuous light-sensitive film deposited on a transparent signal electrode. Magnetic focusing and beam deflection are used with most tubes. However, others use either magnetic or electrostatic fields for these functions. A few types use electrostatic focus and deflection.

The commonly used sizes range from ½- to 4.5-in (13- to 38-mm) diameter, although larger tubes have been made for high-resolution applications. Some of the larger tubes do not drive the video signal directly from the target but instead use a return-beam signal and an electron multiplier similar to the image orthicon.

Antimony Trisulphide Vidicon. The most widely used vidicons employ an antimony trisulphide (Sb$_2$S$_3$) photoconductor. The primary applications for the tube are in closed-circuit industrial systems. The major advantages are variable sensitivity and low cost. It has never been used to any extent in broadcasting studios because the image retentivity, visible as lag or tails on moving objects, and the dark current are excessive, particularly at low light levels. In some noncritical applications such as surveillance systems this is not a drawback. On the other hand, in broadcast telecine service where more than enough light is available to overcome lag and to minimize the dark current, it has been an effective replacement for the iconoscope. However, vidicon color-telecine cameras, with their attendant registration problems of the three or four channels (three colors and luminance), are fast being replaced by the flying-spot scanner.

The photoconductor consists of alternating layers of porous and solid layers of the antimony trisulphide, built up to a thickness of 1 or 2 μm. The operating voltage of the target for optimum performance can vary from a few volts to 100 V. The reason for the variation is twofold: (1) to control the sensitivity and (2) to accommodate manufacturing tolerances. This type of tube suffers from a high residual signal current in the absence of light, called *dark current*. The dark current may range from 1 or 2 nA at a low target voltage to as much as 100 nA at a high target voltage. Dark current also increases drastically with temperature. For every increase of 8°C, the dark current doubles.

Antimony trisulphide is sensitive to light throughout the entire visible spectrum and to a lesser degree to infrared (see Fig. 3.10). The signal output relative to light level is not linear, following approximately a 0.65 power law (see Fig. 3.11). This progressive compression of higher signal-output levels is desirable since it partially compensates for the typical picture tube that compresses blacks and stretches whites to provide a pleasing system transfer characteristic of unity.

The resolution of the tube is quite good, since it is not degraded by the relatively thin photoconductor. In addition, the photoconductor being dark in appearance absorbs most of the incident light, rather than scattering it laterally to reduce the contrast of fine detail.

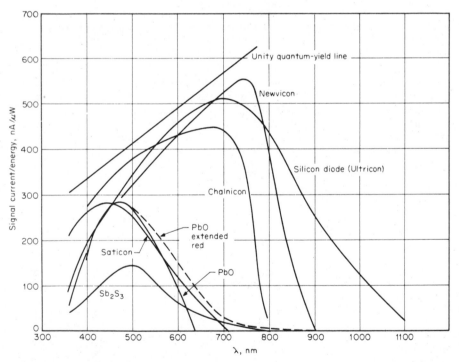

FIGURE 3.10 Absolute spectral response curves of various camera-tube photoconductors. *(Source: Benson, 1986)*

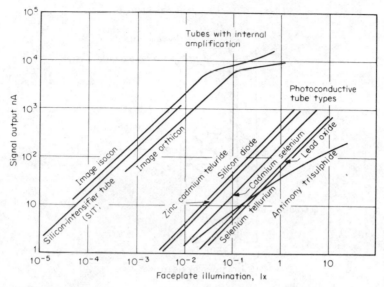

FIGURE 3.11 Light-transfer characteristics of typical camera tubes. *(Source: Benson, 1986)*

Plumbicon. Another type of more complex heterojunction photoconductor that has revolutionized the broadcast camera industry was developed by Philips of the Netherlands. The lead-oxide tubes are manufactured variously as Leddicons, Visticons, or Hi-sensicons by other manufacturers. Unlike the antimony trisulphide photoconductor, the lead-oxide material is a solid-state semiconductor (see Fig. 3.12). It is a porous vapor-grown crystalline layer of lead monoxide, processed during growth to be an *i* type on a transparent *n*-type tin or indium-oxide signal electrode. The *i* type has a nearly equal number of mobile electrons and holes (absence of an electron). The surface of *i*-type layer on the side toward the vacuum and the scanning beam is treated to provide *p*-type conductivity; it will not absorb electrons from the scanning beam.

In operation, light energy from the scene causes a positive-image-charge pattern to be set up on the *ip* junction. When this is scanned by the beam, electrons neutralize the charge and cause a signal current to flow from the *n*-type signal plate.

The dark current of lead-oxide tubes is practically zero (in the order of 1 nA at a typical target voltage of 45). Thus, the resultant very stable black level contributes materially to accurate matching of color camera channels and faithful color reproduction.

Unfortunately the spectral response of conventional tubes in the red region is negligible. Therefore, in color cameras a special red-sensitive tube, generally made by doping the photoconductor with sulphur, is used in the red channel.

Plumbicons have been made in a variety of sizes as well as a variety of combinations of magnetic and electrostatic focus and deflection. The first tube was 1¼ in (30 mm) with magnetic focus and deflection. Improvements in design and manufacturing techniques have resulted in the production of 1-in (25-mm) and

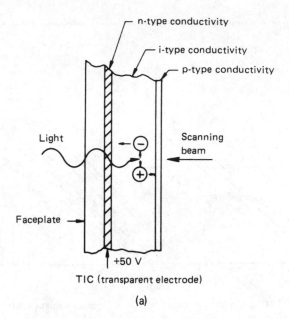

TIC (transparent electrode)

(a)

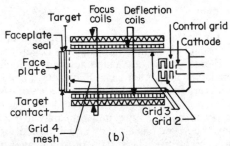

(b)

FIGURE 3.12 Cross-section of a photoconductive camera tube. (*a*) Lead-oxide photoconductor illustrating generation of the output signal. *(Source: RCA Corp., in Benson, 1986)* (*b*) Major components of the tube structure and the external focus and deflection coils. *(Source: Carl Bentz and Jerry Whitaker)*

⅔-in (18-mm) tubes that were more easily adapted to cameras designed for vidicons. The engineering effort and manufacturing improvements have been concentrated on the 1- and ⅔-in formats, rather than the 1¼-in, with the result that the latter are providing performance equal or better than the larger tube.

In the evolution of Plumbicon tube designs, the ⅔-in version with mixed-field electron optics (magnetic focus and electrostatic deflection) has emerged as the leader in the field. The center resolution is excellent, and the corner resolution is superior to that of tubes using both magnetic focus and deflection. This is due in part to the precision of the laser-etched deflection electrodes that are an integral part of the tube.

Further improvement in peripheral resolution is achieved in camera design by the use of dynamic focus where the focus field is modulated by appropriate horizontal- and vertical-scanning waveforms.

The use of a diode electron gun, rather than the conventional triode configuration, reduces the scanning-beam resistance, which in turn shortens the image lag to a time comparable to that of the selenium-based Saticon tube described in the next paragraph.

Saticon. A selenium-based photoconductor of the heterojunction type similar to the lead oxide of the Plumbicon is used in this tube. Two other photoconductive target materials that have been developed and marketed in tubes similar to the Saticon are cadmium selenide in the Chalnicon and zinc selenide in the Newvicon.

Silicon-Diode Photoconductive Target. The silicon-diode structure is the most sensitive of all photoconductive targets. Frequently called the "silicon vidicon," its target is made up of an array of several hundred diodes; a square of about 1000 × 1000 elements is in the 1-in (25-mm) tube.

The target is practically immune from image burn, even in direct sunlight. However, at low lights the lag on moving objects is excessive. In addition, since the light-to-signal transfer characteristic, or gamma, is 1.0, gamma correction is needed to compensate for the picture-tube exponential transfer characteristic of approximately 2.5 in order to enhance detail and gray-scale separation detail in lowlights.

The silicon-diode vidicon is used primarily in closed-circuit monochrome applications where low-light and infrared sensitivity and lack of damage or image burns from very bright images are required.

3.2.4 Single-Tube Three-Color Camera Systems

Until the development of CCD imagers, most low-cost cameras used by consumers, and by many professionals in applications not requiring broadcast quality, utilized a single camera tube to produce a complete television picture. Nearly all these tubes incorporated an array of fine color stripes from which the color information was derived. These tubes are basically vidicons in that they use a photoconductor to detect light and develop the video signal from a readout of the target.

There are two categories of these three-color tubes. One has a single output that provides a luminance signal of restricted bandwidth and color signals as two different high-frequency carriers. The other produces three different color signals from three vertically oriented sets of signal plates, one for each of the red, blue, and green primaries.

Single-Output Signal Tubes. Vertical-striped optical filters are used to generate the color information. The spacing of the stripes is chosen so as to generate a video-carrier signal when scanned horizontally by the electron beam. Yellow stripes are used to generate blue by filtering out the red and green light; cyan stripes generate red by filtering out the green and blue. By arranging the colored stripes at slightly different angles with clear spaces between, two carrier signals are produced that represent red and blue light, respectively. The luminance signal is produced from an addition of the light passing through the filters and the clear spaces between the colored stripes. The individual *R*, *B*, and *G* signals are then encoded in the conventional manner.

Multiple-Output Signal Tubes. The *trielectrode vidicon* has three outputs from three independent sets of signal plates or grids. Each one is backed with an optical filter for the primary colors: red, blue, and green, respectively. The signals from these three sets of grids are then encoded to produce a color video signal.

The Trinicon has two outputs from two sets of interleaved fingers of signal plates positioned behind triplets of vertically oriented red, blue, and green optical filters. There are three filter stripes for each pair of interleaved fingers. The filters are located so that one color-filter stripe straddles the gap between two fingers and the other gap coincides with the junction between the other two color-filter stripes. By a combination of a one-line delay and a subtraction and addition matrixing of the signals, output signals are generated for the three color channels and subsequent encoding.

3.2.5 Solid-State Image Sensors

Early Developments. A solid-state imager sensor consists of a flat array of photosensitive diodes, wherein each diode may be considered a picture element. Solid-state imaging devices using a flat array of photosensitive diodes were proposed as early as 1964 and demonstrated publicly in 1967. The charge voltage of each sensor element was sampled in a horizontal and vertical, or X-Y, address pattern to produce an output voltage corresponding to a readout of the image pixels.

The resolution capability of these first laboratory models did not exceed 180 by 180 pixels, a tenth of that required for television broadcasting applications. Nevertheless, the practicability of solid-state technology was demonstrated.

Charge-Coupled Device (CCD) Storage Technology.[5] In the first camera system, demonstrated publicly in 1967, a video signal was generated by sampling the charge voltages of the elements of the array directly in an x and y (horizontal and vertical) scanning pattern. In the early 1970s a major improvement was achieved with the development of the CCD, or in operation, a charge-transfer device. The photosensitive action of a simple photodiode was combined in one component with the charge-transfer function and metal-oxide capacitor storage capability of the CCD. In the sequence of operation, the photogenerated charges are transferred to a *metal-oxide semiconductor* (MOS) capacitor in the CCD and stored for subsequent readout as signals corresponding to pixels.

Thus, rather than sampling directly the instantaneous charge on each photosensitive picture element, the charges are stored for readout either as a series of picture-scanning lines in the *interline-transfer* system, or in the *frame-transfer* system as image fields.

Interline-Transfer Structure. The early CCD imagers were *interline-transfer* devices in which vertical columns of photosensitive picture elements are alternated with vertical columns of sampling gates (see Fig. 3.13). The gates in turn feed registers to store the individual pixel charges. The vertical storage registers then are sampled one line at a time in a horizontal- and vertical-scanning pattern to provide an output video signal.

---

5. For more detailed information on CCD storage technology, see K. Sadashige, "An Overview of Solid-State Sensor Technology," *J. SMPTE*, **96** (no. 2) (1987): 180–185; A. A. J. Franken and Rao, "Television Camera Tubes and Solid State Sensors for Broadcast Applications," *J. SMPTE*, **95** (no. 8) (1986): 799–801; and Benson, 1986, pp. 11.60–11.65.

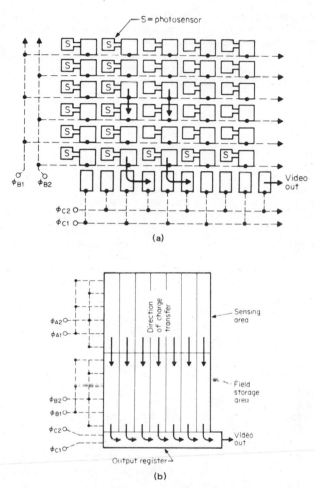

FIGURE 3.13 CCD imager architectures. (*a*) Interline-transfer structure. (*Source: Benson, 1986*) (*b*) Frame-transfer structure. (*Source: Benson, 1986*)

This approach was used in early monochrome cameras and in three-sensor color cameras. It was also used with limited success in a single-tube color camera wherein cyan-, green-, and yellow-stripe filters provided three-component color signals for encoding as a composite signal. The interline system is of only historical interest, and it is discussed here only to provide an understanding of the basic transfer structure of CCD imagers, since currently the frame-transfer system is used in all professional-quality cameras.

Frame-Transfer Structure. This structure differs from the interline system in that at the end of the $\frac{1}{60}$th second field exposure, the entire charge pattern is transferred in parallel through the vertical CCD shift-register columns into the storage register. This action frees the image register to begin sampling the next field.

Concurrently, during this same time interval, the charge packets in the storage register are transferred one row at a time into the output register, through which they are transferred sequentially to the output stage.

When all rows of the storage register have been read out, it is ready to receive the next parallel transfer from the image register, and the entire cycle is repeated for the second field to provide a complete frame of video.

3.2.6 Nonstorage Camera Tubes and Systems

Image-Dissector Tube.   In order to emphasize the importance of the storage feature, the development of a nonstorage electronic pickup and electronic image-scanning device by Philo T. Farnsworth in 1928, called the *image-dissector tube*, is worthy of mention. The electron image composed of electrons emitted by a photocathode surface was deflected by horizontal- and vertical-scanning fields applied by coils surrounding the tube so as to cause the image to scan a small aperture.

In other words, rather than an aperture or electron beam scanning the image, the aperture was stationary and the electron image was moved across the aperture. The electrons passing through the aperture were collected to produce a signal corresponding to the charge at an element of the photocathode at that instant.

The limitation of this invention was the extremely high light level required because of the lack of storage capability to build up the charge level between signal readout by a scanning beam. Consequently, the image dissector found little use other than as a laboratory picture signal source.

Flying-Spot Studio Scanner.   Unsuccessful attempts were made to use pickup devices without the storage capability for studio applications. The most ambitious was the Allen B. Dumont Laboratories' experiment in the 1940s with an electronic "flying-spot" camera. The set in the studio was illuminated with a projected raster frame of scanning lines from a cathode ray tube. The light from the scene was gathered by a single photocell to produce a video signal.

The artistic and staging limitations of the dimly lit studio are all too obvious. Nevertheless, while useless for live pickups, it demonstrates the flying-spot principle that presently has taken over as the universal system for television transmission of motion-picture film and slides.

Flying-Spot Film Scanners.   A more practical application of the flying-spot technique has been in 625-line/50-Hz motion-picture film transmission. For a number of years, and more recently with the advent of the digital time-base corrector for signal processing, the flying-spot technique has been used in 525-line/60-Hz countries. This is accomplished in 50-Hz systems by scanning each film frame twice with the projected image of an interlaced frame of lines on a flying-spot cathode ray tube.

By a fortunate coincidence for 50-Hz countries, film can be run at a rate of 25 frames/s, rather than the standard rate of 24, with no objectionable deleterious effects from the 4 percent increase in speed. Thus, it is possible to transport the film continuously at a 25-Hz frame rate and scan each frame twice by moving the raster between two precisely determined vertical positions. This movement can be accomplished, as in early British designs, by a twin-lens optical projection and shutter system or by a two-position raster shift on the CRT.

For 60-Hz countries the solution is considerably more complex. It is unacceptable to transmit film at a frame rate of 30 Hz, an increase of 20 percent, in order to use the two-position raster system. One alternative is to use a five-position "jump-scan" raster to "chase" the continuously moving film frames. Although a few film scanners of this kind were used for a short period, the tolerances on scanning position and geometry were too tight for a practical operating system.

The advent of cost-effective digital field and frame-store equipment has made possible the development of a universal and mechanically simple flying-spot system that progressively scans each frame of film in a sequential manner. After analog-to-digital signal conversion, the information from a film frame is retained in a two-field video store. The stored video information then is sorted and read out in the proper order and format to produce two interlaced fields of video in either 525-line/60-field or 625-line/50-field systems standards.

The horizontal-scanning speed of the flying-spot cathode ray tube is one-third faster than normal 525-line standards so that a complete film frame is scanned and "written" in storage in the time taken by 1½ television fields. In other words, the length of the write operation is 1½ television fields. Since a ¼₄-s film-frame period is equal to 2½ television fields, there is a full-field gap between write operations.

As shown in Fig. 3.14a, the first line of the write operation is fed to store A, the second to store B, the third to A, the fourth to B, and so on alternatively until a complete frame of 492 lines has been stored. This is slightly more than the 483 active picture lines (assuming 21 lines of vertical blanking).

The television lines now have been sorted and stored into their correct fields. The final step is to read field stores A and B alternately with the proper half-line offset to obtain an interlaced-frame output.

CCD Film Scanners. The film transport for CCD line-scan telecines is functionally very similar to the flying-spot cathode ray tube scanner described in the previous section. However, as in the case of the flying-spot scanner, the practicability of using CCDs for the transfer of motion-picture film to television resulted from the development of low-cost digital-video field and frames stores. Early attempts to use CCDs for 525-line television required either an intermittent film motion, or with continuous motion, an optical repositioning of the film frame to achieve the equivalent of the raster shift used with the flying-spot system to scan the 24-frame film at the 30-frame rate of television. This presented insurmountable mechanical problems in design, as well as in equipment maintenance, to achieve acceptable vertical registration. It was not until the availability of digital-video frame store that the CCD line-scan telecine came into its own. This is accomplished by the use of three video field stores with computer-controlled sequencing of the inputs and outputs.

The film is scanned at a rate of 24 frames/s as it moves by a CCD line sensor at the standard film rate of 24 frames/s. The clock rate for the red-blue-green line sensors is synchronized with the film-frame rate. Referring to Fig. 3.14b, film-frame 1, the odd lines are written into field-store FS_1 while the even lines are switched to write into FS_2. When approximately 60 percent of the frame has been scanned, the readout of FS_1 can start to generate the video lines for the first odd field. The finish of reading field 1 occurs just after the writing of the last line from film-frame 1.

The next (even) television field is obtained by switching to the output of FS_2. This field occurs during the first part of the scanning of film-frame 2. Since FS_2 will be read at a rate faster than the information is written in, by scanning the film, it is possible to begin to write in FS_2 the information scanned from film-

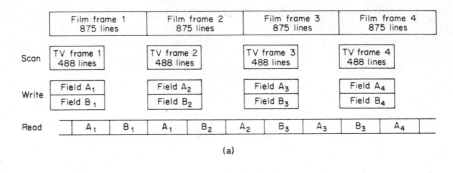

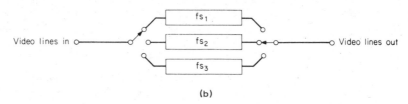

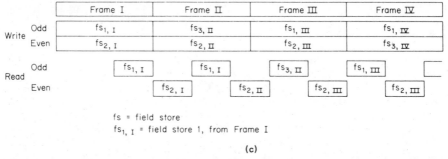

```
fs = field store
fs₁, ₁ = field store 1, from Frame I
```

(c)

FIGURE 3.14 (*a*) Write-read sequence for the 525/60 system. (*Source: Benson, 1988*) (*b*) Diagram of 24-frame sequential scan, 60-field readout in 525/60 system. (*Source: Benson, 1986*)

frame 2. Also, as frame 2 is scanned, the odd lines will be written into FS_3. This leaves the information from film-frame 1 available in FS_1 to be read out to form the second odd television field. This process continues as diagrammed to result in the familiar 3–2 frame conversion from film to television rates used in intermittent projectors.

3.3 IMAGE REPRODUCTION

Each picture of a televised scene is reproduced in a pattern of horizontal-scanning lines identical to that on the photosensitive surface in the television camera. The intensity of the lines as they are reproduced varies in accordance with the brightness of the picture elements of the scene. Black-and-white and color-video display devices using present-day technology can be classified into five different categories as follows:

1. Cathode ray tube (CRT), direct view
2. Large-screen displays, optically projected from a CRT
3. Large-screen displays, projected from a modulated light beam
4. Large-area displays from discrete picture elements
5. Matrixed flat panels of light-emitting picture elements

The CRT remains the dominant type of display for both consumer and professional television applications. The Eidophor and light-valve systems using modulation of a light source have wide application for presentations to large audiences in a theater environment. Stadium displays, which can support a high equipment and installation cost, provide color reproduction of a somewhat reduced resolution. Flat-panel displays are being used in increasing numbers for small-screen personal TV sets. However, the future promises such displays in a size suitable for home viewing from a wall-mounted panel.

3.3.1 Direct-View Picture-Tube Displays

Monochrome Cathode Ray Tubes. From the start of commercial television in the 1940s until the emergence of color as the dominant programming medium in the mid-1960s, virtually all receivers were the direct-view monochrome type. A few large-screen projection receivers were produced, primarily for viewing in public places by small audiences. Initially the screen sizes were 10- to 12-in diagonal. In a few years screen sizes as large as 27 in were produced in relatively small quantities. In the years from 1965, with the universal popularity of color, to the present the trend has been to smaller monochrome screen sizes in compact receivers for personal viewing. This demand has resulted in the use of shorter CRTs with smaller neck diameters and wider beam-deflection angles. Presently screen sizes of 12 in (30 cm) and neck diameters of 0.8 in (20 mm) are typical.

Monochrome tubes consist of the following basic components:

1. *An electron gun*, consisting of a cathode and a filament, and a beam-controlling aperture (G_1). The cathode is heated by the filament to a temperature high enough to cause it to emit a stream of electrons through the controlling aperture. By varying the voltage difference between the cathode and the G_1 aperture, the density of the electron beam can be modulated with a video signal.
2. *An accelerating electrode* (G_2) with a small-diameter defining aperture to limit the diameter of the electron beam.
3. *A focusing aperture*, or a *magnetic focus coil*, to direct the spreading beam of electrons toward a crossover at the phosphor screen, thus producing a small spot.
4. *A vertical- and horizontal-beam-deflection system.* A system of external magnetic coils or internal electrostatic plates to deflect the scanning beam.
5. *A phosphor screen* applied to an optically smooth faceplate. The screen usually is backed with a thin, conductive coating of aluminum. The conductive coating on the rear side of the phosphor screen and the adjacent wall of the

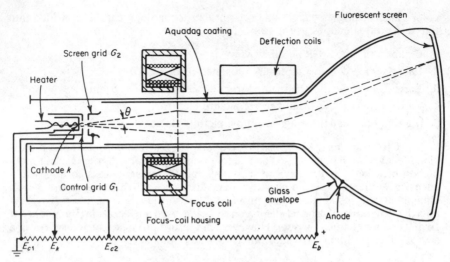

FIGURE 3.15 Generalized schematic of a cathode ray tube with electrostatic focus and deflection.

bulb are maintained at a high potential relative to the cathode to provide an accelerating field for the electron beam. The aluminum coating is thin enough to permit the electron beam to pass through and excite the phosphor screen into the emission of light. The aluminum coating also provides a nearly double output of usable light by reflecting the rays that are emitted from the rear side of the phosphor screen.

In addition to serving as the anode for the electron beam, the aluminum backing provides a gain in brightness and contrast by reflecting the light emitted to the rear from the phosphor to the viewing side of the screen.

A further improvement in contrast for viewing in lighted rooms or in the presence of daylight is achieved, although with some loss in brightness, by the use of a slightly darkened neutral-density faceplate. This results from the room light being attenuated twice, first as it passes through the faceplate and second after it is reflected from the screen and aluminum backing through the faceplate, whereas the light emitted by the screen is attenuated only once.

The component configuration of a typical black-and-white picture tube with electrostatic focus and magnetic deflection is shown in Fig. 3.15.

The horizontal lines of the two fields on a receiver or monitor screen are produced by a scanning electron beam which, upon striking the back of the picture-tube screen, causes the phosphor to glow. The density of the beam, and the resultant brightness of the screen, is controlled by the voltage level of a video signal applied between the controlling aperture and the cathode in the electron gun.

Color Cathode Ray Tubes.   All color-television picture displays synthesize the reproduction of a color picture by *generating* light, point by point, from three fluorescent phosphors, each of a different color. This is called an *additive* system. The *chroma characteristic*, or hue, of each of these color light sources is defined as a primary color. The most useful range of reproduced colors is obtained from

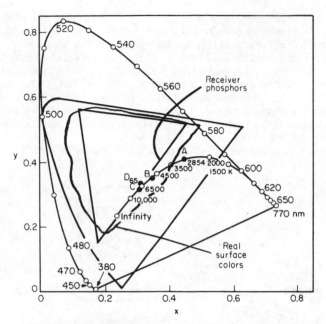

FIGURE 3.16 CIE 1931 chromaticity diagram showing locus of Planckian radiators and CIE standard illuminants *A, B, C,* and *D65*. The color triangle is defined by a standard set of color-television receiver phosphors, compared with the maximum real-color gamut on a *u', v'* chromaticity diagram. Overlaid are the gamut of colors found in real surfaces and from typical receiver phosphors. *(Source: Benson, 1986)*

the use of three primaries with hues of red, green, and blue. A combination of the proper intensities of red, blue, and green light will be perceived by an observer as white.

The balance of intensity among the three primaries can shift the perception of white by the observer between two extremes of warm (reddish) and cold (blue). This can be shown graphically by the presentation on a color triangle as shown in Fig. 3.16.

By further adjustment of the individual intensities, the full range of hues within the limits bordered by the primaries can be reproduced.[6]

By combining pairs of these three colors, three more colors called *complementaries* can be produced. For example, magenta (purple) is produced by red and blue, yellow from red and green, and cyan (aqua) from blue and green.

Shadow-Mask Display Tube. Utilizing this phenomenom of physics, color-television signals were first produced by optically combining the images from three color tubes, one for each of the red, blue, and green primary transmitted

---

6. The fact that white light is composed of the full gamut of colors can be demonstrated in reverse by observing sunlight projected through a prism onto a white screen. The range of colors from the short-wavelength violet, through blue, green, yellow, orange, to long-wavelength red are spread out in sequence because of the increased bending of the short wavelengths going through the nonparallel glass surfaces.

colors. This early *trinescope*, as it was called by RCA, served the purpose of demonstrating the feasibility of color television, but the tubes were too cumbersome and costly to be a practical solution for viewing in the home.

The problem was solved by the invention of the shadow-mask picture tube in 1953. The first successful tube used a triad assembly of electron guns to produce three beams that scanned a screen composed of groups of red, green, and blue phosphor dots. The dots are small enough not to be perceived as individual light sources at normal viewing distances. Directly behind the screen, a metal mask perforated with small holes approximately the size of each dot triad, is aligned so that each hole is behind an *R-G-B* dot cluster.

The three beams are aligned by *purity* magnetic fields so that the mask *shad-*

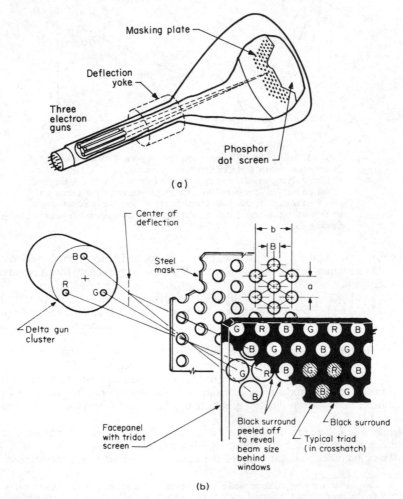

(a)

(b)

FIGURE 3.17 (*a*) Overall schematic of shadow-mask cathode ray color picture tube. (*b*) Delta gun, round-hole mask, and negative guard band tridot screen. Taper of mask holes not shown. *(Source: Benson, 1986)*

ows the green and blue dots from the beam driven by the red signal. Similarly, the mask shadows the red and blue dots from the green beam, and the red and green dots from the blue beam. Fig. 3.17*a* depicts the principal elements of the shadow-mask tube; Fig. 3.17*b* shows the mask and screen assembly in detail.

Figure 3.17*b* shows an enlarged section (not to scale) of a tube with a triad gun assembly showing geometric relationships of the electron beams, the masking plate, and the phosphor dots on the glass faceplate. Note the black graphite surrounding the phosphor dots to eliminate reflection of stray light from the screen.

Slotted-Shadow-Mask Cathode Ray Tube. The principle of the triad gun and shadow-mask design is used in all present-day direct-view picture tubes. However, rather than a triangular configuration of color phosphor dots, the screen is in the form of fine alternating red, green, and blue vertical stripes. The most popular design uses a shadow mask with vertical slots and an in-line assembly of the three color guns as shown in Fig. 3.18. For increased contrast, a black surround is used, similar to that in Fig. 3.17*b*.

Trinitron Cathode Ray Tube. Another popular design is the Sony Trinitron, which uses a grid of wires, rather than a shadow mask, behind the phosphor screen to provide the color separation of the electron beams. In addition, a unique gun design is used to generate all three beams. The three beams, focused by a single electron lens as shown in Fig. 3.19, provide a favorable spot size.

3.3.2 Large-Screen Displays

Projection systems provide television picture displays with a much greater area than is practical to produce for viewing directly on a cathode ray tube phosphor screen. By means of optical magnification, a 5-in diagonal image on a CRT screen can be projected as an expanded display with a diagonal dimension as great as 60-in diagonal on a passive rear or front projection screen.

Direct-viewing CRT screens larger than 30-in diagonal are not commercially practical. The major factor limiting the size of CRTs is the glass strength to withstand the atmospheric pressure on evacuated envelopes. Other important factors are weight—a 30-in (76-cm) tube weighs 88 lb (40 kg)—and the necessary cabinet depth to accommodate a tube with a practical scanning deflection angle. Of no little consideration in a consumer product is the sharp increase in the cost of a cabinet for tubes over 20 or 25 in.

Projection techniques have been used from the earliest days of television to magnify images produced by a rotating Nipkow scanning disk. The major differences between direct-view and projection systems are in the three elements, namely, the image source, the optical elements, and the viewing screen. The electronics are essentially the same for both.

Display Requirements. To be acceptable, a projection system for small viewing audiences must approach or ideally equal the performance of a direct-view screen in terms of brightness, contrast, and resolution. Furthermore, compared to motion pictures that are viewed in a darkened theater, the higher level of ambient light for most television viewing dictates a much higher screen brightness in order to provide adequate contrast and detail in low-light shadow areas. Table 3.3 compares the performance levels of direct-view television receivers and conventional

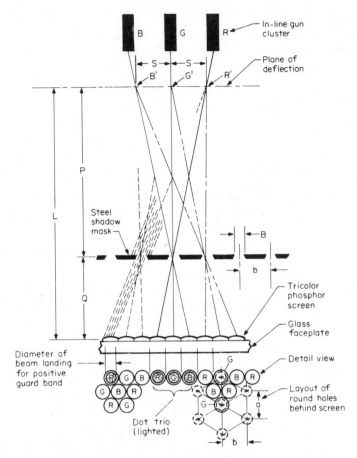

FIGURE 3.18 Shadow mask with in-line guns. Mask has round holes, as in lower view, or vertical slots with stripe sequence as in top row of dots. *(Source: Benson, 1986)*

motion-picture film theaters. The resolution for film theaters is in equivalent television lines (TVL).

Although for large-screen television displays, some compromise in the first two can be accepted in order to achieve a marketable cost, projection displays must excel in resolution because of the tendency of viewers to be positioned considerably closer to the screen than the normal relative distance from the screen to three to eight times the screen height.

Projection Screens. Two basic types of viewing screens are employed, one a diffusing surface where the image is projected onto the viewing side as in conventional film theaters, and the other a translucent plastic or glass plate where the image is projected on the opposite or rear side from the viewer. In both front- and

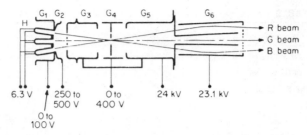

FIGURE 3.19 Electrode arrangement of the Trinitron gun. *(Source: A. Morrell et al.,* Color TV Picture Tubes, *Academic, New York, 1974, in Benson, 1986)*

TABLE 3.3   Television and Film Performance Levels

	Brightness	Contrast	Ambient	Resolution
Receiver	60–120 fL	30:1	5.0 fc	275 TVL
Theater	16–25 fL	100:1	0.1 fc	4800 TVL

rear-projection screens, the directional characteristic of the screen material determines the brightness of the image.

Front-projection screens depend upon reflectivity to provide a bright image, whereas rear-projection screens require high transmission. In either type, the brightness decreases with increased screen size. Thus, for a given projector luminance output (lumens), the viewed brightness varies in proportion to the reciprocal of the square of any linear dimension (width, height, or diagonal) of the screen as follows:

$$B = \frac{L}{A}$$

where B = apparent brightness, cd/m^2
   L = projector light output, 1m
   A = screen viewing area, m^2

To improve the apparent brightness, screens can be designed with directional characteristics. This characteristic is called *screen gain G*, and the equation above becomes:

$$B = G \times \frac{L}{A}$$

Table 3.4 lists some typical front-projection screen types and their gains.

Picture Contrast.   Contrast is the ratio of brightness at maximum and minimum video-signal levels. It is dependent upon the effect of ambient light and stray light in the optical projection system such as lens flare. Figure 3.20 illustrates how a highly reflective front-projection screen reflects ambient light as well as light of the projected image. A similar but lesser effect is shown for a rear-projection

TABLE 3.4   Screen Gains

Type	Gain (cd/m^2)
Flat white paint (magnesium oxide)	0.85–0.90
Semigloss white paint	1.5
Aluminized	1–12
Lenticular	1.5–2
Beaded	1.5–3
Kodak Ektalite	10–15
Scotch-light	Up to 200

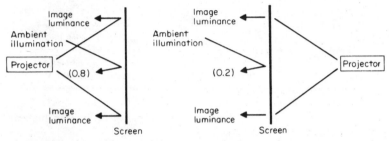

FIGURE 3.20   Effect of ambient light. (*a*) Front projection. (*Source: Benson, 1986*) (*b*) Rear projection. (*Source: Benson, 1986*)

screen. Typical values shown are 0.8 and 0.2, respectively. The ambient light tends to dilute the contrast, although the effect can be reduced somewhat by the use of a highly directive screen, combined with judicious placement of room lighting fixtures.

A rear-projection screen, which is dependent upon high transmission for high brightness, can be designed with low reflectance to reduce the effect of ambient light and thus achieve an increase in contrast by the use of a ribbed lenticular-lens construction on the projector side, and light-absorbing black striping on the viewing surface as shown in Fig. 3.21.

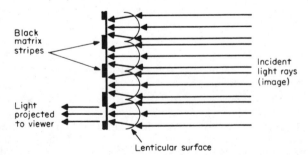

FIGURE 3.21   High-contrast rear-projection screen. (*Source: Benson, 1986*)

This technique focuses the projected light through the lenticular-lens segments onto strips of the viewing surface. The intervening areas are coated with a black nonreflecting material. The lenticular segments usually are oriented vertically to broaden the horizontal viewing angle. The overall result is an efficient transmission, on the order of 60 percent, while absorbing as much as 90 percent of the stray light on the viewing side.

The line structure of lenticular screens is fine enough not to affect the resolution in 525- and 625-line systems. For high-definition television (HDTV) applications, consideration of this possible degradation may require attention.

Projection Optics. For monochrome viewing, a single 5-in tube, operating at an anode voltage of up to 30 kV with either a refractive-lens system or a more efficient spherical-mirror and corrector-lens reflective Schmidt system, provided marginally adequate brightness for viewing under moderate ambient-lighting conditions. However, with the advent of color, single color-tube systems were not found to be practical. Refractive optics were used with less than satisfactory results, primarily because of the low brightness resulting from the inefficiency of the lens system. Shadow-mask color tubes were too large to be used with a more efficient reflective Schmidt optical system. Therefore, it was necessary to resort to more complex light sources and optical systems (Fig. 3.22).

The two most frequently employed three-tube projection optical configurations for color-television image projection are shown in diagram form in Fig. 3.22. The trinescope in 3.22a with its crossed dichroic mirrors for combining of the three-color beams was the system first used to demonstrate 525-line color transmission in the United States. With the introduction of the shadow-mask color tube, the trinescope was retired to the laboratory for service as reproduction standard and reference.

The in-line layout in 3.22b with three tubes, each with its projection-lens system, is the basic system used for all multitube displays. Typical packaging to reduce cabinet size for front and rear projection are shown in c and d, respectively. Because of the off-center positioning of the outboard color channels, the optical paths differ from the center channel, and keystone scanning-height modulation must be used to correct differences in optical throw from left to right. The problem is shown diagrammatically and explained in Fig. 3.23.

One form of the in-line array takes advantage of the relatively large aperture and light efficiency of the Schmidt reflective system by incorporating the optical components in a single tube with the electron optics and the phosphor target. The principal components are shown in Fig. 3.24. Electrons from the electron gun pass through the center opening in the spherical mirror of the reflective optical system to scan a metal-backed phosphor screen. Light from the phosphor (red, green, or blue depending upon the channel) is reflected from the spherical mirror through a corrector lens, which is the faceplate of the tube.

Schmidt reflective optical systems provide a significantly greater efficiency of light transmission over refractive-lens systems. In addition, a greater magnification can be obtained with cost-effective designs. On the other hand, limiting resolution is lower than that of a high-quality, expensive glass projection lens. A comparison of the characteristics is given in Table 3.5.

Light-Valve Color Displays. A number of light-valve systems have been demonstrated. At present two types are used for large-screen television picture presentation. Both employ a system of light-modulating elements and grating patterns

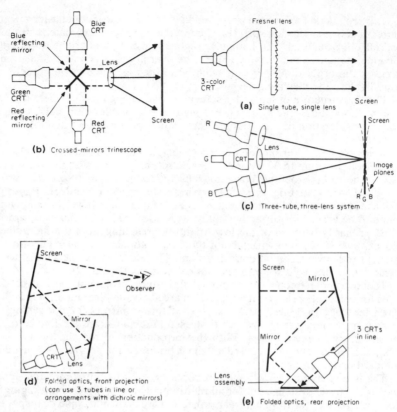

FIGURE 3.22 Projection optical configuration. *(Source: Benson, 1986)*

called *Schlierin optics* for controlling a light source. One is a reflective system developed by the Swiss called the *Eidophor*, and the other is a refractive system developed by General Electric. In a manner similar to film projectors, the systems employ a conventional light source such as a high-intensity xenon lamp, which is modulated by an optical valve between the source and the projection optics. These systems are capable of producing light output of 7000 lm (lumens), compared to a maximum of 200 to 500 for CRT-source projectors. However, because of their relatively high cost, they are used primarily in commercial applications such as theaters, lecture halls, and public displays.

Figure 3.25a is a diagram of the Eidophor monochrome system. Three such systems provide rasters of red, green, and blue primaries which are registered on a screen to produce a color picture. The rays from the lamp (1) pass through a system of lenses (2) and are reflected by a bar-grid of mirrors to the mirrored surface a slowly rotating sphere (5), which is coated with a thin layer of oil. The light rays are reflected from the sphere back through the bars (3) to the projection-lens system and screen. In this return path the bars serve to partially block the light rays. The bars and lenses are aligned so that no light is passed when the liquid surface is undeformed and smooth. When the control layer is deformed, part of the light is defracted and passes through the mirrored bars, and a

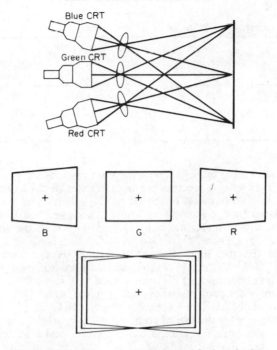

FIGURE 3.23 Three-tube in-line array. Outboard tubes (red and blue) and optical assemblies tilted so that axes intersect axis of green tube at screen center. Red and blue rasters show trapezoidal (keystone) distortion and resulting misconvergence when they are superimposed on the green raster. *(Source: Benson, 1986)*

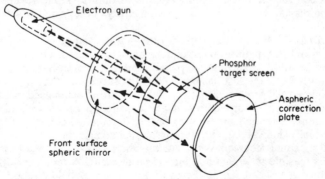

FIGURE 3.24 Projection CRT with integral Schmidt optics. *(Source: Benson, 1986)*

TABLE 3.5 Projection-Lens Performance

Lens	Aperture (f)	Image diagonal (mm)	Focal length (mm)	Magnification (mm)	Response at 300 TVL (percent)
Refractive, glass	1.6	196	170	8	33
Refractive, acrylic	1.0	127	127	10	13
Schmidt	0.7	76	87	30	15

bright spot is then visible on the screen. The deformations are produced by bombarding the oil layer with an electron beam (6) from an electron gun. The electron beam scans the picture area (4) of the oil on the sphere. The scanning beam is velocity-modulated by a constant-frequency ac voltage. The frequency of this modulating voltage determines the size of the raster pixels; the amplitude determines the relative brightness.

The General Electric light-valve system employs the same principle of Schlieren optics as the Eidophor. A schematic diagram is shown in Fig. 3.25b. A deformable oil film on a rotating disk in the light path is modified by an electron beam to produce defraction patterns much in the same way the oil on the Eidophor sphere is distorted by a scanning beam. The G.E. light valve produces color pictures directly by the use of color filters, input slots, and output bars to create three simultaneous, superimposed red, green, and blue images from the same electron beam.

Flood-Beam CRT.   Displays from 30 to 70 ft wide are employed for large audiences in stadiums for coverage of sports and entertainment events. These utilize a cluster of four 1-in or 1.25-in CRTs, one each for red and green and two for blue, for each picture element, to produce adequate brightness for viewing in bright daylight. No scanning of the phosphor screen of each tube is used; hence the nomenclature, "flood beam." Unlike similar-sized displays using incandescent bulbs, there is no lag or tailing on moving objects, or color-hue shift in low-brightness areas of the picture.

A typical 70-ft-wide display contains 150,000 tubes. The power to drive each tube, and its control and power supply circuits, is only 2 W. However, the total power for the display and the terminal equipment is in the order of 400 kW.

Flat-Panel Displays.   Flat-panel displays offer an alternative to the CRT when small screens and thin front-to-back dimensions are desired. Of the techniques demonstrated, namely, gas-discharge panels, thin-film electroluminescence, and liquid crystal displays, the last has been the most successful, having been incorporated in commercial products. Instead of the electron beams used in CRTs that are deflected by magnetic or electrostatic fields, the flat panel is made up of an assembly of either discrete light-emitting or rear-illuminated elements that are selected individually and driven with appropriate video signals.

For small personal receivers with screens where full NTSC resolution is not required, it is feasible to drive each individual element. A receiver with a 3-in-diagonal color screen has been marketed that uses over 90,000 direct-drive transistors. However, for larger sizes where full NTSC resolution is demanded, direct drive to each element, in terms of hardware manufacture, is impractical and

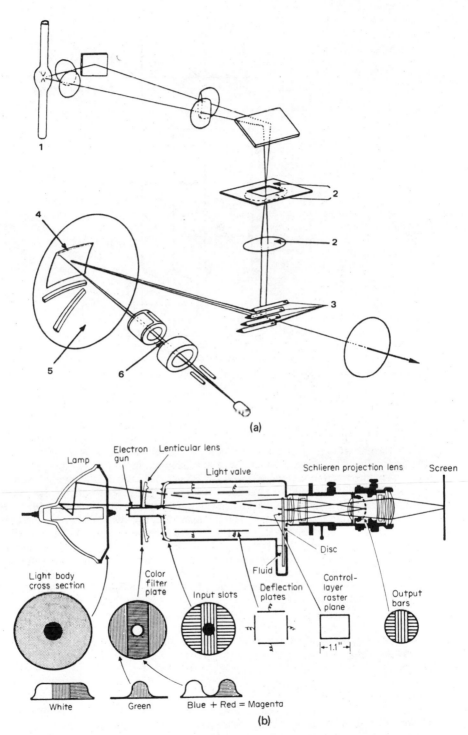

FIGURE 3.25 (*a*) Schematic diagram of Eidophor projector. *(Source: Eidophor, Ltd., Regensdorf, Zurich, Switzerland, in Benson, 1986)* (*b*) Schematic diagram of General Electric single-gun color-television light-valve assembly. *(Source: Benson, 1986)*

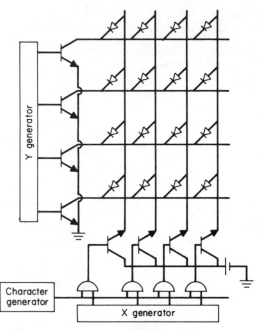

FIGURE 3.26   Matrix addressing. *(Source: Adapted from
S. Sherr,* Electronic Displays, *Wiley, New York, 1979, in
Benson, 1986)*

uneconomical since a 525-line display would require more than 250,000 connections.

The expedient used to solve the problem is termed *matrix addressing*. For matrix addressing, the elements on a panel are arranged in groups of rows and columns. To illustrate the technique, in Fig. 3.26 a 4 × 4 matrix is shown with each row and column connected to all other elements in that row or column. It then is possible to select the drive signals so that when only a row or column is driven, no elements in that row or column will be activated, but when both row and column of a selected element are driven, the sum of the two signals will be sufficient to activate that element. The reduction in the number of connections required, compared to the discrete-element approach, is quite large; the number of connections is reduced from 1 per element to the sum of the rows and columns in the panel. For the 4 × 4 array, the reduction is from 61 to 8, and for a 500 × 500 array, from 250,000 to 1000.

3.4 COLOR SIGNALS: COMPOSITION AND PROCESSING

3.4.1 Color-Signal Encoding

Compatibility Requirements.   In order for the introduction of color-television broadcasting in the United States, and other countries with existing monochrome services, to be economically desirable, it was essential that the transmissions be compatible. In other words, color pictures would be provided with acceptable quality on unmodified monochrome receivers. In addition, because of the limited availability of RF spectrum, another related requirement was the need to fit a little over 2-MHz bandwidth of color information into the 4.2-MHz video bandwidth of existing 6-MHz broadcasting channels with little or no modification of existing transmitters.

This is accomplished using the band-sharing color-signal system developed by the NTSC and by taking advantage of the fundamental characteristics of the eye regarding color sensitivity and resolution.

Harmonic Interleaving of Color Signals. The video-signal spectrum generated by scanning an image consists of energy concentrated near harmonics of the 15,734-Hz line-scanning frequency. Additional lower-amplitude sideband components exist at multiples of 60 Hz (field-scan frequency) from each line-scan harmonic. Substantially no energy exists halfway between the line-scan harmonics, that is, at odd harmonics of one-half line frequency. Thus, these blank spaces in the spectrum are available for the transmission of a signal for carrying color information and its sidebands as shown in Fig. 3.27.

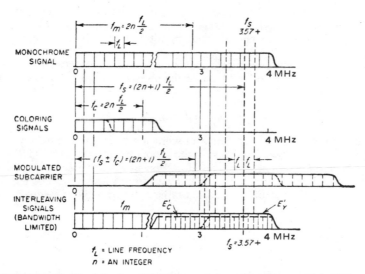

FIGURE 3.27 Band sharing of chrominance information. *(Source: Fink, 1982, in Benson, 1986)*

In addition to the fact that substantially no energy exists in these gaps at one-half line frequency in a normal monochrome video signal, a signal modulated with color information injected at this frequency is of relatively low visibility in the reproduced image because the odd harmonics are of opposite phase on successive scanning lines and in successive frames, requiring four fields to repeat. Furthermore, the visibility of the color-video signal is reduced further by the use of

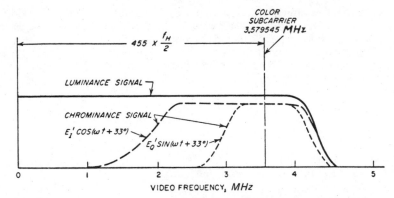

FIGURE 3.28 Luminance and chrominance spectrums. *(Source: Fink, 1982, in Benson, 1986)*

a subcarrier frequency near the cutoff of the video bandpass. The spectrums of the luminance and chrominance signals are shown in Fig. 3.28.

Luminance.   A white image produced by a properly adjusted camera consists of equal red, blue, and green voltages. Similarly, equal voltages of red, green, and blue fed to a properly adjusted monitor or receiver produce a white screen.

On the other hand, luminance signal is composed of components of the three primary colors, red, green, and blue, in the proportions for reference white, E_Y, as follows:

$$E_Y = 0.3E_R + 0.59E_G + 0.11E_B$$

These transmitted values equal unity for white and thus result in the reproduction of colors on monochrome receivers at the proper luminance level. This is known as the *constant-luminance principle*. However, as noted above, on color receivers the individual color-channel gains and background levels must be adjusted for the efficiencies of each of the phosphors in order to achieve white balance of highlights and lowlights.

Chrominance.   The color signal consists of two chrominance components, *I* and *Q*, transmitted as AM sidebands of two 3.579545-MHz subcarriers[7] in *quadrature* (differing in phase by 90°).

The subcarriers are suppressed, leaving only the sidebands in the color signal. Suppression of the carriers permits demodulation of the color signal as two separate color signals in a receiver by reinsertion of a carrier of the phase corresponding to the desired color signal. This system for recovery of the color signals is called *synchronous demodulation*.

I and *Q* signals are composed of red, green, and blue primary color components produced by color cameras or color-signal generators. The usual practice in

---

7. The color-subcarrier frequency is related precisely to the horizontal-scanning frequency by a factor of 455/2. This factor is calculated to provide a low-visibility pattern for the subcarrier components and an RF-signal beat between the subcarrier components and the sound (aural) carrier at 4.5 MHz above the picture (visual) carrier.

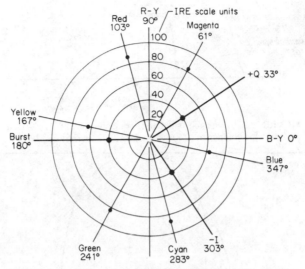

FIGURE 3.29 Electronic Industries Association (EIA) RS189A vectorscope display showing the vector and chroma-amplitude relationships. *(Source: Electronic Industries Association, in Benson, 1986)*

processing the three signals to be transmitted, namely, the luminance signal E_Y and the two chrominance signals E_I and E_Q, is to add the appropriate values of the primaries for each in a resistive matrix. After matrixing, the signals are bandwidth-limited as described in the following subsection.

The phase relationship among the I and Q signals, the derived primary and complementary colors, and the color-synchronizing burst can be shown graphically on a vectorscope cathode ray tube display. The horizontal and vertical sweeps on a vectorscope are produced from $R-Y$ and $B-Y$ sub-carrier sinc waves in quadrature, producing a circular display. The chrominance signal controls the intensity of the display. A vectorscope display of an EIA color-bar signal is shown in Fig. 3.29. A diagram of the corresponding color-monitor display is shown in Fig. 3.30.

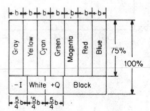

FIGURE 3.30 Electronic Industries Association (EIA) RS189A full-signal displays. Color-picture monitor. *(Source: Electronic Industries Association, in Benson, 1986)*

Color-Carrier Bandwidth Limiting. The chrominance components, I and Q, specified by the NTSC and the FCC, differ in bandwidth to conform roughly to the color-resolution capability of the eye. For example, tests have shown that an observer usually exhibits good perception of red, orange, and cyan. On the other hand, observers confuse purples, blues, and greenish yellow with grays.

Exploiting this characteristic of human vision, bandwidths selected for I- and Q-signal channels are 1.5 and 0.5 MHz, respectively. Colors lying on or near the

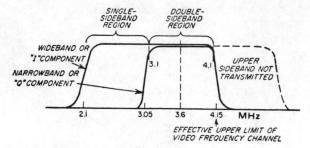

FIGURE 3.31 Vestigial sideband transmission of wide-band chrominance component. *(Source: Fink, 1982, in Benson, 1986)*

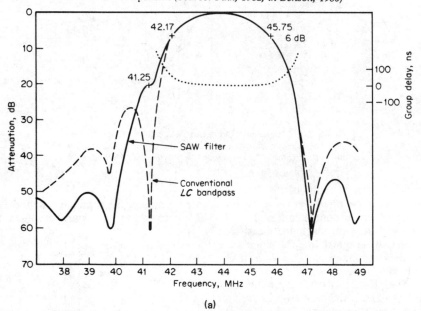

(a)

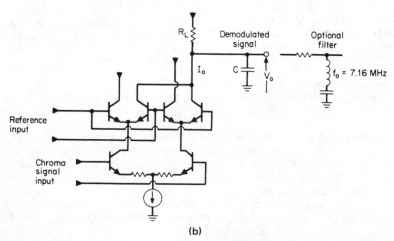

(b)

FIGURE 3.32 *(a)* Basic circuit of synchronous detector. *(Source: Fink, 1982)* *(b)* IC implementation as a chroma demodulator. *(Source: Benson, 1986)*

wide-band I axis are red, yellow, and cyan, while purple and green are near the narrow-band Q axis.

Another difference between the chrominance channels is the method of transmission: The I signal is single sideband, and the narrower Q signal, a double sideband. The single-sideband transmission of the wider-band I signal permits the color subcarrier to be placed nearer the cutoff of the video channel and thus reduces the visibility of color components in the luminance signal to a minimum (see Fig. 3.31).

3.4.2 Color-Signal Decoding

Demodulation.   The chroma signal consists of two amplitude-modulated carriers with a quadrature, or 90°, phase relationship. Each of these two carriers can be recovered individually by means of synchronous detection. A reference subcarrier of the same phase as the desired chroma signal is applied as a gate to the balanced demodulator shown in Fig. 3.32*a*. Only the modulation of the signal in the same phase as the reference signal will be present in the output. Practical implementation of the technique is in the full-wave configuration shown in Fig. 3.32*b* in order to remove components of the reference-switching signal. A low-pass filter may be added to remove second-harmonic components of the chroma signal generated in the process.

Choice of Demodulation Axes.   Over the ±500-kHz spectrum of color subcarrier, the chrominance signal can be demodulated as a double-sideband AM signal on either the I and Q axis or on the $R-Y$ and $B-Y$ axis. The latter permits the use of a simpler matrix to obtain the R, G, and B video-drive signals and consequently has become current practice in many receiver designs. However, this chrominance signal has the drawback of not utilizing the additional resolution provided by the wider bandwidth of I-channel encoding. Nevertheless, the trend in newer designs is to revert to the classic I and Q approach.

REFERENCES

Benson, K. Blair (ed.) (1986). *Television Engineering Handbook*. McGraw-Hill, New York.

Fink, Donald G. (editor-in-chief) (1982). *Electronics Engineers' Handbook*. McGraw-Hill, New York.

CHAPTER 4
AUDIO FUNDAMENTALS

E. Stanley Busby, Jr.
Ampex Corporation (retired), Redwood City, California

4.1 SOUND

Sound is the perception by humans and other animals of variations in air pressure. Very slow variations, as in the change of atmospheric pressure with shifts of weather, are not sensed. Very rapid variations, in the range above about 20000/s, may or may not be sensed, depending on the species. Generally, the smaller the animal, the higher both the lower and upper sensory limits become. Some variations, such as those experienced when descending from a high altitude in an airplane, may be sensed but as *discomfort* as opposed to sound.

The word *sound* is used in some places, principally the United Kingdom, to include television and radio signals created by sound, or those that can be converted to sound. In this chapter, *sound* means the physical variation in pressure of air or other conducting media, while electrical analogs of sound are called *audio*, discussed in Sec. 4.2. Sound has several properties. The rate of pressure variation is given the name *frequency*. The unit of frequency is the hertz (Hz), equal to one complete cycle of intensification and rarefaction of pressure per second. The amplitude of the variations in terms of *maximum change of air pressure* is called the *sound-pressure level*, usually abbreviated SPL. SPL is expressed in decibels (dB) greater than a reference level of 0 dB, approximately equal to the threshold of hearing, defined in Sec. 4.1.6.

4.1.1 Propagation of Sound

Sound spreads away from its source in a manner similar to waves from a disturbance on the surface of a pond, the propagation of radio waves from an antenna, and of light from its source. Sound waves *reflect* from surfaces, are *diffracted* as they pass around solid objects, are *refracted* by variations in speed of propagation through the medium, and can be *focused*. Like electromagnetic waves, sound waves diminish in intensity as they recede from their source, and transmitters (horns) and receptors (microphones) can be made directional. Unlike them, sound waves are deflected by wind and ocean currents. The wavelengths of sound in air and water are within the same dimensions as ultrahigh radio frequencies used in communications. Many of the same transmission and reception techniques used in radio have a direct analog in sound. Parabolic reflectors, for example, are used to focus both microwaves and sound.

When sound originates on a moving object, such as a rolling train, the object's velocity can be a significant fraction of the speed of sound in air. A stationary listener will hear each cycle of pressure variation a little early as the train approaches because its launch point is nearer at the start of each cycle. This is perceived as a higher frequency. After the train passes, the listener will hear each cycle a little later than normal because the launch point is farther away at the start of each cycle and is perceived as a lower frequency. This is called the *Doppler effect*, and it is the principle behind police radar guns. It is also responsible for the shift in light frequency from certain receding stars, called the *red shift*. An interesting outcome occurs when the listener can hear both the direct sound from a departing train and reflections from stationary structures in front of it. Both the decreased and increased frequencies can be heard together, and they make an unusual and complex sound.

The speed of sound through gases ranges from 606 ft/s for heavy gases like carbon disulfide, to 4165 ft/sec for the lightest gas, hydrogen. In air the speed of sound at sea level is given by:

$$c \ (\text{m/s}) = 331.29 + 0.607t \ (°\text{C}) \qquad (4.1)$$

$$c \ (\text{ft/s}) = 1051.5 + 1.106t \ (°\text{F}) \qquad (4.2)$$

where c = the speed of sound
     s = one second
     t = air temperature

The speed of sound in air also varies with altitude, atmospheric pressure, and water vapor content.

The range of the speed of sound in liquids is far smaller than in gases, since most liquids are relatively incompressible. One interesting value is the average speed of sound in seawater, 1.5×10^5 cm/s (4900 ft/s). The density of water changes in a complex way with variations in temperature near the freezing point. An *inversion layer*, a rapid change in temperature versus depth, can reflect or refract an incident sound wave. This can prevent *sonar* (sound navigation and ranging) emissions at the surface from reaching and echoing from a submarine that is beneath the layer.

The speed of sound in solid objects depends not only on the nature of its substance but on its shape and size. In rods and bars, the speed diminishes as the rod or bar becomes larger compared to the wavelength of the sound. The harder metals and glasses, like iron, steel, and crown glass, offer the highest speeds, similar to that in water, while softer metals, like lead and bismuth, are slower by a factor of 4 or 5. Among woods, elm is slower than lead, while oak is faster than many metals. At the bottom is cork, 10 times slower than steel.

4.1.2 Spectrum of Sound

For humans, the total mixture of all perceptible frequencies of air-pressure variation presented to their ears constitutes the *spectrum* of sound. The spectrum can carry, simultaneously, a surprising number of different signals, which the ear can selectively decode. For example, it is easy to imagine a doctor conversing with friends at a table beside a dance floor. Even without watching lips move, he would have no trouble understanding speech while largely ignoring the music. If his beeper emits its characteristic tone, he notices that as well, concludes his conversation without losing anything, and heads for a telephone. The aural decoding of the spectrum is often aided by visual cues and the directionality offered by hearing with two ears (*binaural reception*).

4.1.3 Kinds of Sound

Music. A musical combination of frequencies is one in which they are related to each other ratiometrically by small integers. If either part of the ratio between the frequencies exceeds approximately 13 (depending upon the listener), the result is perceived as *dissonance*, irritating if continuous but often a desirable element of music when followed immediately by a frequency group using small integers. Further, music must consist of an ever-changing presentation of different musical combinations, typically varying in the duration of each change and varying in amplitude as well. The voice can make music, too. Music may also include short bursts of noise, mostly to punctuate the tempo and rhythm of the total presentation.

Voice. Humans, unlike gorillas, have vocal chords and are able to emit pulses of sound pressure by modulating their larynx muscles (vocal chords) while exhaling through the larynx. They can create music or speech: music if the predominant frequencies chosen are related to each other as explained above, or speech otherwise.

Speech also includes bursts of noise, as in pronouncing sibilants, fricatives, and plosives; like ess, eff, tea, and pea. You may consider the human vocal tract as a transmission line excited at the sending end by the pulses emitted by the vocal chords (or by turbulent unpulsed airflow) matched to the impedance of the air by the mouth, depending upon the shape of its opening and, most importantly, modified by the position of a movable ridge on the tongue that causes a reflection to the vocal chords, and a re-reflection toward the mouth. Speech is a very complex mixture of different types of sound. It cannot be accomplished without a tongue.

Noise.   Noise is sometimes thought of as anything that is not music or voice and therefore undesirable. Electrical interference, wind noise in a microphone, and the like are certainly undesirable, but some noises, like crowd noise at a sports broadcast, add credibility to the show. Gunshots are bursts of noise, but what would a cowboy movie be without them?

Noise takes three forms:

1. *Pressure variations:* One type of noise is caused by pressure variations at a single or a few related frequencies, which can even be musically related but which are undesired, like the hum of an electric motor, or are too loud, like a locomotive horn. (Locomotive horns involve 3 to 5 musically related frequencies.)

2. *Random noise:* This kind of noise is made by variations in which no single frequency or group of related frequencies contains the preponderance of acoustic energy, but in which the energy is smoothly spread through the spectrum. The human voice can approximate it by saying "shhhhh." The turbulence created by rapidly moving gases and liquids tends to create random noise. Random noise is useful in measuring the frequency response of a room.

If the energy is uniformly distributed throughout the spectrum, it is said to be *white noise*, a term taken from the optical sciences as being an analog of "white light." If the energy is greater at the low-frequency end of the spectrum and tapers off smoothly toward the high end, it is said to be *pink noise*, an analog of the optical case of a relatively low temperature radiant light source. Pink noise is often used in acoustic measurement, as it closely approximates the spectral distribution of normal speech and music. An example of pink noise is the roar of Niagara Falls.

3. *Impact noise:* In theory, an impact will produce a single change in air pressure resulting in a spectrum of energy that exponentially decreases with increasing frequency. In fact, an impact usually excites resonant vibrations of some sort. A hand clap causes an acoustic resonance in the air between the palms of the hands. A hammer blow on a nail causes compression waves to travel up and down the nail, and along the head of the hammer, and vibrates the structure being nailed into. This periodically expands and shrinks both the nail and the hammer. These and the structure move the air around them, producing sound. The distribution of energy throughout the spectrum for impact sounds varies widely from a "dull thud," to the peal of a church bell, to the jingle of car keys.

Many jurisdictions have enacted legislation to protect their citizens from disturbance by sound sources, mostly designed to assure them a quiet and peaceful home, but also establishing SPL limits for the workplace. The deleterious effects of exposure to high SPLs are a function of the SPL and duration of exposure. The sound levels in jet airplane cockpits or in locomotives or in musical performance groups can temporarily at first, then permanently, decrease the sensitivity of hearing.

In the United States, it is generally accepted that an upper limit of 90 dBA (the A is a weighting curve discussed in Sec. 4.1.6) for an 8-h exposure is acceptable. For each halving of exposure time, a 5-dBA increase is permissible, up to 115 dBA for 15 min. 140 dBA is often taken as an absolute never-exceed level for impact and impulse sounds, such as gunfire and pile-drivers. The judgments made in arriving at these levels are based on permanent impairment of the ability to understand *speech*, so those of us in the broadcasting and recording industry, whose hearing needs to include the full range of frequencies and loudness of music, must be prudent in limiting our exposure to loud sounds of all kinds [1]. Attenuators are available at gun shops and garden supply vendors, and they can be worn over the ears while exposed to heavy noise. They reduce the SPL by 25 to 30 dBA, less if the wearer uses eyeglasses or has thick hair around the ears. Ear plugs are also available.

Above a level of 95 to 140 dBA, varying with the individual, sound is perceived not as *sound*, but as *pain*. The threshold between sound and pain can be a sharp one. For example, the author, having no ear protection, on an aircraft carrier, withstood an airplane launch (with afterburners) with only discomfort. Soon after, during a simultaneous launch of *two* airplanes, he experienced overwhelming and totally disabling pain.

4.1.4 Sources of Sound

Natural Sources. Wind makes sound directly by turbulent flow around obstacles. This effect is most notable during storms. It also rustles leaves, creating a kind of low-level impact noise. Waves at a beach disturb the air in a mostly random way, making a characteristic sound. The turbulent flow around taut wires and ropes can cause them to vibrate, creating a tonal sound.

In the process of communication, people speak, birds sing, insects flap their wings or rub their legs together, whales and porpoises vocalize, and some fishes vibrate a drumlike organ. Male gorillas beat their fists on their breastplate, while the females (who haven't a breastplate) make rhythmic sounds by pounding on a mound of earth. Volcanoes roar, rivers rumble, bees hum...the earth *abounds* in natural sounds.

Manmade Vibrations. Taut strings are used to make music. There are three classes of stringed instruments:

1. *Plucked strings* are deflected from their rest position, then suddenly released. Guitars, harps, lutes, samisens, and sitars are in this class, being directly plucked by fingers, fingernails, or picks held in fingers. Harpsichords are also in this class, but the plucking is indirect, done by a keyboard mechanism struck by the fingers. The spectral balance of the sound depends largely on how close to one end of the string it is plucked. The decay time of a vibrating string depends upon how efficiently its energy is coupled to the surrounding air. This is determined mainly by the design of the box or frame that supports the strings.

2. *Bowed strings* are caused to vibrate by rubbing them with a stretched bundle of hairs coated with a sticky substance. The physical phenomenon of *stick-slip* (the rapid decrease of the coefficient of friction once relative motion has begun) transfers power from the wielder of the bow to the string. Violins, violas, cellos, and double basses fall into this category. The spectral content of the vibrations depends on the dimensions of the box, including the thickness of the box wall, the stiffness of the string support, and the exact position of the (single) support post that couples the top and bottom surfaces of the box.

3. *Hammered strings* in many instruments cause the vibration. The nature of the resulting sound depends in part on how far from one end of the string it is struck and the hardness of the hammer. Pianos and clavichords fall into this class. The energy is transferred to the air mostly by the vibration of the frame that supports the strings. Because the strings are hammered, pianos are usually considered percussion instruments, discussed below.

Percussion Instruments. Drums are characterized by being excited by impact. The chamber of air contained in the drum oscillates at a frequency dependent upon the volume of contained air and the tension and mass of the stretched membrane of the drumhead. The spectral balance of the sound depends upon how near the edge of the drumhead it is struck and the hardness of the drumstick. In the general class of instruments having mechanical vibrations, the transfer of energy from the vibrating membrane of a drum is more efficient than most, resulting in a rapidly diminishing sound amplitude, characterized by a typical rat-a-tat sound. Kettle drums, having a heavy skin and only one radiating surface, are an exception, sounding longer than most drums when struck.

Other percussion instruments include the xylophone and marimba, in which metal pipes or blocks of wood are struck by hammers (usually wooden) to create musical sounds.

Vibrating Air Columns. The kinds of air-column instruments differ mainly in the manner in which pressure variations are introduced into the column:

Voice. Human, mammalian, and avian voices directly excite variation in air pressure by modulating the flow of air through their throat (a kind of tube), using a pair of stretched muscles between the lungs and the mouth. Aquatic mammals and some fishes also "sing," probably by transferring air-pressure variations to the surrounding water through their skin.

The human voice in particular is capable of an exquisite range of aural sensation, when combining words, musically uttered, with other musical instruments and enhanced by gestures. The human voice has also been used to imitate certain musical instruments, epitomized by the development of the bebop style of music.

Pipes and Whistles. Blowing air across the top of a tube at the proper angle will excite an oscillation within it, producing a mostly tonic sound having a predominant frequency proportional to the length of the tube and dependent on whether the tube is closed or open at the unblown end. A whistle is a pipe with a specially shaped orifice in the side of the pipe near the blown end. Air is blown *through* a whistle. Examples range from a police whistle to organ pipes. The fundamental (lowest) frequency depends on tube length and whether the far end is closed.

The spectral character of the tube's output depends on the material from which it was fashioned, whether or not it is tapered, the distance from the blowing port to the orifice, and the ratio of diameter to length.

Organ pipes are mostly made of metal, but some are wooden, usually of square cross-section. Metal tubes are usually cylindrical, but some increase in diameter toward the unblown end (*trompette*), while others decrease in diameter toward the unblown end and are usually capped. Some organ pipes modulate the airflow by a vibrating spring-metal flapper, excited into oscillation by the airflow in a manner analogous to reed instruments; thus these organ pipes are indeed called *reeds* even though the vibrating element is metallic. (See "Reeds," below.)

If holes are drilled in the tube, the effective length of the tube can be varied by using the fingers or finger-operated flappers, to cover or uncover the holes. Examples include the flute, recorder, saxophone, and ocarina. A sliding piston can be inserted in the tube to produce the "pennywhistle," a child's musical toy. Sounds from closed tubes, like many organ pipes and the pennywhistle, emanate entirely from the orifice near the blown end.

Reeds. Reed instruments vary the airflow in a tube by the use of a thin vibrating closure at the blown end that "almost" closes that end. Typically, the resonant frequency of the tube is varied by fingering a set of valves that cover or uncover holes along the length of the tube. The resulting character of sound depends on the stiffness and mass of the reed, the placement of the mouth upon it, and the muscles of the mouth.

There are two classes of reed instruments. One class uses one reed (usually literally made from a carved bit of reed), mounted near a stationary part of the tube. The other uses two bits of reed symmetrically opposed, very much like a flattened cigarette filter. The first class is characterized by clarinets, certain organ pipes, and saxophones while the second class includes oboes, bassoons, and English horns.

It is important to remember that much of the sound emanating from a fingered tube comes from the open holes, and *not* the end of the tube. Microphones are best placed *above* the instrument and not necessarily in line with the open end.

Horns.   Horns are pipes (mostly metal) whose effective length is either fixed, as in certain antique instruments, or varied by fingered valves that insert extra lengths of tubing between the mouthpiece and the bell. The bell is the radiating opening at the unblown end. This flared radiating opening offers a better impedance match between the source and the air than most wind instruments, so horns can be *loud*. Excitation of the exponentially tapered tube is through the use of a shaped metal mouthpiece, called an *embouchure*, against which the mouth is pressed with pursed lips. The embouchure is a resonant chamber, excited by blowing air through vibrating lips. The entire system of horn, embouchure, and muscles of the lips form an oscillating system producing pulses of air into the input end of the tube. Horns include French horns, cornets, bugles, trumpets, tubas, Sousaphones, double-belled euphoniums, and trombones. Modern horns are coiled to reduce their maximum dimension while maintaining their acoustic length. Two exceptions are the antique trumpet and the alpenhorn. The trombone is unique in that the length of the tube is varied by the use of a single smoothly sliding extension. It can produce frequencies *between* the usual harmonically related increments. Among the horns, the trombone does the best WA-WA.

4.1.5 Sensitivity of the Ear

The ability of humans to discern the loudness of sounds of various frequencies depends on the sound-pressure level (SPL). Figure 4.1 shows how the human ear perceives loudness at various frequencies. The lowest curve shows the SPL called the *threshold of sound*, below which the average ear cannot hear a tone. At other SPLs it can be clearly seen that the frequency response "flattens" as the SPL increases. The frequency response and sensitivity of an individual can vary from zero (totally deaf) to 20 dB or so greater than average. The sensitivity to high frequencies normally diminishes with age. Trauma and disease can decrease sensitivity. Trauma includes frequent exposure to loud sounds. Deafness among operators of loud machinery, such as airline pilots and railroad engineers, is well known.

Not only does the frequency re-

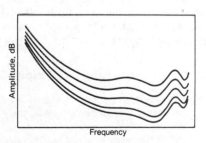

FIGURE 4.1 Fletcher-Munson curves of hearing versus frequency. *(Source: Blair K. Benson (ed.),* Audio Engineering Handbook, *McGraw-Hill, New York, 1988)*

sponse of the ear change with amplitude, but if presented a sudden loud sound, the ear's sensitivity is diminished *and remains so for a short time*, on the order of 50 ms. Some digital audio transmission techniques take advantage of this fact by using coding methods of variable resolution, and therefore variable quantization noise, with the choice based on the loudest recent digital sample. Called *near-instantaneous companding*, it reduces the amount of data to be transmitted.

4.1.6 Measurement of Loudness

The unit of loudness is the *phon*, which is the SPL, at any frequency, that is perceived by listeners to be equal to the loudness of a 1-KHz tone of that same SPL. Measuring loudness by this method is an expensive proposition, involving panels of listeners, and it is therefore seldom done.

Loudness can be adequately estimated by measuring SPL directly, using a calibrated microphone and amplifier and inserting a filter that discriminates against those frequencies at which the ear is least sensitive, i.e., having a response that is approximately the inverse of Fig. 4.1. Three curves are offered for American Standard SPL meters:

The *A* curve, derived from the 40-dB equal-loudness curve of Fig. 4.1, is for measuring sounds up to 55 dB*A* above threshold.

The *B* curve, derived from the 70-dB equal-loudness contour, is for SPLs between 55 and 85 dB*B*.

The *C* curve (flat response) is for loud sounds above 85 dB*C*.

It is customary to indicate which curve was used by appending its letter to the decibel reading, for example, 45 dB*A*.

Most instruments offer the *A* and *C* curves, while noise-abatement laws tend to reference only the *A* meter. Typically, instruments have a switch to select the response time of the meter. "Fast" will deflect the meter as rapidly as possible, while "slow" allows the meter to stabilize in about 1 s. Obviously, "slow" discriminates against impulse noise. Noise-abatement laws usually specify slow response.

Small, portable battery-powered noise meters are available that can be worn by workers. The meters periodically measure the SPL and store it in digital memory. At the end of the workers' shift, the recorded data is processed to provide a printed history of their exposure to sound.

The unit of SPL is the pascal, equal to a pressure of one newton per square meter. (The newton is equal to 10^5 dyn, or 0.1020 kg.) The reference level for SPL measurements is 20 μp approximating the threshold of human hearing, using the *A* weighting curve. Table 4.1 shows the SPL created by some typical sources.

4.1.7 Frequency Range of Various Sources

The narrowest frequency range of all are those sources related to the ac power supply or that are otherwise fixed in frequency. Examples are locomotive whistles and pigeons.

Among humans, and the sounds generated by them, the narrowest frequency range is the voice, spanning a range of almost 4 octaves. The pipe organ can produce the lowest frequency and challenges those that produce the highest. The strings span just over 7 octaves. Blown pipes cover just over 5 octaves.

The highest frequencies are created by mechanical vibrations, such as

TABLE 4.1   Sound-Pressure Level (SPL) Created by Some Typical Sources

Source	SPL (dBA)
Jet engine at 10 m	150*
Threshold of pain	130
SST takeoff at 500 m	120†
5-chime locomotive horn at 100 ft	117
Rock music, amplified	110
Chain saw at 1 m	100
Lawn mower at 1.5 m	90
Jet, window seat, takeoff	82–88
Automobile, 15 ft	70
Automobiles in traffic, 15 m	70
Conversation, 1 m	55–70
Author's computer, 2 ft	54
Residence, suburban	40–50
Recording studio	20
Quiet whisper, 1.5 m	20
Quiet breathing	10
Reference	0

*Painful and damaging.
†Painful for some.

a squeaking door or a chirping insect. The lowest frequencies are created by the longest organ pipes, on the order of 16 IIz. For more detail, see *Reference Data for Engineers* [2].

Ordinarily, the spectral distribution of energy of the human voice and the other instruments that humans play is such that the average energy at frequencies above about 3 kHz becomes smaller as the frequency increases. One technique that makes use of this distribution is to selectively boost higher frequencies before recording or transmitting them, then attenuate them upon reproduction or reception, which also attenuates any noise encountered in the process. The boost is called *preemphasis*, and the attenuation is called *deemphasis*. Their use is mandatory in analog tape recording and FM and TV audio transmission. On digital recordings, their use is optional. Most digital recordings are not emphasized. Some sounds of the modern repertoire are electronically generated, in which case the distribution of energy in the spectrum is entirely in the hands of the performer, who, using a *synthesizer*, is capable of producing sounds at any frequency and any amplitude. Using preemphasis with synthetic music invites overload of the recording or transmission path, causing distortion.

4.1.8   Reinforcement of Sound

In earlier times, *sound reinforcement* was called "public address." Before electronic amplifiers, it was accomplished by *steam*. The speaker spoke loudly into a diaphragm that modulated the flow of steam through a pipe, which was the input to a horn, which addressed the crowd.

In large auditoriums, the sound made by performers on stage must be amplified so that those in the audience far away from the stage, perhaps observing it with optical magnifiers, can enjoy the sound as if they were closer to the stage. The frequency response of an auditorium varies greatly from the ideal and is sharply dependent upon population density, since people are an efficient absorber of sound. The frequency response of an auditorium can be compensated by adjusting the frequency response of the amplifier used.

In the case of very large auditoriums, the propagation time of sound from the stage to the audience may require that the reinforcing sound be electronically delayed, so that it arrives at the listener's seat at the same time as the sound from the stage itself. Audio delay is generally accomplished by taking periodic samples of the microphone input, converting these to numeric digital values, passing them through a digital memory of appropriate size, then reconverting them to an analog signal.

The ability of the human brain to tolerate sounds that arrive earlier or later than the imaged events that caused them is asymmetric. Sounds that "lag" the picture by about 40 ms can be tolerated, but when sound "leads" the picture by about 20 ms, the brain senses the result as abnormal. This is not surprising if you consider that when watching someone speaking from 22 ft away, their sound is delayed by 20 ms. The brain has experienced this delay and compensated for it. Large movie theaters tend to adjust their sound-to-picture timing so that proper synchronism is presented to those seated at one-third the way from the front row to the back row.

Synchronization problems can occur when video-frame synchronizers are used on video signals transmitted via satellite. If more than one synchronizer is in the video path and no compensating audio delay is applied, the difference can be disturbing. If the video delay path varies, the synchronizer adjusts by repeating or discarding an entire frame of video. Instant adjustments cannot be made in the audio path without causing notice. There are digital audio devices made to accommodate these time shifts by *slowly* increasing or decreasing the delay by one TV frame time. For further information, see *Reference Data for Radio Engineers* [2], and Floyd E. Toole, *Audio Engineering Handbook* [3].

4.2   AUDIO

As used in this book, *audio* is the electrical, magnetic, or optical analog of what was once, or is yet to be, a physical variation of pressure in air or other media, called *sound*, discussed in Sec. 4.1 above.

4.2.1   Conversion of Sound to Audio

Sound waves in air or in liquids are usually converted to an electric signal through the use of a thin diaphragm vibrated by the pressure waves of the sound and converted to an electric signal in several ways:

Telephone.   The classic telephone transmitter, or mouthpiece, contains a packet of carbon granules (usually anthracite coal) that is impacted by a piston coupled to the diaphragm. The variation of pressure on the granules alters the packet's electrical resistance and therefore the current through it. Over short distances the current variations are sufficient to excite a telephone receiver without amplification.

Moving-Coil Microphone.   A diaphragm is attached to a coil of wire positioned between the poles of a strong magnet. Movement of the coil generates a voltage. In one implementation, the coil is a linear half turn of an extremely light conductor, which is flat and very thin. The conductor and the diaphragm therefore are one. It is positioned between two long, parallel magnetic pole pieces. Vibrated by the air, its movement generates a voltage.

Many inexpensive intercom systems use a small loudspeaker as a microphone for transmission, and the same loudspeaker for reproduction.

Capacitor Microphone.   A diaphragm plated with a microscopically thin conductive material is spaced close to a parallel conductive surface, forming a capacitor. The capacitor is charged to a potential of 12 to 48 V through a series resistor. The energy stored in a capacitor is proportional to the product of voltage and capacitance and tends to remain constant. If the capacitance varies because the sound pressure changes the spacing of the capacitor plates, then the voltage varies so as to maintain a constant product. This voltage variation becomes the output signal. Capacitor microphones can be very small, with diaphragm areas on the order of 10 to 40 mm^2. They are therefore especially useful in applications in which the microphone must be visually unobtrusive.

Many audio mixing boards and preamplifiers supply the charging potential, called a *phantom supply*. If the microphone cable is a single shielded conductor, the voltage is applied to the inner conductor through a resistor. If the cable is a shielded twisted pair, the voltage is applied to both conductors equally, using either two equal resistors or a center-tapped transformer. The output impedance of capacitor microphones is inherently high, and it is customary to use a field-effect transistor at or near the microphone to convert to low impedance. Some implementations convert to *balanced* low impedance, a desirable feature if the cable lengths are long as they might be in, for example, a large music hall or church.

Piezoelectric Microphones.   A diaphragm is coupled to a bar of ceramic material, typically lead zirconate titanate, which has the property of generating a voltage when deformed. The output impedance is high. Once in favor because it could supply an adequate signal level directly to the grid of a vacuum-tube amplifier, its advantage diminished when the cost of gain (in semiconductor amplifiers) became infinitesimal compared to the cost of maintaining a vacuum tube.

Contact Microphones.   Using any of the above techniques, a suspended mass inside an enclosure fastened to a vibrating solid can be coupled to an element that will generate a voltage. Contact microphones are sometimes used on guitars, pianos, and other stringed instruments and are probably applicable on any instrument in which sound is radiated by the overall vibration of the frame of the instrument. Wind instruments seem to be contraindicated. In one application, the electric guitar, the microphone is directly excited by the vibration of steel strings in the magnetic field of the microphone.

4.2.2 Amplification

Usually, audio signals must be amplified before they are converted to sound. In the case of a microphone output signal, the design of the amplifier is dominated by the requirement for low-noise low-amplitude amplification and is called a *pre*amplifier. An amplifier used for driving a loudspeaker is designed for low distortion at high power-output levels and is called a *power* amplifier. All amplifiers, especially power amplifiers, are capable of introducing distortion, usually *harmonic* distortion, in which the frequencies of the spurious artifacts are integer multiples of the fundamental frequency of the input signal.

4.2.3 Preemphasis and Deemphasis

In transmission paths that ordinarily exhibit flat frequency response, it is often desirable to boost the higher frequencies of the audio signal as it enters the system and to equally attenuate it at the receiver. This attenuates any high-frequency noise introduced along the path. It is especially appropriate for FM broadcasting, in which the noise spectrum introduced at the receiver is concentrated at the higher modulation frequencies. The use of preemphasis/deemphasis tends to "flatten" the noise characteristic of an FM system. Since every FM transmitter must preemphasize and every receiver must deemphasize the signal, it is necessary to *standardize* the *degree* of emphasis/deemphasis.

Figure 4.2*a* shows a typical *preemphasis* circuit, while 4.2*b* demonstrates its inverse, a *deemphasis* circuit. The response of both circuits is generally expressed in terms of the product of *R* and *C*, as shown in Fig. 4.2, and is often spoken of as the *time constant*, although there is no direct connection between the charge-discharge time of an *RC* circuit and its frequency response *other* than that both involve the product of *R* and *C*. The frequency at which the boost is 3 dB is given by:

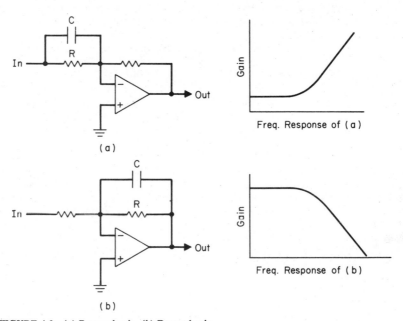

FIGURE 4.2   (*a*) Preemphasis. (*b*) Deemphasis.

$$F_{3dB} = \frac{1}{2\pi fT} \tag{4.3}$$

The boost at any frequency is given by:

$$\text{Gain (dB)} = 10 \log_{10}(1 + 2\pi fT)^2 \tag{4.4}$$

where f = frequency, in Hz
π = 3.14159
$T = R \times C$

AM broadcasting uses no preemphasis. FM and TV audio in the United States uses a T of 75 μs, while FM and TV audio in Europe uses a T of 50 μs. Compact disks (CDs) and other digital audio recordings sometimes employ preemphasis prior to being digitized. If preemphasis is used, a data bit in the digital signal stream is dedicated to signaling this fact, so the reproducer will know whether to deemphasize or not. Figure 4.3 shows typical circuitry and response shapes. The maximum boost, in decibels, is given by:

$$\text{Gain}_{dB} = 20 \log_{10}\left(1 + \frac{R_2}{R_1}\right) \tag{4.5}$$

The response is stated in terms of two time constants, such that:

$$T_1 = R_1 \times C \quad \text{and} \quad T_2 = (R_1 + R_2) \times C$$

The values most used on compact disks, if used at all, are T_1 = 50 μs and T_2 = 15 μs. Approximately, the larger time constant determines the frequency at which boost starts, and the smaller number the frequency at which the boost stops increasing.

4.2.4 Equalization

Equalization is the introduction of a frequency and/or phase-response adjustment into a transmission path that is *not* flat, so as to *make it flat*. Long transmission

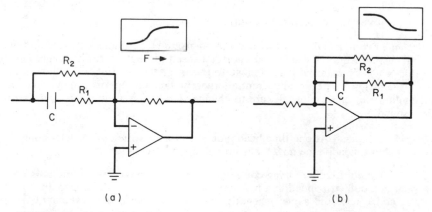

FIGURE 4.3 (*a*) Preemphasis for compact disk. (*b*) Deemphasis for compact disk.

lines tend to lose response at high frequencies. A circuit at either end of the line, or even distributed between the two ends, having a response that is the *inverse* of the line's response, can make the overall path response flat. Unlike preemphasis and deemphasis, equalization has no word for its inverse, such as "de-equalization."

Analog audio-tape reproducers have a natural response, due to the inductive nature of the reproduce head, which rises 6 dB with every doubling of recorded frequency. At some frequency approaching the upper limit of the system, various high-frequency losses become significant, and above that frequency predominate. Figure 4.4 shows the general shape of the reproduce response and the response of the equalizer. It is typical to standardize the response of the reproduce equalizer and make adjustments in the record path to produce an overall flat result. Expressed as a time constant, as in Sec. 4.2.4 above, the reproduce curves for analog audio tracks on video recorders range from 10 to 35 μs. Audio recorders use thicker tape coatings and a wider range of tape speeds and reproduce curves ranging from 25 to 130 μs. Analog disk recordings at 33 and 45 r/min expect a 75-μs reproduce curve.

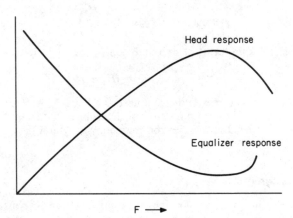

Head response

Equalizer response

F ⟶

FIGURE 4.4 Tape playback equalization.

4.2.5 Conversion of Audio to Sound

Telephone Receiver. A coil of wire is wound around a permanent magnet and the signal current from the distant mouthpiece passed through it. A diaphragm made of thin magnetic material is held close to the magnet, resulting in its flexure toward the magnet. The signal current alternately aids and opposes the permanent magnetic field, causing the flexure to change, producing a vibration, and therefore sound.

Headphone Types. At one time headphones were constructed like telephone receivers, but a headphone now takes one of the following forms:

1. *Moving coil:* Like a moving-coil microphone in reverse, a diaphragm is coupled to a coil suspended in a magnetic field. Passing the audio signal through the coil causes the signal magnetic field to react with the permanent field,

pushing and pulling the coil and vibrating the diaphragm, making sound. A low-impedance coil (less than 30 Ω) is useful when the headphone is used with a circuit designed to drive a loudspeaker. A high-impedance coil is more appropriate in a recording or broadcasting studio because it might be accidentally connected across a transmission line in active use. The limit to high-frequency response is the mass of the diaphragm and coil.

2. *Piezoelectric:* A bar of piezoelectric ceramic is coupled to a diaphragm and the audio signal, amplified to supply a relatively high voltage (but low current) applied across two faces of the bar. The bar flexes according to the applied voltage and moves the diaphragm, producing sound. The limit to high-frequency response is the mass of the mechanism.

3. *Capacitive:* A very light, plated diaphragm is brought close to a fixed conductive plane. A relatively high dc potential is applied to the capacitor formed by the diaphragm and the conductive plane. Unlike charges attract, so the diaphragm is deflected toward the plane. The audio signal, amplified to a high voltage, is superimposed on the dc potential, alternately making the total potential, and therefore the deflection, vary above and below the average, producing sound. Since the diaphragm is extremely light, the high-frequency response is greatly extended.

Headphone Construction. The physical construction of headphones has taken several forms:

1. A single earpiece, with or without a microphone boom, that holds a single microphone in front of the wearer's mouth. These are used by telephone and control tower operators for verbal communications in relatively quiet surroundings.

2. Dual earpieces, with or without a microphone boom, but with no special attention taken to exclude external noise. In some studios, program audio is supplied to one ear and instructions from the director to the other. In others, the audio signal is a stereo pair, and in others both ears are supplied with the same monaural signal.

3. Dual earpieces each with a large conformable soft ring, which completely surrounds each ear and provides an acoustic seal against the sides of the skull. In this way, outside sound interference is attenuated from 20 to 30 dB. These are especially useful when it is necessary to make critical audio-quality judgments in a moderately noisy environment such as a video-tape-recording room, or to hear radio instructions in a very noisy airplane cockpit. One shortcoming is a frequency-response null between 5 and 15 kHz due to an acoustic resonance of the air in the sealed chamber containing the earphone and the ear itself.

4. Dual earpieces with low-density foamed plastic pads to hold the earphones slightly away from the ear and that supply no acoustic seal against the skull isolate the wearer from the environment and do not exhibit an acoustic null.

Use of headphones became particularly popular among commuters, but because their use inhibits the detection of acoustic alarms, horns, sirens, and the like, many jurisdictions have declared their use illegal for motor vehicle drivers. The use of extremely high sound-pressure levels in motor vehicles also overshadows these warnings, but this practice has yet to be addressed by legislators.

There are even *null* headphones, called *attenuators*, that supply no other function than to shield one's ears from loud sounds which might impair hearing. They are used by leaf-blowing gardeners, locomotive engineers, gunners, airport and flight deck crew, and others exposed to high SPLs.

Loudspeakers.   These are, at the minimum, large headphones that many people can hear at once. In the audio and television industry, each is more likely to be an array of loudspeakers, each of a different dimension, and each designed to efficiently radiate its own portion of the audio spectrum. These multiple loudspeaker systems fall into two classes:

1. The frequency-wise distribution of acoustic energy to the several radiators is determined by passive networks of resistors, capacitors, and inductors, usually contained within the enclosure housing the radiators. The input to each set of speakers is the output of a single amplifier.

2. The distribution of energy to the radiators is selected by a low-level network applied to the signal *prior* to power amplification, with separate power amplifiers used for each frequency range. The terms *bi-amp, tri-amp*, and *quad-amp* identify the number of amplifiers needed for each channel.

Loudspeakers have been constructed that are analogs of capacitor microphones and earphones and ribbon microphones [14, 15]. *Tweeters* (high-frequency loudspeakers) have been made by ionizing a gas (typically helium) in a chamber using a supersonic ac voltage. The supersonic voltage is amplitude modulated by the audio signal and causes alternate heating and cooling of the gas. The chamber is impedance-matched to the atmosphere by a horn.

Most loudspeakers are conical configurations of paper or plastic constructed with a cylindrical extension at the small end, which is positioned within the circular gap of a strong permanent magnetic field. On the extension is wound a "voice coil." The driving amplifier causes an audio current to flow in the coil. The coil's magnetic field reacts with the permanent field to push and pull the cone, producing sound.

Audio and acoustic engineers have produced a great volume of mathematics treating the design and application of loudspeakers. Audiophiles have produced a great volume of mystique concerning the choice of loudspeakers, enclosures, speaker location, the chemistry of the preparation of the copper for wiring, and the metallurgy of connector contacts. The author of this chapter suggests: for loudspeakers that will be heard by others (who might have better ears than yours), judge by the specification sheets, but for audio to be heard by *you*, listen to them, preferably in the location in which they will be used, or in a room with similar dimensions, check your wallet, and decide. It is not unlike choosing a bottle of wine for dinner.

4.2.6   Wire Transmission of Audio

Balanced/Unbalanced.   Electronic equipment for the home tends toward *single-ended* or *one-wire transmission*, employing a single shielded conductor. This is adequate for the transmission distances found in a home. Professional equipment generally uses *balanced* or *push-pull transmission*, by means of a pair of twisted conductors having an overall conductive shield. This greatly reduces interference from extraneous electromagnetic fields.

Source Impedance. Many years ago the source impedance of an audio signal for wire transmission was standardized at 600 Ω, approximately the impedance of the open-wire lines of the era. Frequency-response equalizers (see Sec. 4.2.4) were designed whose performance depended strongly upon that impedance. Even though the characteristic impedance of the twisted pairs in modern multiconductor cables is approximately 100 Ω, the 600-Ω standard still survives. Today it is the practice, within an installation, to provide a low source impedance (30 Ω or less) and a high termination or input impedance. This allows many receivers to be fed by one source with no need to adjust levels. In addition, if by accident a 600-Ω load is connected to the source, the resulting 0.4-dB reduction in level is not catastrophic. When transmitting to the telephone company, it is wise to ask whether *you* must supply a 600-Ω source impedance, or *they* will.

Amplitude Levels. In the following discussion, the reference level of 0 dBm refers to the voltage developed when 1 mW is dissipated in a 600-Ω resistor. The "standard level" within an installation is established by the system's designer and ranges from a low of about −24 dBm to a high of +8 dBm, the standard level when transmitting to the telephone company for distribution. This choice was made by noting that signal peaks are typically 10 dB greater than indicated on most meters and that levels of greater than +18 dBm cause excessive crosstalk into adjacent telephone subscribers' lines.

Measurement of Levels. In the United States the unit of audio level is the *volume unit*, or VU. For a single-frequency tone, 0 VU is the same as 0 dBm. The VU meter has a semilogarithmic scale, brought about by a nonuniform spacing of the magnet's poles in which the moving coil rotates. The VU meter contains a rectifier. The meter is never connected directly across an active signal line. It is customary to use a series resistor of 3.6 kΩ, minimum; else the nonlinear loading of the line by the rectifier will distort the signal. The ballistics or time response of the meter is closely specified and controlled and would be adversely affected by being driven from a low-impedance source. In Europe, audio is measured with a peak program meter (PPM) specified in EBU-3205 and CCIR Recommendation 468. The signal is processed with a logarithmic amplifier, and the signal level is presented on a linear scale. By means of a storage capacitor, the indicator will respond fully to a tone burst as short as 10 ms and gradually return to the steady-state level in 2 or 3 s. A PPM for use in the United States, proposed by the IEEE, currently is under review by industry [16]. Home tape recorders and recording studio mixing boards often employ a bar graph indicator that combines both averaging and peak-reading characteristics. The bar graph, using light-emitting diodes or a plasma-discharge device and having no mass, has no problem with ballistics and indicates instantaneously. It is typical to have the uppermost or rightmost element of the bar graph remain lit for a few seconds as an indication of the peak amplitude attained. Both linear and logarithmic versions are used.

Audio signals vary greatly in their ratio of peak to average amplitude, with a church organ, for example, having a small ratio, and a snare drum having a large one. The use of an averaging instrument, like the VU meter, tends to maintain a constant average level, and therefore a constant signal-to-noise ratio, but allows peaks that are too short to indicate on the meter to be clipped or otherwise distorted. The PPM, on the other hand, tends to hold the worst-case distortion constant, while the signal-to-noise ratio varies with the peakedness of the program. The average difference in the displayed amplitude of the VU and PPM instruments is generally accepted to be 9 dB.

Connectors.   Home electronics, having short distances of interconnection, mostly use the RCA *phonoplug*, a single-ended device with a single inner conductor and outer shield, while a few use the XLR 3-pin connector that provides for two conductors and a shield, even if only one inner conductor is used. The XLR connector is described in the Electrotechnical Commission Publication 268-12 (1975), *Circular Connector for Broadcast and Similar Use.* The standard usage is as follows:

1. An output connector fixed on an equipment shall use male pins with a female shell. The corresponding cable-mounted connector will thus have female pins with a male shell.

2. An input connector fixed on an equipment shall use female pins with a male shell. The corresponding cable-mounted connector shall use male pins with a female shell.

3. The pin usage is:
   Pin 1: Cable shield and/or signal ground
   Pin 2: "Low" side of signal, or signal ground
   Pin 3: "High" side of signal

This same connector is specified for the transmission of bit-serial digitized audio, in which case the polarity of pins 2 and 3 is unimportant [4].

4.2.7   Recording of Audio Signals

Analog.   Audio signals are recorded in an analog fashion on magnetic tape [5], on optical tracks of motion picture film [6], on grooved analog disk [7], and in an FM fashion on video disk [13]. In the case of magnetic recordings, the important adjustments are bias amplitude, record level, and equalization to attain flat response. If Dolby [8] noise reduction is employed, record-level adjustments become critically important, while if DBX [9] noise reduction is used, frequency-response adjustments become critical.

Digital.   Once having been digitized, the numeric representation of the signal can be recorded on compact disks [10], digital audio [11] or television [12] recorders or on computer data disks, which especially renders it useful for facile editing. An important datum to record is whether or not preemphasis was applied before digitization. All digital recording and transmission formats provide for this, and one [4] permits the transmission of a wide range of other channel information as well, including two time codes, source and destination codes, and more.

4.2.8   Reproduction of Audio Signals

Analog.   Most analog reproducers use well-established standards of frequency-response equalization. Standard tape speeds are 30, 15, 7½, 3¾, 1⅞, and ¹⁵⁄₁₆ inches per second (ips). Adjustment of playback response is accomplished by the use of a constant-amplitude oscillator coupled to a coil of a few turns of wire temporarily taped or clipped to the reproduce head. The frequency response of the reproduce circuitry is adjusted until it most closely approaches the ideal. An alternative is to pass the oscillator's output through a frequency-response shaping network having a response that is the inverse of the standard response. The reproducer should then be adjusted for flat response.

Calibration tapes and films are available that have recordings at several frequencies and at a standard magnetization level. In addition, they have a short-wavelength (high-frequency) section to facilitate the adjustment of the reproduce head azimuth angle. Calibration tapes tend to deteriorate with usage, so it is suggested that a new one be used *twice* to adjust two recorders, then a copy made of it using those two recorders. Play the copy and record the magnitude of any deviations from standard response. Use the copy and its deviation record for routine maintenance. When the copy wears out, make a new one. Standard response analog disks are also available for servicing phonographs.

Digital.   Digital playback includes recovering the numeric value of each sample as well as overhead data that is used to detect and correct errors due to random noise and tape dropouts. Adjustments are limited to establishing the best record current and playback equalization that result in the lowest digital error rate. Playback equalization in a digital reproducer has nothing whatever to do with the frequency response of the audio signal. It does compensate for variations in head responses, tape coating formulation, and surface. Generally, both frequency and phase response are adjustable.

Of the two current digital video recorder formats, the D1 (component) format 12 uses high-coercivity (850 Oe) gamma-ferric-oxide tape while the D2 (composite) format uses iron-particle tape. Both formats use the same cassette, so beware of the opportunity to introduce the wrong kind of tape into a recorder.

In both formats, the four audio channel records are merged into the video path and written on tape using the video heads and the video data rate. The audio data is written in bursts, with two identical bursts for each channel, for a total of eight bursts. There are separations between each burst to allow for editing individual channels. The D1 format writes the audio records near the centerline of the tape, while the D2 format writes the audio near the edges of the tape.

Computer Disk.   A rapidly growing segment of the digital audio business offers editing systems that store and recall digital samples using computer disks. They offer rapid selection of program sectors without having to wait for a tape machine to spool through a reel to find them. Most of these systems can accept digital samples at four popular rates:

1. 48 kHz, which can be derived from either the NTSC or PAL horizontal-scanning rate. The lowest common frequency of the two systems is 2.25 MHz. 2.25 MHz/144 is 15,625 Hz, the horizontal-scanning rate of the PAL and SECAM transmission standards, while 2.25 MHz/143 is the NTSC color horizontal-scanning rate. The standard sampling rate for the D1 and D2 digital video recorders is 48 kHz.

2. 44.1 kHz, the standard for compact disks, which is derived from either the 50-Hz field-scan rate of the PAL system, or the 60-Hz vertical-scan rate of North American *monochrome* TV.

3. 44.056 kHz, the standard for some video recorders adapted for recording audio samples, results when those recorders are locked into the NTSC color TV vertical rate of 60/1.001 or 59.94 Hz.

4. 32 kHz, used primarily in Europe, can be derived as in rate 1, above.

4.2.9 Synthetic Audio Signals

Many audio signals, including both music and speech, are created electronically, having never existed as sound. The *electric organ* of the 1940s was one of the first. Modern devices can create an enormous range of sound. They can "copy" a sound, such as a frog croaking, then dynamically modify it to agree with a musical scale, allowing a musician to "play" frogs on an electronic keyboard.

Analog. Synthetic audio can be treated like any other source of sound, with one exception. In systems that employ preemphasis, like FM radio, TV audio, and some CDs, the usual assumption that the input energy at high frequencies is small can no longer be made. A performer using an electronic synthesizer can easily generate high-energy, high-frequency signals that can overload a system using preemphasis.

Digital. Sometimes synthetic audio is directly generated in the digital domain. In ordinary systems that digitize analog signals, the input spectrum is carefully limited by a low-pass filter before digitization to prevent audible artifacts from being generated that fall inside the audio passband. In the case of direct generation, the generating algorithm must ensure that it does not create any representations of audio having a frequency higher than 20 kHz.

REFERENCES

1. Charles Martiez and Samuel Gilman. "Results of the 1986 AES Audiometric Survey." *Jour. AES*, **36** (no. 9) (September 1988).

2. Howard W. Sams & Co., Inc. "Electroacoustics." *Reference Data for Radio Engineers*, 6th ed., Chap. 37, 1975.

3. Floyd E. Toole. "Principles of Sound and Hearing." *Audio Engineering Handbook.* McGraw-Hill, New York, 1988, Chap. 1.

4. "Recommended Practice for Digital Audio Engineering. . . Serial Transmission Format for Linearly Represented Digital Audio Data." *ANSI S4.40–1985.*

5. E. S. Busby Jr. "Magnetic-Tape Recording and Reproduction." *Audio Engineering Handbook.* McGraw-Hill, New York, 1988, Chap. 10.

6. Ronald E. Uhlig. "Film Recording and Reproduction." *Audio Engineering Handbook.* McGraw-Hill, New York, 1988, Chap. 12.

7. Gregory A. Bogantz and Joseph C. Rude. "Analog Disk Recording and Reproduction." *Audio Engineering Handbook.* McGraw-Hill, New York, 1988, Chap. 8.

8. Ray Dolby, David P. Robinson, and Leslie B. Taylor. "Noise Reduction Systems." *Audio Engineering Handbook.* McGraw-Hill, New York, 1988, Chap. 15.

9. Leslie B. Taylor. "dBx-TV Noise Reduction." *Audio Engineering Handbook.* McGraw-Hill, New York, 1988, Chap. 15, Sec. 15.7.3.

10. Hiroshi Ogawa, Kentaro Odaka, Masanobu Yamato, and Toshi T. Doi, technical consultant. "Digital Disk Recording and Reproduction." *Audio Engineering Handbook.* McGraw-Hill, New York, 1988, Chap. 9.

11. P. Jeffery Bloom, Guy W. McNally, Leonard Sherman, and Jerry Whitaker. "Digital Audio." *Audio Engineering Handbook.* McGraw-Hill, New York, 1988, Chap. 4.

12. "Proposed American Standard for Component Digital Video Recording," *SMPTE 224M*, and "Proposed American National Standard Digital Television Tape Recorder for Composite Digital Video Recording," *SMPTE V16*.

13. Robert A. Castrignano. "Video Disk Recording and Reproduction," *Television Engineering Handbook*. McGraw-Hill, New York, 1986, Chap. 16.

14. Katsuake Satoh. "Other Types of Loudspeakers." *Audio Engineering Handbook*. McGraw-Hill, New York, 1988, Chap. 7, Sec. 7.7.1.

15. J. A. M. Nieuwenddijk. "Compact Ribbon Tweeter/Midrange Loudspeaker," *Jour. AES.*, **36** (no. 10) (October 1988).

16. Randall Hoffner. "Audio Program Metering in the 1980s: The Work of the IEEE Audio Measurements Subcommittee," *SMPTE Jour.*, **98** (no. 8) (August 1989), pp. 590–593.

CHAPTER 5
DIGITAL FUNDAMENTALS

E. Stanley Busby, Jr.
Ampex Corporation (retired), Redwood City, California

Strictly speaking, a *digit* is one of your fingers or toes. The word *digital*, as an adjective describing calculating machinery, is appropriate because early calculators used and displayed the well-known set of 10 Arabic numerals. It would have been better if the early mechanical computers had been described as "decimal," but they weren't. Keep this in mind as you read on, and remember that all operations inside a digital device are *binary*, with each wire and pin having only *two* states. Table 5-1 equates various words used to describe the two states. If the words in the columns are reversed, the wire or pin is said to be "low active" or "negative logic."

TABLE 5.1

State	Opposite
On	Off
High	Low
1	0
Active	Inactive
On	$\overline{\text{on}}$ (The bar means "not.")
True	False
Positive	Zero

5.1 COMBINATIONAL LOGIC

When the inputs to a logic circuit have only one meaning for each, the circuit is said to be *combinational*. The devices tend to have names reflecting the function they will perform, like AND, OR, exclusive OR, *latch, flip-flop, counter*, and *gate*. To perform the logic of the statement "If the record button and the play button are pressed at the same time, the record mode is entered," an AND gate will be used to provide an output *only* if both its inputs are active.

5.1.1 Symbols

Logic circuits are usually documented by the use of schematic diagrams. For simple devices, the *shape* of the symbol tells the function it performs, while the presence of small bubbles at the points of connection tell whether that point is high or low when the function is being performed.

More complicated functions are shown as rectangular boxes. Figure 5.1 shows a collection of symbols. Note the use of the prefix N to the word describing the function when the output of the device is "low" when the function is being performed. "Exclusive OR" means the output is active if *either, but not both*, of the inputs are active. For a two-input device, this means that a high output indicates the inputs are different, while a low output indicates sameness. In the case of more than two inputs to an exclusive OR gate, the output is active if the number of active inputs is odd.

The clocking input to memory devices and counters is indicated by a small triangle at (usually) the inside left edge of the box. If the device is a *transpar-*

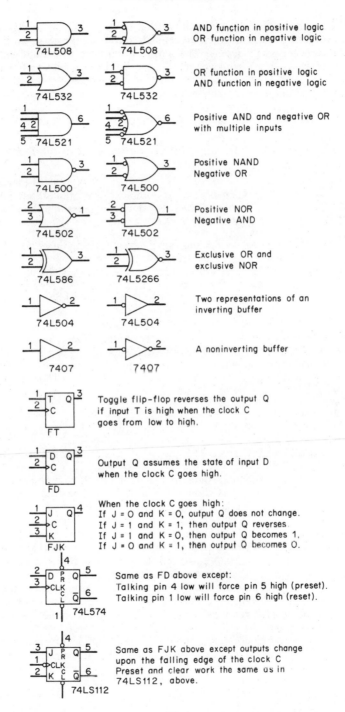

AND function in positive logic
OR function in negative logic

OR function in positive logic
AND function in negative logic

Positive AND and negative OR
with multiple inputs

Positive NAND
Negative OR

Positive NOR
Negative AND

Exclusive OR and
exclusive NOR

Two representations of an
inverting buffer

A noninverting buffer

Toggle flip-flop reverses the output Q
if input T is high when the clock C
goes from low to high.

Output Q assumes the state of input D
when the clock C goes high.

When the clock C goes high:
If J = 0 and K = 0, output Q does not change.
If J = 1 and K = 1, then output Q reverses.
If J = 1 and K = 0, then output Q becomes 1.
If J = 0 and K = 1, then output Q becomes 0.

Same as FD above except:
Talking pin 4 low will force pin 5 high (preset).
Talking pin 1 low will force pin 6 high (reset).

Same as FJK above except outputs change
upon the falling edge of the clock C
Preset and clear work the same as in
74LS112, above.

FIGURE 5.1 Symbols used in digital system block diagrams.

ent latch, the output follows the input while the clock input is active, and the output is "frozen" when the clock becomes inactive. A flip-flop, on the other hand, is an edge-triggered device. The output is allowed to change only upon a transition of clock input from low to high (no bubble) or high to low (bubble present).

Three types of flip-flops are shown in Fig. 5.1:

A *T* (*toggle*) device will reverse its output state when clocked while the *T* input is active.

A *D* flip-flop will allow the output to assume the state of the *D* input when clocked.

A *J-K* flip-flop is more complicated. If both *J* and *K* inputs are inactive, the output does not change when clocked. If both are active, the output will toggle as in *T*, above. If *J* and *K* are different, the output will assume the state of the *J* input when clocked, similar to *D* above.

The origin of the designation *J-K* flip-flop is an interesting bit of history: Early flip-flops were designed using a dual-triode vacuum tube mounted on a plug-in module. The pins of the module were named for the letters of the alphabet. The particular pins that carried the logic inputs to the device were pins *J* and *K*.

Flip-flops, latches, and counters are often supplied with additional inputs used to force the output to a known state. An active *set* input will force the output into the active state, while a *reset* input will force the output into the inactive state. Counters also have inputs to force the output states. There are two kinds of these: *asynchronous*, in which the function (preset or clear) is performed immediately, and *synchronous*, in which the action occurs on the next clock transition. Usually, if both preset and clear are applied at once, the clear function outranks the preset function.

5.2 SEQUENTIAL LOGIC

Also known as a *state machine*, sequential logic can perform combinational logic with fewer parts. An example is shown in Fig. 5.2. It is desired that an alarm sound if an airplane's throttle is retarded while the wheels are not extended. A 4-bit group of clocked flip-flops is connected to a read-only memory containing 256, 8-bit bytes. The 8-bit input is an address pointing to one of the bytes to be output. Assume that one of the inputs is connected to a switch that is actuated when an airplane's wheels are fully retracted (Up). Assume that another switch is actuated when the wheels are fully extended (Dn). Assume another switch that is actuated when the throttle is retarded (Ret). Four of the outputs are connected to the input of the four flip-flops and determine their output state upon the next clock cycle. A fifth output is connected to a sonic alarm. Assume that the initial condition of the flip-flops are all zeros. From the figure, you can see that if the throttle is not retarded, we don't care about wheels. If the throttle is retarded and the wheels are NOT down, then the alarm should be set. If the alarming situation is corrected, then return to 0000 and keep testing in case the mistake is made again. The circuit shown in Fig. 5.2 is capable of testing four external items and driving four outputs, in 16 steps.

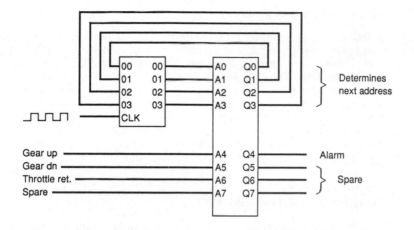

FIGURE 5.2 Typical sequential-logic diagram. X indicates "don't care."

This State	Throt	Wheels	Next State
0000	On	X	0000 (normal, continue)
0000	Ret	Dn	0000 (normal, continue)
0000	Ret	Up	0001 (sound alarm)
0000	Ret	Not dn	0001 (sound alarm)
0001	On	X	0000 (return to normal)
0001	Ret	Dn	0000 (return to normal)
0001	Ret	Up	0001 (continue alarm)
0001	Ret	Not dn	0001 (continue alarm)

5.3 FAMILIES OF LOGIC DEVICES

Early electrical logic was performed using vacuum tubes, mechanical relays, and motor-driven cams and switches. Early solid-state logic was done with a collection of germanium transistors and diodes. The discussion below is limited to *integrated* circuits, a collection of transistors, usually silicon, packaged into a single container.

5.3.1 Resistor-Transistor Logic (RTL)

RTL is mostly of historic interest. It used a 3.6-V positive power supply, and it was not very compatible with the families that came later. The packages were round with a circular array of wires (not pins) for circuit board mounting. Inputs were applied to the base of a transistor, and the transistor was turned on directly by the input signal if it was high. An open input could usually be considered as an "off" or "0."

5.3.2 Diode-Transistor Logic (DTL)

RTL was followed by the popular DTL, mounted in a DIP (dual in-line package). It had 14 or 16 stiff pins arranged in two parallel rows 0.3 in apart with the pins on 0.1-in centers. For simple devices, like a two-input NAND gate, four gates were packaged into one DIP. The stiff pins made possible the use of sockets. An internal resistor attached to the positive 5.0-V supply turned on the input transistor. Input signals were applied through diodes such that if an input signal were low, it "stole" the resistor's current, and the transistor turned off. It is important to remember that a disconnected DTL or TTL input is a logic high. The DTL output circuit was pulled low by a transistor and pulled up to +5 V by an internal resistor. As a result, fall times were faster than rise times.

5.3.3 Transistor-Transistor Logic (TTL)

TTL, like DTL, supplies its own turn-on current but uses a transistor instead of a resistor. The inputs do not use diodes but instead use multiple emitters on an input transistor. The output is pulled down by one transistor and pulled up by another. Defined below are the various family variations on this basic design, using the 7400 device, a two-input NAND gate:

7400	The prototype
74L00	A *l*ow-power version, but slow
74S00	(*S*chottky) fast, but power-hungry
74LS00	*L*ow power and *s*peed, too
74AS00	*A*dvanced *S*chottky
74ALS00	Like LS, but improved
74F00	*F* for *f*ast

All variants can be used in the presence of the others, but doing this complicates the design rules that determine how many inputs can be driven by one output. The dividing line between an input high and an input low is about 1.8 V. A high output is guaranteed to be 2.4 V or greater, while an output low will be 0.8 V or less.

5.3.4 NMOS and PMOS (*n* and *p* Metal-Oxide Semiconductors)

MOS logic devices use field-effect transistors. The initial letter tells whether the device uses *n*- or *p*-type dopant on the silicon. At low frequencies, MOS devices are very frugal in power usage. Early MOSs were fairly slow, but smaller and smaller conductor sizes have reduced on-chip capacity and therefore charging time.

5.3.5 Complementary MOS (CMOS)

A very popular logic family, the CMOS uses both *p*- and *n*-type transistors. At direct current, input currents are almost zero. Output current rises with frequency since the output circuit must charge and discharge conductors and the capacity of the inputs it is driving. The early CMOS was fairly slow when powered with a 5-V supply, but it sped up when powered at 10 or 15 V. Modern microscopic geometry produces CMOS parts that challenge TTL speeds while using less power.

The input decision level of the CMOS is nominally midway between the positive supply and ground. The logic state of an open input is indeterminate. It can and will wander around depending on which of the two input transistors is leaking the most. Unused inputs *must* be returned either to ground or the supply rail. CMOS outputs, unlike TTL, are very close to ground when low and very close to the supply rail when high. The CMOS can drive TTL inputs, but in a 5-V environment, the CMOS decision level of 2.5 V is too close to the TTL guaranteed output high. The solution is an external pull-up resistor between the output pin of the TTL part and the supply rail.

Early CMOS parts had their own numbering system (beginning at 4000) that was totally different from the one used for TTL parts. Improvements in speed, etc., have spawned subfamilies that seem to tend toward a return to the use of the 7400 convention; for example, 74HC00 is a *h*ighspeed *C*MOS part.

5.3.6 Emitter-Coupled Logic (ECL)

ECL has almost nothing in common with the families previously discussed. Inputs and outputs are push-pull. The supply voltage is negative in respect to ground at −5.2 V. Certain advantages accrue from this configuration:

1. Due to the push-pull input output, inverters are not needed. To invert, simply reverse the two connections.
2. The differential-amplifier construction of ECL input and output stages causes the total current through the device to be almost constant.
3. The output voltage swing is small and, from a crosstalk standpoint, is opposed by the complementary output.
4. To drive a balanced transmission line does not require a "line-driver" since an ECL output (with some resistors) *is* a line-driver.
5. Mainly because the transistors in ECL are *never* saturated (maximum conduction), they operate at maximum speed. Early ECL was power-hungry, but new ECL gate-array products are available that will toggle at 1.2 GHz without running hot.

5.3.7 Programmable Devices

As was shown in Sec. 5.2, memories are used to perform logic. Generally called *ROM* (*read-only memory*), their different varieties are defined below. It is typical for equipment manufacturers to update their equipment by sending their customers a new set of ROMs to plug in.

ROM: The contents of the device are decided at the time of manufacture, before being packaged, using a mask to decide where conductors will be plated on.

PROM (Programmable ROM): These devices are preprogrammed with all zeros or all ones, then packaged. The user decides what data he or she wants in the PROM and enters this data into a PROM programmer. This circuit supplies enough current to the selected memory locations to cause the microscopic fusible link that defines the one or zero to melt, thus "burning in" the data. The process cannot be reversed, and the part cannot be reprogrammed.

EPROM (Erasable PROM): These parts are also programmed by the end user, but the data is stored as capacitive charges near the gate of an FET transistor. The charges will remain for at least 10 years, but the part may be reused by exposing it to intense ultraviolet (UV) light. The parts are equipped with a quartz window so the light can reach the surface of the silicon. EPROM devices are often shipped in new equipment, with the windows covered by a part-number label to further keep UV light out. Ultraviolet light EPROM erasers are available that will erase an EPROM in from 5 to 50 min.

EEPROM (Electrically Erasable PROM): This device is like the EPROM above, but it does not require UV erasure.

PLD (Programmable Logic Device): These are hybrids between ROMs and PROMs. Portions of the device are configured at the time of manufacture, then packaged. They contain a number of flip-flops and gates and an array of criss-crossed conductors. At the intersections of the conductors, the user may program an optional connection between the conductors to build the desired circuit. As might be inferred from Sec. 5.3.7, several implementations are sold. Some may be programmed only once, by the blowing of fusible links, others may be erased like EPROMS and EEPROMS and reused, while others may be programmed by "downloading" instructions from external memory each time power is applied, or at any other time it is desired to change the circuit. Each connection point is made by a transistor controlled by a flip-flop that is either set or cleared by the input instructions. It is not possible to selectively program portions of the device. It must be totally programmed each time.

5.3.8   Size Scale of Digital Circuit Packages

The term *small-scale integration (SSI)* includes those packages containing, for example, a collection of four gates, a 4-bit counter, a 4-bit adder, and any other item of less than about 100 gate equivalents. *Large-scale integration (LSI)* describes more complex circuitry, such as an asynchronous bit-serial transmitter-receiver, or a *DMA (direct memory access)* controller, involving a few thousand gate equivalents. *Very large scale integration (VLSI)* represents tens of thousands of gate equivalents, like a microprocessor or a graphics controller. LSI and VLSI devices are typically packaged in a larger version of the DIP package, usually with the two rows spaced 0.6 in, and having from 24 to 68 pins.

Other Packages.   Many devices are available in dual in-line packages designed to be soldered to the surface of a circuit board rather than using holes in the circuit board. The pin spacing is 0.05 in. Another surface-mount device is a square device called a *leadless chip carrier.* Its contact spacing is also 0.05 in, and there is an equal number of contacts along each edge of the square. As many as 84 contacts may be had. Sockets are available for these packages, but once the package is installed, a special tool is required to extract it. Yet another large-scale package is called the *pin-grid array*, with pins protruding from the bottom surface of a flat, square package in a row-and-column "bed-of-nails" array. The pin spacing is 0.1 in. A little more than 200 pins may be had. Extraction tools are available for these packages as well. For exact dimensions and outline drawings, see Texas Instruments, *2-μm CMOS Standard Cell Data Book* [1].

5.4 REPRESENTATION OF NUMBERS AND NUMERALS

A single bit, wire, or flip-flop in a binary system can have only two states. When a single bit is used to describe numerals, by convention those two numerals are 0 and 1. A group of bits, however, can describe a larger range of numbers. Conventional groupings are identified below.

5.4.1 Nibble

A nibble is a group of 4 bits. It is customary to show the binary representation with the least significant bit on the right. The *least significant bit* (*LSB*) has a decimal value of *1* or 0. The next most significant bit has a value of *2* or 0, and the next, *4* or 0, and the *most significant bit* (*MSB*), *8* or 0. The nibble can describe any value from binary 0000 (= 0 decimal) and 1111 (= 8 + 4 + 2 + 1 = 15 decimal), inclusive. The 16 characters used to signify the 16 values of a nibble are the ordinary numerals 0 through 9, followed by the letters of the alphabet A through F. The 4-bit "digit" is called a *hexadecimal* representation.

Octal.   An earlier numbering scheme, octal, used groupings of 3 bits to describe the numerals 0 through 7. Used extensively by the Digital Equipment Corporation, it is falling out of use, but it is still included in some figures for reference.

5.4.2 Byte

A byte is a collection of 8 bits, or 2 nibbles. It can represent numbers (a *number* is a collection of numerals) in two ways:

> Two *hexadecimal* digits, the least significant representing the number of 1s, and the most significant the number of 16s. The total range of values is 0 through 255 (FF).

> Two *decimal* digits, the least significant representing the number of 1s, and limited to the range of numerals 0 through 9, and the most significant representing the number of 10s, again limited to the range 0 through 9.

The use of 4 bits to represent decimal numbers is called *binary-coded decimal* (*BCD*). The use of a byte to store two numerals is called *packed BCD*. The least significant nibble is limited to the range of 0 through 9, as is the upper nibble, thus representing 00 through 90. The maximum value of the byte is 99.

5.4.3 Word

A word, usually a multiple of 8 bits, is the largest array of bits that can be handled by a system in one action of its logic. In most personal computers, a word is 16 bits, or 2 bytes, or 4 nibbles. Larger computers use words of 32 and even 64 bits in length. In all cases, the written and electrically mapped representation of the numeric value of the word is either hexadecimal or packed BCD, as described above.

5.4.4   Negative Numbers

When a byte or word is used to describe a *signed* number (one that may be less than zero), it is customary for the most significant bit to represent the sign of the number, 0 meaning positive and 1 negative. The representation is called *two's complement*. To negate (make negative) a number, simply show the number in binary, make all the zeros into 1s, and all the 1s into zeros, and then *add* 1.

5.4.5   Floating Point

In engineering work, the range of numerical values is tremendous and can easily overflow the range of values offered by even 64-bit systems. Where the accuracy of a computation can be tolerably expressed as a *percentage* of the input values and the result, *floating-point* calculation is used. One or two bytes are used to express the *characteristic* (a power of 10 by which to multiply everything), and the rest used to express the *mantissa* (that fractional power of 10 to be multiplied by). It is commonly called *engineering notation* (Table 5.2). For example:

$$4967 = 10^4 \times 0.4967 \tag{5.1}$$

5.5   MICROPROCESSORS

A microprocessor is a collection of binary logic that can perform certain operations on a nibble (a 4-bit device) or a byte (an 8-bit device) or other device up to 32 bits. Most personal computers are 16-bit devices, although internally, a 16-bit microprocessor can often do some 32-bit tricks. All microprocessors contain at least the following sets of logic.

5.5.1   Program Counter

Usually at least a 16-bit counter, a program counter furnishes an address to the off-chip memory when fetching an instruction. Instructions come in various sizes, but the processor knows by looking at the first byte or word of an instruction how many more times it must fetch. The address bus is an output from the processor.

5.5.2   Storage

Storage is a collection of small memories, often configurable in length. The memories are called *registers*, and they sometimes have special assigned purposes. Usually all registers may be output to the address bus or to the data bus.

TABLE 5.2 Number and Letter Representations

Decimal	Hexadecimal	Octal	Binary	
0	0	0	0000	
1	1	1	0001	
2	2	2	0010	
3	3	3	0011	
4	4	4	0100	
5	5	5	0101	
6	6	6	0110	
7	7	7	0111	
8	8	10	1000	
9	9	11	1001	
10	A	12	1010	
11	B	13	1011	
12	C	14	1100	
13	D	15	1101	
14	E	16	1110	
15	F	17	1111	
81	51	121	01010001	$16 \times 5 + 1$
250	FA	372	11111010	$16 \times 15 + 10$
+127	7F	177	01111111	(signed)
−1	FF	377	11111111	(signed)
−128	80	200	10000000	(signed)

5.5.3 Accumulator

An accumulator is a special register capable of numerically adding its contents to one of the other registers or to a number fetched from memory. The result is retained in the accumulator. Subtraction is accomplished by negating the number being subtracted, then adding. In addition (no pun intended) the accumulator can perform the following instructions:

1. *Increment:* Add 1 to its contents.
2. *Decrement::* Subtract 1 from its contents.
3. *Multiply by 2:* This is done by shifting all bits one bit position to the left.
4. *Divide by 2:* This is done by shifting all bits one position to the right.

5.5.4 Test

This action puts into a separate "status" register the state of the contents of the accumulator. One bit is set if the contents are zero, another if the contents are negative, another if the last action resulted in an overflow, or carry, etc.

5.5.5 Compare

A comparison involves negating one of the two numbers being compared, then adding them and testing the result. If the test shows zero, the two numbers are equal. If not, the test reveals which of the two is greater than or less than the other, and the appropriate bits in the status register are set.

5.5.6 Jump

The orderly progression of the program counter may be interrupted and instructions fetched from a *new* location in memory, usually based upon a test or a comparison, for example, "If the result is zero, jump to location X and begin execution there; if the result is positive, jump to Y and begin execution there; else keep on counting." This ability is probably the most powerful asset of a computer.

5.6 ANALOG-TO-DIGITAL CONVERSION

Any signal may be converted into a series of numbers representing the amplitude of the signal. The conversion process is called *sampling*, and it is usually done at a constant rate. If certain rules are followed, the series of numbers may be used to reconstitute the original signal to some degree of accuracy.

5.6.1 Sampling Rate

The sampling rate, even in analog sampling systems, is crucial. Figure 5.3a shows the spectral consequence of a sampling rate that is too low for the input bandwidth, 5.3b shows the result of a rate equal to the theoretical minimum value, which is impractical, and 5.3c shows typical practice. The input spectrum *must* be limited by a low-pass filter to greatly attenuate frequencies near one-half the sampling rate and above. The higher the sampling rate, the easier and simpler the design of the input filter becomes. An excessively high sampling rate is wasteful of transmission bandwidth and storage capacity, while a low but adequate rate complicates the design and increases the cost of input and output analog filters. Early digital audio recordings were made at a sampling rate of 50 kHz. Later, recordings on modified video recorders were made at 44.1 or 44.056 kHz. Even later, 48 kHz was established as a standard rate for professional recording and finds its primary use on digital video recorders [2, 3]. The video sampling rate for *component* digital-video recorders [2] is 13.5 mHz for the luminance signal and half that, 6.75 mHz, for each of the two color difference signals. These frequencies are multiples of 2.25 mHz and therefore may be derived from the horizontal-scan rate of either the NTSC 525-line system or the PAL or SECAM 625-line systems. *Composite* digital-video recorders [3] typically sample at four times the color-subcarrier frequency. This frequency is greater than is needed, but many color-processing tasks become simpler, considering that the color components are modulated on the subcarrier in quadrature. By adjusting the phase of the sampling frequency relative to the phase of the color encoder subcarrier reference, separation of chroma and luminance in the digital domain is made easier.

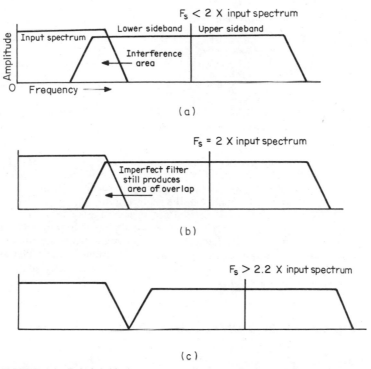

FIGURE 5.3 Relationship between sampling rate and bandwidth. (*a*) Sampling rate too low for input spectrum. (*b*) Theoretical minimum sampling rate (F_s) requires theoretically perfect filter. (*c*) A practical sampling rate, using a practical input filter.

5.6.2 Digital-to-Analog Conversion

Each digital number (typically a 16-bit signed integer for audio or an unsigned 8-bit integer for video) is converted to a corresponding voltage and stored in a capacitor until the next number is converted. Figure 5.4 shows the resulting spectrum. The energy surrounding the sampling frequency must be removed, and an output low-pass filter is used to accomplish that. One cost-effective technique used in compact disk players is called *oversampling*. A new sampling rate is se-

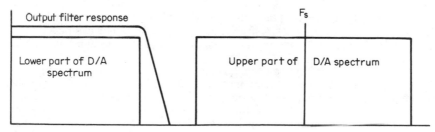

FIGURE 5.4 Output filter response requirements, ordinary D/A converter.

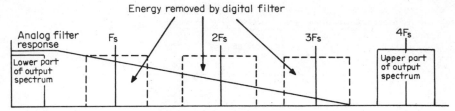

FIGURE 5.5 Oversampling simplifies the requirements of the output filter.

lected that is a whole multiple of the input sampling rate. The new rate is typically two or four times the old rate. Every second or fourth sample is filled with the input value, while the others are set to zero. The result is passed through a digital filter that distributes the energy in the real samples among the empty ones and itself. The resulting spectrum (for a 4X oversampling system) is shown in Fig. 5.5. The energy around the 4X sample frequency must be removed, which can be done simply and cheaply because it is so distant from the upper band edge. The response of the output filter is chiefly determined by the digital processing and is therefore very stable with age, in contrast to a strictly analog filter, whose component values are susceptible to drift with age.

5.7 PROCESSING OF DIGITAL SAMPLES

5.7.1 Filtration

A filter response equivalent to an analog filter can be achieved by digital processing. Two types of digital filters are generally appropriate for processing digital samples. The first, called *finite-impulse response* (*FIR*), accumulates N samples prior to and after the current 1 and multiplies each by a coefficient, adds each product to a total, and outputs the aggregate sum. This approach has the advantage of linear phase, or uniform delay. The second, which is applicable when correcting for an analog response adjustment such as preemphasis, offers both amplitude and phase adjustment, and is called *infinite-impulse response* (*IIR*). The number N is typically around 10 for video filters and 96 for audio filters.

5.7.2 Mixing, Editing, and Cross-Fading

Editing a video signal is a simple matter of beginning a new recording at a time corresponding to the vertical-blanking interval. An abrupt switch between two *audio* signals, however, is not desirable, as strong click sounds can occur at the moment of switch, depending on the instantaneous phase and amplitude of the two edited signals.

Two methods of performing an audio edit exist. In the first, the outgoing samples are multiplied by a steadily diminishing coefficient, while the new samples are multiplied by a steadily increasing coefficient, such that the sum of the two coefficients totals 1. The results of the two multiplications are added and output. This is called a *cross-fade* and usually occupies a time of 5 to 100 ms. Another, simpler approach is to multiply the outgoing samples by a steadily decreasing coefficient, such that at the moment of switch, the coefficient is zero. The incoming samples are then multiplied by a steadily increasing coefficient until the coefficient reaches 1. In this way, the potential editing transient is "covered up" by silence. This technique is sometimes called a *V edit*. The time taken to perform it is typically 32 to 128 sample periods.

5.8 DIGITAL GENERATION OF SYNTHETIC SIGNALS

Video and audio test signals can be generated in the digital domain. Video synthetic signals include color bars, zone plates, and computer-generated graphics. Audio synthetic signals encompass test tones, musical approximations to real instruments, and sounds derived from sampling real sounds but processed so that they can be "played" at any frequency, usually using a piano-like keyboard. If signals are generated by direct mathematical algorithm, it is important that the algorithm not generate frequency components that would have been removed by the usual analog input low-pass filter. For example, character generators, or "titlers," require significant processing to smooth out the jagged edges of a sloping line as in the sides of the letter A.

5.9 DIGITAL ERRORS

When a digital signal is transmitted through a noisy path, errors can occur. Early methods to deal with this from the transmitted data generated one or more digital words, using check sums, cyclic redundancy checks, and the like, and appending the result at the end of a block of transmitted data. Upon reception, the same arithmetic was used to generate the same results, which were compared to the data appended to the transmission. If they were the same, it was unlikely that errors occurred. If they differed, an error was assumed to have occurred, and a retransmission was requested. Thus the method performed *only* error detection. In the case of digital tape recording, retransmission is not possible, and methods are employed that not only *detect* but *correct* errors.

5.9.1 Error Detection and Correction

Given a string of 8-bit bytes, additional bytes can be generated using Galois field arithmetic and appended to the end of the string. The length of the string and the appended bytes must be 256 or less, since 8 bits can have no more than 256 different states. If 2 bytes are generated, upon playback 2 "syndrome" (symptom) bytes are generated. If they are zero, there was likely no error. If they are nonzero, then after some fancy arithmetic, 1 byte "points" to the location of the damaged byte in the string, while the other contains the 8-bit error pattern. The error pattern is used in a bit-wise exclusive OR function upon the offending byte, thus reversing the damaged bits and correcting the byte. With 4 check bytes, 2 flawed bytes can be pinpointed and corrected; with 6, 3 can be treated; and so on. If the number of bytes in the string is significantly less than 256, for example, 64, the error-detection function becomes more robust since, if the error pointer points to a nonexistent byte, it may be assumed that the error-detection system itself made a mistake. It can happen.

Errors in digital-video recorders fall into two classes: *random* errors brought on by thermal random noise in the reproduce circuitry, and *dropouts*, long strings of lost signal due to tape imperfections. The error detection and correction system of digital-video recorders is designed to cope with both types of errors. Figure 5.6 shows how data can be arranged in rows and columns, with separate check bytes generated for each row and each column in a two-dimensional array. The data is recorded (and reproduced) in row order. In the example in Fig. 5.6, it can be seen that a long interruption of signal will disrupt every tenth byte. The row corrector cannot cope with this, but it is likely that the column corrector can since it "sees" the burst error as being spread out over a large number of columns.

The column corrector, if taken alone, can correct $N/2$ errors where N is the number of check bytes. Given knowledge of which *rows* are uncorrectable by the row corrector, then N errors can be corrected. Generally, the row (or "inner") corrector corrects errors caused by random noise, while the column (or "outer") corrector takes care of burst errors. In both digital-video recorder formats, the digital audio has its own column generator and corrector, but it is combined with the video data stream prior to the row generator. On playback the inner corrector treats both audio and video.

Generally, error detection and correction schemes have these characteristics:

1. Up to a threshold error rate, all errors are corrected.
2. If the error rate is greater than the above first threshold, the system will flag the blocks of data it is unable to correct. This allows other circuits to attempt to *conceal* the error.
3. Above an even higher error rate, the system will occasionally lie, reporting that all is well when it isn't.

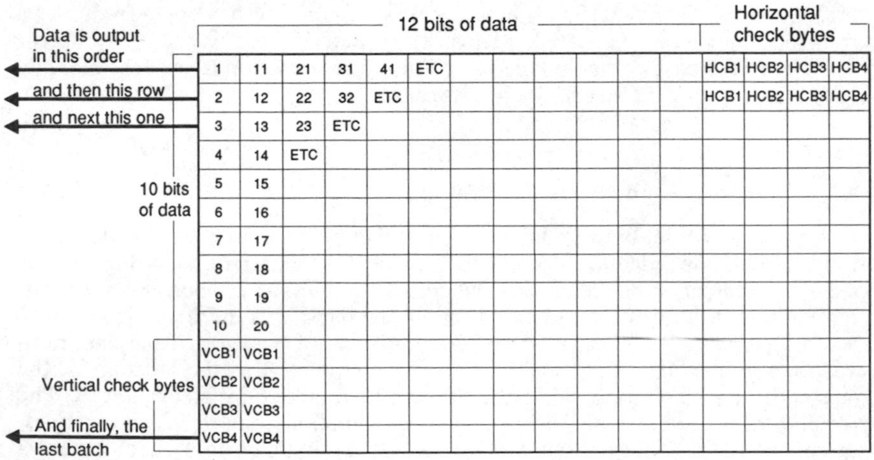

	12 bits of data												Horizontal check bytes			
1	1	11	21	31	41	ETC							HCB1	HCB2	HCB3	HCB4
2	2	12	22	32	ETC								HCB1	HCB2	HCB3	HCB4
3	3	13	23	ETC												
4	4	14	ETC													
5	5	15														
6	6	16														
7	7	17														
8	8	18														
9	9	19														
10	10	20														
VCB1	VCB1															
VCB2	VCB2															
VCB3	VCB3															
VCB4	VCB4															

Data is output in this order, and then this row, and next this one — 10 bits of data. Vertical check bytes. And finally, the last batch.

FIGURE 5.6 An example of row and column two-dimensional error-detection coding.

Most recorders have some form of readout to indicate the level of activity in the error detection and correction apparatus. A level approaching the limit of the system to fully correct is a clear indication that it is time to make a copy, verify it, and then discard the worn tape.

5.9.2 Error Concealment

When the error-correction system is overloaded and error-ridden samples are identified, it is typical to calculate an estimation of the bad sample. In video, samples that are visually nearby and that are not corrupted are used to calculate an estimate of the damaged sample. The estimate is then substituted for the unusable sample. In the recording process, the video data samples are scrambled in a way that maximizes the chance that a damaged sample will be surrounded by good ones. The scrambling algorithm is different for the component and the composite recording standards.

In the case of audio, the samples are scrambled such that failure of the correction system is most likely to result in every *alternate* sample being in error. Replacement of a damaged audio sample consists of summing the previous (good) sample and the following (good) sample and dividing by 2. If the error rate becomes unreasonable, then the last good sample is simply repeated, or "held." If the error rate gets very high, such that many adjacent samples are damaged and uncorrectable, it is common practice to mute the channel.

Video error concealment is roughly 10 times more effective than audio concealment, probably because the eye is an integrator and the ear is a differentiator.

5.10 TRANSMISSION OF DIGITAL SIGNALS

5.10.1 Byte-Wide Digital-Video Signals

Transmission of digitized video samples can take place as byte-serial (8 twisted pairs plus 1 pair for clock), at 27 Mbytes/s, 13.5 Mbytes for the luminance, and 13.5 Mbytes for the two-color difference signals. The signal levels are compatible with the emitter-coupled logic (ECL) family. The connectors, pin assignments, source, and receiving impedances are given in "Bit-Parallel Digital Interface for Component Video Signals" [2].

5.10.2 Bit-Serial Digital-Video Signals

Work done by Thompson-CSF [4] and later tested at CBC Montreal and still later standardized by the European Broadcasting Union [5] created a bit-serial transmission standard for component digital-video signals. The byte rate is the same as in Sec. 5.10.1, above, but each byte is used with a PROM-based look-up table to locate a 9-bit sample. The set of 9-bit samples was chosen to provide plenty of transitions from 1 to zero and zero to 1. This ensures that the appropriate clocking rate can be generated at the receiving end while at the same time reduces the low-frequency energy. This simplifies the design of cable equalizers. The receiver accumulates each group of 9 bits and uses a look-up table to recover the original sample values. The standard is intended to be used with the same coaxial cables used for transmission of analog-video signals and has been tested over a distance of 160 m.

An Important Note.   There is significant energy at the clocking rate of 243 mHz (27 × 9) and at half that (121.5 mHz). These two frequencies are the international distress frequencies for aircraft. When servicing equipment using this standard, it is vitally important to maintain the integrity of shields, line filters, and the like.

5.10.3   Byte-Wide Digital-Audio Signals

The creation of the D2[2] video-audio recording format introduced a simplified byte-wide interface. Four bytes are used for each of the four channels. The first 2 bytes contain the 16 bits of the audio sample, and the other 2 carry ancillary information about the sample and the channel. The byte rate is therefore 48 kHz (the sampling rate) × 4 (bytes) × 4 (channels) = 768 kHz. The connector is the same as is used for byte-wide video transmission, and each bit employs, as in the bit-parallel video standard, a twisted pair compatible with the ECL family. Two clocks are supplied, each on a twisted pair, one a square wave at 768 kHz, for clocking in each byte, and the other at the sampling rate, 48 kHz, to identify the first sample of the group of four.

5.10.4   Bit-Serial Digital-Audio Signals

Several formats exist, as described below.

PC-1610 Format.   The Sony PC-1610 is an early implementation of digital-audio recording. It uses a specialized signal system that "packages" a stereo pair of 16-bit audio samples between television sync pulses so that the result is palatable to a helical-scan video recorder. The digital input-output consists of one set of wires for each channel plus a sample-rate clock. The data is sent in *nonreturn-to-zero* (*NRZ*) form. Each channel employs 32 bits per sample, 16 for the sample bits, and another 13, largely squandered, for ancillary information, consisting of a parity check bit, a bit to signal whether or not preemphasis was employed, and terminated by one cycle of a square wave having a duration of three bit periods used as a synchronization sequence. The preemphasis bit is transmitted once every 256 samples, with the phase of the synchronization sequence inverted to signal that event. The sample data is transmitted most significant bit first.

AES/EBU Format [6].   This format accommodates two channels, either stereo or two separate channels. Sixty-four bits are used per sample period. Each group of 32 bits treats one channel, with the left channel (*A*) first. The first 4 bits of the 32 may be optionally the least significant 4 bits of a 24-bit sample, or auxiliary information. Which usage of the first 4 bits is being made is signaled by a data path discussed later. The next 4 bits are the least significant 4 bits of a 20-bit sample. If the sample resolution is less than 20 bits, the unused bits are set to zero. The next 16 bits are occupied by sample bits, least significant first. The last 4 bits contain 1 bit for each channel and for each sample:

1. A validity bit, to indicate whether or not the associated sample is of doubtful accuracy, i.e., was interpolated.
2. A user bit, totally given over to the user of the signal path.
3. A channel status bit, one of 196 per "block." During the period of a block, a great deal of information may be sent, including preemphasis usage, the usage of the first 4 bits, source and destination codes, two time codes, and an error check byte.
4. An even-parity bit, encompassing all the previous 31 bits, and of course, itself.

Each transmission of 28 bits is begun with a preamble using 4 bit-periods of time. Three preambles are specified: one that marks the start of the sequence of 196 samples as well as the start of a pair of samples, one that otherwise marks the start of a pair of samples, and one that marks the start of the second (channel B) sample. Each preamble violates the channel coding rules to render it unique and never duplicated by the data stream. The bit rate is 64 times the sample rate. The electrical connection is balanced and uses the same XLR connector used for analog audio.

Rotary-Digital Audio Transport (R-DAT) Format. This format, in the temporal sense, is very similar to the AES/EBU format, except for a largely ineffective copy-protection bit, and other reassignments within the channel status 196-bit group. The bit rate is the same, but the connector is a single-ended RCA phonoplug, and the voltage level is much smaller. The transmission order remains least significant first. This format is also supplied with some compact-disc players.

5.11 DIGITAL DIAGNOSTICS

The most general form of digital troubleshooting is *signature analysis* in which a 16-bit shift register with feedback is presented with a binary signal at a test point in a circuit and is shifted through the register for some prescribed and known number of clock cycles (usually large). The contents of the shift register are then examined to see if they are the same as when a healthy circuit was examined. If they are not the same, a fault is likely affecting the point being examined. If they are the same, there is, in the face of purely random errors, 1 chance in $2^{16} = 65,536$ that the system reported no problem when, in fact, there was one. Typical faults are more deterministic, consisting of a bus driver output stuck low or two bus wires shorted by a lump of solder. The chances of these types of faults going undetected is very much smaller than 1 in 65,536.

With the application of some labor, typical faults can be simulated by, for example, shorting two points on a circuit board together, running a signature analysis, and writing down the resulting erroneous signature, which becomes one more item in a dictionary of likely errors.

Some specific applications of this and another technique are discussed in Chap. 7, digital-video recorders.

REFERENCES

1. Texas Instruments. *2-μm CMOS Standard Cell Data Book*, 1986, Chap. 8.
2. "Bit-Parallel Digital Interface for Component Video Signals." *SMPTE RP*, **125** (1984).
3. Robert Boyer, JeanLuc Grimaldi, Jacques Oyaux, and Jacques Vallee. "Serial Interface Within the Digital Studio." *J. Soc. Motion Pict. Telev.* (November 1984).
4. Publication Tech 3247-E, Technical Centre of the EBU. Brussels (1985).
5. *Recommended Practice for Digital Audio Engineering—Serial Transmission Format for Linearly Represented Digital Audio Data*. ANSI, S4.40 (1985).
6. Hewlett Packard Application Note 222-2. *Application Articles on Signature Analysis* (undated).

CHAPTER 6
TELEVISION AND VIDEO EQUIPMENT

K. Blair Benson
Television Technology Consultant, Norwalk, Connecticut

Carl Bentz
Intertec Publishing Corporation, Overland Park, Kansas

6.1 STUDIO AND FIELD CAMERAS

6.1.1 Design Requirements

The major components of all television cameras, from the simple black-and-white surveillance type to the most sophisticated broadcast system with computer-assisted control and memory, may be classified into several interconnected functional blocks leading from the conversion of light picked up from a scene to a video signal suitable for display on a picture monitor or receiver. The assembly of the blocks is called a *television camera chain*. Variations among different designs result primarily from differences, if any, in degree of operator control, mono-chrome or color-signal output, picture-quality requirements, and cost considerations.

Camera designs may be classified broadly in increasing degree of complexity and cost as follows: (1) monochrome security surveillance, (2) amateur photography, (3) educational and industrial programming, (4) broadcasting and cable-TV electronic news gathering (ENG), (5) electronic field production (EFP), and (6)

Portions of this chapter have been adapted from K. Blair Benson, ed., *Television Engineering Handbook*, McGraw-Hill, New York, 1986.

broadcasting studio production. These applications frequently overlap; for example, many industrial productions use equipment and techniques equal to those of network broadcasting. On the other hand, broadcasters often use consumer-type hand-held cameras for ENG in order to gain access to restricted areas, or to allow greater camera mobility and range of camera shooting angles.

The range in complexity is reflected in the cost, which may vary from over six figures for the most elegant broadcast camera to under $100 for a black-and-white surveillance vidicon or CCD package. In addition, the very important consumer market presents another set of cost and performance requirements which are limited rigorously by the marketplace.

Nevertheless, a similarity in basic designs will be found among cameras in all the categories listed above. Therefore, for an explanation of the function and operation of the major elements of a television camera, the following is a description of a typical color camera system. The basic functional components, shown in Fig. 6.1, are outlined below.

1. The *optical* section of the TV camera consists of a lens and other optical devices. The optics collect light from the scene and prepare that light for application to the light-sensitive pickup devices.

2. The *pickup* section converts light energy into an electrical signal. The conversion, whether through vacuum-type camera tubes or charge-coupled devices (CCDs), is accomplished under the control of an organized deflection or scan-

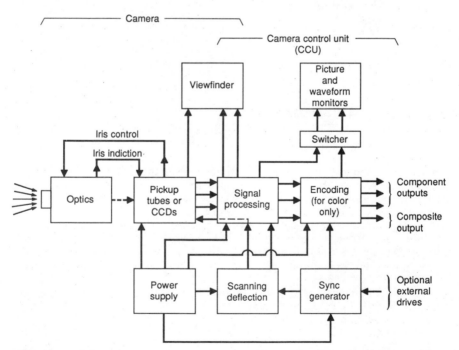

FIGURE 6.1 Block diagram of a color camera showing basic functional components. All or some of the CCU functions may be contained in the camera for applications such as broadcast and cable ENG and home video.

ning plan. Preamplifiers in this block boost the minute electrical output of the sensors to a more usable level.

3. The *processor* section accepts signals from the pickup devices and applies various corrections to them. Gamma correction controls the linearity of the transfer from light to electric energy by camera sensor and circuitry. Gamma may be considered a predistortion of the video signal to compensate for a nonlinear distortion that occurs in the human eye as we view the monitor. Clipping reduces signal overshoot, particularly on fast luminance changes from dark to light or bright to dark. Clamping controls dc voltage effects by using a specific portion of the video signal as a reference level. Enhancement improves the apparent frequency response of the system by increasing level transitions from low to high and high to low. Enhancement is possible in both the vertical and the horizontal direction. Peak white and black level controls in this section limit the blackness of the signal.

4. The *encoder* combines processed video with synchronizing (sync) pulses to form a composite signal.

5. The *deflection* or scanning and *sync* sections include a self-contained synchronizing pulse generator. In some cases, the sync generator may receive reference information from another source, in which case it genlocks to the source. The sync generator creates a number of sync pulses to drive the deflection or scanning operation in the image sensors. The sync pulses are used in creating the composite video output of the encoder.

6. A *power supply* develops all voltage levels needed for the other sections of the camera. Voltage regulation is often needed to ensure that the output signal is not affected by changes in the input power. Such changes could produce variation in the brightness, contrast, and size of the picture, if a picture is even created.

Signal monitors required in a camera chain will depend on the application of the camera. For example, a remotely operated surveillance camera does not require a monitor at the camera location, but consumer camera-recorders (camcorders) or those used in video production require a viewfinder attached to the camera so that the operator can see what is included in the picture. Camera chains involved in production and broadcast facilities include additional video monitor (picture, waveform, and vector) units, to allow technicians to maintain proper camera operation.

6.1.2 Functional Components

Optics. The optics in a three-channel color camera are a combination of an objective lens and spectral filtering systems to provide color-separated images on the photosensitive surfaces of the red, green, and blue pickup tubes or charge-coupled devices. Alternatively, in a single-tube CCD camera, the color-separation component is not required. The objective lens may be a simple fixed-focal-length lens or a multielement zoom lens with a wide range of focal-length adjustment.

Cameras used for professional production and broadcast use, and by more ambitious amateurs, can be fitted with any of several lens assemblies, according to the application of the camera. A typical studio lens has a relatively short focal length with a small zoom ratio. Because of the distance between the camera and the talent in the studio, high magnification is not required, but a greater wide-angle view may be desired. Systems for ENG and other hand-held uses have somewhat longer focal lengths with greater zoom capability.

Special optical equipment is needed for cameras for outside broadcasting, such as sporting or other media events. Closeups are often desired, but the separation between the camera and the subject is too long for a standard lens to produce the picture preferred. In such cases the camera can be fitted with lenses that have longer focal lengths and higher zoom ratios. In the Winter Olympics of 1988, a lens with a zoom ratio of greater than 50:1 was used for the first time in broadcasting to obtain closeup views of participants in skiing events. The camera location was a mountainside position across the valley from the ski slope, from which the operator could get a picture of any portion, or all, of the ski course.

During some of the first space shuttle landings, a lens with a very large zoom ratio was used to sight the craft as it began its descent from orbit to the California landing strip. When using exceptionally long focal length lenses, the stability of the camera and lens support structure becomes increasingly important.

Lens systems for industrial applications often include servo-controlled zoom and focus adjusts, particularly for surveillance systems. The operator in the surveillance center can remotely control the camera's pan and tilt attitude as well as the lens focus and zoom. With cameras intended for consumer use, the number of lens options is limited. Most consumer TV cameras are not designed to allow lens changes. However, attachments that provide focus on extreme closeups are frequently available.

From a practical point of view, camera size, the focal length of the lens, and the event being televised determine whether a camera support is needed. In the studio setting, larger cameras are commonly used, and a mounting system with a pedestal on rubber wheels provides sufficient mobility for the action. Smooth movement and image stability with relatively static shots are usually more important than high mobility.

The intent of sports and news coverage is to track action for the home viewer. Many ENG camera operators prefer lenses with a zoom range of about 14X. This length allows reasonable closeup work from the sidelines of the football field, for example, without overly magnifying the instability inherent in handheld operation. An important factor in ENG activity is maneuverability, suggesting at least handheld (and often even wireless camcorder or microwave) operation of the camera. Tripods and monopods are valuable as long as tracking of fast motion is not essential.

As the zoom range of the lens increases, the need for stable support for the camera and lens also increases. Any unintentional movement of the camera-lens system will be magnified by the current zoom ratio setting of the lens. As was seen during the 1988 Olympics, the ability to track a skier down the entire slalom course, from a facing mountainside, with the additional capability of zooming in on that skier's face required a special tripod support system to keep even minor erratic movement of the lens and camera completely out of the picture. In some cases, when the lens mass is much greater than that of the TV camera, it is not uncommon to use a pedestal or tripod support for the lens, while the smaller, lighter-mass camera is merely attached to the lens.

An adjustable iris in the lens assembly controls the intensity of the light from the scene being televised which is transmitted to the photoelectric sensors in the camera. The effect is the same as that of a lens on a still or motion-picture film camera. The lens is rated according to its transmission efficiency with the iris opened to the maximum diameter. This is called the maximum *speed* of the lens.

The terms *fast* and *slow* for lens speed come from film photography, where a faster lens has more light-gathering power and thus requires less the time for proper exposure of the photographic emulsion.

The light-gathering power of a lens is stated in terms of its f number. This is calculated from the ratio between the effective diameter of the lens opening and its focal length (see Fig. 6.2). For example, the specification of a lens as an $f/4.5$ means that its focal length is 4.5 times its effective diameter. Thus, it can be seen that the smaller the f-number of a lens, the greater the effective diameter D, and thus the greater the light-gathering power: $f/4.5 = D$.

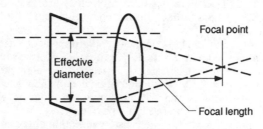

Focal point

Effective diameter

Focal length

FIGURE 6.2 The speed of a lens relates the effective diameter and focal length as the f/stop number.

The *focal length* is defined as the distance from the center of the lens to the point where the image from a distant object, or in optical terms "from infinity," will be in focus, as shown in Fig. 6.2. The determination of the focal length can be demonstrated with a simple magnifying glass by projecting an image of the sun or a relatively distant lamp on a white card. The distance from the lens to the card is the focal length of the lens.

A number of accessories for the lens may be available. (See Fig. 6.3.) Range extenders and diopters are used to change the magnification capability of the basic lens. (Diopter is the term for the magnification rating. It is equal to the focal length in meters.) Functionally, the range extender may be considered part of the movable zoom group. Physically, however, the range extender is inside the lens assembly, but is not a part of the zoom group. Instead, it can be moved into po-

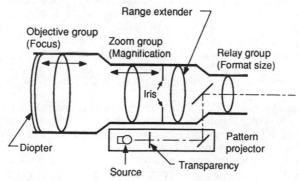

Range extender

Objective group (Focus)

Zoom group (Magnification

Relay group (Format size)

Iris

Diopter

Source

Transparency

Pattern projector

FIGURE 6.3 Zoom lens, showing the *diopter* magnifying lens used as a range extender, the rear-mounted range extender, and the test-pattern projector accessory.

sition in the optical path without a change in the zoom setting. In contrast, diopters are additional elements that can be attached to the front of the lens assembly to change the magnification range.

An option usually found in only the more expensive broadcast lenses is the pattern projector. This projector contains one or more small lamps and a test pattern transparency. When used periodically to check the registration of the camera, this accessory bypasses much of the normal light path through the lens. With a mirror, the image of the transparency is projected to the light preparation elements and then to the image sensors, allowing accurate positioning of the sensors. In lenses for CCD cameras, pattern projectors are essential, because the positioning of CCD sensors is fixed during camera manufacture and cannot be changed in the field.

The alignment of individual lens elements and groups is critical. In most cases, alignment is accomplished with lasers during the original assembly. For that reason, manufacturers insist that changes or repairs to the elements of the optical path should be attempted only by trained optical technicians who have proper equipment. Attempts to adjust the alignment may introduce severe chromatic or other aberrations and restrict the resolving power of the lens.

During maintenance of lenses, one must realize that there is a risk of introducing dust or other contaminants into the controlled environment inside the lens. The surfaces of some elements are in focus, and dust particles become visible in the picture. Ingress of dirt into the mechanism of the zoom lens increases the chance for abrasive wear and eventual difficult operation. In addition, because all high-quality lenses are manufactured with a transparent nonreflective coating to improve contrast by reducing light scattering, extreme care must be taken not to use any abrasive cleaning tissues or solutions.

The real maintenance requirements of most lens systems are limited, but one is very important. The outside of the assembly and accessible areas under any protective housing should be kept clear of dust, dirt, and moisture. Screwdriver adjustments are provided through small snap-out caps on portable systems or under the housing of larger studio assemblies. These adjustments control parameters for the electrically controlled functions with the lens system and circuitry.

Color Separation. Light from the scene being televised is imaged on the targets of the photosensitive tubes or charge-coupled devices (CCDs) by the objective lens. The separate images for the red, blue, and green primary colors are derived by means of a split-cube separation optical system. Dichroic coatings on the surfaces of the cube provide spectral transmission characteristics which result in green light passing straight through the cube on the optical axis to the tube or CCD feeding the green channel. The red and blue rays are reflected to the red and blue channels on opposite sides of the optical axis. Spectral trimming filters are placed ahead of each of the photosensitive tube or CCD surfaces to improve the color saturation.

The first commercial color cameras used a dichroic-mirror arrangement instead of a cube for color separation. This approach was discarded in favor of the cube because of deterioration of the dichroic coating from exposure to air and because of possible damage to the optical elements from the frequent cleaning found necessary to remove the accumulation of dust and air pollutants.

Another disadvantage of mirrors, compared with a cube, is the need for a relay lens to accommodate the longer optical path. This not only results in a larger optical assembly, but adds one or two *f*-stops to optical efficiency. The prism optical system of a three-tube color camera is shown in Fig. 6.4.

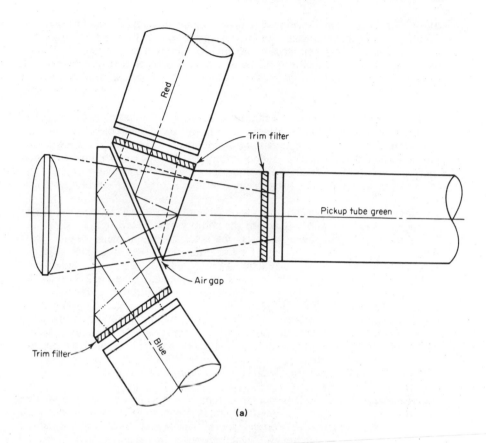

(a)

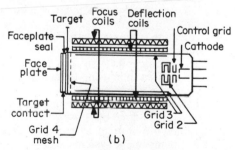

(b)

FIGURE 6.4 (*a*) Prism optical system of a three-tube color camera. (*Source: K. Blair Benson, ed.,* Television Engineering Handbook, *McGraw-Hill, New York, 1986*) (*b*) A cutaway view of a camera tube showing the major parts of the tube structure.

Video Amplification. The relatively low-level video-output signals from the pickup tubes are amplified by preamplifiers and corrected by high-peaking circuits for the high-frequency response loss in the coupling circuits between the pickup tubes and the preamplifiers. At this point, parabola- and sawtooth-correction signals are added to the video signal to compensate for shading errors in the pickup tubes and spurious flare light in the optical system.

This is followed in the processing circuits by line-by-line clamping at blanking level, image enhancement for loss in sharpness because of aperture losses in the optics and pickup tubes, and gamma correction for the linear light-to-video transfer characteristic of the tubes or CCDs.

Gamma Correction.   Gamma is a film term adopted for video transmission characteristics. In algebraic terms, film gamma is the average exponent of the light-versus-film density curve. The gamma in television cameras is the exponent of light versus the video signal; in video amplifiers, it is the exponent of the signal input versus output. Vidicon-type camera tubes and CCDs have a gamma of 1, which means that the video output signal varies directly in proportion with light intensity.

Gamma correction is used in television cameras to provide an increase in gain for low-light signals to compensate for the compression of low light by CRTs and thus obtain an overall linear transfer characteristic from scene brightness to picture-display-tube light output. In fact, the FCC Rules and Regulations (Subpart E of Part 73) state, "The gamma-corrected voltages . . . are suitable for a color picture . . . having a transfer gradient (gamma exponent) of 2.2 associated with each primary color." However, because most picture tubes have a gamma greater than the FCC value of 2.2, in order to provide a pleasing picture with good low-light detail under home-viewing conditions in lighted rooms, a camera gamma somewhat less than 1/2.2 (4.5) is generally used. In some cameras an empirical *black-stretch* circuit is to operate only with low light. In addition, because the increased gain in blacks increases the noise level, black-stretch circuits are narrowbanded to improve the signal-to-noise ratio in blacks and grays.

Image Enhancement.   The lens and camera pickup tubes both contribute to a reduction in the resolution of fine detail, which is apparent as a lack of image sharpness. This loss is called an *aperture effect* because it may be described as an enlargement of picture elements. In the direction of horizontal scan, the picture elements are widened; in the vertical dimension, the number of lines representing a picture element is increased.

Horizontal aperture correction corrects for the horizontal loss by boosting the high-frequency video components with a zero phase-shift network. Such a network, unlike a resistor-capacitor combination, does not introduce any timing shift with an increase in amplitude at an increasing frequency.

Vertical aperture correction is accomplished by delaying the video signal by one horizontal line and subtracting the sum of the undelayed signal and the signal delayed by two lines. Since this is done on a per-field (267.5 lines) basis, in the interlaced picture of 525 lines the vertical correction is + and − two lines. A block diagram of an image-enhancement system using one-line delays is shown in Fig. 6.5.

Color Enhancement.   An electronic equivalent (Fig. 6.6) of the photographic process called *masking* is used to correct for lack of color saturation (color intensity) in camera optics, camera pickup tubes and devices, and CRT phosphors. Electronic masking subtracts calculated values of the other two color signals in resistive matrices for each color channel. A masking-matrix block diagram for a single channel (green) is shown in Fig. 6.6.

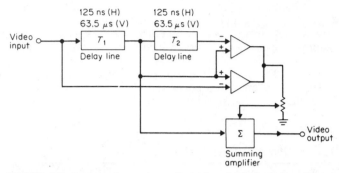

FIGURE 6.5 Image enhancement using +1 and −1 single-line delays. *(Source: K. Blair Benson, ed., Television Engineering Handbook, McGraw-Hill, New York, 1986)*

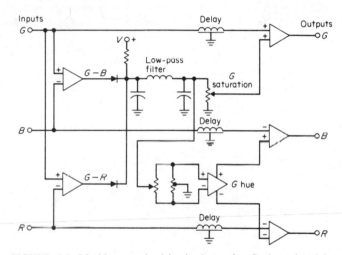

FIGURE 6.6 Masking matrix (circuit shown for *G* channel only). *(Source: K. Blair Benson, ed., Television Engineering Handbook, McGraw-Hill, New York, 1986)*

Black-Level Control. The three color signals, after gamma correction, contain horizontal- and vertical-scan blanking signals inserted in the pickup tubes or devices, but no adjustment has been made for the level of peak-black signals to blanking level. Furthermore, the width of the blanking signals is narrower than the system blanking, and the vertical-blanking interval does not have the required equalizing and serrated vertical-sync pulses. This processing is accomplished in the signal-processing stage, where system blanking is added and the combined signals are clipped to establish black level.

The *clipping level*, relative to peak blacks and camera blanking, is controlled by the camera-control operator through the blanking-level control. This control, which sets the blackest portion of the video signal, usually has a large range in clipping level. This permits operators to match cameras and to provide artistic effects by stripping low-light portions of the video signal.

Alternatively, automatic black-level control circuits may be used to detect black peaks in the video signal and set the clipping level at a preset value.

Signal Processing. The corrected and processed color signals are mixed with *system blanking* and clipped to produce three color signals for encoding into a composite 525-line television signal.

6.2 VIDEO-SIGNAL DISTRIBUTION

6.2.1 Distribution Amplifiers

Video signals are distributed by video distribution amplifiers (DAs). These are wide-bandwidth amplifiers designed to drive the low-impedance, unbalanced (75-Ω nominal) coaxial cables used in television facilities. Video DAs typically provide bridging (high-impedance) inputs with two paralleled input connectors to allow the input signal to be *looped through* to additional equipment and ultimately terminated in 75 Ω. They also normally provide multiple, isolated 75-Ω source-terminated outputs to drive distribution cables for one or more destinations. A simplified block diagram of a typical DA is shown in Fig. 6.7.

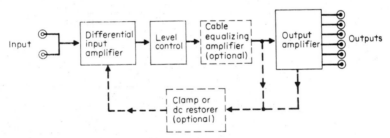

FIGURE 6.7 Simplified block diagram of a distribution amplifier. *(Source: K. Blair Benson, ed.,* Television Engineering Handbook, *McGraw-Hill, New York, 1986)*

There are three basic types of DAs, designated by their intended use as video DAs, subcarrier DAs, and pulse DAs. Video DAs are designed to accommodate standard 1-V (nominal) peak-to-peak (p-p) composite or 0.7 V (nominal) p-p noncomposite video signals. Some video DAs may be used interchangeably as subcarrier DAs because they have sufficient headroom to accommodate the 2-V p-p (nominal) terminated or 4-V p-p unterminated subcarrier signal levels. Pulse DAs typically accommodate up to 4-V p-p (nominal) pulse signal inputs and may provide shaped output pulses with controlled rise and fall times.

Subcarrier and pulse DAs may be further categorized as being linear or regenerative. Linear DAs linearly amplify the signal, whereas regenerative DAs replace the original sync, burst, and blanking portions of the signal with a regenerated version.

Differential Inputs. Differential input circuits provide balanced input imped-ances with respect to ground and amplify only the difference between the signals. Thus, *common-mode* (in-phase) signals, such as power-line hum caused by equip-ment ground loops and other noise and interference components commonly en-countered on long cable runs, are automatically canceled.

Most modern DAs now provide differential inputs in order to avoid hum and noise pickup problems. It is not uncommon in typical installations to encounter several volts of common-mode hum on a 1-V p-p video signal. (In one installa-tion, nearly 100 V of common-mode hum was present on a DA input owing to a power-line ground loop between two buildings.) In some installations clamp-on ammeters attached to coaxial cables have measured several amperes of ground-loop current passing through the shield. In severe cases, power-line ground loops have been known to melt video cables. In installations where such conditions may exist, DA input specifications should be examined carefully to avoid damage to the DA and distortion of the video signal.

Probably the most serious problem that occurs in specifying DA inputs results from using a nondifferential input in a noisy studio environment. Unfortunately, nondifferential-input DAs will amplify any hum and noise that are present as shield currents, in addition to amplifying the video signal. If a clamping DA is used to remove the hum and noise already added by a nondifferential-input DA, further irreversible distortion is added to the signal (even the most carefully de-signed clamping circuits add some distortion). The problem is that once hum and noise become additive (are amplified by a nondifferential-input DA), they are nearly impossible to remove without causing further problems. The solution is to always specify a differential-input DA for noisy signal environments to minimize video hum and noise in the first place, and to use a clamping DA to remove hum only when absolutely necessary. Generally, if there is any question about how clean the electromagnetic environment of the installation is, the cost of specifying a differential-input DA is so small that it is worthwhile insurance. Most modern DAs provide differential inputs. In any case, installations should be designed us-ing good engineering practices. Single-point grounding of video equipment and coaxial cables with a separate, clean video grounding system helps to eliminate unwanted hum and noise at the source that might otherwise cause hum and noise currents in coaxial cable shields.

Level Control. Although DAs normally operate at the standard video levels of 1 or 0.7 V p-p, most DAs provide some form of selectable and/or variable level control to compensate for nonstandard input levels or line losses. The gain con-trol typically has an adjustment range of from -2 to $+3$ dB and is normally ac-cessible from the front panel.

Cable Equalization. Distribution amplifiers usually make provision for video ca-ble loss equalization. The distributed capacitance of long cable runs degrades fre-quency response, particularly at the higher frequencies. Consequently, many DAs will accommodate optional plug-in equalizer networks which compensate for distributed high-frequency rolloff effects by selectively boosting the high-frequency gain of the amplifier. Typically, cables are equalized when their lengths exceed 50 to 100 ft (depending on cable type) and high-frequency losses begin to be excessive. Standard equalizing DAs will compensate for up to 1000 ft (304 m) of Belden 8281/9231, Western Electric WE724, or equivalent cable.

It is important to realize that distribution amplifiers are designed to equalize video cables, not RF cables. If proper video cables are not used, correct equalization may not be possible. RF cables are used for applications such as CATV where low loss at VHF or UHF frequencies is required. These cables often use a silver-coated steel center conductor (for strength and efficient RF transmission) and an aluminum shield which exhibit severe low-frequency rolloff characteristics below 1 MHz and are nearly impractical to equalize for video. It should be noted that the amount of cable loss specified for these cables applies to operation at RF frequencies only. Video transmission cables should have a full copper shield and a copper center conductor to provide a low dc-loop resistance (the series resistance of the shield and center conductor). This type of cable (for example, Belden 8281/9231, Western Electric WE724, or equivalent) ensures proper response across the video band.

Both fixed and variable equalizers are generally available for use with DAs. Fixed equalizers compensate for a fixed cable length and are typically specified in 50-ft (15.2-m) factory-adjusted increments. An advantage of fixed equalizers is that the parameters are generally calculated by computers and checked by sophisticated equipment at the factory, and so it is not necessary to be concerned with the complex equalization parameters that are being compensated for. (Proper equalization involves more than simple frequency-response correction.) In applications where fixed equalizers can be used, selection simply involves choosing the value for the related cable length. Another advantage of fixed equalization is that there is no variable-equalization control that must be adjusted with test signals and measuring equipment. A factor often ignored with variable equalizers is that varying the equalization changes both the frequency response and subcarrier timing or delay through the DA (often considerably, depending on cable length). A disadvantage of fixed equalizers is that cable length must be known before installation and must not be subject to change. If this is not the case, a variable equalizer should be specified.

Long-Cable Equalization.   Special long-cable-equalizing DAs will compensate for up to 3000 ft (914 m) of Belden 8281/9231, WE724, or equivalent cable. Long-cable DAs often provide some form of lightning protection. Long-cable DAs may employ as many as three specially designed cascaded equalizer stages (depending on cable length) to provide the correct equalization at up to 3000 ft (914 m). It is important to note that because cable loss is not linear with distance, three 1000-ft equalizing DAs placed in series will not properly equalize 3000 ft of cable. When an equalizing DA is selected, the total distance to be equalized must be specified.

Fiber optics offer a unique alternative to resolving the long-cable-equalization problem. For example, equalizing a mile of RG/59 cable, which exhibits about 45 dB of loss at subcarrier frequency, would not be very practical. A fiber optic communications system, such as the Grass Valley Group Wavelink, provides a 10-MHz broadcast-quality video or a 6-MHz video plus audio FM transmission link which does not need to be equalized, is totally unaffected by electromagnetic interference and hum pickup, and is highly cost-effective for long-cable runs. The cost of suitable fiber cable is less than half the cost of coaxial cable (and is still decreasing). Thus, the additional cost of terminal equipment is often more than offset by the lower cost of fiber compared with coaxial cable alone.

Clamping.   Clamping is a video-processing operation that provides a line-by-line correction of the video blanking or sync tip level to a fixed dc reference voltage. The primary application of clamping amplifiers is to: (1) reduce additive low-frequency noise and hum, and (2) minimize dc-level bounce upstream of the video switching system when switching between synchronous video sources.

Clamping also increases the dynamic range of amplifiers by reducing the swing in peak levels with APL changes. Clamping DAs can usually accept either composite video inputs (self-driven mode) or noncomposite inputs (external sync mode). Clamping is also used for restoring the dc component of the video signal prior to processing circuitry such as clipping, blanking insertion, and gamma correction.

DC Restoration.   DC restoration, as opposed to clamping, provides a slow return of the blanking or sync tip level to a fixed dc reference voltage. Because of their long time constants, dc restorers will not respond to rapid changes, such as noise, hum, or sync pulses, and cannot make the rapid corrections possible with clamping DAs. Nevertheless, they are useful in restoring the dc component of the video without introducing the distortion caused by clamping.

The dc restorer is usually used as an anti-bounce circuit to reduce the effects of multiple ac couplings in large systems without introducing the noise-translation characteristics of line-by-line, or so-called hard, clamps. DC restorers can operate on either composite or, when a sync-gating signal is provided, noncomposite signals.

Delay Adjustment.   Some distribution amplifiers provide for signal timing or delay adjustment over a 60- to 500-ns range by a combination of switchable steps and a continuously variable vernier. This allows system signal delays to be matched for a variety of video applications.

Automatic Video Delay Adjustment.   A recent development in delay distribution amplifiers is the introduction of automatic delay capability. Even though stable delay DAs employing high-quality delay lines are now available, it is still possible for overall system delay to shift by as much as $+15$ ns as a result of daily temperature effects. Generally, an automatic-delay DA is one whose delay automatically varies as the function of an error signal derived by monitoring the difference in timing between reference subcarrier and burst.

An automatic-delay DA will adjust the timing of source signals to ensure precise color-phase timing at the input to a video switcher or processing facility, to eliminate undesirable hue shifts. Generally, a range of $\pm20°$ (NTSC) or $+24°$ (PAL) of automatic video delay correction is provided with less than 1° (NTSC) or 1.5° (PAL) of phase error.

Since the amplifier varies the total video delay (shifts the luminance, chrominance, sync, and burst simultaneously) to compensate for overall system timing errors, it does not affect the critical ScH phase relationship.

DA Design Specifications.   Good engineering practice for video system design dictates that distribution amplifier specifications be an order of magnitude better than system requirements. This rigorous demand results from the fact that a video signal may be routed through 10 or more DAs in a single installation. A typical network program may pass through literally hundreds of DAs from the originating point in a camera to a local transmitter. Signal degradation or attenuation can result from system interconnections, such as uncompensated loop-through connections, and reflections from improper sending and receiving end terminations. The manner in which DA performance parameters are measured should reflect actual, dynamic operating conditions. For instance, differential gain change through the amplifier should be measured during a 10 to 90 percent APL bounce test. Similarly, a low-frequency bounce test should be used to ensure proper amplifier ac coupling. Other important parameters often not specified

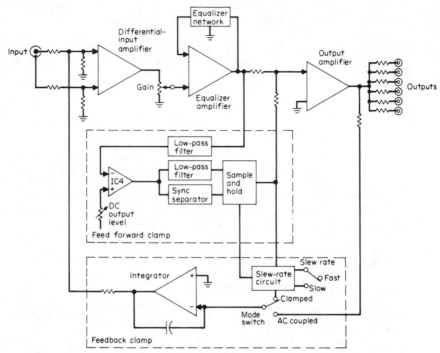

FIGURE 6.8   Block diagram of a clamping distribution amplifier. *(Source: K. Blair Benson, ed., Television Engineering Handbook, McGraw-Hill, New York, 1986)*

by manufacturers are cable frequency-response degeneration and clamping distortion, which was previously discussed.

All clamp circuits introduce a certain amount of distortion in the video at the time interval of the clamp pulse. To prevent distortion during the vertical interval, one clamping DA design uses both feedforward and feedback circuits which operate only on sync tips to minimize such distortion. A block diagram is shown in Fig. 6.8.

6.2.2   Processing Amplifier Designs and Applications

Video signals are processed by high-performance video amplifiers (proc amps) which regenerate the sync, blanking, and subcarrier portions of the video signal. Video processors are typically used to improve noisy video signals, to provide stable and standard video inputs to video tape recorders (VTRs), studio systems, network distribution services, and transmitters. In addition, special proc amps are used for more complex processing applications such as image enhancement, gamma correction, and automatic white- and black-level control. Figure 6.9 is a simplified block diagram of a typical video processing system.

Early Designs.   In the early period of monochrome television, the major concern in signal transmission was the problem of variations in sync level when switching

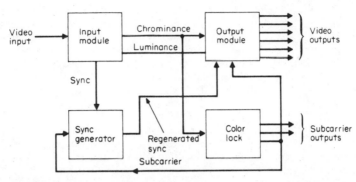

FIGURE 6.9 Simplified block diagram of a video processor. *(Source: K. Blair Benson, ed., Television Engineering Handbook, McGraw-Hill, New York, 1986)*

among different signal or program sources. Excessive sync levels would cause transmitter overload; low levels would result in low-contrast, excessively dark pictures on dc-restored receivers. To correct the problem, the first processing amplifiers were used as sync (amplitude) stabilizing amplifiers (stab amps). These used *stretch-and-clip* sync processing.

Current Designs. The stretch-and-clip systems were succeeded by regenerative video processor designs which removed the entire blanking interval (including sync, color burst, and setup) from the incoming video signal and replaced it with an accurate, adjustable blanking interval from a local sync generator that was locked to the incoming signal. The main advantage of this approach is that an incorrect amplitude, or missing, sync or burst is removed automatically and accurate sync and burst reinserted; in fact, the entire vertical interval will be reinserted, if necessary.

In summary, video processing amplifier designs presently provide the following features: (1) regeneration and amplitude control of sync, burst, and blanking, with accurate clamping of the blanking level, (2) color-black generation and automatic switching between station generator signals, (3) subcarrier-to-horizontal (ScH) phasing, (4) adjustable blanking width, with soft and/or hard white-level clipping, (5) high slew rate capability, fast sync-lockup time, (6) phase-linear separation of chroma and luminance, (7) deletion of vertical interval noise, (8) common-mode rejection, (9) cable equalization capability, (10) vertical interval reference (VIR) correction capability, and (11) remote-control capability.

Noise Reduction. Probably the most critical application of video processors is in improving noisy video signals. (Some video processors are designed to operate only in clean environments and therefore do not improve noisy signals.) To do this, it is necessary to separate and regenerate sync reliably and to remove (clip) any luminance noise spikes that go below the video blanking level. Noise spikes can occur during video time owing to microwave signal fade-out, VTR drop-outs, or spurious random transients. If these negative excursions are not removed, downstream equipment can malfunction by processing and using the noise pulses as sync pulses.

Other video processor requirements are: (1) regeneration of subcarrier (burst), control of burst level, and correction of burst axis offset errors, (2) accurate detection and clamping of the original video blanking level and adjustment of the pedestal setup portion of the dc picture axis, (3) accurate clipping of the video at the maximum white level to limit the signal amplitude to downstream equipment such as VTRs or television transmitters, (4) removal of noise pulses that occur during the vertical interval, (5) automatic gain control (AGC) correction of video, chroma, burst, and setup level, and (6) generation of controlled rise time pulse outputs for sync, blanking, V and H drive, burst flag (PAL), color-frame identification, and PAL identification pulses.

A desirable optional feature is provision for signal outputs with controlled rise time. Another desirable proc amp option is an external reference which eliminates horizontal picture movement resulting from timing errors at the input to a studio switching system.

Video-Tape-Recording Requirements.   Another important application of video processors is in improving switching system output signals or signals into a VTR. (Typically the output of a switcher is processed before it is applied to a VTR.) In order to perform properly in this application, the processor must be able to maintain a constant sync amplitude between different video sources that are selected by the switcher. This is particularly important for VTRs, since many base their AGC-video action on sync amplitude. The processor must also be able to maintain a consistent color-burst amplitude for different video input signals. This is also very important for VTRs, since tape machines often reference video head equalization to burst level; if the burst level is allowed to change as video sources are switched, the overall VTR frequency response could be affected.

The processor must also maintain uniform sync and subcarrier timing for multiple video inputs to minimize VTR time-base errors. A typical amplifier with this feature inserts new sync, blanking, and subcarrier and allows manual or automatic switching between source and station sync based on programmable nonsynchronization timing windows. This allows automatic switching to station sync if a nonsynchronous source is selected. In addition to normal sync clipping, the external reference is also useful for clipping video excursions below the blanking level to ensure that the dynamic signal range of VTRs is not exceeded. In some master-control switching systems a video AGC mode of operation may be desirable to maintain bright pictures without manual attention.

Transmitter-Input-Signal Processing.   In order to perform adequately in a transmitter application, the video processor must provide all the functions required for processing noisy signals, since transmitter signals may be subject to fade-out and resulting noise problems from microwave studio-to-transmitter links (STLs) or terrestrial conditions. The processor must also maintain constant-amplitude sync, blanking, and peak video and must clip negative-going luminance spikes to prevent overload shutdown of the transmitter. White-level clipping is also required to limit peak luminance to 100 IRE units and prevent modulation of the transmitter through zero carrier. To ensure that peak luminance is not exceeded for all luminance and chroma conditions, it is important that the processor separately process the luminance and chroma signals.

Processors are also used as sync generators at the transmitter to accurately lock test equipment to the transmitter signal for off-air testing. For transmitter use, a processor should provide the capability of increasing setup. The luminance clip level must also be adjustable to satisfy FCC requirements of 7.5 IRE units in the United States and 0 in PAL and SECAM countries. The processor should be capable of establishing FCC blanking levels and widths. (Throughout the signal path it is often desirable to maximize the picture-signal horizontal duration by operating all facilities at narrow blanking and allowing the video processor at the transmitter to insert the correct blanking width before transmission.)

6.2.3 Routing Switchers

Routing switchers may differ considerably in configuration to suit a variety of applications. They range from simple single-bus input selectors to complex master-grid systems with many inputs and outputs. They may switch program and monitoring video and audio signals, intercommunication signals, control and data signals, or any combination thereof. Control of the switching function may be centralized or distributed, performed locally or remotely, and may follow any number of different types of operational logic.

To clarify the descriptions that follow, some basic terms are included with their definitions:

Source: Any device which outputs a (video/audio) signal.

Destination: Any device which accepts a (video/audio) signal as an input.

Switching system: An assembly of hardware whose function is to establish a signal path between one of several sources and one or more destinations.

Crosspoint: A circuit element which acts as a switch, allowing information to pass only in the on state.

Matrix: An array of crosspoints represented schematically on horizontal and vertical axes, with inputs along one axis and outputs along the other. This term is often used in reference to a complete switching system but may also be used to describe a single subsection or collection of subsections (usually on circuit cards) within a complete system.

Switching bus: A group of crosspoints along a common output row.

Switching-Bus Routing Switchers. This is the simplest type of routing switcher. It may be nothing more than a passive device consisting of a mechanically interlocked multiple-pushbutton switch assembly. However, most broadcast applications demand advanced features (e.g., remote control, vertical-interval switching) and performance (e.g., high input return loss, low crosstalk) which passive devices cannot provide. Therefore, solid-state designs are used for all but the most elementary applications.

Single-bus switchers are generally used where it is necessary to select one of a relatively small number of dedicated sources at a given destination. For example:

1. Studio camera-match monitoring
2. Transmitter signal-path monitoring (video in, modulator out, IPA diode out, PA diode out, demod out)
3. Edit-bay record VTR input selector
4. Sync generator gen-lock source selector
5. Studio camera viewfinder input selector

The relative simplicity of a single-bus routing switcher when compared with a master control or production switcher, for example, equates directly to a high mean time before failure (MTBF) and a low mean time to repair (MTTR). Therefore, it is common practice to use them as *emergency bypass* switchers on the output of master control. A limited number of often-used sources (network, studio outputs, and VCRs) appear, along with the output of the master control switcher, as inputs to the bypass switcher. The output of the bypass switcher, in turn, feeds the STL. Under normal circumstances, the output of the master-control switcher is selected on the bypass switcher. Should the master-control switcher fail, the bypass switcher is used to keep the station on the air.

Single-bus routing switchers are frequently used to extend the input capabilities of a production or master-control switcher. Large numbers of sources are common in modern television facilities, and quite often a substantial percentage of them are needed in studio control rooms or master control. Since the number of inputs on production and master-control switchers is generally limited to a maximum of about 20, it is not uncommon to find that a single unit simply does not have enough. An excellent solution to this problem is to connect the most often used sources directly, reserving one input for a single-bus routing switcher *preselector*. This makes a number of additional sources, equal to the number of inputs on the routing switcher, available to the production or master-control switcher. Of course, it is necessary to account for the delay through the preselector and take appropriate steps to preserve system timing.

The functional circuit blocks used in most single-bus routing switchers are very similar to those employed in larger multiple-bus designs. Thus, the single-bus system is a good starting point for a close look at basic routing switcher design and operation. Figure 6.10 is a block diagram of a 10-input, single-bus. The functional blocks employed in the video and audio sections are similar.

Quite often, single-bus switchers are used as building blocks to make a multiple-bus switcher. This is usually done by looping cables through the inputs of several single-bus units. While this will work, performance is compromised, mainly because of the effects of uncompensated looping on system input return loss.

Input return loss is an important factor in any video system component. Input return loss is the ratio of the power delivered by the signal source to the power reflected back to the source, expressed in decibels. A high input-return-loss figure is important, because the reflections caused by mismatch can affect video system transient and frequency response. The best way to ensure a high input-return-loss figure is to make the input impedance of the component purely resistive and exactly equal to the impedance of the source. This assumes, of course, an *ideal* cable impedance between them.

Even the best grades of coaxial cable are less than perfect, to some degree. More important, they become even less perfect (as transmission lines) if they are looped through several pieces of equipment. For this reason, this practice should be used with caution. The best system performance will always be realized in systems where looping is not used. When looping cannot be avoided, the cables should be kept as short as possible, and equipment should be selected which has been properly designed to compensate for looping inputs. Unless the equipment includes an input compensation network, it will probably introduce a significant impedance discontinuity when attached to the middle of a section of transmission line.

The *crosspoint* and *bus* act as a multiposition switch to allow the operator to select one of many sources. The crosspoint must therefore behave as much like a perfect switch as possible; that is, it must exhibit zero transmission loss in the on

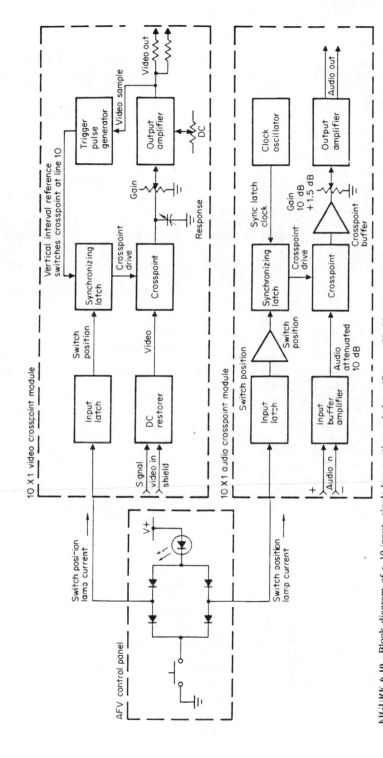

FIGURE 6.10 Block diagram of a 10-input, single-bus routing switcher. (*Source: K. Blair Benson, ed., Television Engineering Handbook, McGraw-Hill, New York, 1986*)

state and infinite transmission loss in the off state. With most solid-state devices, low transmission loss in the on state is usually not a problem. High off-state loss can be a problem, especially at high video frequencies; therefore, a T configuration is often employed.

A T crosspoint consists of three solid-state switch elements. Two are connected in series and are placed in the signal path. The third is connected between the junction of the other two and ground. When the crosspoint is selected, the two series elements are turned on and the shunt element to ground is turned off. The signal is able to flow with little or no attenuation. When the crosspoint is not selected, the series elements are turned off and the shunt element is turned on. Thus, the signal path is opened and any residual signal leakage through the first series element is diverted to ground.

The *output amplifier* serves primarily to buffer the output of the bus to the following circuitry and system. Usually, a gain-trim adjustment is provided. However, since a routing switcher is normally regarded as a unity-gain device, the output gain trim has only enough range to compensate for minor system variations.

The output amplifier may also include provisions for output cable-loss equalization. This is similar to the equalization function which may be included in the input amplifier.

Crosspoint control schemes vary in complexity depending on the number of crosspoints being controlled and the desired control functions. Once again, the single-bus case provides a good model, since control is limited in terms of both crosspoints and functions. Generally speaking, the control of single-bus routing switchers is restricted to the selection and tally of a relatively small number of crosspoints in one dimension (a single switching bus).

If the number of crosspoints is 10 or less, wire-per-crosspoint control is usually employed. A control line is provided for each crosspoint, and shunting any line to ground causes the corresponding crosspoint to be selected. Crosspoint status (confirmation of actual occurrence of the switch) is provided by the switcher control system in the form of a sustained ground on the control line after the temporary ground is removed.

Simplicity is the primary advantage of wire-per-crosspoint control. Control panels may be constructed using momentary-action normally open pushbuttons with LED or incandescent indicators. A properly executed wire-per-crosspoint control system should accommodate a wide variety of inputs for selection, while providing a status signal acceptable to a wide variety of devices. A good design will support control/status by external logic as well as switches and indicators.

Since it is possible to ground more than one line at a time, wire-per-crosspoint control systems must include logic to inhibit simultaneous selections. Priority encoding is usually used, resulting in selection of the highest-numbered input.

When the number of crosspoints being controlled exceeds 10, it becomes attractive to use some form of coded-control selection which reduces the number of wires between the switcher and the control panel. Typically, binary or binary-coded decimal (BCD) coding is employed.

Regardless of the method used to control crosspoint selection, a means of timing the switch point is required. Nearly all modern video switchers switch during the vertical interval. Since the inputs to routing switchers are often nonsynchronous with respect to each other, the question arises as to which vertical interval should be used. A popular solution, especially on single-bus switchers, is to strip sync from the output of the switcher and use it to phase-lock a trigger-pulse generator operating near the vertical rate. The switch will then be timed to the vertical interval of the last-selected input. If there was no signal present at the last-selected input, the switch will occur randomly.

Multiple-Bus Routing Switchers.   It is natural to think of a multiple-bus routing switcher as a group of single-bus switchers with common (looped) inputs. This approach can be, and in fact is, used, but there are a number of reasons why it is a makeshift one at best. The most serious technical limitation, the effects of looping on performance, has already been discussed. The difficulty in implementing full-matrix xy control is yet another. But the most compelling reason for avoiding this approach is its inefficiency, in terms of both cost and routing density. A truly efficient and cost-effective multiple-bus routing switcher is a difficult design exercise involving a number of tradeoffs, such as reliability versus packaging efficiency, price versus performance, and flexibility versus complexity.

Multiple-bus routing switchers are generally used when it is necessary to make a fairly large number of sources (usually 20 or more) available to multiple destinations. In many instances, a *master-grid* concept is employed, where virtually all sources in the facility appear as inputs to a large matrix with many switching buses. Alternatively, smaller matrices may be employed, each performing a specific function, such as VTR input selection, production switcher input preselection, and production switcher iso-bus switching.

The master-grid concept provides the greatest flexibility, since all sources are available at each destination. System timing is greatly simplified, since the length of program paths is predictable. Efficiency is sacrificed, however, because some of the crosspoints (the ones representing unused or seldom-used source-to-destination connections) are not required.

The alternative approach, employing a number of *functional-area matrices*, eliminates the efficiency problem since each matrix is fed only by the required sources. Thus, there are fewer redundant crosspoints. Flexibility is sacrificed, however, because the seldom-used sources are not available, except by patching them temporarily. Preservation of system timing is also difficult, unless strict rules are observed regarding program paths through multiple matrices.

Both approaches can be appropriate, but the decision as to which is best for a given facility can be determined only after a careful study of present and future needs, operating philosophy, and budget.

It should be fairly obvious that controlling a multiple-bus routing switcher is a more complicated task than controlling a single-bus routing switcher. This is true not only because of the increased number of crosspoints, but also because a unique crosspoint in a single bus switcher can be specified by a single number (x, the input number). In a multiple-bus switcher, two numbers must be used (x, the input number, and y, the output-bus number).

Operationally, a multiple-bus routing switcher is generally controlled in one of two ways: on a dedicated bus-by-bus basis, or full-matrix xy control. Quite often, both methods are employed. For example, a given system may include several buses for VTR input selection with control panels for these buses located at the VTRs. At the same time, an xy panel would be provided at a central maintenance/monitor location. Under normal circumstances, the VTR operator would control his or her own bus. In some cases, several buses are controlled using a single control panel.

Although xy control is used primarily for maintenance purposes, it does have other applications. An xy control panel can be used for *emergency* control of a bus whose dedicated control panel has failed. Small routing switchers which are being used as patch panel replacements may use a single xy panel as the primary means of control. Regardless of how they are used, *xy* control panels should be applied with caution, since an inadvertent operator error has the potential of causing a selection to be made on a bus other than the desired one.

Routing-switcher control-system technology has become quite sophisticated. Most contemporary designs are microprocessor based. The choices as to basic control system architecture closely parallel those of basic signal system architecture. Instead of the choices being output-oriented versus matrix-oriented, they are distributed versus centralized. Choice criteria are similar in both cases and primarily relate to the importance of reliability and serviceability.

Control Panels.   Routing switcher manufacturers' literature typically includes a dazzling array of types of control panels. Selecting a panel for any given application can be greatly simplified if three primary criteria are considered: function, form, and features.

What control function is required? There are three possibilities: single-bus, multibus, and *xy*. If the routing switcher has multiple levels (e.g., video, audio 1, audio 2, time code), the question arises whether independent control of each level is necessary or whether all switching operations will be married.

There are numerous ways to enter and display data. Data entry methods include individual pushbuttons, key-pad pushbutton arrays, and thumb-wheel switches. Data display methods include indicator lamps (incandescent or LED) and multiple-segment numeric and alphanumeric displays.

1. *Button-per-Source:* A button-per-source control panel has one or more rows of individual pushbuttons. Each button is equipped with an illuminating indicator to show the operator which source is active. The unit shown in Fig. 6.10 is designed to accept film legends with the name of the source in each button. This enables the operator to select sources rapidly. Typical applications for this type of panel include production switcher iso-buses and camera match monitor selector. There are, of course, physical limitations on the number of buttons that can be mounted on a reasonable-sized control panel; button-per-source panels are generally constrained to deal with a maximum of 30 or so sources. If the switching system has more inputs than this, the control panel must provide some means for designating which subset of sources it controls.

2. *Thumb wheels:* The thumb-wheel control panel consists of a multidigit thumb-wheel switch, a take button, and a multidigit numeric display. The major limitation of thumb wheels is their slow data entry speed. Their main advantage is that they do not require much space.

The most popular type of general-purpose control panel for routing switchers is the *key pad*. Key pads provide access to a large number of sources without requiring much space. A key pad is *natural* for data entry, a characteristic no doubt attributable to the widespread use of electronic calculators and pushbutton telephones.

Summary.   Routing switcher systems can be characterized in many different ways: size, architecture, and control philosophy, to name just a few. Selecting a particular system requires a careful analysis of the application. Factors to consider include performance and reliability, flexibility, expansion capability, operating features, and control requirements.

Alternatives to the classic rectangular crosspoint array have begun to emerge and should be considered where they offer a significant advantage in terms of performance and/or cost.

6.2.4 Production Switchers

The portion of the studio video system that handles video production includes equipment that switches, modifies, or creates video for the purpose of enchancing program material supplied by cameras, VTRs, or other outside sources. Typical video production devices and their interfaces are shown in the video system block diagram in Fig. 6.11.

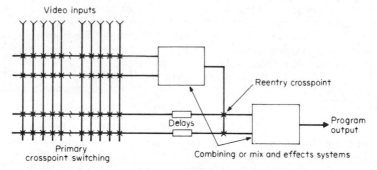

FIGURE 6.11   Block diagram of a video switching system. *(Source: K. Blair Benson, ed.,* Television Engineering Handbook, *McGraw-Hill, New York, 1986)*

Video production has been heavily affected by recent developments in digital control and digital video. While the basic functions of studios have remained the same, several new devices have changed video production considerably. These devices, video switching systems, switcher memory systems, and digital video effects systems, will be discussed in this section. Video-tape editors also play an important role in video production, and the expanding interface between switchers, editors, and other production devices is also discussed.

Basic Functions.   A video production switching system (*switcher*) is one of the primary devices used to produce a television program. Production switchers are essential to all live operations and most postproduction situations. The main function of a production switcher is to either switch or cut between two video sources, or combine them in a variety of ways. The principal methods used to combine video are (1) mixing, (2) wiping, and (3) keying.

A production switcher consists of two main sections: (1) an input selection matrix which provides the input switcher functions, and (2) video mixing amplifiers, usually called the *mix and effects* (*mix/effects*) *system*, which provides the combining functions. Improvements in production switchers have been quite rapid in recent years. Early systems could only switch or mix video sources using an input selection matrix and a simple mixing amplifier controlled by a lever arm. This selected the proportions of the two video signals being combined and added the result to form the output signal. In Europe, video switchers were patterned after audio mixers, where the amplitude of each input was controlled by a separate control (knob), allowing two or more signals to be combined simultaneously. This approach is called the *knob-a-channel* system.

As technological advances provided high-speed switching transistors and integrated circuits, wipe and keying functions were added to production switchers, usually in the form of separate two-input combining amplifiers in addition to the mixing amplifier. A selection of different wipe patterns was provided to allow the operator to produce a variety of wipe effects between one picture and another. These units also provided the ability to key or switch from one picture to another under the control of a video or external input signal. Keying is typically used for adding captions or titles to the picture.

Currently produced switchers usually provide all the above functions of mixing, wiping, and keying in a single mix/effects (M/E) amplifier. Controls for the functions are usually grouped together to allow wipe patterns to be used to mask out parts of keys, while mixing in and out of the resultant combination. The largest and most sophisticated production switchers usually have two or three mix/effects systems, each with four to five input rows that can produce multilevel keys and utilize a large number of wipe patterns.

The configurations of production switchers vary in a number of ways to accommodate production requirements. The design variations generally encountered are as follows.

Number of Inputs. The number of inputs usually is between 8 and 32 video channels. The number is determined by the requirements of the application, such as field, studio, production, and postproduction. Field or mobile applications often require a small panel size and fewer inputs than a studio or postproduction switcher. When a very large number of sources is available, a preassignment or routing switcher may be employed to preselect the sources, such as video tape or film, to limit the inputs to the production switcher's input matrix.

Number of M/E Systems. The number of mix/effects (M/E) systems varies considerably and depends largely on the complexity of the production, the number of effects needed simultaneously, and the extent of postproduction required. There is often less need for multiple-effects production for mobile use than for live and postproduction operations. In live operations more than one M/E system is useful so that one can be used while a second is being set up. Usually M/E systems can be cascaded or reentered one into another to allow production of complex source combinations.

M/E Functions. Most mix/effects systems provide several basic functions, which include:

1. *Mix:* An additive combination of two video sources (usually summing to unity).
2. *Mix transition:* A change from one video source to another using a mix operation (duration greater than one field).
3. *Wipe:* A switch occurring during the active video at specific points on the raster to produce a pattern between two video sources (switching is usually controlled by an internal waveform generator).
4. *Wipe transition:* A change from one video source to another using a wipe operation (duration greater than one field).

5. *Key:* A switch occurring during active video between two video sources controlled by a video or video-related signal.

M/E Features. A great variety of operating features are associated with these modes. Some examples are listed below:

1. *Mixing:*

> Manual and auto transitions
> Nonadditive mixing (ability to combine two pictures so that only the brightest elements of each are present in the result)

2. *Wiping:*

> Number of patterns
> Rotational patterns
> Bordering modes, hard or soft, and colored edges
> Direction modes
> Modulation modes
> Positioner modes
> Aspect change
> Multiple patterns

3. *Keying:*

> Title keys (or liminance keys)
> Source selection
> Invert mode
> Insert selection (video or colorfill)
> Bordering selection
> Chroma keys
> Source selection
> Key features
> Key separate or hold modes
> Encoded chroma keys

Often these features are provided as add-on options to enable the user to select a certain degree of customizing. Other common options are a quad split system, a final or downstream keying (DSK) system, and additional switching rows for utility use. Each of these features will be described later in more detail.

The availability of each of these features to the operator is largely dependent on the design of the control panel and its logic. To get the most effective use from a system, the control panel layout must provide a logical arrangement of controls for the operator. Grouping controls by functions, providing easily identifiable repeating patterns for similar functions (each effects system should look like the others), and making often-used functions such as keys, wipes, and mixes easy to find help make a switcher easy to use.

Effects Memory Systems. With the increased capabilities of larger switchers, the number of operating changes that are required between operations has increased. This has led to the introduction of mix/effects memory systems, usually microprocessor controlled, that enable the operator to store panel settings for later recall. This allows the operator to go from one setting to another quickly and to utilize the full capability of the production switcher. The memory systems usually operate independently on each mix/effects system, providing the ability to recall and preview the control settings before taking them to the switcher's output.

Two general types of effects memory systems are in use:

1. *Learn-mode programming:* The operator sets up the desired effect in the normal manner and then learns (enters) the effect into the user memory register. A more permanent storage medium, such as flexible magnetic disc storage, is often provided as well.

2. *User programming:* The operator uses a keyboard to enter source and mode selections, for example, into memory locations, and can recall these on the switcher to preview and air the result.

The memory system for switcher control also makes it easier to interface with editing computers for postproduction editing by providing access to the full switcher operations, and the editing computers are easier for the operator to change than the editor listing for a particular editing sequence. The ability to interface to editing computers is usually by means of a serial or parallel interface between two microcomputer systems. The serial mode often provides access to other peripheral devices used to feed production switchers, such as still stores and character generators. This can provide full interaction with editing computers.

6.2.5 Digital-Video Effects

Digital-video effects or *digital effects* are created by digitizing the video-effects signal so that it can be stored, retrieved, and manipulated in a digital format. Digital effects include all video effects that can be produced by moving the picture on the screen, changing its size, or giving it motion. To produce these effects, the individual picture elements must appear on the screen at locations different than in the original picture. This can be accomplished only by writing the picture into a portion of the memory that represents the normal raster, then reading the processed picture from the memory at a later time. A digital video-effects device can be a stand-alone unit that adds special effects to one or more video signals, or it can be an integrated unit that is part of the production switcher, much as wipe and key special effects are integrated.

Features. Similar digital effects are possible whether or not digital effects are an integrated part of a switcher, although more creative flexibility can be achieved with an integrated system. Some common terms that describe these effects are:

Compression
Expansion
Reversal
Splitting
Decay
Freeze
Resolution or data-bit reduction
Perspective
Rotation

A digital-effects unit can be separated, for purposes of discussion, into two parts—the digital-effects processor and the digital-effects controller. A typical digital-effects system consists of the control panel, controller electronics to generate the required control signals to define the effect, and the digital video processor, which uses the control signals to control the video signal in the digital domain, producing the desired effect.

The digital-effects processor digitizes the analog television signal and performs all the high-speed processing that is required for digital manipulation of the picture (compression, decay, rotation, etc.). The digital-effects controller typically provides the human interface between the switcher and the digital-effects processor. In some cases, the controller also provides the system integration of the digital effects. The effects controller normally sends the data required to control the effect (effects mode, size, rotation, etc.) to the effects processor once per field to control the digital effect. The digital-effects processor and the effects controller may or may not be physically separated; however, the operations are quite different.

System Considerations.   Digital effects can be used in a wide variety of operational environments. For example, on-air and postproduction are clearly very different environments that both use digital effects. The following is a list of system considerations whose relative importance will depend on the operational environment:

1. *Video input source selection:* How are the sources selected? Are they selected by a separate routing switcher, or is the selection interpreted with the effects controller? How many input sources are required?

2. *Digital-effects video and key output:* Normally, the digital-effects output is connected to the input of a production switcher to be combined with other video sources as part of the final picture. To accomplish this, a key signal must also be supplied by the digital-effects processor, and the key signal must be in a form that is usable by the switcher. Many digital-effects processors provide the capability of keying the transformed picture onto a matte background. Without a production switcher, the final picture would be limited to the matte background.

3. *Control delegation:* To save costs, it is sometimes desirable to share one digital-effects processor among several studios or postproduction suites with the ability to delegate controls to one studio or suite when needed.

4. *Video-tape editor control:* In postproduction applications it is necessary to control digital effects from a video tape editor. This control is normally provided through a serial interface.

Stand-Alone Digital-Effects System.   A controller for a stand-alone system provides the human interface to the digital-effects processor and would not normally include interfaces to other devices. There are two basic types of human interface to the controller—a menu-directed interface and a dedicated interface.

In the menu-directed interface, the human interface inputs are *soft* keys, i.e., switches and analog controls whose functions change depending on which mode or *menu* is selected. The human interface output may include a CRT display which would also change with each menu selected. The advantage of the menu system is that many control functions can be accommodated with a small number of switches, potentiometers, and display devices. Its disadvantage is that it may take some time (i.e., several menu selections) to get access to any particular operation.

The dedicated interface provides an input device (pushbutton switch, potentiometer, etc.) and a display indicator for each operation. The advantages and disadvantages are the opposite of those of the menu system. Quick access to all functions is provided at the cost of a large number of control devices and the accompanying large control panel.

Most systems are a combination of the two basic types. The mix of menu-directed and dedicated interface functions must be evaluated on the basis of what would work best in the particular operational environment.

Integrated Digital-Effects System.   An integrated system with its digital data and control paths (video paths are not shown) is shown in Fig. 6.12. The major components of the system are:

1. *Mix/effects system:* Each mix/effects system and the flip-flop system of the switcher are controlled.
2. *Digital-effects unit:* This component consists of the digital-effects processor and the digital-effects controller.

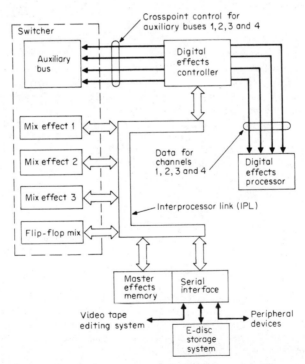

FIGURE 6.12 Block diagram of an integrated special-effects system. *(Source: K. Blair Benson, ed.,* Television Engineering Handbook, *McGraw-Hill, New York, 1986)*

3. *Master effects memory:* This component learns (stores) and recalls all the control parameters of the switcher and digital-effects unit.

4. *Serial interface adapter:* This provides communication with external equipment, such as a video-tape editor, disk storage system, and other peripheral devices.

5. *Auxiliary video crosspoint bus:* This component selects the video input to the digital-effects unit. The selection can be made either manually or by the digital-effects unit.

These components communicate with one another via a high-speed data bus called the *interprocessor link* (IPL). The IPL is a significant factor in the integration of the digital-effects unit, the switcher, and peripheral devices.

The integrated system provides the following capabilities in addition to those found in a stand-alone system:

1. *Video input sources:* These can be selected by the digital-effects units, which allows changing the input source to become part of the overall effects, for example, changing the input source when the picture goes to zero size or changing the input to produce a sequence of frozen pictures each from a different source.

2. *Video-tape editors:* These can control both the switcher and the digital-effects unit from one serial port.

3. *Effects memory functions:* Effects memory functions can include both the switcher and digital-effects unit. (The effects memory stores the condition of all the pushbutton and analog controls in memory; it can be recalled as required to accurately repeat the digital-effects operation.)

4. *Additional functions:* By using the video output from the digital-effects unit as an input to the switcher system, additional functions can be performed, for example, on-air tally and improved key- or wipe-tracking operations. (*Key or wipe tracking* is the ability to make the compressed picture fit into a key or wipe shape and follow or track that shape as it moves.)

Specifications. Digital video effects are difficult to specify in concrete terms since the objective of an effects device is to produce by electronic means effects heretofore created on film and to generate new and different effects. However, experienced operators can determine whether the effects are easily repeatable, are easy to set up, and produce acceptable picture quality. Certain qualities of the effects controller are subtle and should be verified, such as whether screen movements appears smooth, whether controls have good resolution, and whether automatic transitions start and stop smoothly.

CHAPTER 7
VIDEO-TAPE AND FILM REPRODUCTION

K. Blair Benson
Television Technology Consultant, Norwalk, Connecticut

E. Stanley Busby, Jr.
Ampex Corporation (retired), Redwood City, California

7.1   VIDEO-TAPE RECORDING AND REPRODUCTION

The first commercially practical video-tape recorder was developed by a team of Ampex engineers headed by Charles Ginsburg. The recorder was used initially in November 1956 by the CBS Television Network for time-zone delay of evening news programs. The new magnetic-tape medium provided a picture quality significantly superior to the kinescope process in use at the time without the time-consuming and expensive film laboratory processing. Within a short time the use of video tape expanded to the storage of all types of programming for later release, first of uninterrupted recordings and within a year for complex edited productions.

Previous attempts to adapt audio-recording techniques using stationary heads to record wide-band video signals were unsuccessful because of the exceedingly high tape speed required and the accompanying difficulty in maintaining uniform video-signal levels. These problems were solved by two important developments, one mechanical and the other electrical:

First, a very high writing speed for the magnetic head relative to the tape is achieved by the use of a rotating-head drum to scan across the tape, while the tape moves at a normal audio speed of 15 ips or less. This permits the use of proven designs for tape transport mechanisms and for longitudinal recording of audio signals.

Second, in a manner similar to FM radio broadcasting, the video signal is recorded as frequency modulation of a carrier signal. Thus, amplitude variations and nonlinearities in the record-playback process are not transferred to the playback signal, and in fact, limiting or clipping amplifiers can be used to remove spurious interference.

The progress that has been made in the video-tape field since that first broadcast in 1956, while remarkable, is evolutionary rather than the result of technological breakthroughs and may be attributed primarily to improvements in video heads and magnetic tape, the advent of solid-state components, and the use of digital processing.

7.1.1   Record and Reproduce Process

The basic elements of a magnetic-tape recorder with a stationary head are shown in Fig. 7.1. A magnetic tape is pulled in the direction indicated by the arrow across the erase and recording heads by the capstan and the take-up reel drive motor. The magnetic coating contacts the heads in sequence, first the erase head, then the video head in the scanner assembly, and last the audio and control-track heads.

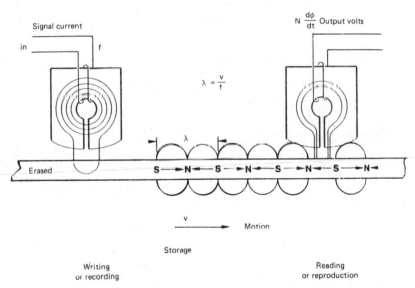

FIGURE 7.1   Fundamental recording and reproduction process. *(Source: Benson, 1986)*

Recording Sequence.   The erase head demagnetizes the tape before recording by exposing the magnetic particles in the coating to a high-frequency field that is several times greater than the maximum magnetization property of the particles. As the tape is drawn past the erase head, the erase field gradually decreases to zero, thus leaving the coating in a demagnetized state with the magnetic fields of the particles oriented in random directions.

Next the tape moves in contact with the recording head and across the gap where the magnetic field from the head forces the magnetic particles into alignment with the lines of force from the head. The number of magnetic particles forced into alignment as the tape leaves the *recording zone* of the head depends upon the strength of the magnetic field. This determines the amplitude and direction of the magnetically recorded signal.

The magnetization of the basic recording action described is not linear with respect to the head current. However, linear magnetization can be obtained by adding a high-frequency ac-bias current to the signal current. This scheme is used in audio recorders to provide a linear magnetization characteristic and reduce amplitude distortion. In video recorders, since the signal is recorded as frequency modulation, the carrier is recorded directly without ac bias.

Recorded-Signal Wavelength.   During each cycle of signal there are two reversals of the magnetic field. The measured distance occupied by one cycle of the recorded signal on the tape is called the *wavelength*. This is indicated in Fig. 7.1 by the distance between two like poles, namely, north (north-seeking) to north. The wavelength, in turn, is proportional to the relative speed between the *head* and *tape* and inversely proportional to the frequency of the recorded signal. This is equal to 2 bits in digital-data terms.

With stationary heads used for audio-signal recording, the relative speed is equal to that of the tape traveling through the tape transport. The relative speed for video recording is much higher because the head is mounted in a rapidly rotating cylinder. Consequently, even though the tape is moving at a slow speed of 15 ips or less, a very high video writing speed is achieved.

Packing Density.   The number of bits, or reversals of the magnetic field, per measure surface distance on the tape is the *linear* packing density. The linear packing densities and recording characteristics for several magnetic-tape recording applications are tabulated for comparison in Table 7.1.

TABLE 7.1   High- and Medium-Density Recording Applications and Characteristics

Recorder	Tape speed [ips (m/s)]	Maximum frequency	Lin. packing dens. [bits/in (bits/cm)]
Type-C video	1000 (25.4)	15MHz	30,000 (11,811)
Consumer VCR	220 (5.6)	7MHz	64,000 (25,197)
Cassette audio	1⅞ (0.0476)	20kHz	21,300 (8,386)
Professional audio	15–30 (0.38–0.76)	20kHz	2,700 (1,063)

Longitudinal and Transverse Recording.   When the magnetization is oriented in the direction of the relative motion between the head and tape, the process is referred to as *longitudinal recording*. *Transverse recording* exists when the recording is oriented at right angles to the relative head-to-tape motion. From these definitions, longitudinal magnetization patterns are produced by both rotary- and stationary-head recorders. Therefore, stationary-head recorders should be identified as such, rather than as longitudinal recorders.

Perpendicular Recording.   If the magnetization is aligned vertically to the surface of the tape, it is called *perpendicular* or *vertical* recording. This requires that the north and south poles of the recording head be on opposite sides of the tape and is not used for conventional video or audio recording.

Saturation Recording.   The recording process applies a changing magnetic field, in accordance with an electric signal, to a recording head as the head moves across the tape in video recording (the tape moves across a fixed head in audio recording). The playback signal will increase nonlinearly with an increase in the recording signal up to a saturation point, above which the playback signal will decrease. Furthermore, as the recording signal is increased to a maximum and then returned to zero, the magnetization of the tape does not also return to zero but a *remnance* magnetization is retained.

This is shown by the curve, known as a *hysteresis loop*, of magnetization force H versus magnetization M in Fig. 7.2. The dashed line indicates the increase, and decrease to zero, of the applied magnetic force. The remnance is M_r. Thus, the magnetic recording and playback amplitude characteristic is nonlinear and, in addition, is distorted by the value of the remnance magnetization existing as the polarity of the magnetizing force is reversed.

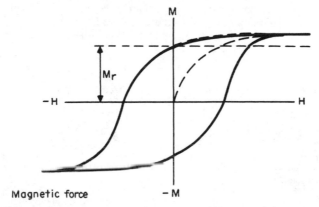

FIGURE 7.2   Magnetic force, magnetization hysteresis loop for a typical tape.

In video FM signal recording, as in digital signal recording, linear amplitude reproduction is not required. Instead, a binary on-off signal is all that is required, since the signal information is contained in the frequency in FM, or timing in digital, of the magnetization reversals. This permits the use of a constant-amplitude FM signal that is optimized for maximum playback amplitude with the particular tape-head combination in use.

Bias Recording.   Unlike FM video recording, in audio recoding exact linearity between the record and playback signals is necessary to meet the low-distortion requirements for high-quality sound reproduction. This is accomplished in both video and audio tape recorders by superimposing a large-amplitude, high-frequency signal well above the audio spectrum on the audio signal. The bias signal can be thought of as a high-frequency switching signal that magnetizes the particles half the time in one direction and half the time in the other. As a result, the remnance magnetization components in effect cancel.

Frequency-Modulated (FM) Video-Signal Recording.   The full color signal is recorded directly in most VTRs intended for broadcast use. This technique is called *direct color recording.* The video signal is recorded as a frequency-modulated carrier with sync tips clamped at a frequency substantially above the nominal video bandpass of 4.2 MHz and the peak-white video modulation extending to 10 MHz. This separation is necessary to avoid visible herringbone beat patterns between the video-modulation signal and the FM carrier. However, in order to maintain the precision of the color-subcarrier frequency and the phase of its color modulation components required for faithful color reproduction, extremely accurate *time-base correction* (TBC) is necessary. In consumer-type recorders, which are limited in video-bandwidth capability because of the lower head-to-tape writing speed, the need for a system video bandwidth to accommodate the color subcarrier in a direct FM-recording system and costly TBCs is avoided by the use of the "color-under" system. Color-under recording is described in a following subsection.

The reference frequencies specified in the Society of Motion Picture and Television Engineers Recommended Practices (SMPTE RPs) for the frequency-modulated luminance components of the 1-in direct-recording Types B and C, and the principle ¾-in and ½-in color-under, Types E and F, 525-line system are listed in Table 7.2. The complete specifications for these and other video-tape systems are listed in Sec. 7.1.2.

TABLE 7.2 Reference Frequencies for FM Video-Tape Recording

Recording system	Sync	Blanking	Peak white
1-in types B and C	7.06	7.9	10.0
¾-in type E	3.8	*	5.4
½-in type F	3.1	*	4.5

*Not specified.

Color-Under Recording.   Rather than recording the luminance and chrominance signals directly, the composite video signal is fed through a low-pass filter to separate the luminance signal from chrominance and to limit the high-frequency luminance components to about 2.5 to 3.0 MHz before frequency modulation. In addition, as shown in Table 7.2, the FM carrier frequency and deviation are less than that of 1-in direct-video recorders.

The chrominance AM subcarrier is bandpass filtered to provide a 0.5 MHz-wide double-sideband signal for amplitude modulation of subcarrier at about 688 kHz. The 688-kHz AM color subcarrier is added to the FM luminance carrier to provide a composite-signal for recording. The spectrum of the recording signal is shown in Fig. 7.3.

The advantages of the system are that time-base errors from variations in head-to-tape velocity cause very little change in color-subcarrier phase and resultant color fidelity. For professional applications such as broadcasting where more stringent requirements are dictated for editing and for program integration, as well as by the FCC, use of a TBC is necessary.

On the other hand, trade-offs in picture quality are: (1) higher chroma-noise because of the low level of chroma signal necessary to reduce beat patterns with the FM-chroma signal to an unnoticeable level, and (2) the limited luminance resolution resulting from the reduced luminance bandwidth, compared to direct composite-signal recording.

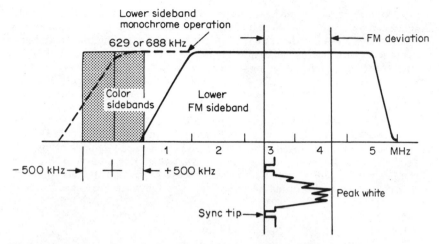

FIGURE 7.3   Color-under spectrum of FM luminance and AM chrominance carriers. To make room for down-converted color sidebands, part of the FM luminance lower sideband is removed. *(Source: McGinty, 1979)*

7.1.2   Recording-Head and Tape Configurations

Type C 1-in Open-Reel Recorders.   The SMPTE Type-C format is a helical scan of one video field by a single head rotating in a fixed drum at an angle of approximately 2½° to the edge of the tape. The layout of the drum and the placement of the heads is shown in Fig. 7.4.[1]

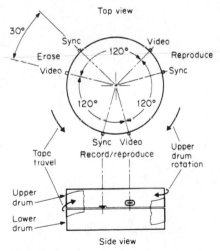

FIGURE 7.4   SMPTE Type-C head layout. *(Source: Fink and Christiansen, 1982, in Benson, 1986)*

---

1. ANSI C98.18M and C98.19M.

The minimum number of heads required for picture recording and reproduction are one for erase and a second to serve for both recording and playback. The *C* format specification provides a total of six heads. In some implementations of the format, a *sync* head is included with each video head to provide a contiguous vertical-sync signal during the signal-dropout time when the video head moves from the tape first contacting the drum to the tape leaving the drum. This head, incidentally, is used also for time-code editing pulses and other digital information signals.

A second set provides a playback signal while recording and for electronic editing. The third set provides the erase function for all helically recorded signals.

Stationary heads are provided for erase, recording, and playback of longitudinal tracks for audio signals, tape-speed control, and editing time code. Figure 7.5 is the layout of a typical transport showing these heads and the tape path. The layout of the Type-C recorded magnetic tracks is shown in Fig. 7.6*a*.

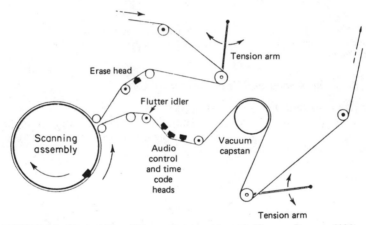

FIGURE 7.5   Typical Type-C format transport layout. *(Source: Benson, 1986)*

Video-Cassette Recorders.   The Type-E ¾-in VCR format was introduced by Sony in 1972 as the U-matic system. Figure 7.6*b* is a layout of the Type-E recorded magnetic tracks. This led to the development of a variety of lower-cost, easy-to-operate systems and opened the door for widespread use of video recording and playback in the consumer marketplace. A large number of ½-in formats with much longer playing time than the 1 h of the U-matic have emerged. Their salient characteristics are listed in the following section.

7.2   VIDEO-TAPE-RECORDING FORMATS: A GLOSSARY

All video recorders have at least one analog recording track running along one edge of the tape. Audio recording on video transports is so similar to audio-only recording that the subject is included in Chap. 10 under audio recording.

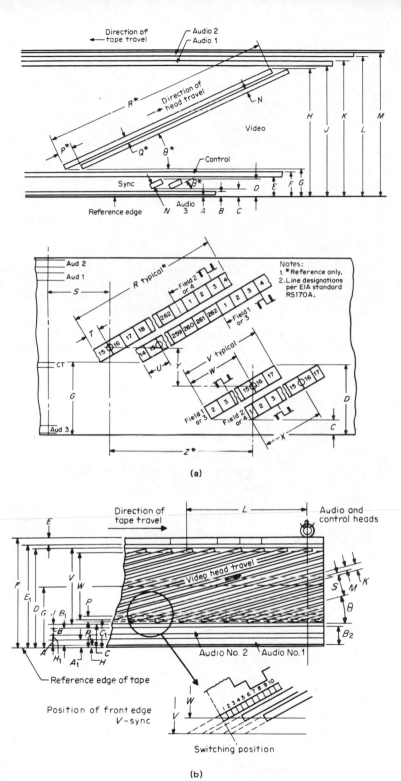

FIGURE 7.6 Layout of recorded magnetic tracks. (a) Type-C 1-in. (*Source: Fink and Christiansen, 1982, in Benson, 1986*) (b) Type-E ¾-in. (*Source: Fink and Christiansen, 1982, in Benson, 1986*)

In all cases below, Oe is the abbreviation for oresteds and i/s is the abbreviation for inches per second. Numbers given in parentheses are for either of the European 625-line systems.

Where the term *cobalt-modified* is used, it describes two possible applications of cobalt [1]. In one process cobalt-doped gamma-ferric oxide creates crystals of a matrix of cobalt and iron oxide, while the other process causes cobalt to be adsorbed onto the surface of the iron-oxide particles. Cobalt-modified tape and chromium-dioxide formulations can be made to provide roughly equivalent results.

7.2.1 Quadruplex

The first successful video recorder, the quadruplex, employs a scanner whose heads move perpendicularly to the tape. The tape is held in a cupped configuration by a vacuum tape guide whose radius of curvature is slightly less than the radius of the scanner. The same video and audio heads are used for recording as well as playback.

Although quadruplex recorders are no longer in production, their description is included here because many are still in use and there is a small, but rapidly decreasing market for used and refurbished machines.

Signal system:	FM analog, composite video
Peak-white frequency:	10 MHz (9.3)
Black-level frequency:	7.9 MHz (7.8)
Chroma carrier:	N/A
Scanner type:	Transverse, 4 heads
Media type:	2 in wide, gamma-ferric oxide, 350 Oe, particles oriented transversely
Tape consumption:	75 ft/min (78)
Head-to-tape speed:	1550 i/s (1600)
Track width:	10 mil
Track pitch:	15 mil
Track length:	1.818 in
Track angle:	89.43°
Azimuth angle:	Perpendicular to plane of scanner
Scanner diameter:	2.064 in
Wrap angle:	115°
Scanner rotation:	240 r/s (250)

7.2.2 SMPTE Type B

This format was introduced first by Fernseh in Germany, but is used in other parts of the world, including the United States. Like the quadruplex format, it is segmented, meaning that it takes more than one head pass to record one TV field. There are 5 or 6 scans per field. Unlike most formats, the longitudinal control-track head is incorporated into the fixed half of the scanner, as opposed to being mounted in a separate head assembly. This scanner is also used in some data recorders manufactured in the United States.

Signal system:	FM analog, composite video
Peak-white frequency:	10.00 MHz (8.90)
Black-level frequency:	7.90 MHz (7.68)
Chroma carrier:	N/A
Scanner type:	Helical, 4 heads
Media type:	1-in-wide, cobalt-modified gamma-ferric oxide, 650 Oe
Tape consumption:	243 mm/s
Head-to-tape speed:	47.4 m/s
Track width:	0.16 mm
Track pitch:	0.20 mm
Track length:	411 mm
Track angle:	14.3°
Azimuth angle:	Perpendicular to plane of head
Scanner rotation:	150 r/s
Scanner diameter:	50.33 mm
Wrap angle:	180°

7.2.3 SMPTE Type C

This format is an outgrowth of the SMPTE Type A, promulgated by Ampex. Type A used 1-in tape and was similar to Type C except that Type A had audio tracks of unequal widths. Initially developed jointly by Ampex and Sony, other manufacturers have produced compatible Type-C recorders. Initial deliveries to Europe used the same carrier frequencies as the quadruplex standard, but the 625-line standards were later altered and are shown below in parentheses.

Signal system:	FM analog composite
Peak-white frequency:	10 MHz (8.90)
Black-level frequency:	7.9 MHz (7.68)
Chroma carrier:	N/A
Scanner type:	Helical, 3 heads
Media type:	1-in-wide, cobalt-modified gamma-ferric oxide, 650 Oe
Tape consumption:	48 ft/min (40)
Head-to-tape speed:	1009 i/s (841)
Track width:	5.1 mil
Track pitch:	7.2 mil (6.0)
Track length:	16.1718 in
Track angle:	2°, 24 min
Azimuth angle:	Perpendicular to the track

Scanner diameter:	5.35 in
Wrap angle:	330°

7.2.4 SMPTE Type-E U-matic

This is the first of the formats to use a cassette and the first to exploit the technique of color-under, in which the chroma component of the signal is translated to a lower frequency and recorded directly, as in anhysteretic, or bias recording, with the luminance FM carrier providing the bias. Introduced in 1972 by Sony, the format is in wide use at this writing not only for video recording but in a modified form for digital audio recording (see Chap. 10).

Signal system:	FM composite-luminance video, color-under chrominance
Peak-white frequency:	5.4 MHz
Sync-tip frequency:	3.8 MHz
Chrominance carrier:	688 kHz
Scanner type:	Helical, 4 heads, 2 tracks per revolution
Media type:	¾-in-wide, 1.1-mil-thick, cobalt-modified gamma-ferric oxide, 650 Oe
Tape consumption:	18.75 ft/min
Head-to-tape speed:	410 i/s
Track width:	3.35 mil
Track pitch:	5.39 mil
Track angle:	4° 57'
Azimuth angle:	±7° from perpendicular
Scanner diameter:	4.34 in
Wrap angle:	180°

7.2.5 SMPTE Type-G Beta

This format, developed for home recording, has been expanded in performance both in the home and in broadcasting (see Sec. 7.2.18, "Betacam SP"). For home recording in the United States, it is losing favor. Two modes are supported. These are identified as Mode I and Mode II. In the first, there is a guard band between tracks; in the second there is no guard band.

Signal system:	FM composite-luminance video, color-under chrominance
Peak-white frequency:	4.8/4.8 MHz (Mode I/Mode II)
Sync-tip frequency:	3.5/3.5 MHz
Chroma carrier:	688 kHz
Scanner type:	Helical, minimum of 2 heads
Media type:	½-in-wide, 0.8-mil-thick, cobalt-modified gamma-ferric oxide, 650 Oe

Tape consumption:	40/20 mm/s
Head-to-tape speed:	6.9 m/s
Track width:	29.2 μm/29.2 μm/19.5 μm
Track pitch:	58.5 μm/29.2 μm/19.5 μm
Track length:	115 mm
Track angle:	5° 10'
Azimuth angle:	± 7° from perpendicular
Scanner diameter:	2.9 in
Wrap angle:	180°

7.2.6   SMPTE Type-H VHS

Developed by Japan Victor Company and Matsushita Electric Co., Video Home System (VHS) has become the predominant home-video-recording format in the United States and is very popular in the rest of the world, primarily due to its greater recording time.

Signal system:	FM composite luminance, color-under chrominance
Peak-white frequency:	4.4 MHz
Sync-tip frequency:	3.4 MHz
Chroma carrier:	629 kHz
Scanner type:	Helical, minimum of 2 heads
Media type:	½-in-wide, 0.8-mil-thick, cobalt-modified gamma-ferric oxide, 650 Oe
Tape consumption:	33.35 mm/s
Head-to-tape speed:	229 i/s
Track width:	2.3 mil
Track pitch:	2.3 mil
Track length:	97 mm
Track angle:	6° 58'
Azimuth angle:	± 6° from perpendicular
Scanner diameter:	62 mm
Wrap angle:	180°

7.2.7   Super VHS (S-VHS)

Signal system:	FM composite luminance, color-under chrominance
Peak-white frequency:	7.0 MHz
Sync-tip frequency:	5.4 MHz
Chroma carrier:	629 kHz
Scanner type:	Helical, minimum of 2 heads
Media type:	½-in-wide, 0.8-mil-thick, cobalt-modified gamma-ferric oxide, 850 Oe

Tape consumption:	11.12 mm/s
Head-to-tape speed:	5.80 m/s
Track width:	19.3 μm
Track pitch:	19.3 μm
Track length:	97 mm
Track angle:	6.00°
Azimuth angle:	±6° from perpendicular
Scanner diameter:	62 mm
Wrap angle:	180°

7.2.8 High-Band Beta-III Mode (Beta-Hi-fi)

Signal system:	FM composite luminance, color-under chrominance
Peak-white frequency:	5.6 MHz
Sync-tip frequency:	3.4 MHz
Chroma carrier:	688 kHz
Scanner type:	Helical, minimum of 2 heads
Media type:	½-in-wide, 0.8-mil-thick, cobalt-modified gamma-ferric oxide, 650 Oe
Head-to-tape speed:	6.9 m/s
Track width:	19.5 μm
Track pitch:	19.5 μm
Track length:	115 mm
Track angle:	5° 10'
Azimuth angle:	±7° from perpendicular
Scanner diameter:	74.5 mm
Wrap angle:	180°

7.2.9 Extended-Definition Beta-Hi-fi (ED-Beta)

This format introduces stereo audio on FM carriers around 1.5 mHz. The performance is better than most direct recording methods. However, the FM audio is recorded at the same time as the video since it is written by the video heads.

Signal system:	FM composite luminance, color-under chrominance
Peak-white frequency:	9.3 MHz
Sync-tip frequency:	6.8 MHz
Chroma carrier:	688 kHz
Scanner type:	Helical, minimum of 2 heads
Media type:	½-in-wide, 0.8-mil-thick cobalt-modified gamma-ferric oxide, 650 Oe
Tape consumption:	13.33 mm/s

Head-to-tape speed:	6.9 m/s
Track width:	19.5 μm
Track pitch:	19.5 μm
Track length:	115 mm
Track angle:	5° 10'
Azimuth angle:	±7° from perpendicular
Scanner diameter:	74.5 mm
Wrap angle:	180°

7.2.10 8-mm

This format, formalized by the IEC Standards Committee, includes stereo FM audio carriers around 1.5 MHz, written by the video heads.

Signal system:	FM composite luminance, color-under chrominance
Peak-white frequency:	5.4 MHz
Sync-tip frequency:	4.2 MHz
Chroma carrier:	743.4 kHz
Scanner type:	Helical, minimum of 2 heads
Media type:	8-mm wide, 13-μm-thick metal particle, 1450 Oe
Tape consumption:	14.34 mm/s
Head-to-tape speed:	3.76 m/s
Track width:	20.5 μm
Track pitch:	20.5 μm
Track length:	62.6 mm
Track angle:	4° 53'
Azimuth angle:	±10° from perpendicular
Scanner diameter:	40.0 mm
Wrap angle:	180°

7.2.11 8-mm High-Band

This format is promulgated by the industry 8-mm Video Conference. It also includes a pair of FM audio carriers at about 1.5 MHz.

Signal system:	FM composite luminance, color-under chrominance
Peak-white frequency:	7.7 MHz
Sync-tip frequency:	5.7 MHz
Chroma carrier:	743.4 kHz
Scanner type:	Helical, minimum of 2 heads
Media type:	8-mm-wide, 10-μm-thick, enhanced metal particle, 1450 Oe
Tape consumption:	14.34 mm/s

Head-to-tape speed:	3.76 m/s
Track width:	20.5 μm
Track pitch:	20.5 μm
Track length:	62.6 mm
Track angle:	4° 53'
Azimuth angle:	± 10° from perpendicular
Scanner diameter:	40.0 mm
Wrap angle:	180°

7.2.12   Improved-Enhanced Definition Television Systems (IDTV/EDTV)

This format, developed by Hitachi Denshi Ltd., is described as follows:

Signal system:	FM composite luminance, color-under chrominance
Peak-white frequency:	5.0 MHz
Sync-tip frequency:	4.0 MHz
Chroma carrier:	629 kHz (includes high-frequency luminance components)
Scanner type:	Helical, minimum of 2 heads
Media type:	½-in-wide, 13.0-μm thick, cobalt-modified gamma-ferric oxide 650 Oe
Tape consumption:	33.35 mm/s
Head-to-tape speed:	5.80 m/s
Track width:	58.0 μm
Track pitch:	58.0 μm
Track length:	97.0 mm
Track angle:	5° 58'
Azimuth angle:	± 10° from perpendicular
Scanner diameter:	62.0 mm
Wrap angle:	180°

7.2.13   Multiple Sub-Nyquist Sampling Encoding (MUSE)

The MUSE format, developed by Nippon Hoso Kyokai (NHK), is intended for the recording of *direct broadcasts from satellite* (*DBS*) high-definition (1125/60) transmissions to the home. Audio is recorded longitudinally.

Signal system:	FM composite luminance, time-compressed chrominance
Peak-white frequency:	18.5 MHz
Sync-tip frequency:	12.5 MHz
Chroma carrier:	N/A

Scanner type:	Helical, 4 scanning segments per field
Media type:	½-in-wide, 10-μm-thick, metal-particle or evaporated metal, 1450 Oe
Tape consumption:	67 mm/s
Head-to-tape speed:	23.2 m/s
Track width:	29.0 μm
Track pitch:	29.0 μm
Track length:	97.0 mm
Track angle:	6°
Azimuth angle:	±6° from perpendicular
Scanner diameter:	62.0 mm

7.2.14 Adaptive Dynamic-Range Digital Encoding

This is a digital recording format proposed by Sony. It is one of the first to embed signals, written by the video heads, that are intended to be read by a longitudinal head. Typical of many modern recording formats, the track *width* is wider than the track *pitch*. This form of recording is possible only by using opposing azimuth angle offsets on alternate tracks. The data is scrambled before encoding, and a Reed-Solomon error detection and correction code is appended during recording.

Signal system:	Rate-reduced digital, composite
Scanner type:	Helical, segmented with 2 tracks per field
Media type:	8-mm, 10-μm thick, evaporated metal, 1150 Oe
Tape consumption:	21.5 mm/s
Head-to-tape speed:	7.5 m/s
Track width:	18.5 μm
Track pitch:	15.0 μm
Track length:	62.6 mm
Track angle:	4° 53'
Azimuth angle:	10° from perpendicular
Scanner diameter:	40.0 mm

7.2.15 Discrete Cosine-Transform Data Reduced

This format is proposed by N. V. Phillips Co. of the Netherlands. It is intended for digital home recording.

Signal system:	Digital, data-rate reduced by coding the composite video signal
Scanner type:	Helical, 2 channels, 3 tracks per field
Media type:	8-mm, 10-μm thick, metal-particle, 1450 Oe
Tape consumption:	26.0 mm/s
Head-to-tape speed:	9.4 m/s

Track width:	(2) 10.0 μm
Track pitch:	20 μm
Track length:	62.6 mm
Track angle:	4° 53'
Azimuth angle:	±10° from perpendicular
Scanner diameter:	40.0 mm

7.2.16   8-mm Hadamard Transform-Interfield Sub-Nyquist Sampling-Vector Quantization Data Reduced

This format, proposed by Matsushita, uses two-dimensional Reed-Solomon error detection and correction as do the two professional formats, but it is intended for home recording.

Signal system:	Digital, data reduced, composite video
Scanner type:	Helical, 2 channels, 1 track per field
Media type:	8-mm, 10-μm-thick, evaporated metal, 1150 Oe
Tape consumption:	14.34 mm/s
Head-to-tape speed:	3.75 m/s
Track width:	20.5 μm
Track pitch:	20.5 μm
Track length:	62.6 mm
Track angle:	4° 53'
Azimuth angle:	±10° from perpendicular
Scanner diameter:	40.0 mm

7.2.17   Half-Inch Hadamard Transform-Interfield Sub-Nyquist Sampling-Vector

This format is a half-inch version of the 8-mm format in Sec. 7.2.16 proposed by Matsushita. The modulation code is an improved 8–10 code coupled with a two-dimensional Reed-Solomon error detection and correction scheme.

Signal system:	Digital, data reduced, composite video
Scanner type:	Helical, 2 channels, 2 tracks per field
Media type:	12.7-mm, 10-μm-thick, evaporated metal, 1150 Oe
Tape consumption:	110 mm/s
Head-to-tape speed:	5.8 m/s
Track width:	9.7 μm
Track pitch:	9.7 μm
Track length:	97.0 mm
Track angle:	6°

Azimuth angle: ±6° from perpendicular

Scanner diameter: 62.0 mm

7.2.18   Betacam Superior Performance (SP)

This format was developed for professional applications. It does not record a composite video signal but instead records the luminance and two-color components thereof. Note also that preemphasis, in the frequency sense, is supplemented with nonlinear emphasis in the amplitude sense. This calls for careful control of video-signal amplitudes at the point where the nonlinear treatment is applied. The luminance channel is recorded on one channel while the chrominance components and two stereo audio channels are recorded on the other. The FM audio channels are centered on 310 kHz and 540 kHz.

Signal system: FM, component video

Luminance peak white: 7.7 MHz

Luminance sync tip: 5.7 MHz

Scanner type: Helical, 1 track per field

Media type: 12.7-mm-wide, 14-μm-thick, metal particle, 1450 Oe

Tape consumption: 118.52 mm/s

Head-to-tape speed: 7.07 m/s

Track width: 42.0 μm

Track pitch: 84.5 μm

Track length: 115.0 mm

Track angle: 4.67°

Azimuth angle: ±15° from perpendicular

Scanner diameter: 74.49 mm

7.2.19   M-II

This format, developed by Matsushita, is somewhat similar to the Sony SP described above. It, too, uses amplitude nonlinear compression-expansion as well as amplitude-frequency preemphasis. The format allocates space for *pulse-code modulation* (PCM) for digital audio recording. As in the Beta-SP format above, luminance is recorded on one track, and chrominance and two audio subcarriers are recorded on the other. The two audio FM subcarriers are centered on 400 and 700 kHz.

Signal system: FM, component video

Luminance peak white: 7.7 MHz

Luminance sync tip: 6.2 MHz

Scanner type: Helical, 1 track per field

Media type: 12.7-mm-wide, 13-μm thick, metal particle, 1450 Oe

Tape consumption: 67.69 mm/s

Head-to-tape speed:	7.09 m/s
Track width:	42.0 μm
Track pitch:	84.5 μm
Track length:	118.25 mm
Track angle:	4.25°
Azimuth angle:	±15° from perpendicular
Scanner diameter:	76.00 mm

7.2.20 8-mm Broadcast Electronic News Gathering/Electronic Field Production (ENG/EFP)

This format was developed by Hitachi Denshi of Tokyo. From the information available at this writing, it appears to be an 8-mm version of the two formats above.

Signal system:	FM, component video
Luminance peak white:	7.2 MHz
Luminance sync tip:	4.63 MHz
Chrominance:	6.3 MHz, maximum
Scanner type:	Helical, 1 track per field
Media type:	8-mm-wide, 13-μm-thick, metal particle, 1450 Oe
Tape consumption:	120.68 mm/s
Head-to-tape speed:	7.00 m/s
Track width:	36.0 μm
Track pitch:	78.0 μm (both tracks)
Track length:	62.6 mm
Track angle:	2° 12'
Azimuth angle:	Perpendicular to plane of heads
Scanner diameter:	76.00 mm

7.2.21 SMPTE/EBU D1

This is a digital component format with a sampling rate that can be derived from the horizontal-scan rate of either NTSC color or Pal or Secam. The luminance component is sampled at 13.5 MHz, and the two color-difference signals are sampled at 6.75 MHz each for a total of 27 megasamples of 8 bits each per second. This format is particularly useful when a recorded signal is to be used later for chroma-keying. The same sample rate is used for 525- and 625-line systems. The input data is scrambled to optimize error concealment, and a two-dimensional Reed-Solomon error detection and correction code is appended. The data is written on two tracks per head pass. Up to 20 bits per sample of 4 digital audio channels are written by the video heads. (See Chap. 10 for details.)

Signal system:	Digital, component video
Scanner type:	Helical, 6 tracks per field

Media type:	19-mm-thick, 13- or 16-μm-thick, cobalt-adsorbed gamma-ferric oxide, 850 Oe
Tape consumption:	286.58 mm/s
Head-to-tape speed:	35.3 m/s
Track width:	35.0 μm
Track pitch:	45.0 μm
Track length:	62.6 mm
Track angle:	5.4°
Azimuth angle:	Perpendicular to plane of heads
Scanner diameter:	75- or 96-mm diameter

7.2.22 SMPTE D2

This format was introduced by Ampex and Sony. It is not formalized at this writing, but equipment made to its format has been manufactured and sold, and format details have been furnished to SMPTE and EBU. This format uses a different tape than D1, above, but uses the same cassette, so beware of tape type mixups. Like the D1 format, the video heads write four channels of digital audio with up to 20 bits per sample. The location of the audio records on the tape is different, however. See Chap. 10 for details.

Signal system:	Digital, composite, 4 × Fsc
Scanner type:	Helical, 6 tracks per field (8 tracks for Pal or Secam)
Media type:	19-mm-wide, 13-μm thick, metal particle, 1450 Oe
Tape consumption:	131.8 mm/s
Head-to-tape speed:	27.3 m/s (30.44)
Track width:	39.1 (35.2) μm
Track pitch:	39.1 (35.2) μm
Track length:	150.78 mm (150.71)
Track angle:	6.12° (6.13)
Azimuth angle:	±15° from perpendicular
Number of channels:	2
Tracks per field:	6 (8)
Error-correcting code:	Two-dimensional Reed-Solomon code of 64 + 4 and 85 + 8

7.2.23 Three-Dimensional DPCM Data-Reduced 8-mm

This format, in development, is intended for ENG and medical X-ray recording.

Signal system:	Digital, composite, reduced to 30-Mb per channel (with overhead)
Scanner type:	Helical, 2 tracks per field
Media type:	8-mm-wide, 10-μm-thick, metal particle, 1150 Oe

Tape consumption:	57.4 mm/s
Head-to-tape speed:	7.5 m/s
Track width:	20.5 μm
Track pitch:	20.5 μm
Track length:	62.6 mm
Track angle:	4° 53'
Azimuth angle:	± 10° from perpendicular
Scanner diameter:	40 mm
Number of channels:	2
Tracks per field:	2
Modulation method:	Miller squared
Error correction code:	Reed–Reed-Solomon

7.2.24   NHK/SMPTE Type J, High Definition

This format is proposed for recording high definition (1125/60).

Signal system:	FM, analog, component
Luminance peak white:	20.23 MHz
Luminance sync tip:	14.28 MHz
Scanner type:	Helical, 3 tracks of Y, R-Y, B-Y, or R, G, B during one turn of the scanner
Media type:	25.4-mm-wide, 30-μm-thick, gamma-ferric oxide, 650 Oe
Tape consumption:	483.12 mm/s
Head-to-tape speed:	25.9 m/s
Track width:	280.0 μm
Track pitch:	357.0 μm
Track length:	414.81 mm
Track angle:	2° 32'
Azimuth angle:	Perpendicular to plane of scanner
Scanner diameter:	134.6 mm

7.2.25   High-Definition Component

This format is proposed by Mitsubishi for recording high-definition signals on half-inch tape. The chrominance components are time-compressed.

Signal system:	FM, analog, component
Luminance peak white:	19.5 MHz
Luminance sync tip:	12.5 MHz
Scanner type:	Helical, 2 tracks, 3 passes per field

Media type:	12.74-mm-wide, 10-μm-thick, evaporated metal, 1450 Oe
Tape consumption:	100 mm/s
Head-to-tape speed:	17.4 m/s
Track width:	29.0 μm
Track pitch:	29.0 μm
Track length:	97.0 mm
Track angle:	6.00°
Azimuth angle:	6.00° from perpendicular

7.2.26 High-Definition Digital

This format is intended for studio production work in a 1125/60 environment. Using six tracks per field, this machine lays down 648 Mb/s/s.

Scanner diameter:	134.63 mm
Signal system:	Digital, component
Encoding method:	Run length limited, 8-to-8-bit mapping
Error-correction code:	Two-dimensional Reed-Solomon
Scanner type:	Helical, 6 tracks per field
Media type:	25.4-mm-wide, 19-μm-thick, cobalt-adsorbed gamma-ferric oxide, 850 Oe
Tape consumption:	488.0 mm/s
Head-to-tape speed:	52.0 m/s
Track width:	27.5 μm
Track pitch:	40.0 μm
Track length:	410.0 mm
Track angle:	2° 34'
Azimuth angle:	Perpendicular to plane of scanner
Scanner diameter:	134.63 mm

7.2.27 High-Definition Digital, Sony

This format, proposed by Sony, is intended for production in a 1125/60 environment. Using 8 tracks per field, this machine lays down 1.188 Gb/s.

Scanner diameter:	134.63 mm
Signal system:	Digital, component
Encoding method:	Run length limited, 8-to-8-bit mapping
Error-correction code:	Two-dimensional Reed-Solomon
Scanner type:	Helical, 8 tracks per field
Media type:	25.4-mm-wide, 13-μm-thick, metal particle, 1450 Oe

Tape consumption:	805.0 mm/s
Head-to-tape speed:	52.5 m/s
Track width:	34.0 μm
Track pitch:	37.0 μm
Track length:	410.0 mm
Track angle:	2° 34'
Azimuth angle:	Perpendicular to plane of scanner
Scanner diameter:	134.63 mm

7.2.28 M-III

This format, proposed by Matsushita, is intended for digital recording in the field.

Scanner diameter:	76 mm
Signal system:	Digital, component
Encoding method:	Run length limited, 8-to-10-bit mapping
Error-correction code:	Two-dimensional Reed-Solomon
Scanner type:	Helical, 6 tracks per field
Media type:	12.7-mm-wide, 10-μm-thick, metal particle, 1450 Oe
Tape consumption:	96.7 mm/s
Head-to-tape speed:	21.0 m/s
Track width:	20.0-μm
Track pitch:	20.0-μm
Track length:	389 mm
Track angle:	4.25°
Azimuth angle:	±15° from perpendicular ·
Scanner diameter:	134.63 mm

7.2.29 Data Cassette Recording System (DCRS)

This data recording format, promulgated by Ampex, is a distant outgrowth of the quadruplex transverse TV recorder, the first to gain popularity. Unlike quadruplex, however, there is no vacuum guide. The tape is in a cassette.

Tape:	1 in, 0.57-mil base thickness, 650 Oe
Track width:	1.35 mil
Track pitch:	1.50 mil
Tape usage:	5.5 i/s
Write speed:	2500 i/s
Scanner diameter:	1.574 in
Track angle:	89.91°
Tracks per revolution:	6

Input data rate: 200 Mb/s maximum, configurable by the user

7.2.30 ANSI ID1/MIL Standard 2179

This data-recording format is a direct outgrowth of the SMPTE/EBU D1 video-recording format. As much of the D1 format was retained as possible, including the cassette, but customer requirements caused large changes to be made in the signal-processing path. Data recordists are faced with a need to record slowly and replay rapidly (as in down-loading from an orbiting vehicle) or to do the opposite when the data processing equipment is limited in speed. TV recordists, on the other hand, have a highly standardized signal to record in which is embedded a wealth of timing information useful during playback.

Tape:	0.75-in, 0.37-mil base thickness, cobalt-modified gamma-ferric oxide, 850 Oe
Track width:	1.57 mil
Track pitch:	1.77 mil
Tape usage:	Varies with choice of speed
Write speed:	1192 i/s at 200 Mb/s, varies with bandwidth
Scanner diameter:	2.95 in
Track angle:	5.4005°
Tracks per revolution:	2, 4, 6, or 8
Input data rate:	100, 200, 300, or 400 MHz, depending on heads

7.3 MECHANICAL ADJUSTMENTS

7.3.1 Tape Tension

Modern machines use tape that is much thinner than tape formulations of 35 years ago. Thin tape is more difficult to edge-guide because its column stiffness is less as it wraps around a guide post. The force required to steer the tape around a post also depends on how tightly it is wrapped around it, which is governed by tape tension, especially in the case of rotating guides. Edge damage is the usual cause of death of tapes, especially cassette tapes. Measure tape tension with an instrument called a *Tentelometer*. It is capable of measuring tape tension with tape in motion or stationary. {See Tentel Corp. [2].}

7.3.2 Motor Torque

Many recorders, especially small helical-scan units, have elements that generate or cause (as in brakes) *torque*, as opposed to *tension*. {See Tentel Corp. [2].}

7.3.3 Tape Guides

Some helical-scan machines have adjustable entrance and exit tape guides. Their accurate adjustment is crucial in obtaining tape interchangeability. A reference tape is required and can usually be obtained from the machine's manufacturer. The object of the adjustment is to obtain a signal recovery from the heads that is uniform in amplitude. It is especially important that accurate adjustments be made on machines that are used to edit tapes made elsewhere or to record tapes that will be edited elsewhere.

7.3.4 Head-Tip Projection

While tip projection is not an *adjustment*, it is a useful *measurement* that leads to an estimation of the remaining life of the head tip. Tentel Corp. [2] is a source of appropriate instrumentation.

7.3.5 Spindle Height

This distance is critical in cassette VTRs since the tape pack is "outside" the interior tape path elements. If the platform that supports the cassette is not just right, edge damage and binding cassettes result. Tentel Corp. [2] is a source of instrumentation to measure this critical dimension.

7.3.6 Control-Track Head Position

Many helical-scan recorders have the ability to adjust the position of the control-track head. This requires a standard reference tape, usually obtained from the machine's manufacturer. This adjustment would normally need to be done only if the control-track head were replaced or if the transport were damaged. It is especially important to standardize the control-track gap position when recorders will be editing tapes made elsewhere or when recording tapes that will be edited elsewhere.

7.4 ELECTRICAL ADJUSTMENTS

7.4.1 Analog Carrier Frequencies

On machines that will edit other tapes or generate tapes to be edited elsewhere, it is important that the FM carrier frequencies representing peak white, black level, and tip of sync be maintained. Some recorders can automatically establish a standard frequency at black level, while others can automatically establish a standard frequency at tip of sync. In installations that contain a mixture of machines, it is important not only to maintain all carrier frequencies but to maintain standard video-to-sync ratios. These ratios are for NTSC 0.714 video/0.286 sync, and for Pal or Secam, 0.7 video/0.3 sync. The objective of all this is to prevent, at the point of an edit, a sudden jump of carrier frequency at either black level *or* tip of sync.

7.4.2 Analog Equalization

This is a playback adjustment. It is more critical in the case of quadruplex and SMPTE Type-B formats since it must be very precise to avoid "banding," or visible artifacts as a function of vertical scanning in the output picture.

7.4.3 Record Current

Record current in both analog and digital formats is usually adjusted to maximize the recovered signal at some particular wavelength. In the case of analog formats, record current affects frequency response, while in the case of digital formats, record current affects only the digital error rate.

7.4.4 Reproduce Equalization

In an analog situation, reproduce equalization is adjusted to provide flat reproduction of a standardized recording. In a digital situation, reproduce equalization is adjusted for minimum digital error rate. Some recorders automate the adjustment of digital reproduce equalization. At this writing, none automate the adjustment of record current.

7.4.5 Record-Reproduce Levels

Maintenance of input-output levels is, like motherhood, GOOD. It is especially good in the formats that use (and require) amplitude preemphasis.

References

1. Robert H. Perry. "Magnetic Tape," *Video Engineering Handbook*. McGraw-Hill, New York, 1986, Sec. 15.2, p. 15.52.
2. Tentel Corp., 1506 Dell Avenue, Campbell, CA 95008.
3. "A Complete Guide to Media and Formats." Ampex Magnetic Tape Division, April 1988.

7.5 MOTION-PICTURE FILM EQUIPMENT

7.5.1 The History of Film in Television

Film has been a major source of program material for television broadcasting since its beginning in the 1930s. With the advent of commercial broadcasting, by means of kinescope recording, it was the only medium for the storage and subsequent delayed playback of live television studio productions. Unfortunately, the image quality of "taking a picture of a picture" on a cathode ray tube left much to be desired. The shortcomings were a nonuniform gray scale causing loss of detail in shadow areas, excessive noise from film grain, and a fixed grain pattern from the CRT phosphor. With the introduction of color, these problems became more acute and were compounded by the additional requirement of stable and faithful hue and saturation reproduction.

These problems were solved for programming originating on television cameras by the invention and implementation of videotape recording in 1956. However, since over half of all television programming originates from motion-picture film, significant improvements in the transfer of film to television were necessary to provide a picture quality comparable to the competing "live-on-tape" programming.

7.5.2 Frame-Rate Conversion

Unsuccessful attempts were made to use flying-spot and CCD scanners, which had become the standard in Europe for 625-line systems. However, the problem in maintaining image stability in "chasing the raster" to convert from film's 24 frames/s to 525-line television's 30 frames/s was insurmountable, since the raster must be moved in a repeating sequence to five different positions. Alternatively, on 625-line/50-Hz systems film is run at 25 frames/s with the raster sequenced between two positions and the slightly faster transmission rate considered not objectionable. It was not until many years later when the digital frame store was available, which permitted continuous-motion film in a scanner and a progressive rather than "jump" scan, that flying-spot and CCD scanners became practical on the 525-line system. (See Sec. 3.2.6.)

2–3 Pulldown.   The 6-Hz difference of 24 to 30 in frame rate between film and television was solved by the storage capability of the vidicon and the use of a film projector modified for a pulldown time that was shorter than normal compared to a conventional theater projector. This increased the time the film is in the gate and permitted the use of a shutter with a longer light-application time.

In addition, the pulldown timing is modified to expose 4 film frames every 5 TV frames. As shown in Fig. 7.7a, film frames 1 and 3 are timed to expose *two* TV film frames; 2 and 4 are timed to expose *three* TV frames—thus, a 2–3 pulldown. A standard projector time cycle is shown in Fig. 7.7b for reference.

The illustration is drawn with the light application for four frames and television scanning cycle of five frames in synchronism. In practice, it has been found that nonsynchronous operation of the film motion and television scanning is practical if the exposure time is at least 40 percent of the television scan. Otherwise a flicker effect between the two frame rates will occur.

7.5.3 Pulldown Mechanisms

In order to achieve the greatest film-positioning accuracy in the taking camera, since the film is running at a uniform rate, a claw pulldown is utilized to advance the film frame to frame. In addition, a registration pin is inserted into a sprocket perforation for each frame to ensure very high frame-to-frame consistency.

Claw.   One widely used technique in reproducing intermittent projectors, to advance the film, is to employ a two- or three-tooth claw which is periodically pushed horizontally toward the film so that the claw teeth engage sprocket holes and then the claw is moved vertically to advance the film in the gate. Both motions are accomplished by cam-driving the mechanisms against spring

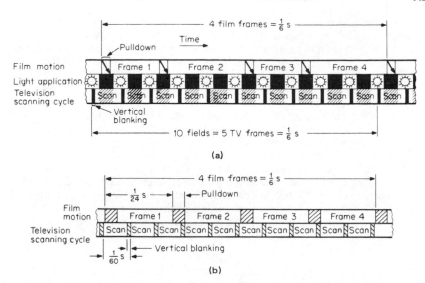

FIGURE 7.7 Projector time cycles. (a) Long-application television projector. (*Source: Benson, 1986*) (b) Standard projector. (*Source: Benson, 1986*)

loading. Since the film motion entails a 3–2 relationship, the cams have two lobes and cause two frame advance cycles per revolution of the driving shaft. Although two and sometimes three teeth are present on the claw assembly, in general only one tooth surface engages a sprocket hole edge for driving the film. In the event of a damaged sprocket hole one of the other teeth will usually have a good sprocket hole to engage and thus provide insurance that the film will continue to advance through the projector. The two-lobe cam which drives the claw for advancing the film must be very precisely made so that the stopping point of the claw is precisely the same for each lobe. This is to ensure that there will be no spurious variation in the vertical positioning of successive film frames in the gate.

Mechanically Controlled Sprocket. A second approach to moving the film intermittently at the gate of a projector is to use a sprocket which is rotated by an angle suitable to advance the film one frame pitch per cycle of film advance. Such a drive is shown in Fig. 7.8. The most commonly used sprocket for 16-mm film contains four teeth, while that most commonly used for 35-mm film contains 16 teeth. The sprockets are usually rotated through the proper incremental angles by means of a Geneva drive, which is shown in Fig. 7.8. Sprocket pulldown systems are more difficult to design and fabricate than is the claw pulldown system because, in addition to requiring very precise cycle-to-cycle accuracy of the

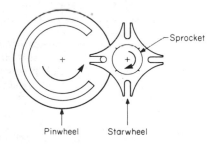

FIGURE 7.8 Geneva sprocket drive. (*Source: Benson, 1986*)

Geneva drive, the tooth-to-tooth spacing on the sprocket becomes critical with respect to vertical registration of the frames in the projector gate. However, because of the greater film mass and in an effort to minimize stress on the driving contact points of 35-mm film, the sprocket drive is universally used for 35-mm projectors.

Servocontrolled Sprocket. In the mid-1970s a sprocket pulldown utilizing a servocontrolled stepping motor for rotating the sprocket was introduced. This approach permits a fine degree of control in the stopping point of each succeeding tooth on the sprocket. This approach also lends itself readily to running the projector at widely different speeds.

The servo stepping-motor drive of sprocket pulldowns has resulted in the capability of operating the associated projector over a wide range of speeds. One projector, the RCA FR-35, can be adjusted at an incremental film rate from 0 to 48 frames/s. The servocontrolled projectors operate independently from the power line and thus can be locked to television vertical sync if so desired.

7.5.4 Telecine Cameras

The signals from a telecine camera's three photoconductive pickup tubes must be passed through several processing circuits before they are ready for encoding into signals that are suitable for transmission. Figure 7.9 is typical of the processing chains found in photoconductive telecine cameras. There are three such cascades of circuits, substantially identical; for simplicity, Fig. 7.9a shows only one of those three.

The signal from the photoconductive tube passes first into a *preamplifier*, which has the task of either providing gain for the low-level signals received from the pickup tube or altering the input time constant so that the tube and circuit capacity cannot substantially affect the circuit bandwidth. If the latter task is the designer's choice, the preamplifier is designed to have a very low input impedance and to operate as a transresistance amplifier (see Fig. 7.9b).

Clamping and Black Level. The preamplifier output usually passes through a combined *clamping* and *black-level correction* circuit, which allows for the compensation of undesired backgrounds arising from dark current or bias light. One popular means for such correction consists of adding to the signal a blanking pulse of variable amplitude and polarity, thereby allowing the dc component of the active scan to be varied slightly to achieve this compensation. Alternately, the reference voltage for the clamp may be made variable, thus achieving the desired black-level correction.

Once black level has been set and clamping accomplished, the signal is passed to a *flare-correction* circuit. In this circuit, the signal is integrated (by means of a low-pass filter having a cutoff below frame rate) to determine the overall illumination of the scene. The resulting signal is fed back to the black-level control to compensate for the stray illumination arising from light reflections in the optics path and from the tube surfaces.

Shading. Shading is accomplished next in the typical circuit cascade. Shading correction is of two types. The first, called *axis shading*, adds to the signal a combination of sawteeth and parabolas, at both horizontal and vertical rates. The sec-

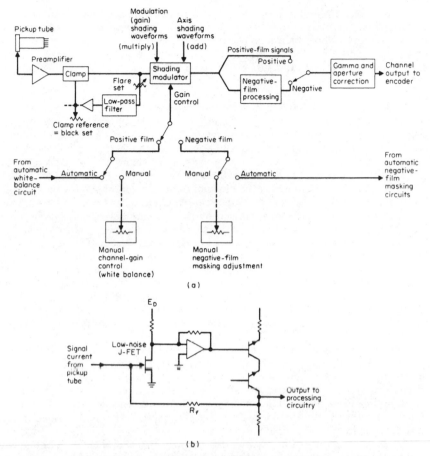

FIGURE 7.9 Photoconductive telecine camera system. (*a*) Block diagram of one color-signal channel. (*Source: Benson, 1986*) (*b*) Typical preamplifier circuit. (*Source: Benson, 1986*)

ond type, called *gain shading*, multiplies the signal by a different combination of horizontal and vertical sawteeth and parabolas. If both types of shading are provided, the operator is usually offered four control knobs for each type, so that independent adjustments of both axis shading and gain shading may be made. Axis shading permits the operator to correct for uneven dc conditions across the picture plane, such as might be caused by dark currents. Gain shading, on the other hand, permits correction of uneven illumination from the projector or uneven sensitivity in the photoconductive pickup tube.

Gain Control. Since the multiplicative (gain) shading in a modern telecine camera is usually done through four-quadrant multipliers, it is economical to use this same multiplier as a dc-controlled, remotable *gain control*. As Fig. 7.9a shows, the lead which controls the multiplier gain is switched, with the switch position depending upon whether the telecine chain is handling positive film or negative film. For positive film, the gain control goes to a (remote) potentiometer labeled either *channel gain* or *white balance*. Since Fig. 7.9 depicts only one of the three channels, there will be three such (remote) gain controls—one each for red, blue, and green—and the operator may use these three to adjust the three channel gains to achieve equal outputs from each channel for a white signal, thus achieving white balance.

White Balance.   Most modern cameras will also include a servomechanism for the automatic adjustment of white balance. A switch allows the (remote) gain control to be replaced by a servogenerated direct current which can automatically adjust channel gain to achieve white balance. In most telecine cameras, this automatic action is achieved by holding the green channel gain to a fixed value and allowing the servo to vary red and blue channel gains to achieve white balance. The circuitry that detects the channel behavior and generates the red and blue control signals is called *auto-white balance* circuitry and is described below.

When negative film is being handled by the telecine chain, the switch in the multiplier gain control (see Fig. 7.9a) is switched to the negative-film position, and the channel gain is then controlled by a potentiometer labeled *negative-film masking*. The gain controls of the three channels can then be set to different values, to compensate for the orange or yellow mask that is added to negative film to minimize crosstalk between film dyes in the printing process. Although this masking coating is of no use in television-signal processing, being present chiefly to improve positive prints made from the negative film, it is still present in all standard negative film and must be compensated for by circuitry in the telecine chain.

Initial compensation for the orange mask is usually done optically, by the addition in the optical path of cyan filters, such as one or two Kodak CC40s. Such optical filters reduce red-channel output, making it more nearly equivalent to green. The balance of the compensation is then accomplished by the individual negative-film masking potentiometers just discussed, and sometimes by additional channel gain controls in the negative-film path, which is described in the next paragraph.

By referring again to Fig. 7.9a, it can be seen that the shaded and compensated signal is then passed to a pair of paths—one for *negative-film* signals and one for *positive-film* signals. A switch at the output of the alternate paths allows the operator to choose the appropriate path for the film in use at the time. The positive path is a direct connection; the positive-film signal from the preceding circuitry is ready for use as it stands. The negative-film signal, however, must be processed further before it can be encoded for transmission.

Cross-Channel Masking.   The above discussion covers only one channel of a telecine camera and extrapolates to the entire camera by suggesting that the other two channels are identical, except, perhaps, for the green-only aperture compensation. There is another aspect of the camera that may be shown only as an interaction among the three channels.

Consider what would happen if a film camera photographs a color-bar pattern from a kinescope. Assume that the color-bar generator had been perfectly set up and that the picture-tube color temperature is specially set to match the color-temperature rating of the film. The resulting film, if placed in a telecine camera, would *not* produce proper color bars at the camera output, because the picture-tube phosphors, the film dyes, and the telecine camera taking filters do not have the same spectral shapes. Consequently, the color-bar signal from the telecine camera would evidence errors in both hue and saturation.

To correct these errors, it is necessary to cause deliberate crosstalk among the three primary-color channels. Carefully calculated amounts of green signal and red signal, for example, are caused to contaminate the blue channel; similarly, green is contaminated by red and blue, and red, by blue and green. The contaminating colors may be either negative or positive, as required. This overall arrangement is usually called *linear matrixing*, but the name *linear masking* or simply *masking* is also seen in the literature. The last two names can be confused with the masking process described above.

Since dye sensitivity is not a linear function, some telecine manufacturers have chosen to attempt compensation by performing *nonlinear matrixing*. In such an arrangement, the crosstalk signals are passed through logarithmic amplifiers before they are added to the neighboring channels.

Other manufacturers chose to perform corrections of this nature by circuits that operate in the domain of the color difference signals—I and Q or $R - Y$ and $B - Y$.

7.5.5 Flying-Spot Scanners

In addition to continuous-motion film transports as described in Sec. 3.2.6, flying-spot scanners utilize extensive electronics to scan film horizontally, transduce the optical image to electric signals, process the component color signals, and produce suitably encoded color-video output. Digital video-signal storage makes possible great flexibility in scanning and improved quality in the generated video signal.

Block Diagram.   A functional block diagram of the video-signal channels in a flying-spot scanner is shown in Fig. 7.10. Only the green channel is shown in its entirety since the red and blue channels are identical and confining the diagram to the green channel only makes it easier to follow. The video signals are nonstandard analog signals up to the digiscan block. Within this block the analog signals are converted to digital for further processing and are delivered at its output as reconverted analog signals at the standard scanning rates for encoding in the encoder. Since basic video amplification and color encoding are functions that are common to essentially all telecine cameras, they will not be discussed further here.

Light-Spot Source.   The light spot for flying-spot scanning is generated by a special cathode ray tube which utilizes a very fine grain phosphor and a high-resolution scanning beam and which operates at high voltage to generate a very high light output. It should be noted that when operating at anode voltages of 20 to 30 kV, the cathode ray tube is very likely to emit X-rays that exceed the recommended maximum permissible level of 0.5 mR/h as measured at a distance 50 m from any part of its external surface. Suitable shielding and operating precautions should be taken when using this equipment. Since the light is generated by electrons exciting a phosphor, there is a continuation of light output after the electrons have stopped exciting a particular area of the phosphor. This persistence or afterglow requires electronic compensation at a suitable point in the signal channel.

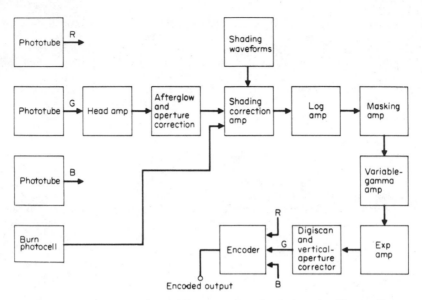

FIGURE 7.10 Block diagram of flying-spot video-signal channels. (*Source: Benson, 1986*)

7.5.6 CCD Line Scanners

The combination of two developments, the line-scan CCD photosensor and the digital video and frame field store, has led to the development of line-scan CCD telecines. This telecine is based on capstan-drive continuous-motion film transports. The vertical component of the film scanning is generated entirely by motion of the film. The sequentially scanned signals are subsequently stored in digital video stores and then read out in line-interlaced form. In the case of the 525/60 television system the 24-frame film rate is converted to the 60-frame television rate by means of this same digital video store system. Two commercially available CCD line-scan telecines, the Bosch and the Marconi, are typical of CCD telecine scanners.

Block Diagram.   The functional block diagram shown in Fig. 7.11 is typical of the video-signal system. The *RGB* signals, following generation by the line-scan CCDs, are amplified and processed in analog form and subsequently A/D converted for storage in the digital frame stores. As generated, these signals are line sequential *RGB* signals. In the readout process from the digital stores the signals are converted to line interlaced fields. After suitable matrixing to recover *RGB* in addition to the *Y* signals, they are encoded to the standard NTSC output signal. In the Marconi line-scan telecine, the *RGB* signals, as generated, are converted to the digital form early in the video channel with gamma correction and masking signal processing being carried out in the digital domain.

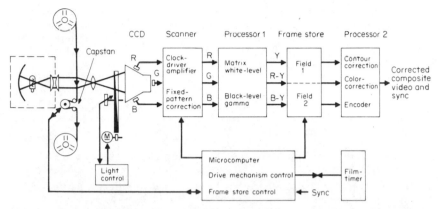

FIGURE 7.11   Functional block diagram pf CCD scanner video systems *(Source: Robert Bosch CmbH, Fernseh Group, in Benson, 1986)*

7.6 ELECTRONIC EDITING

7.6.1 Time-Code Editing

The need for fast and accurate conjunction of video signals from two or more sources is met by the recording of a digitally coded signal on each auxiliary audio channel. Promulgated by the Society of Motion Picture and Television Engineers, it is called the *Time and Control Code*. Each frame of a recording is uniquely defined by an 80-bit coded signal. To identify a frame, 32 bits are used, expressed in hours, minutes, seconds, and frames. 32 bits are reserved for undefined purposes, and a 16-bit sequence defines the frame boundary and includes a bit sequence from which the direction of tape movement can be discerned. The channel coding is *biphase mark* in which each bit cell is delineated by a flux reversal, and a binary 1 is defined by an extra midcell transition. Since the frame rate of NTSC color television is $30 \times 1/1.001$ Hz, the indicated time falls behind real clock time at the rate of several seconds per hour. To compensate for this, two frame counts are skipped over each minute except every tenth minute. This adjustment, which is optional, is called *drop frame*. It is not necessary in 25-frame systems, whose frame rate is exact.

Vertical-Interval Time Code Editing.   The time-code signal is recoverable over a wide range of tape speeds, about one-fifth normal speed to 50 times normal. Even so, helical-scan recorders offer the capability of reproducing a single field while the tape is stationary, obviating recovery of the time-code longitudinal recording. To provide frame identification in this case, vertical-interval time code (VITC) may be injected into the video-signal path during recording. One iteration of the code occupies one horizontal interval and is typically recorded on two nonadjacent intervals during vertical blanking.

On most helical scan recorders, VITC may be recorded using either the dedicated sync head, which records only the vertical sync interval, or the main video head, which records all other video, including much of the vertical blanking interval. If recorded on the sync channel, the advantage is obtained that time code may be recorded independently of the video, perhaps at a later time than the video. A disadvantage is that the video head, which is often able to move to adjacent tracks, may not be scanning the track identified by the sync channel. If recorded on the video channel, there can be no doubt that it accurately identifies the frame, but it must be applied at the time of video recording, with no possibility of independently altering it later.

In both time codes, numbers are expressed in binary-coded decimal (BCD), using 4 bits per digit. Since the tens of frames and tens of hours digits never exceed 2, and the tens of seconds and tens of minutes digits never exceed 5, 6 bits of the 32 time-code bits are not needed to express time. Several of these are used to convey other information:

1. In NTSC, whether drop-frame time correction is being used in the counting sequence.

2. Whether the counting sequence has been phased to reflect a particular horizontal-to-subcarrier timing relationship.

3. Whether the user bits contain coded alphabetic characters.

4. A parity bit, used to cause the total number of transitions, after coding, to be even. This is a necessary, but not sufficient, requirement to edit a time-code track without loss of data during subsequent decoding.

The complete time-code specifications may be found in Chap. 17.

7.6.2 Sync-to-Subcarrier Phasing Requirements

The color-phase relationship between H-sync and color subcarrier reverses every other frame of two vertical fields. This parameter of video is of no significance in television viewing; it is of great significance during video-tape editing.

Color-Frame Editing. A typical example of an NTSC edit where the field sequence recorded on the tape with both playback machines locked to the edit-room sync generator is illustrated below:

Edit

Video *A*   Video *B*

Field no. 1234 1234 12 1234 1234 1234 123

At the edit point during playback, since the tape machine must remain locked to the edit-room generator, and since the subcarrier must match the reference subcarrier in phase, the recording tape machine is forced to do one of two things: either shift the horizontal phase by one-half cycle of subcarrier timing (140 ns), which is the normal playback mode of operation, or unlock the servo system and slide the tape one frame to realign the tape video to the same color frame as the edit room. This is operation of the machine in a color-frame playback mode. If the 140 ns shift occurs, and if the edit occurred at a time when there is a portion of the B video which is identical to the A video, a very noticeable shift will occur. Also, since the output processor of the tape machine is adding new sync and new blanking, sync appears to remain stationary, and the picture moves either right or left. The tape machine may be inserting a new horizontal-blanking interval in the video, and when the picture shifts, a portion of active picture may be blanked, causing a widening of blanking and a narrowing of the active picture.

In cases where a program undergoes an extensive amount of editing and where color-frame editing is not followed, if the tape machine blanking width is set for the maximum standard (10.6 µs), the growth of blanking can be serious enough to exceed FCC tolerances. For example, after five edits there is a possibility of as much as a 1-µs increase in blanking and an equivalent horizontal picture shift.

The only solution to this problem is to ensure that all video on the tape has a consistent and stable sync-to-subcarrier phase relationship. For the sake of uniformity and interchangeability of tapes among machines, RS170A states the preferred sync-to-subcarrier relationship. If all video which can be recorded is timed and properly ScH phased, and if the tape machine is referenced to a stable and properly phased source, the tape machine color-frame editor will provide contiguous color-frame timing.

A video processor ahead of the tape machine can alternate color-frame editing problems. The processor at the output of a switcher, or at the input to a tape machine, can be operated with external sync and subcarrier to ensure consistent ScH-phased signals to the tape machines. If the processor is fed mistimed sources, it can still shift the picture and widen blanking before recording; however, the processor will prevent any picture shift during playback.

An operating practice is outlined in the EIA RS170A specifications that prevents blanking growth when improperly timed sources are passed through an externally referenced processor. The practice includes recommended blanking widths at various locations in a facility. It uses sources with narrow blanking and gates wide blanking in just before transmission to the consumer. Running narrow blanking internal to the facility ensures that there will be ample picture available to meet FCC specifications for over-the-air broadcast even if several non-color-frame edits, or passes of mistimed video through an externally referenced processor, occur.

7.6.3 In-Machine Editors

The advent of microprocessors greatly increased the density of logic functions available within a video recorder. Much of this logic capability is applied to editing requirements. Two modes of editing are typically offered, INSERT and ASSEMBLE. The INSERT mode implies that some previously recorded element is being totally replaced. Here, a control track recording is assumed to exist, and the recorder uses it to control the tape position during the recording, just as it would in playback. In the ASSEMBLE mode, it is assumed that a new recording is being added at the end of an old one, and a new control track is recorded, phase-coherent with the old one. In helical-scan recorders, erasure of the tape into which the new video recording is to go is provided by a *flying erase head*, mounted on the rotating scanner, and located just ahead of the record head, to pave the way. Many video recorders include the ability to reproduce time code and synchronize their tape movement with an external reference. When the specified frame is approached, all the precursory actions are taken to cause a splice to occur at the specified frame number and to cease at another specified frame number. Typically, video and audio splice points are separately definable.

Communications with a video recorder that is a peripheral of an editing system are accomplished with a high-speed serial interface. Data which are time-related are transferred within a frame interval. The video recorder is expected to be able to position its tape at a specified frame, synchronize its tape movement with other recorders, begin recording at a specified frame, and cease recording at still another specified frame. Many other machine functions may be executed via the serial interface.

The serial interface may either be an external accessory to the recorder or, as is increasingly the case or, be contained within the recorder's electronics.

REFERENCES

Benson, K. Blair (ed.) (1986). *Television Engineering Handbook*. McGraw-Hill, New York.

Fink, D. G., and D. Christiansen (eds.) (1982). *Electronic Engineers' Handbook*, 2d ed. McGraw-Hill, New York.

McGinty, Gerald P. (1979). *Videocassette Recorders*. McGraw-Hill, New York.

CHAPTER 8
TELEVISION TRANSMISSION

Jerry Whitaker
Editorial Director, Broadcast Engineering *Magazine*

Carl Bentz
Intertec Publishing Corporation, Overland Park, Kansas

Portions of this chapter were adapted from: (1) John T. Wilner, "Television Transmitters" in *Television Engineering Handbook*, K. Blair Benson, ed., McGraw-Hill, New York, 1986, Chap. 7. (2) Oded Ben-Dov and Krishna Praba, "Television Antennas," *Television Engineering Handbook*, K. Blair Benson, ed., McGraw-Hill, New York, 1986, Chap. 8.

8.1 FCC SYNCHRONIZING WAVEFORMS RULES AND REGULATIONS

The synchronizing waveform shown in Fig. 8.1 is from FCC Rules and Regulations (Part 73.699, Fig. 6) and was in effect in 1983. The waveform is specified in terms of the modulating signal that must be applied to an ideal transmitter. Such a transmitter is one with perfectly linear modulation characteristics and with the specified vestigial sideband attenuation and envelope delay.

The envelope of the signal radiated is identical with this waveform for all the components lower than 0.75 MHz. The higher-frequency components are reduced in amplitude to half that shown, by the removal of the lower-frequency sideband.

When the signal of Fig. 8.1 is used to modulate a transmitter with the ideal characteristics, the radiated envelope will be that shown except that the HF components will be reduced in amplitude and advanced in phase with respect to the lower-frequency components. The effect would be to reduce the burst by 6 dB and to reduce the interval following the horizontal pulse, shown in the drawing as $0.006H$. After this signal has been amplified by a receiver having the specified attenuation and delay, the envelope will be identical to that shown in Fig. 8.1.

The division of amplitude range between the video modulation and the synchronizing signal should be such that relatively simple circuits will be able to hold satisfactory synchronism. That is, the picture should be stably synchronized when the signal is so weak that it is difficult to tell from visual inspection of the raster whether or not the picture is present. The synchronizing signal amplitude equal to one-quarter of the peak carrier amplitude was chosen for monochrome television and was continued for color transmission. Experience has shown that the value is an appropriate one.

The following considerations led to the specification of the duration of each portion of the signal: Referring to the detailed drawing of the horizontal pulse (5), the interval preceding the pulse is used to stabilize the conditions at the blanking level. If the video modulation were continued until the time of start of the horizontal pulse, the timing of the pulse would be affected by the modulation content. This would be particularly serious in a narrow-band receiver.

The duration of the pulse is 7½ percent of the horizontal period. The total period at the blanking level provided for receiver retrace is $0.165H$ minimum or $0.145H$ after the start of the horizontal pulse. The time specified for the horizon-

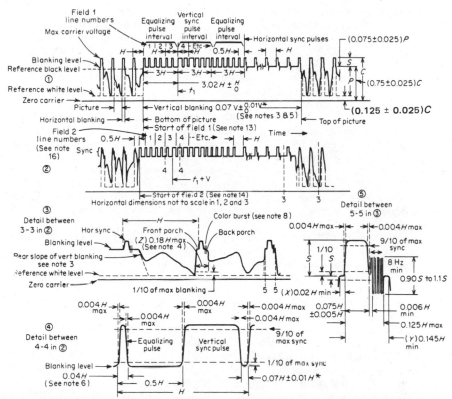

FIGURE 8.1   Television synchronizing waveform for color transmission. *Notes:* (1) H = time from start of one line to start of next line. (2) V = time from start of one field to start of next field. (3) Leading and trailing edges of vertical blanking should be complete in less than 0.1H. (4) Leading and trailing slopes of horizontal blanking must be steep enough to preserve minimum and maximum values of (x + y) and (z) under all conditions of picture content. *(5) Dimensions marked with an asterisk indicate that tolerances are permitted only for long time variations and not for successive cycles. (6) Equalizing pulse duration shall be between 0.45 and 0.55 of the horizontal synchronizing pulse duration. (7) Color burst follows each horizontal pulse but is omitted following the equalizing pulses and during the broad vertical pulses. (8) Color bursts to be omitted during monochrome transmission. (9) The burst frequency shall be 3.579545 MHz. The tolerance on the frequency shall be ±10 Hz with a maximum rate of change not to exceed 0.1 Hz/s. (10) The horizontal scanning frequency shall be 2/455 times the burst frequency. (11) The dimensions specified for the burst determine the times of starting and stopping the burst but not its phase. The color burst consists of amplitude modulation of a continuous sine wave. (12) Dimension P represents the peak excursion of the luminance signal from blanking level but does not include the chrominance signal. Dimension S is the synchronizing pulse amplitude above blanking level. Dimension C is the peak carrier amplitude. (13) Start of field 1 is defined by a whole line between first equalizing pulse and preceding H sync pulses. (14) Start of field 2 is defined by a half line between the first equalizing pulse and the preceding H sync pulses. (15) Field 1 line numbers start with the first equalizing pulse in field 1. (16) Field 2 line numbers start with second equalizing pulse in field 2. (17) Refer to text for further explanations and tolerances. (18) During color transmissions, the chrominance component of the picture signal may penetrate the synchronizing region and the color burst penetrates the picture region. *(Source:* FCC Rules and Regulations, Part 73.699, Fig. 6, 1983. *In* Television Engineering Handbook, *K. Blair Benson, ed.,* McGraw-Hill, New York, 1986)

tal retrace is a compromise between two conflicting requirements. The duration of the video modulation interval should be as large as possible, since the amount of information that can be transmitted is directly proportional to its duration. The retrace time should be as long as possible to simplify the receiver scanning requirements. The blanking interval duration must be sufficient to include the complete receiver retrace. The 17.5 percent duration was selected as a compromise between simple receiver retrace requirements and the maximum time for the transmission of video modulation.

The chrominance synchronizing burst must be received without interference from the chrominance information present during video modulation. A guard interval of $0.02H$ is provided. The difference in envelope delay for the low frequencies and the burst frequency will advance the burst in the radiated signal with respect to the horizontal pulse by approximately 0.17 μs, or approximately $0.003H$. A guard interval is therefore needed preceding the burst. The remaining interval, $0.039H$ minimum, is available for the transmission of the burst. This corresponds to 8⅞ cycles at 3.58 MHz. The specifications call for a minimum of 8 cycles.

In November 1977 the Electronic Industries Association (EIA) issued Tentative Standard RS170A, "Color Television Studio Picture Line Amplifier Output." This tentative standard on synchronization waveforms was in effect in June 1983.

To facilitate measurements, video voltage levels are expressed in IRE units, where sync level equals 40 IRE units, blanking level 0 IRE units, and reference-white level 100 IRE units. When applied to a linear television transmitter, these levels correspond to 100 percent carrier, 75 percent carrier, and 12.5 percent carrier, respectively, which are the FCC specifications.

Critical pulse measurements during the horizontal-blanking interval and preceding the first portion of the vertical sync pulse are expressed in microseconds.

One innovation in the drawing is the showing of four color fields of a color signal. Field 1 is defined by a particular subcarrier to horizontal-sync timing relationship; all other fields then follow accordingly.

FCC rules specify the parameters of the radiated signal. It has been recommended that the target values and tolerances shown in EIA Drawing RS-170A be maintained in television programs as they leave the video signal source to other broadcasters or to transmitter plants. In practice, maintaining the various timing values within the tolerances specified will facilitate compliance with FCC rules.

Compliance with these EIA-recommended values and tolerances is voluntary. However, it is expected that with the adoption of these values as industry standards for analog television signal sources, most broadcasters will begin to use these recommended values and that equipment manufacturers will design new equipment so that eventually all video signal sources will be able to follow these specifications.

8.2 TRANSMISSION EQUIPMENT

Television transmitters are classified in terms of their operating band, power level, final-tube type, and cooling method. The transmitter is divided into two basic subsystems: the *visual section*, which accepts the video input, frequency-modulates (FM) a radio frequency (RF) carrier, and amplifies the signal to feed the antenna system; and the *aural section*, which accepts the audio input, frequency-modulates a separate RF carrier, and amplifies the signal to feed the antenna system. The visual and aural signals are combined to feed a single radiating system.

Television transmitters in the United States operate in three frequency bands: low-band VHF (channels 2 through 6), high-band VHF (channels 7 through 13), and UHF (channels 14 through 83). (UHF channels 70 through 83 currently are assigned to mobile radio service.) Because of the wide variety of operating parameters for television stations outside the United States, this chapter will focus primarily on TV transmission as it relates to the United States.

Maximum power output limits are specified by the Federal Communications Commission (FCC) for each type of service. The maximum effective radiated power (ERP) for low-band VHF is 100 kW; for high-band VHF it is 316 kW; and for UHF it is 5 MW. The ERP of a station is a function of transmitter power output (TPO) and antenna gain. ERP is determined by multiplying these two quantities together.

The second major factor that affects the coverage area of a TV station is antenna height, known in the broadcast industry as height above average terrain (HAAT). HAAT takes into consideration the effects of the geography in the vicinity of the transmitting tower. The maximum HAAT permitted by the FCC for a low- or high-band VHF station is 1000 ft (305 m) east of the Mississippi River, and 2000 ft (610 m) west of the Mississippi. UHF stations are permitted to operate with a maximum HAAT of 2000 ft (610 m) anywhere in the United States (including Alaska and Hawaii).

The ratio of visual output power to aural power can vary from one installation to another, although the aural is typically operated at between 10 and 20 percent of the visual power. This difference is the result of the reception characteristics of the two signals. Much greater signal strength is required at the consumer's receiver to recover the visual portion of the transmission than the aural portion. The aural power output is intended to be sufficient for good reception at the fringe of the station's coverage area, but not beyond. It is of no use for a consumer to be able to receive a TV station's audio signal but not the video.

Two classifications of low-power TV stations have been established by the FCC to meet certain community needs: translators and low-power television stations.

- *Translators* are low-power systems that rebroadcast the signal of another station on a different channel. Translators are designed to provide "fill-in" coverage for a station that cannot reach a particular community because of local terrain. Translators operating in the VHF band are limited to 100 W power output (ERP), and UHF translators are limited to 1 kW.

- *Low-power television* (LPTV) is a service recently established by the FCC to meet the special needs of particular communities. LPTV stations operating on VHF frequencies are limited to 100 W ERP, and UHF stations are limited to 1 kW. LPTV stations originate their own programming and can be assigned by the FCC to any channel, as long as full protection is afforded against interference of a full-power station.

8.2.1 Transmitter Design Considerations

Each manufacturer has a particular philosophy with regard to the design and construction of a broadcast TV transmitter. Some generalizations can, however, be made with respect to basic system design. Transmitters can be divided into categories based on the following criteria.

Output Power.   When the power output of a TV transmitter is discussed, the visual section is the primary consideration. Output power refers to the peak power of the visual section of the transmitter (peak of sync). The FCC-licensed ERP is equal to the transmitter power output minus feedline losses times the power gain of the antenna.

A low-band VHF station can achieve its maximum 100 kW power output through a wide range of transmitter and antenna combinations. A 35-kW transmitter coupled with a gain-of-4 antenna would do the trick, as would a 10-kW transmitter feeding an antenna with a gain of 12. Reasonable parings for a high-band VHF station would range from a transmitter with a power output of 50 kW feeding an antenna with a gain of 8, to a 30-kW transmitter connected to a gain-of-12 antenna. These combinations assume reasonable feedline losses. To reach the exact power level, minor adjustments are made to the power output of the transmitter, usually by a front-panel power-trim control.

UHF stations that want to achieve their maximum licensed power output are faced with installing a very high power (and very expensive) transmitter. Typical pairings include a transmitter rated for 220 kW and an antenna with a gain of 25, or a 110-kW transmitter and a gain-of-50 antenna. In the latter case, the antenna could pose a significant problem. UHF antennas with gains in the region of 50 are possible but not advisable for most installations because of the coverage problems that can result. High-gain antennas have a narrow vertical radiation pattern, which can reduce a station's coverage in areas near the transmitter site. Whatever way is chosen, getting 5 MW ERP is an expensive proposition. Most UHF stations, therefore, operate considerably below the maximum permitted ERP.

At first examination, it might seem reasonable and economical to achieve licensed ERP using the lowest transmitter power output and highest antenna gain possible. Other factors, however, come into play that make the most obvious solution not always the best solution. Factors that limit the use of high-gain antennas include:

• Effects of high-gain designs on coverage area and signal penetration
• Limitations on antenna size because of tower restrictions, such as available vertical space, weight, and windloading
• Cost of the antenna

Final-Stage Design.   The amount of output power required of a transmitter will have a fundamental effect on system design. Power levels dictate whether the unit will be of solid-state or vacuum-tube design; whether air, water, or vapor cooling must be used; the type of power supply required; the sophistication of the high-voltage control and supervisory circuitry; and many other parameters.

Tetrodes are generally used for VHF transmitters above 5 kW and for low-power UHF transmitters (below 5 kW). As solid-state technology advances, the power levels possible in a reasonable transmitter design steadily increase. As of this writing, all-solid-state VHF transmitters of 30 kW have been produced.

By and large, however, the workhorse of VHF broadcasting is still the tetrode. Transmitters built around such devices are reliable, easy to service, and relatively straightforward from the design standpoint.

In the realm of UHF transmitters, the klystron reigns supreme. Klystrons use an electron-bunching technique to generate high power (55 kW from a single tube is not uncommon) at microwave frequencies. They are currently the first choice for high-power service.

Klystrons, however, are not particularly efficient. A stock klystron with no special circuitry might be only 40 to 50 percent efficient, depending on the type of device used. Various schemes have been devised to improve klystron efficiency, the best known of which is beam pulsing. Two types of pulsing are in common use today: mod-anode pulsing, a technique designed to reduce power consumption by the device during the color burst and video portion of the signal (and thereby improve overall system efficiency); and annular control electrode (ACE) pulsing, which accomplishes basically the same thing by incorporating the pulsing signal into a low-voltage stage of the transmitter rather than a high-voltage stage (as with mod-anode pulsing). Experience has shown the newer ACE approach—and other similar designs—to provide greater improvement in operating efficiency than mod-anode pulsing, and better reliability as well.

Still another approach to improving UHF transmitter efficiency involves an entirely new class of vacuum tube: the Klystrode. The Klystrode is a unique device that essentially combines the cathode/grid structure of the tetrode with the drift tube/collector structure of the klystron. As of this writing, several high-power (60 kW or greater) transmitters utilizing Klystrode technology have been put into on-air service. Data taken from these systems indicates a significant efficiency improvement over conventional klystron designs.

Modulation System. A number of approaches may be taken to amplitude modulation of the visual carrier. Current technology systems utilize low-level, intermediate-frequency (IF) modulation. This approach allows superior distortion correction, more accurate vestigial sideband shaping, and significant economic advantages to the manufacturer.

8.2.2 Elements of the Transmitter

A television transmitter can be divided into four major subsystems:

- The exciter
- An intermediate power amplifier (IPA)
- A power amplifier (PA)
- A high-voltage power supply

Figure 8.2 shows the audio, video, and RF paths for a typical design. In this diagram, the exciter subsystem consists of:

- The video tray
- The exciter-modulator tray
- The RF processor tray

Depending on the design of the transmitter, these sections may be separate units or simply incorporated into the exciter itself. A separate power supply section (not shown in Fig. 8.2) supplies operating voltages to the various subassemblies of the transmitter.

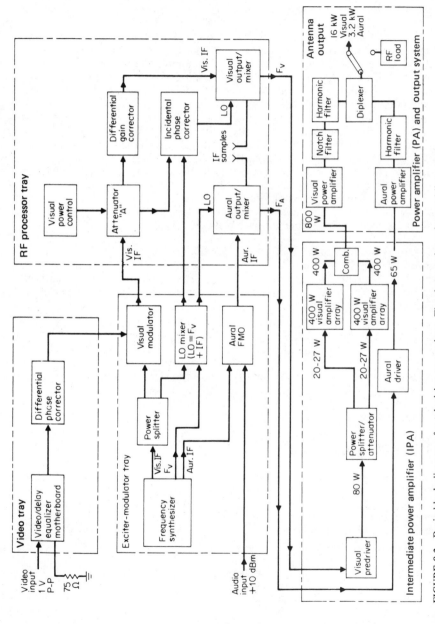

FIGURE 8.2 Basic block diagram of a television transmitter. The three major subassemblies are the exciter, the IPA, and the PA. A power supply system (not shown) provides operating voltages to all sections and high voltage to the PA stage.

8.3 EXCITER SUBSYSTEM

The input video signal is first fed into a video processing amplifier (proc amp) and equalizer. The proc amp performs the following functions:

- Video input amplification
- Group delay equalization/correction
- Video clamping and hum cancellation
- Phase delay compensation
- White clipping
- Sync regeneration

The processed video is converted to amplitude modulation of an intermediate frequency on a carrier generated by a master oscillator. Major circuits include:

- The frequency synthesizer (or crystal oscillator)
- The visual-aural modulator
- The vestigial sideband filter

The modulated visual IF signal is processed for distortion corrections and subsequent RF transmission in the following stages:

- Power amplifier linearity corrector
- Driver amplifier linearity corrector
- Incidental phase modulation corrector
- RF up-converter
- Channel bandpass filter

8.3.1 Group Delay Equalizer-Corrector

The group delay equalizer is designed to introduce video delays to offset and correct for delays produced in the transmitter PA stage and notch diplexer, and delays inherent in typical TV receivers. Each delay equalizer contains up to 5 *all-pass* delay circuits, each designed to correct a specific frequency band. Figure 8.3a shows a simplified active all-pass network.

All-pass networks are constant-impedance circuits that allow composite delay curves to be made by cascading sections. The result is an overall delay curve that complements the transmitter's delay characteristics. Such circuits can independently correct phase and amplitude variations with a minimum of interaction.

The circuit shown in Fig. 8.3a is a second-order phase rotator in which 360° of phase rotation is achieved. There are also first-order phase rotators, in which either C or L is missing, and the maximum phase rotation is 180°.

8.3.2 Video Clamp

The video clamp is required to hold the *back porch* level constant, regardless of changes in the average picture level. A sample of the input signal is sent to the clamp circuit, where a clamp-pulse generator produces an output that is coinci-

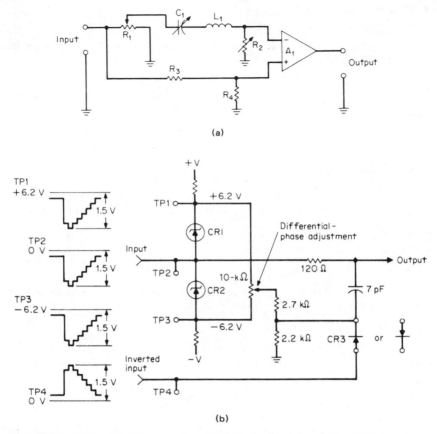

(a)

(b)

FIGURE 8.3 (*a*) Simplified diagram of an active all-pass network, used in a TV transmitter for group delay correction. In a typical system, five such circuits would be cascaded to provide the necessary range of correction. (*Courtesy Harris Corp. In K. Blair Benson, ed.,* Television Engineering Handbook, *McGraw-Hill, New York, 1986*) (*b*) Simplified diagram of a differential phase corrector and a set of representative input waveforms. The inverted input at TP4 requires an inverting amplifier, not shown. (*Courtesy RCA Corp. In K. Blair Benson, ed.,* Television Engineering Handbook, *McGraw-Hill, New York, 1986*)

dent with the back porch of the video signal. A separate signal from the output of the video processing system, after removal of the color subcarrier, is applied to a tip-of-sync detector. These two signals are processed to produce a control voltage that holds the back porch level constant. In the event of a loss of video, the circuit will automatically set to blanking level, preventing overdrive of subsequent amplifiers.

8.3.3 Differential Phase Corrector

Differential phase correction is performed on a *precorrection basis* to compensate for the distortion that *will be* produced in the IPA and PA stages of the transmitter. The corrector functions by generating signals of opposite phase delay that will, when added to the input video signal, cancel subsequent unwanted differential phase distortion. Figure 8.3b shows a simplified diagram of a differential phase corrector. In a typical implementation, seven such diode gating circuits would be used. This type of configuration permits up to 14° correction.

An examination of Fig. 8.3b will help to provide understanding of the concept involved. The 10-kΩ differential phase adjustment potentiometer produces a bias of up to 6.2 V on zener diodes CR1 and CR2. Depending on the bias value, the zener diodes will conduct over a portion of the video signal, with a correction signal being added to the output through the 7-pF capacitor. An inverted input signal, produced by an amplifier not shown on the diagram, is added to the output via diode CR3.

8.3.4 White Clip

A white-clip circuit is used to prevent negative-going video modulation from dropping below 12.5 percent of reference white. If the video is below 12.5 percent, overmodulation and high incidental (intercarrier) phase modulation (ICPM) will occur. Audio buzz may also be encountered in home receivers.

The white-clip circuit acts on the active video signal and keeps it from reaching reference white. In so doing, it prevents the transmitter from overmodulating and degrading the audio.

8.3.5 Frequency Synthesizer

The carrier signal for the visual and aural outputs of a TV transmitter begins with a master oscillator. This oscillator can be either crystal-based or generated by a frequency synthesizer. The latter is by far the most common today.

A frequency synthesizer is built around a phased-locked-loop (PLL) circuit that accurately controls the operating frequency of a free-running L/C oscillator. The PLL uses a highly stable crystal oscillator (mounted in a temperature-controlled oven) or a temperature-compensated crystal oscillator (TCXO) as a reference signal source. Depending on the design, a single frequency synthesizer circuit may be used to cover the entire VHF or UHF operating band. Any single operating frequency can be selected by setting DIP switches on the synthesizer circuit board.

The heart of a frequency synthesizer is the PLL, shown in Fig. 8.4 in simplified form. In this example, a 5-kHz reference clock is fed to the phase detector, where it is compared with a sample of the final IF carrier, which has been divided down to 5 kHz. If the two signals differ in phase or frequency, an error voltage proportional to the difference is generated and applied to the voltage-controlled oscillator (VCO). This error voltage forces the VCO to move back to the desired operating frequency.

The output of the VCO feeds a string of digital counters/dividers that determine the operating frequency of the system. By selecting different "divide-by" configurations, the operating point of the VCO can be made to change.

The output signal of the frequency synthesizer at visual IF is applied to the visual modulator. A second visual output plus the visual carrier is used for incidental phase correction and up-conversion. A similar signal is also sent to the aural up-converter. Although the aural system of a television transmitter is es-

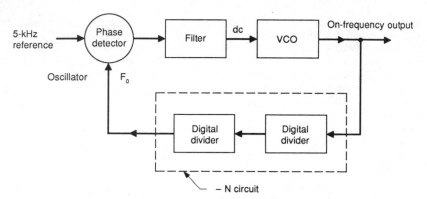

FIGURE 8.4 Block diagram of a phase-locked-loop (PLL) frequency synthesizer. This type of circuit is commonly used to generate the operating frequencies in a TV transmitter exciter. *(Courtesy RCA Corp. In K. Blair Benson, ed.,* Television Engineering Handbook, *McGraw-Hill, New York, 1986)*

sentially independent from the visual (in most designs), the aural carrier must be accurately locked to the visual.

The need to keep the visual and aural carriers locked together is a function of the basic television system design. In the NTSC system, the color subcarrier is an odd multiple of half the line scanning rate, making the visibility of the subcarrier minimal on monochrome receivers. Developers of the NTSC system determined that optimum performance would be gained if the beat between the subcarrier and the aural carrier was also an odd multiple of half the line scanning rate, rendering it minimally visible. To meet this requirement, aural carrier separation must be a multiple of the horizontal scanning rate. Therefore, although the visual and aural sections of a transmitter can be considered as separate systems, the carrier frequencies of the two systems must be linked together and maintained to within tight tolerances.

8.3.6 Visual-Aural Modulator

The visual modulator receives a precorrected video signal from the processing system and uses it to modulate a carrier from the IF synthesizer. Mixer designs vary from one manufacturer to another, but all perform essentially the same function. Figure 8.5 shows a modulator built around a balanced mixer. The resulting amplitude-modulated signal is produced with double sidebands. A downstream filter in the modulator removes most of the lower sideband, resulting in a vestigial sideband signal (as described in Sec. 8.3.7).

The aural modulator functions in essentially the same fashion as an FM radio broadcast exciter. The peak deviation of the FM aural carrier is 25 kHz (FM broadcast utilizes 75-kHz deviation). Audio input is preemphasized to match the standard 75-μs preemphasis curve specified by the FCC. In most 625-line countries, 50 μs preemphasis is used, with 50-kHz deviation.

The aural modulator signal frequency modulates an IF carrier output by the master frequency synthesizer and then is routed to the aural up-converter.

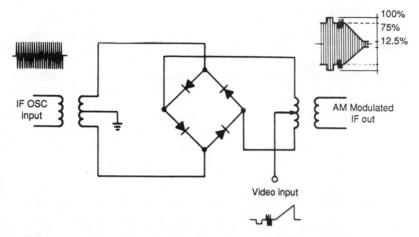

FIGURE 8.5 Simplified schematic diagram of a balanced mixer. This design is typically used in the visual modulator section of a TV exciter.

8.3.7 Vestigial Sideband Filter

The modulated visual IF signal is band-shaped in a vestigial sideband filter, typically a surface-acoustic-wave (SAW) filter. Envelope delay correction is not required for the SAW filter because of the filter's uniform delay characteristics. Envelope delay compensation may, however, be needed for other parts of the transmitter.

The SAW filter provides many benefits to transmitter designers and operators. A SAW filter requires no adjustments and is stable with respect to temperature and time. SAW filters are made of a substrate about 2 mm thick and generally use a piezoelectric crystal such as lithium tantalate (other crystalline structures are also used successfully). The dimensions of the filter elements are determined by the desired frequency-response characteristics of the crystal.

SAW filters have exceptionally steep skirts at band edges and can provide up to 60 dB attenuation to out-of-band frequencies. Insertion loss for a SAW filter is about 25 dB, making it necessary to use low-noise amplifiers to maintain a satisfactory signal-to-noise ratio for the overall system.

A color notch filter is required at the output of the transmitter because imperfect linearity of the IPA and PA stages introduces unwanted modulation products (see Sec. 8.6.1).

8.3.8 Linearity Corrector

The linearity corrector predistorts the modulated IF signal to compensate for transmitter differential-gain distortion. This is accomplished by compressing or stretching the nonlinear portions of the transfer characteristic of an IF signal amplifier. The approach is similar to the method used for gamma correction in a video amplifier.

8.3.9　ICPM Corrector

The incidental phase modulation corrector cancels ICPM distortion generated within the visual RF path. A sample of the modulated visual carrier is detected to develop a video signal that contains all delay and differential-phase precorrections. This signal is filtered and then used to phase-modulate the local oscillator feed to the visual up-converter. This modulation is equal to and opposite in phase with the undesired ICPM.

Once correction has been obtained, envelope and synchronous demodulators in home receivers should produce essentially the same output waveforms (see Sec. 8.3). This attribute is important for the growing numbers of stereo receivers being purchased by consumers.

8.3.10　Up-Converter

The modulated IF signal is mixed with an output from the local oscillator that is the sum of the IF carrier and the operating carrier frequency. Sum and difference frequency signals are produced in the mixing process, and a final carrier difference signal is extracted through the use of a bandpass filter. The visual output of the mixer (*up-converter*) at the final carrier frequency is sent to the IPA driver.

A similar process occurs in the aural signal path, as shown in Fig. 8.2.

8.4　INTERMEDIATE POWER AMPLIFIER

The intermediate power amplifier (IPA) is used to develop the power output necessary to drive the power amplifier stages for the aural and visual systems. A low-band, 16- to 20-kW transmitter typically requires about 800 W of RF drive, and a high-band, 35- to 50-kW transmitter needs about 1600 W. A UHF transmitter utilizing a high-gain klystron final tube requires about 20 W of drive, while a UHF transmitter utilizing a Klystrode needs about 80 W.

Because the aural portion of a television transmitter operates at only 10 to 20 percent of the visual power output, the RF drive requirements are proportionately lower.

Virtually all transmitters manufactured today utilize solid-state devices in the IPA. Transistors and other semiconductors are preferred because of their inherent stability, reliability, and ability to cover a broad band of frequencies without retuning. Present solid-state technology, however, cannot provide the power levels needed by most transmitters in a single device. To achieve the needed RF energy, devices are combined using a variety of schemes.

A typical "building block" for a solid-state IPA provides a maximum power output of approximately 200 W. In order to meet the requirements of a 20-kW, low-band VHF transmitter, a minimum of four such units would have to be combined. In actual practice, some amount of *headroom* is always designed into the system to compensate for component aging, imperfect tuning in the PA stage, and device failure.

Most solid-state IPA circuits are configured so that in the event of a failure in one module, the remaining modules will continue to operate. If sufficient headroom has been provided in the design, the transmitter will continue to operate without change. The defective subassembly can then be repaired and returned to service at a convenient time.

Because the output of the RF up-converter is about 10 W, an intermediate amplifier is generally used to produce the required drive for the parallel amplifiers. The individual power blocks are fed by a splitter that feeds equal RF drive to each unit. The output of each RF power block is fed to a hybrid combiner that provides isolation between the individual units. The combiner feeds a bandpass filter that allows only the modulated carrier and its sidebands to pass.

The inherent design of a solid-state RF amplifier permits operation over a wide range of frequencies. Most drivers are broadband and require no tuning. Certain frequency-determined components are added at the factory (depending on the design), but from the end user's standpoint, solid-state drivers require virtually no attention. IPA systems are available that cover the entire low- or high-band VHF range without tuning.

Advances continue to be made in solid-state RF devices. New developments promise to extend substantially the reach of semiconductors into medium-power RF operation. Coupled with better devices are better circuit designs, including parallel devices and new push-pull configurations. Another significant factor in achieving high power from a solid-state device is efficient removal of heat from the device itself.

8.5 POWER AMPLIFIER

The power amplifier raises the output energy of the transmitter to the desired RF operating level. As noted in Sec. 8.2.1, solid-state devices are increasingly being used through parallel configurations in high-power transmitters. The vast majority of television transmitters in use today, however, still utilize vacuum tubes. The workhorse of VHF television is the tetrode, which provides high output power, good efficiency, and good reliability. In UHF service, the klystron is the standard output device for transmitters above 10 kW.

Tetrodes in television service are operated in the class B mode to obtain reasonable efficiency while maintaining a linear transfer characteristic. Class B amplifiers, when operated in tuned circuits, provide linear performance because of the flywheel effect of the resonance circuit. This allows a single tube to be used instead of two in push-pull fashion. The bias point of the linear amplifier must be chosen so that the transfer characteristic at low modulation levels matches that at higher modulation levels. Even so, some nonlinearity is generated in the final stage, requiring differential gain correction (as discussed in Sec. 8.3.8). The plate (anode) circuit of a tetrode PA is usually built around a coaxial resonant cavity, which provides a stable and reliable tank circuit.

UHF transmitters using a klystron in the final output stage must operate class A, the most linear but also most inefficient operating mode for a vacuum tube. The basic efficiency of a nonpulsed klystron is approximately 40 percent. Pulsing, which provides full available beam current only when it is needed (during peak of sync), can improve device efficiency by as much as 25 percent, depending on the type of pulsing used. (As discussed in Sec. 8.2.1, recent development of the Klystrode promises to radically change the efficiency penalty now faced by UHF broadcasters.)

Two types of klystrons are presently in service: integral-cavity and external-cavity devices. The basic theory of operation is identical for each tube, but the mechanical approach is radically different.

The major elements of a klystron are shown in Fig. 8.6. A heated cathode emits electrons that are accelerated from a gun assembly into a drift tube. The electrons finally encounter the collector (operating at a potential of about 24 kV relative to the cathode) and are returned to the high-voltage power supply. Positioned around the drift tube are a series of cavities (typically four or five) that act as the resonant circuits of the device. Magnet coils associated with the cavities aid in focusing the electron beam.

In the integral-cavity klystron, the cavities are built into the klystron to form a single unit. In the external-cavity klystron, the cavities are outside the vacuum envelope and bolted around the tube when the klystron is installed in the transmitter.

A number of factors come into play in a discussion of the relative merits of integral versus external cavity designs. Primary considerations include operating efficiency, purchase price, and life expectancy.

The PA stage includes a number of sensors that provide input to supervisory and control circuits. Because of the power levels present in the PA stage, sophisticated fault-detection circuits are required to prevent damage to components in the event of a problem either external to or inside the transmitter. An RF sample, obtained from a directional coupler installed at the output of the transmitter, is used to provide automatic power-level control.

The transmitter block diagram of Fig. 8.2 shows separate visual and aural PA

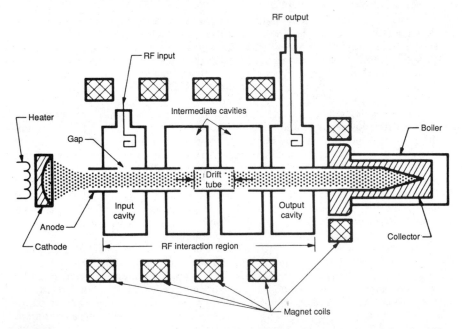

FIGURE 8.6 Basic mechanical structure of a klystron. The cavities shown surrounding the drift tube can be external to the tube (an external-cavity klystron) or internal to the tube (an integral-cavity klystron). The "boiler" shown on the diagram is part of the tube's heat removal system.

stages. This configuration is normally used for high-power transmitters. Low-power designs often use a combined mode in which the aural and visual signals are added prior to the PA. This approach offers a simplified system, but at the cost of additional precorrection of the input video signal.

PA stages often are configured so that the circuitry of the visual and aural amplifiers is identical. While this represents a good deal of "overkill" insofar as the aural PA is concerned, it provides backup protection in the event of a visual PA failure. The aural PA can then be reconfigured to amplify both the aural and the visual signals, at reduced power.

The aural output stage of a television transmitter is similar in basic design to an FM broadcast transmitter. Tetrode output devices generally operate class C, providing good efficiency. Klystron-based aural PAs are used in UHF transmitters.

8.6 COUPLING-FILTERING SYSTEM

The output of the aural and visual power amplifiers must be combined and filtered to provide a signal that is electrically ready to be applied to the antenna system. The primary elements of the coupling and filtering system of a TV transmitter are:

- The color notch filter
- The aural and visual harmonic filters
- The diplexer

In a low-power transmitter (below 5 kW), this hardware may be included within the transmitter cabinet itself. Normally, however, it is located external to the transmitter.

8.6.1 Color Notch Filter

The color notch filter is used to attenuate the color subcarrier lower sideband to the −42 dB requirements of the FCC. The color notch filter is placed across the transmitter output feedline, as shown in Fig. 8.2. The filter consists of a coax or waveguide stub tuned to F_c = 3.58 MHz below the picture carrier. The Q of the filter is high enough so that energy in the vestigial sideband is not materially affected, while still providing high attenuation at 3.58 MHz.

8.6.2 Harmonic Filters

Harmonic filters are used to attenuate out-of-band radiation of the aural and visual signals to ensure compliance with FCC requirements. Filter designs vary depending on the manufacturer, but most are of coaxial construction utilizing L and C components housed within a prepackaged assembly. Stub filters are also used, typically adjusted to provide maximum attenuation at the second harmonic of the operating frequency of the visual carrier and the aural carrier.

8.6.3 Diplexer-Combiner

The filtered visual and aural outputs are fed to a hybrid diplexer where the two signals are combined to feed the antenna. For installations that require dual antenna feedlines, a hybrid combiner with quadrature-phased outputs is used. Depending on the design and operating power, the color notch filter, aural and visual harmonic filters, and diplexer may be combined into a single mechanical unit. For the purposes of this discussion, however, they will be considered separately.

An RF combiner is used to add two RF power sources, such as transmitters or amplifiers, and present a constant impedance to match each source. One RF unit must be phase-delayed by 90° in order to cancel out the inherent quadrature splitting by the combiner. Figure 8.7 shows a simplified power combiner.

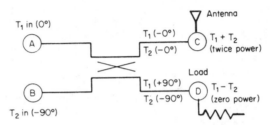

FIGURE 8.7 Operating principles of a hybrid combiner. This circuit is used to add two identical signals at inputs *A* and *B*. *(Courtesy Micro Communications, Inc. In K. Blair Benson,* Television Engineering Handbook, *McGraw-Hill, New York, 1986)*

The combiner accepts one RF source and splits it equally into two parts, one arriving at output port C with 0° phase (no phase delay; it is the reference phase), and the other delayed by 90° at port D. A second RF source connected to input port B, but with a phase delay of 90°, will also split in two, but the signal arriving at port C will now be in phase with source 1 and the signal arriving at port D will cancel, as shown in Fig. 8.7.

Output port C, the summing point of the hybrid combiner, is connected to the load (the antenna). Output port D is connected to a resistive load to absorb any residual power resulting from slight differences in amplitude and/or phase between the two input sources. If one of the RF inputs should fail, then half of the remaining transmitter output would be absorbed by the resistive load at port D.

The hybrid combiner works only when the two signals being mixed are identical in frequency and amplitude, and when their relative phase is 90°. Such a device, therefore, will not work to combine the aural and visual portions of a transmitter into the load. The hybrid combiner, however, serves as the building block of the notch diplexer, which accomplishes the desired function.

Figure 8.8 illustrates the basic concept of a notch diplexer, which combines the aural and visual RF signals to feed a common antenna system and provides a constant impedance load to each section of the transmitter.

The notch diplexer consists of two hybrid combiners and two sets of reject cavities. The system is configured so that all of the energy from the visual transmitter passes to the antenna (port D), and all of the energy from the aural transmitter passes to the antenna. The phase relationships are arranged so that the in-

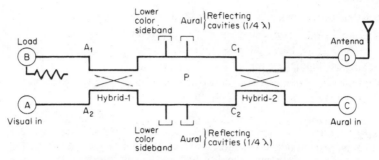

FIGURE 8.8 Functional diagram of a notch diplexer, used to combine the aural and visual outputs of a television transmitter for application to the antenna. *(Courtesy Micro Communications, Inc. In K. Blair Benson,* Television Engineering Handbook, *McGraw-Hill, New York, 1986)*

put signals cancel at the resistive load (port B). Because of the paths taken by the aural and visual signals through the notch diplexer, the amplitude and phase characteristics of each input do not change from the input ports (port A for the visual and port C for the aural) and the antenna (port D), thus preserving signal purity.

8.7 TRANSMITTER PERFORMANCE CHARACTERISTICS

Many shortcomings in the received picture and sound resulting from reflections and terrain limitations cannot be controlled by the broadcaster. However, other problems, such as poor resolution, color fringing, hue shift, luminance shift, and poor audio quality can be prevented through proper equipment adjustment. Among the critical performance parameters causing these types of degradation are:

- Incidental phase modulation (ICPM)
- Differential phase and gain
- Frequency-response errors
- Envelope delay
- Quadrature distortion
- Luminance-to-chrominance shift
- Audio frequency response errors and distortion

8.7.1 Signal Distortions

There are two broad classifications of signal distortions in television systems: linear and nonlinear. Linear distortions occur independent of picture level, while nonlinear distortions vary with the amplitude of the video signal.

Linear distortions usually occur in systems with incorrect frequency response and are divided into four general categories:

1. Short-time distortions affecting horizontal sharpness
2. Line-time distortions causing horizontal streaking
3. Field-time distortions causing errors in low-frequency response and problems with vertical shading of the picture
4. Long-time distortions in the dc-to-extremely-low frequency range causing flicker and slow changes in picture brightness

Nonlinear distortion occurs with changes in the average picture level (APL). Luminance nonlinear distortion occurs when the gain of the system varies as the picture changes from black to white. Differential gain distortion is the change in amplitude of the chrominance signal as luminance amplitude shifts from black to white. Differential phase distortion is the change in color subcarrier phase as the luminance signal varies in amplitude.

Incidental carrier phase distortion is the undesired phase modulation of the visual carrier and its color subcarrier caused by amplitude modulation of the visual carrier. Incidental carrier phase modulation (ICPM) makes it appear as though the picture and color subcarrier are being phase-modulated. In other words, the normal AM picture and color subcarrier have an FM component.

ICPM results in noise in the audio section of the receiver and color distortion in the received picture. Audio noise is usually observed when bright-level picture content, such as the output of a character generator, is being transmitted. Buzz occurs because aural detectors use the difference between the picture and sound carriers to produce the aural intermodulation carrier. If the picture carrier is phase-modulated as a function of visual modulation, changing separation between the two carriers will result in unwanted aural frequency modulation, or buzz.

Differential gain distortion occurs when the amplitude of the chrominance signal changes with respect to the amplitude of the luminance signal. This causes a change in chroma saturation as the luminance varies between black and white levels.

Group delay is a distortion that results when a group of related frequencies differ in "arrival time" after passing through a circuit. Causes of group delay include narrow-band filters in the transmission system (the vestigial sideband filter, aural sound notch, and color notch filter) and bandwidth restrictions at the consumer's receiver.

High-frequency (HF) group delay distortion affects luminance vertical edges, usually in the form of overshoots. Low-frequency (LF) group delay distortion affects primarily the position of the chroma with respect to the accompanying luminance. Extreme group delay over the entire passband can produce the "funny-paper effect," in which the color components are shifted with respect to the monochrome outline.

8.8 TRANSMISSION ANTENNA SYSTEMS

Broadcasting is accomplished by the emission of coherent electromagnetic waves in free space from one or more radiating antenna elements that are excited by modulated RF currents. Although, by definition, the radiated energy is composed of mutually dependent magnetic and electrical vector fields, conventional practice in television engineering is to measure and specify radiation characteristics in terms of the electrical field only.

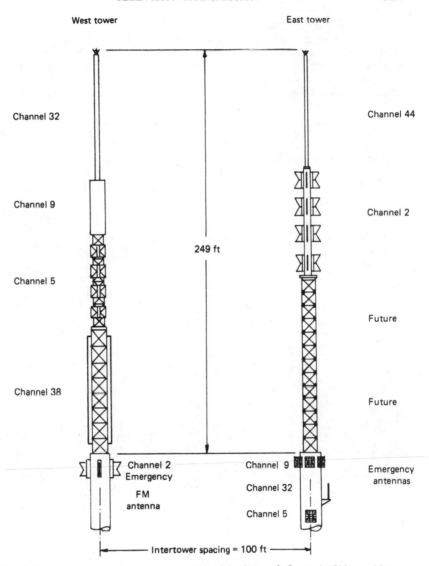

West tower   East tower

Channel 32   Channel 44

Channel 9   Channel 2

249 ft

Channel 5

Channel 38   Future

                                                        Future

Channel 2   Channel 9   Emergency
Emergency   antennas

FM   Channel 32
antenna
                                   Channel 5

⊢— Intertower spacing = 100 ft ——⊣

FIGURE 8.9 Twin tower antenna array atop the John Hancock Center in Chicago. Note how antennas have been stacked to overcome space restrictions.

The field vectors may be polarized, or oriented, horizontally, vertically, or circularly. Television broadcasting, however, has used horizontal polarization for the majority of installations worldwide. More recently, interest in the advantages of circular polarization has resulted in an increase in this form of transmission, particularly for VHF channels.

Both horizontal and circular polarization designs are suitable for tower-top or side-mounted installations. The latter option is dictated primarily by the existence of a previously installed tower-top antenna. On the other hand, in metropolitan areas where several antennas must be located on the same structure, either a stacking or candelabra-type arrangement is feasible. Figure 8.9 shows an example of antenna stacking on top of the John Hancock Center in Chicago, where six TV antennas, four standby antennas, and numerous FM transmitting antennas are located. Figure 8.10 shows a candelabra installation atop the Mt. Sutro tower in San Francisco. The Mt. Sutro tower supports eight TV antennas on its uppermost level. A number of FM transmitting antennas and two-way radio antennas are located on lower levels of the structure.

Another approach to TV transmission involves combining the RF outputs of two stations and feeding a single wide-band antenna. This approach is expensive and requires considerable engineering analysis to produce a combined system that will not degrade the performance of either transmission system. In the Mt.

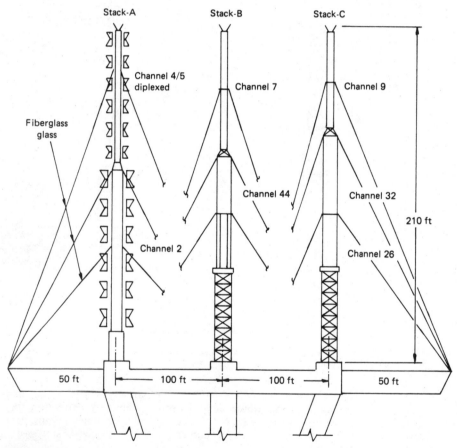

FIGURE 8.10 Installation of TV transmitting antennas on the candelabra structure at the top level of the Mt. Sutro tower in San Francisco. This installation makes extensive use of antenna stacking. (*Source: K. Blair Benson,* Television Engineering Handbook, *McGraw-Hill, New York, 1986*)

Sutro example, it can be seen that two stations (channels 4 and 5) are combined on a single transmitting antenna.

8.8.1 Top-Mounted Antenna Types

The typical television broadcast antenna is a broad-band radiator operating over a bandwidth of several megahertz with an efficiency of more than 95 percent. Reflections from the antenna and transmission line back to the transmitter must be small enough to introduce negligible picture degradation. Furthermore, the gain and pattern characteristics of the antenna must be designed to achieve the desired coverage within acceptable tolerances. Tower-top, pole-type antennas designed to meet these requirements can be classified into two categories: resonant dipoles and multiwavelength, traveling-wave elements.

The primary considerations in the design of a top-mounted antenna are the achievement of a uniform omnidirectional azimuth field and minimum wind-loading. A number of different approaches have been tried successfully. Figure 8.11 illustrates the basic mechanical design of the most common antennas.

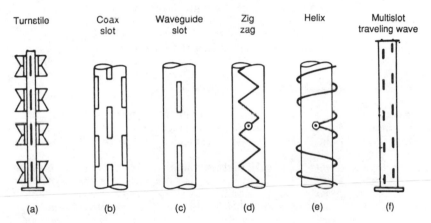

| Turnstile | Coax slot | Waveguide slot | Zig zag | Helix | Multislot traveling wave |

(a)   (b)   (c)   (d)   (e)   (f)

FIGURE 8.11 Various antennas used for VHF and UHF broadcasting. All designs provide linear (horizontal) polarization. Illustrated are: (*a*) turnstile antenna, (*b*) coax slot antenna, (*c*) waveguide slot antenna, (*d*) zigzag antenna, (*e*) helix antenna, and (*f*) multislot, traveling-wave antenna.

Turnstile Antenna.   The turnstile antenna was the earliest and most popular resonant antenna for VHF broadcasting. The antenna is made up of four batwing-shaped elements mounted on a vertical pole in a manner resembling a turnstile (see Fig. 8.11*a*). The four "batwings" are, in effect, two dipoles fed in quadrature phase. The azimuth-field pattern is a function of the diameter of the support mast. The pattern is usually within 10 to 15 percent of a true circle.

The turnstile antenna is made up of several layers, usually 6 for channels 2 through 6 and 12 for channels 7 through 13. The turnstile is unsuitable for side-mounting, except for standby applications in which coverage degradation can be tolerated.

Coax Slot Antenna.   The coax slot antenna, commonly referred to as the *pylon antenna*, is the most popular top-mounted unit for UHF applications (see Fig. 8.11*b*). Horizontally polarized radiation is achieved by using axial resonant slots on a cylinder to generate RF current around the outer surface of the cylinder. A good omnidirectional pattern is achieved by exciting four columns of slots around the circumference, which is basically just a section of structurally rigid coaxial transmission line.

The slots along the pole are spaced approximately one wavelength per layer, and a suitable number of layers are used to achieve the desired gain. Typical gains range from 20 to 40. By varying the number of slots around the periphery of the cylinder, directional azimuth patterns can be achieved.

Waveguide Slot Antenna.   The waveguide slot antenna is a UHF waveguide slot (Fig. 8.11*c*) that is a variation on the coax slot antenna. The antenna is simply a section of waveguide with slots cut into the sides. The physics behind the design is very complicated, but the end result is the simplest of all antennas. This is a desirable feature in field applications, because simple designs translate to long-term reliability.

Zigzag Antenna.   The zigzag antenna is a panel array design that utilizes a conductor routed up the sides of a three- or four-sided panel antenna in a "zigzag" manner (see Fig. 8.11*d*). With this design, the vertical current component along the zigzag conductor is mostly canceled out, and the antenna can effectively be considered as an array of dipoles. With several such panels mounted around a polygonal periphery, the required azimuth pattern can be shaped by proper selection of feed currents to the various elements.

Helix Antenna.   The helix antenna is a variation on the zigzag that accomplishes basically the same goal using a different mechanical approach. The classic helix design is shown in Fig. 8.11*e*. Note the center feed point shown in the illustration.

VHF Multislot Antenna.   In a VHF multislot antenna, which is similar mechanically to a coax slot antenna, the radiator consists of an array of axial slots on the outer surface of a coaxial transmission line, as shown in Fig. 8.11*f*. The slots are excited by a traveling wave inside the slotted line. The azimuth pattern is typically within 5 percent of omnidirectional. The antenna is generally about 15 wavelengths long.

8.8.2   Circularly Polarized Antennas

Circular polarization holds the promise for many broadcasters of improved penetration into difficult coverage areas. There are a number of points to weigh in the decision to use a circularly polarized (CP) antenna, not the least of which is that a station will have to double its transmitter power in order to maintain the same ERP if it installs a CP antenna.

Circularly polarized antennas have been developed for both resonant and traveling-wave antenna designs. The traveling-wave configuration is essentially a side-fire helical antenna supported on a pole, as illustrated in Fig. 8.12*a*. A suitable number of helixes around the pole provide azimuth pattern circularity. This type of antenna is especially suited to applications on channels 7 to 13 because only 3 percent pattern and impedance bandwidth are required. Circular polarization on channels 2 to 6, where the bandwidth is approximately 10 percent, typically incorporates resonant dipoles around a support pole.

Circularly polarized traveling-wave antennas for UHF applications have gained a good deal of interest in recent years, but the economics of generating large amounts of RF power at UHF frequencies has limited growth in this area. Figure 8.12b shows the basic design of a slot CP antenna for UHF broadcasting.

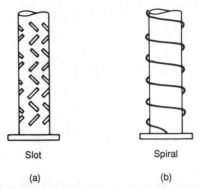

Slot   Spiral

(a)   (b)

FIGURE 8.12 Two circularly polarized traveling-wave antenna designs. (a) Slot CP antenna. (b) Spiral CP antenna.

8.8.3 Side-Mounted Antenna Types

Television antennas designed for mounting on the faces of a tower must meet the same basic requirements as a top-mounted antenna—wide bandwidth, high efficiency, predictable coverage pattern, high gain, and low windloading—plus the additional challenge that the antenna must work in a less than ideal environment. Given the choice, no broadcaster would elect to place a transmitting antenna on the side of a tower rather than the top. The laws of physics are against you when side mounting is required. There are, however, a number of ways to solve the problems presented by side mounting. Antenna manufacturers have developed products that take the guesswork out of side mounting. Some of the more common antennas designed for side mounting include the following:

Butterfly Antenna. The butterfly antenna is essentially a batwing panel developed from the turnstile radiator. The butterfly is one of the most popular panel antennas used for tower face applications. It is suitable for the entire range of VHF applications. A number of variations on the basic batwing theme have been produced, including modifying the shape of the turnstile-type wings to rhombus or diamond shapes. Another version utilizes multiple dipoles in front of a reflecting panel.

For circularly polarized applications, two crossed dipoles or a pair of horizontal and vertical dipoles are used. A variety of cavity-backed crossed-dipole radiators are also utilized for CP transmission.

The azimuth pattern of each panel antenna is unidirectional, and three or four such panels are mounted on the sides of a triangular or square tower to achieve an omnidirectional pattern. The panels can be fed in phase, with each one centered on the face of the tower, or fed in rotating phase with the proper mechanical offset. In the latter case, the input impedance match is considerably better.

Directionalization of the azimuth pattern is realized by proper distribution of the feed currents to individual panels in the same layer. Stacking layers provides gains comparable with top-mounted antennas.

The main drawbacks of panel antennas are (1) high windloading, (2) a complex feed system inside the antenna, and (3) restrictions on the size of the tower face, which determines to a large extent the omnidirectional pattern of the antenna.

UHF Side-Mounted Antennas. Utilization of panel antennas in a manner similar to those for VHF applications is not always possible at UHF installations. The high gains required for UHF broadcasting (in the range of 20 to 40, compared with gains of 6 to 12 for VHF) require far more panels with the associated complex feed system.

The zigzag panel antenna described in Sec. 8.8.1 has been used for special omnidirectional and directional applications. For custom directional patterns, such as a cardioid shape, the pylon antenna can be side-mounted on one of the tower legs. Many stations, in fact, simply side-mount a pylon-type antenna on the leg of the tower that faces the greatest concentration of viewers. It is understood that viewers located behind the tower will receive a poorer signal, however, given the location of most TV transmitting towers—usually on the outskirts of their licensed city or on a mountaintop; this practice is sometimes acceptable.

8.9 SATELLITE COMMUNICATIONS SYSTEMS

8.9.1 The Beginning of the Space Age

In October 1957, when the Soviet Union launched Sputnik I, most people were unaware of Arthur C. Clarke or of a suggestion he had made in 1946. Then, in 1958, when the U.S. space program began, there was more interest in the race to space for military purposes than in developing an orbiting satellite communications network. Several more years would pass before satellites were to form such a global communications network.

Clarke's suggestion had two parts. First, if a relay satellite were placed into orbit around the earth at a specific altitude (22,235 statute miles) and with a specific velocity (6789 statute miles per hour), that object would appear to be suspended over a point on the earth's surface. Second, if a group of three such relay satellites were placed in such a geosynchronous or geostationary orbit and spaced equally (at 120° longitude from one another) around the earth, they could provide a relay service for almost instantaneous communications to just about every point on earth. Each would cover 42.5 percent of the earth's surface with some overlapping coverage. The only areas not served would lie above 81° north latitude and below 81° south latitude. Earth-based stations communicating to these satellites could be fixed in position and would not require antennas capable of tracking satellites. (See Fig. 8.13.)

In 1946, the world lacked rocket engines with enough power and guidance systems with enough accuracy to lift a satellite from the earth's surface and into the required circular orbit. By the late 1950s, however, advances in rocketry had led to engines that could place a spacecraft into a highly elliptical, pseudostationary orbit. (See Fig. 8.14.) For example, a satellite, traveling in an orbit based on calculations first carried out by seventeenth-century astronomers and mathematicians, would speed rapidly by the earth at perigee (at an altitude of 500

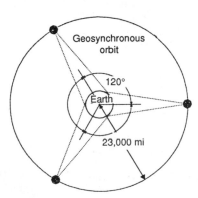

FIGURE 8.13 Clarke's geosynchronous orbit could carry three relay satellites, each covering 42.5 percent of the globe with some overlap.

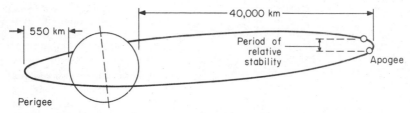

FIGURE 8.14 An elliptical, pseudo-stationary orbit places a satellite in a position of relative stability for several hours, during which it can be used as a relay station with only limited tracking requirements.

km). At apogee, 40,000 km, the satellite would appear to be relatively stationary for a period of several hours and during that time could be used as a relay station. The drawback to a satellite in such an orbit was the need for an earth station antenna with tracking capabilities for transmission and reception.

Commercial Satellite Communications. The first commercial satellite operation occurred on July 10, 1962, when television pictures were first beamed across the Atlantic Ocean through the Telstar 1 satellite. The launch vehicle, however, was still unable to place the spacecraft into a stationary position. Then, three years later, with millions of dollars routed to the development of rockets, INTELSAT saw its initial craft, Early Bird 1, launched into a geostationary orbit—a rapidly growing communications industry was born. In the same year, the USSR inaugurated the Molnya series of satellites, which traveled in more elliptical orbits, to better meet the needs of that nation and its more northerly position. The Molnya satellites are placed in orbits that are inclined about 64° relative to the equator, with an orbital period half that of the earth.

The geostationary orbit at 23,000 mi altitude has a limited number of open parking places for new satellites today. In fact, in an effort to make more satellite locations available, spacing between existing craft was reduced amid outcries that serious interference would result. For more channel capacity in the crowded area, a second band of frequencies, the Ku band (11–14 GHz), was initiated in addition to the original C band (4–6 GHz). Neither of these changes has resulted in a significant increase in interference. The possible introduction of 40-GHz systems for additional channel capacity is also not expected to produce any noticeable detrimental effects.

In this discussion, attention is directed toward the operation of the earth and space segments of a satellite communications system. For additional and more technical material on this topic and on the mechanics of geosynchronous and pseudo-geosynchronous orbits and placement of a spacecraft into such orbits, see the bibliography at the end of this section.

8.9.2 The Satellite Communications Link

A satellite relay system involves three sections. On the ground is an up-link transmitter station, which beams signals to the satellite. The satellite, as the space segment of the system, receives, amplifies, and retransmits the signals back to the earth. The down-link receiving station completes the system and with the up-link forms the earth segment. (See Fig. 8.15.)

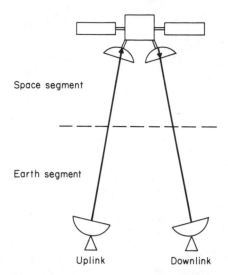

Space segment

Earth segment

Uplink   Downlink

FIGURE 8.15 The satellite communications link consists of the up-link and down-link, which form the earth segment, and the satellite, which forms the space segment.

Two frequency bands are assigned for standard satellite communications. C band operates on 6-GHz frequencies for transmission of information to the satellite. The electronics package aboard the craft converts the signals to 4-GHz frequencies for the return to earth. At much higher frequencies, the Ku band up-link transmits around 14 GHz, with conversion at the satellite to down-link channels at 12 GHz. Many satellites provided service for either C or Ku band. Now, some craft include electronics for both frequency bands. For military and maritime uses, other frequencies are also in use.

Because of the frequencies used, satellite communications are considered a microwave radio service. This introduces some concerns with regard to satellite-link operations, including the following.

1. A line-of-sight path between transmitter and receiver is required.

2. Attenuation of signals resulting from meteorological conditions, such as rain and fog, is particularly serious for Ku-band operation, but less troublesome for C-band operation.

3. Arrangements for shielding the antennas from terrestrial interference include siting the antenna in a depression in the terrain, either constructed or natural, or building a wall of material that is opaque to microwave energy around the antenna. Any foliage at the antenna site threatens possible attenuation effects if it extends into the beam path of the antenna or through scattering of the signal energy by individual leaf surfaces.

4. A received signal strength, based on the inverse-square law, requires highly directional transmitting and receiving antennas, which, in turn, need significant accuracy of aiming. Most antennas are parabolic in form and can be moved for best signal reception. The development of the phased array introduced a steerable radiation pattern from a relatively fixed, flat-surface antenna primarily for Ku-band use.

5. The effects of galactic and thermal noise sources on low-level signals require electronics for satellite service with exceptionally low noise characteristics.

The Satellite. It is convenient to discuss the satellite communications link by starting with its middle point, the satellite. By understanding the relay station requisites for receiving and transmitting, it is easier to realize the requirements placed on the up-link and down-link.

Like all relay stations, the communications spacecraft contains antennas for receiving and retransmission. A low-noise amplifier (LNA) boosts the signal from the antenna before frequency conversion for transmission. Then, a high-power amplifier (HPA) feeds the signal to a directional antenna with a radiation pattern designed to cover a prescribed area of the earth. The electronics receive power generated with solar cells and buffered with storage batteries, which arc recharged by the array of solar cells. The batteries are the only source of power when the satellite passes into the earth's shadow at certain times of the year. Finally, a guidance system stabilizes the attitude of the craft, because as it rotates around the earth, the craft would otherwise begin to tumble and its antennas would no longer point toward the earth. As a part of the guidance system, the craft also contains rocket engines for station keeping—maintaining its position in the geostationary arc. The overall mass and volume of all components of the craft must be limited, however, largely because of launch considerations.

Antennas. The antenna structure of a communications satellite consists of several antennas. One receives signals from earth. Another transmits those signals back to earth. The transmitting antenna may be constructed in several sections to carry more than one signal beam. Finally, a receive-transmit beacon antenna provides communication with a ground station to control the operation of the satellite. The control functions include turning parts of the electronics on and off, adjusting the radiation pattern, and maintaining the satellite in its proper position.

The design of the complex antenna system for a satellite relies heavily on the horizontal and vertical polarizations of signals to keep incoming and outgoing information separated. Multiple layers in carefully controlled thicknesses of reflecting materials, which are sensitive to signal polarizations, can be used for such purposes. Also, multiple-feed horns can develop more beams to earth. Antennas for different requirements may combine several antenna designs, but nearly all are based on the parabolic reflector, because of two properties of the parabolic curve:

1. Rays received by the structure, which are parallel to the feed axis, are reflected and converged at the focus.
2. Rays emitted from the focus point are reflected and emerge parallel to the feed axis.

Special cases may involve spherical and elliptical reflectors, but the parabolic is most important.

Transponders. From the antenna, a signal is directed to a chain of electronic equipment known as a *transponder*. This unit contains the circuits for receiving, frequency conversion, and transmission. Some of the first satellites (or birds, as they are often called) launched for INTELSAT and other systems contained only one or two transponder units. However, initial demand for satellite link services forced satellite builders to develop more economical systems with multiple-transponder designs. Such requirements meant refinements in circuitry that would allow each receiver-transmitter chain to operate more efficiently (less power draw). Multiple-transponder electronics also meant a condensation of the physical volume needed for each transponder with as small as possible an increase in the overall size and mass of the satellite to accommodate the added equipment.

Generally, satellites placed in orbit now have 12 or 24 transponders in C-band units, each with 36-MHz bandwidths. However, variations in the number of transponders and in their operating bandwidths exist. Some Ku-band systems use fewer transponders with wider bandwidths, but others, particularly those designed for use in Europe, have 40 transponders of 27-MHz bandwidth. Identifi-

TABLE 8.1 Identification of the Transponders, Described by a Frequency Plan for C Band

#1. 3720	#2. 3760	#3. 3800	#4. 3840
#5. 3880	#6. 3920	#7. 3960	#8. 4000
#9. 4040	#10. 4080	#11. 4120	#12. 4160

cation of the transponders is described by a frequency plan for C band, as shown in Table 8.1.

Users of satellite communication links are assigned to transponders generally on a lease basis, although it is possible to purchase a transponder. The cost, however, is prohibitive for most customers. Assignments usually leave one or more spare transponders aboard each craft, allowing for standby capabilities in the event a transponder should fail. Some current satellites have additional unused transponders, because lessees have been unable to afford the costs of operation.

By assigning transponders to users, the satellite operator simplifies the design of up-link and down-link facilities. An earth station controller can be programmed for the transponders of interest. For example, the TV station may need access to several transponders from one satellite. The operator enters only the transponder number (or carrier frequency) of interest. The receiver handles retuning of the receiver and automatic switching of signals from the dual-polarity feed horn on the antenna. Controllers can also be used to move an antenna from one satellite to another automatically, but most broadcast facilities have more than one antenna for multiple satellite reception.

Each transponder has a fixed center frequency and a specific signal polarization according to the frequency plan. For example, all odd-numbered transponders might use horizontal polarization, while even-numbered ones might be vertically polarized. Excessive deviation from the center carrier frequency by one signal does not produce interference between transponders and signals because of isolation provided by cross-polarization. This concept extends to satellites in adjacent parking spaces in the geosynchronous orbit. Center frequencies for transponders on adjacent satellites are offset in frequency from those on the first craft. An angular offset of polarization affords additional isolation. The even and odd transponder assignments are still offset by 90° from one another. As spacing is decreased between satellites, the polarization offset must be increased to reduce the potential for interference.

While the discussion has centered on the down-link frequency plan for a C-band satellite, up-link facilities follow much the same plan, except that up-link frequencies are centered around 6 GHz. A similar plan for Ku-band equipment uses up-linking centered around 14 GHz and down-link operation around 12 GHz.

Powering Sources. Providing power to the electronics aboard the spacecraft is a major engineering challenge. Nuclear power has not been utilized because many questions regarding integration of a small reactor into a satellite system remain unsolved. Fortunately, constant evolution of the storage battery has provided the space program with various types of renewable power devices.

During the launch phase, control circuitry obtains power from rechargeable batteries on the craft. After the final orbital position is achieved, the control station activates the deployment of arrays of solar cells. When the solar panels are functional, all operational power is derived from the panels, as well as current to recharge the batteries for backup during times of eclipse, caused when the earth blocks illumination of the panels from direct sunlight.

The solar cells most commonly used are silicon with boron or arsenic doping to create *pn* junctions. The light-to-electricity conversion efficiency of this material is approximately 10 percent. Aluminum or gallium arsenide cells with approximately 18 percent efficiency are under development. The array is created by connecting cells in series to achieve a larger voltage and a number of series groups in parallel for greater current capability. An array of 2000 cells covering 1 m^2 produces approximately 100 W.

With any type of photovoltaic material, the output of a cell is determined by a number of factors. The primary factors include:

1. The surface area of the cell exposed to light
2. The average solar flux per unit area (light intensity)
3. The temperature of the cell material
4. The variation of the light source from a position perpendicular to the cell surface

The initial capabilities, as well as the end-of-life capacity, of a solar power source must take these factors into account and allow for them by designing in accordance with worst-case conditions. One such consideration is the proper positioning of the solar array when operation of the satellite is initiated. Positioning remains important at all times: The variation of solar energy with the seasons varies approximately 8 percent between the solar equinoxes and solstices, because of the varying angle of incidence of the sunlight. This seasonal change produces a variation in temperature as well, and over time, the bombardment by proton radiation from solar flares reduces the efficiency of electric current production.

Two types of solar panels are in general use. One is a flat configuration arranged as a pair of deployable panel wings that extend to either side of the satellite body. In the other type the solar cells are attached to the barrellike body of the satellite. (See Fig. 8.16.)

Each approach has advantages and disadvantages. On the winglike panels, approximately 18,000 cells are used to maintain a 1200-W output capability at end-of-life. The other configuration uses 80,000 cells mounted on a 2.16-m-diameter cylinder that is 5.5 m in height. The wing approach is the more efficient of the two, with an end-of-life (7 years) rating between 15 and 25 W/kg. The spinning cylindrical array is rated at 6 to 10 W/kg. The wing keeps all cells illuminated at all times, producing more electric current. However, all cells are subject to a higher rate of bombardment by space debris and by solar proton radiation and to a higher average temperature (approximately 60°C) because of constant illumination.

The rotating-drum design places a small portion of all cells perpendicular to

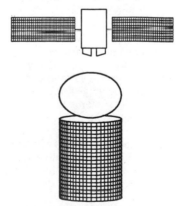

FIGURE 8.16 Two types of solar cell arrays are commonly used.

the sun's rays at any one time. Cells that are not perpendicular to the light source produce power as long as they are illuminated, but the amount is reduced and depends on the angle of light incidence. The individual cells are not all exposed in one direction simultaneously and are less likely to be struck by space debris. Also, because cells on the drum are not exposed to sunlight at all times, their average temperature is significantly less, approximately 25°C, which extends their operating lifetime.

The output current, voltage, and power from the solar panels undergo variation, depending on the conditions. Without a means of controlling or regulating the power, the operation of the electronics package would also vary. An unregulated power bus is simpler and takes less of the allowable mass, or mass budget, of the craft. System electronics circuitry must include on-board regulators to maintain consistent RF levels. The alternative is to use bus regulation with more complex circuitry, a somewhat lower output power, and reduced reliability because of the additional components in the system.

Some variation can be accommodated by the storage batteries in either case. The most serious need for batteries following insertion into orbit of the geosynchronous satellite occurs approximately 84 days per year. The earth passes between the sun and the satellite, causing an eclipse. The blackout period can last as long as 70 min within one day. During these periods the need for a rechargeable supply is essential. Nickel-cadmium batteries have played a major role in powering such spacecraft.

Charging facilities allow for a constant $C/10$ charge rate (C = total battery capacity in ampere-hours or the charge delivered in 1 h; $C/10$ is a charging current of one-tenth that amount). When nearly full charge is reached, the current is cut back to a $C/50$ rate to maintain the charge. Silver-cadmium, nickel-hydrogen, and silver-hydrogen combinations provide higher energy/mass unit values, but at a higher initial cost. The silver-hydrogen cell increases the storage capacity by a factor as much as five times over Ni-Cad cells. Reliability dictates redundancy, with batteries typically connected in parallel to maintain the needed voltages and currents in the event that one or more cells in a battery fail.

Power to the electronics on the craft requires protective regulation to maintain consistent signal levels. Most of the equipment operates at low voltages, but the final stage of each transponder chain ends in a high-power amplifier or HPA. The HPA of C-band satellite channels may include a traveling-wave tube (TWT) or a solid-state power amplifier (SSPA). Ku-band systems rely on TWT devices at present. Traveling-wave tubes, similar to their earth-bound counterparts, klystrons, require multiple voltage levels. The filaments operate at 5 V, but beam-focus and electron-collection electrodes in the device require voltages in the hundreds and thousands of volts. To develop such a range of voltages, the power supply includes voltage converters.

From these voltages, the TWT units produce output powers in the range of 8.5 to 20 W. Most systems are operated at the lower end of the range for increased reliability and greater longevity of the TWTs. In general, lifetime is assumed to be 7 years.

At the receiving end of the transponder, signals coming from the antenna are split into separate bands through a channelizing network, which directs each input signal to its own receiver, processing amplifier, and HPA. At the output, a combiner brings all the channels together again into one signal to be fed to the transmitting antenna.

8.9.3 Control of Satellites

The electronics and attitude control section of the satellite performs tasks of turning transponders on and off, as directed from the ground station, adjustments of antennas to keep the beams focused to form the footprint of signal levels designated for each satellite system, and controlling the small package of rocket engines in the craft to maintain the correct parking position in the geosynchronous orbit. Much of the attitude control is microprocessor-controlled, but the ground station can override on-board equipment to perform necessary corrections.

Communications with the satellite from the ground often use other frequencies in addition to those relayed through the transponders. A 148- to 149.9-MHz up-link and 136- to 138-MHz down-link in VHF or a 2.025- to 2.120-GHz up-link and 2.2- to 2.3-GHz down-link in S band are common. Control information and telemetry data reporting on the status of the satellite system to the ground station are transmitted as encoded tones. Because of the proximity of satellites, strict protocols are needed to avoid one satellite from responding when the neighboring "bird" is being interrogated. Protection against erroneous control signal reception is improved through narrow-band reception or wide-band reception combined with spread-spectrum transmission and digitally encoded data.

Positioning and Attitude Control.   Maintaining the position of the spacecraft in its orbit is called *station keeping*. With the altitude of the craft at about 23,000 mi, some motion of the craft is possible without requiring major redirecting of earth station antennas. In most cases, a window of sides approximately 0.1° in length as seen from earth (equivalent to 75 km) is considered acceptable. If the craft moves outside the window, reception S/N (signal-to-noise) is degraded.

Movement of a craft from its assigned location is caused by forces of the surrounding environment. One significant factor is gravitational effects. It is the normal pull of gravity upon the fast-moving craft (centrifugal force) that keeps the satellite in its orbit. The force of gravity has been found to vary at some locations around the equator. As a result, satellites positioned near the longitudes of gravitational anomalies become advanced or retarded from their correct location, and corrections are periodically required.

Other factors to be accounted for are the gravitational attractions of the moon and sun. The pull of the moon is approximately three times stronger than that of the sun. Most of the error to the orbit caused by these bodies is perpendicular to the equatorial plane, that is, in the inclination of the orbit with respect to the equator. Slight errors in the longitude of the craft result from solar and lunar attractions, but these are compensated with a change in the dimensions of the orbit.

Lesser factors that disturb the position of the spacecraft and may require occasional corrections include:

1. The pressure on the solar power panels by the solar radiation flux
2. The force caused by radio frequency power radiated from the satellite
3. The magnetic field of the earth
4. The self-generated torques related to antenna displacements, the solar arrays, and changing masses of on-board fuel supplies

The results of meteorite impacts on spacecraft are considered, but have not been experienced significantly by any satellite to date.

Attitude of the spacecraft is defined as the combination of pitch, roll, and yaw. These are motions that may occur without causing a change in the orbital position of the satellite. (See Fig. 8.17.) *Pitch* is the rotation of the craft about an axis perpendicular to the plane of the orbit. *Roll* is rotation of the craft about an axis in the plane of the orbit. *Yaw* is rotation about an axis that points directly toward the center of the earth. Satellite attitude is determined by the angular variation between satellite body axes and these three reference axes.

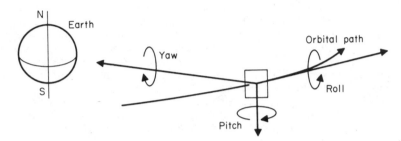

FIGURE 8.17 Attitude of the spacecraft is determined by pitch, roll, and yaw, rotations around three reference axes.

Attitude control involves two functions. First, it is necessary to rotate that part of the satellite that must point toward the earth around the pitch axis. Rotation must be precisely one revolution per day or 0.25° per minute. Second, the satellite must be stabilized against torque from any other source. To perform these functions, numerous detectors measure the sense of the satellite with respect to earth horizons, the sun, specified stars, gyroscopes, and radio frequency signals. Attitude correction systems may involve a control loop with the ground control station or may be designed as a closed loop by storing all required attitude reference information within the satellite itself.

Relative to the first function listed, satellites can be spin-stabilized gyroscopically by rotating the body of the satellite in the range of 30 to 120 times per minute. This effectively stabilizes the craft with respect to the axis of the spinning and is used with cylindrical-type satellites. Although a relatively simple approach, it has a major drawback—the antenna would not remain pointed toward earth if it were mounted on the rotating portion of the craft. The solution is to de-spin the section holding the antenna at one revolution per day about the pitch axis to maintain a correct antenna heading.

8.9.4 The Up-Link

The ground-based transmitting equipment of the satellite system consists of three sections: base band, intermediate frequency (IF), and radio frequency (RF).

The base-band section interfaces various forms of incoming signals with the design of the satellite being used. Signals provided to the base-band section may already be in a modulated form with modulation characteristics—digital, analog, etc.—decided or determined by the terrestrial media bringing signals to the up-link site. Depending on the nature of the incoming signal—voice, data, or video—

some degree and type of processing will be applied. In many cases, multiple incoming signals can be combined into a single up-link signal through multiplexing.

When the incoming signals are in the correct format for passage through the satellite, they are applied to an FM modulator, which converts the entire package upward to an intermediate frequency of 70 MHz. The use of an IF section has several advantages. First, a direct conversion between the base band and the output frequency presents difficulties in maintaining frequency stability of the output signal. Second, any mixing or modulation step has the potential of introducing unwanted by-products. Filtering in the IF may be used to remove spurious signals resulting from the mixing process. Third, many terrestrial microwave systems include a 70-MHz IF section. If a signal is brought into the up-link site by terrestrial microwave, it becomes a simple matter to connect the signal directly into the IF section of the up-link system.

From the 70-MHz IF section, the signal is converted upward again, this time to the output frequency of 6 GHz (or 14 GHz) before application to a high-power amplifier (HPA). Several amplifying devices are used in HPA designs, depending on the power output and frequency requirements. For the highest power level in the C or Ku band, klystrons are employed. Different devices are available with pulsed outputs ranging from 500 W to 5 kW and a bandwidth capability of 40 MHz. This means that a separate klystron is required for each 40-MHz-wide signal to be beamed upward to a transponder.

Another type of vacuum power device for HPA designs is the traveling-wave tube (TWT). While similar in some aspects of operation to a klystron, the TWT is capable of amplifying a band of signals at least 10 times wider than the klystron. Thus, one TWT system can be used to amplify the signals sent to several transponders on the satellite. With output powers from 100 W to 2.5 kW, the bandwidth capability of the TWT offsets its much higher price than the klystron.

Solid-state amplifiers, based on field-effect transistor (FET) technology, can be used for both C- and Ku-band up-link HPA systems. The power capability of solid-state units is limited for C band (5–50 W) and for Ku band (1–6 W). The cost of the devices is high, but the wide-band capabilities are good and reliability is excellent. These devices are useful in amplifying signals directed at more than one transponder.

Antennas.   The seemingly limited output of the HPA, when applied to a parabolic reflector antenna, experiences a high degree of gain (dB$_i$), when referenced to an ideal isotropic antenna. For example, large reflector antennas approximately 10 m in diameter offer gains to 55 dB, increasing the output of the 3-kW klystron or TWT amplifier to an effective power level of 57–86 dB W. Smaller reflector sizes (6–8 m) may be used, with observation of certain restrictions with regard to creation of interference. For these smaller sizes, gain will be somewhat less. With reflector sizes to 30 m, such as those for international services, gains are higher by a few dB.

Several configurations of parabolic antenna, used for satellite communications service, are illustrated in Fig. 8.18. The prime focus type, with a single parabolic reflector, places the signal source in front of the reflector precisely at the focal point of the parabola. Antennas of larger sizes commonly employ a feed horn supported with a tripod arrangement of struts. Because the struts, feed horn assembly, and waveguide coupling to the feed horn are located in the transmitted beam, these components are designed with as little bulk as possible, yet are strong enough to withstand weather conditions.

If we move the feed horn and its support to the side of the reflector and out of

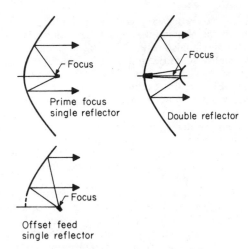

FIGURE 8.18 Satellite receiving-transmitting antennas are of three general types: prime focus, single reflector; offset feed, single reflector; and double reflector.

the radiated beam, we have a single-offset reflector type of antenna. The reflector maintains the shape of a section of a parabola; the closed end of the curve is not included. The feed horn, while still located at the focal point of the curve, points at an angle from the vertex of the parabola shape.

The third antenna configuration is the double reflector. The primary reflector is parabolic, while the subreflector surface, mounted somewhat in front of the focal point of the parabola, is hyperbolic. One focus of the hyperbolic reflector is located at the parabolic focal point, while the second focal point of the subreflector defines the position for the feed horn. Signals reflected from the hyperbolic subreflector are spread across the parabolic prime reflector, which directs them as a parallel beam toward the satellite.

The two-reflector antenna provides several advantages over a single-reflector type. These include the following:

1. The overall front-to-back dimension of the two-reflector system is shorter, simplifying its mounting structure and decreasing its windloading.
2. Placing the subreflector closer to the main reflector means less spillover signal because signals are not directed as close to the edge of the main reflector.
3. When used in receive mode, a smaller level of noise is experienced, because spillover from sky radiation is usually much less than terrestrially generated noise.
4. The accuracy requirement of the reflector surfaces is not as stringent as with a single-reflector type of structure.

The antenna for transmission of signals to the satellite can be used to receive signals from the satellite. To provide this capability, a directional switch or coupling is added to prevent the signal from the transmitter HPA from directly entering the receiver. These switching devices, called circulators, use waveguide characteristics to create a signal path linking the transmitted signal to the antenna feed horn, while a received signal path from the feed horn to the receiver input exists simultaneously.

8.9.5 The Down-Link

Satellite receiving stations, like up-link equipment, interface ground-based equipment to the satellite. To serve this function, the earth station, as a TVRO (television receive only) or a TR (transmit-receive) facility, consists of the receiving antenna with a low-noise amplifier (LNA). A 4-GHz (or 12-GHz) tuner, a 70-MHz IF section, and a base-band output stage complete the necessary equipment package.

Antennas for ground-based receiving stations fall into several categories. First, the antenna used for transmitting to the satellite can also be used as a receiving antenna. Unlike terrestrial two-way radio systems, where one station transmits and then the other, the satellite system participates in both functions simultaneously. This fact explains the offset between up-link and down-link frequencies. To make operation possible, the circulator separates and directs the transmitted and received signals to the correct locations.

Receiving antennas for commercial applications, such as radio/TV networks, CATV networks, and special services or teleconferencing centers, range from 7 to 10 m in diameter for C-band operation, while the Ku-band units can be smaller. Antennas for consumer and business use may be even smaller, depending on the type of signal being received and the quality of the signal to be provided by the down-link installation. If the signals being received are strictly digital in nature, smaller sizes are sufficient. If TV signals are of interest and "broadcast quality" is required, a larger-size reflector is suggested.

Antenna size for any application is determined somewhat by the type of transmission, the band of operation, the location of the receiving station, typical weather in the receiving station locale, and the quality of the output signal from the down-link. As indicated previously, digital transmissions allow a smaller main reflector to be used, because digital systems are designed with error correction. The data stream periodically includes information to check the accuracy of the data and, if errors are found, to make corrections. If errors are too gross for error correction, the digital circuitry provides error concealment techniques to hide the errors. Greater emphasis is placed on error correction for applications involving financial transactions or life-critical data, such as might be involved with a manned space flight. Absolute correction is less critical for entertainment-type programming functions, such as TV and audio. The nature of the application also helps to determine if the antenna must be strictly parabolic, or if one of the spherical types, generally designed for consumer use, will be sufficient.

Regarding the frequency band of operation, the lower the signal frequency, the larger the antenna reflector must be for the same antenna gain. However, there are compensating factors such as the required output signal quality, station location, and local environment. Generally, the gain figure and directivity of larger reflectors are greater than those of smaller reflectors. In short, the size of the reflector depends on the level of signal that can be received reliably at a specific location under the worst possible conditions. Antenna gain must boost the RF signal to a level that can be accommodated by the receiving electronics.

One of the most critical sections of the receiver is the low-noise amplifier (LNA) or low-noise converter, the first component to process the signal following the antenna. The amplifier should not add noise to the signal, because once added, the noise cannot be removed. LNAs are rated by their noise temperature, usually a number around 211 K. This is equivalent to −62°C. The cost of LNAs increases significantly as the temperature figure goes down.

Following the LNA, the receiver tuner converts signals to a 70-MHz IF frequency. As with the up-link equipment, an output at 70 MHz is useful to connect to terrestrial microwave equipment if desired. A second conversion takes video, audio, or data to base-band signals in the form most convenient for those using the communications link.

8.9.6 Signal Formats

The signal transmitted from the up-link site or from the satellite is in the form of frequency modulation (FM). In FM modulation, the center or carrier frequency is rapidly changed above and below its assigned value by the modulating information. The rate at which the frequency alternates above and below the center is determined from the frequency of the modulating signal. How far the modulated signal changes or deviates from the center frequency is based on the amplitude of the signal to be modulated.

In radio transmitters, the bandwidth of the information to be modulated onto a carrier must be controlled within certain limits. If an audio signal with components to 15 kHz is applied to an AM transmitter, everything greater than about 10 kHz must be removed with a bandpass filter. If the information above 10 kHz modulated the AM carrier, it would cause interference with AM radio stations in adjacent channel assignments. The same situation occurs with FM radio except that components greater than 15 kHz are filtered out. For satellite use, the transmitted bandwidth should remain within 36 MHz.

Certain limits are placed on up-linked signals to avoid excessive bandwidth conditions. For example, a satellite channel for television typically contains a single video signal and its associated audio. Audio is carried on one or more subcarriers. To develop the composite signal, each audio channel is modulated onto its subcarrier frequency. Then the subcarriers and the main channel of video are applied as modulation to the up-link carrier. The maximum level of each component is carefully controlled to avoid overmodulation.

In the case of telephone relay circuits, the same concept of subcarriers is used. A number of individual voice circuits are combined into groups. These groups can then be multiplexed to subcarriers through various digital methods. The result is as many as 1000 telephone conversations operating simultaneously through the satellite. No interference occurs between any of them as long as levels are carefully controlled to avoid overmodulation by any individual signal.

The majority of the transponders serve specific organizations. Each of the major TV networks in the United States, for example, has specific transponders assigned on certain satellites. Specific member stations of those networks are designated as up-link sites for use of "back-haul" transponders to send signals to the network's central headquarters. If station A has material to be sent to headquarters, a schedule is set up for when that station can use the transponder. Stations may send signals to the transponder only according to that schedule. If two stations attempt to transmit at the same time, that is, to double-illuminate the transponder, the signal returned to earth will be an unusable combination from the two up-links. Occasional errors in scheduling do occur, but the system works reasonably well.

In the case of radio stations, a different system has been devised. Because there are many more licensed radio stations, a different approach allows multiple users of a transponder simultaneously. In effect, the 36-MHz bandwidth of a transponder is subdivided into smaller channels that are well suited to high-quality audio. (Audio transmitted by most stations is limited to 15 kHz, while TV signals contain components reaching to 5 MHz.) Each prospective user of the transponder is assigned to a channel carrier for this system, called *single channel per carrier* (SCPC).

Each of the single channels in the satellite transponder in this system is processed as a component of a wider-bandwidth signal. The only effect the transponder has on a group of single channels operating at any one time is to reduce the level of any channels that are exceptionally high. In effect, the output level of all signals is controlled by the overall number of signals passing through the unit at the time. Without this effect, one significantly higher-level signal would cause intermodulation interference with all other active channels.

Bibliography

Bleazard, G. B. *Introducing Satellite Communications*. Halsted (Wiley) Press, New York, 1985.

Johnson, R. C., and H. Jasik. *Antenna Applications Reference Guide*. McGraw-Hill, New York, 1987.

Freeman, R. L. *Reference Manual for Telecommunication Engineers*. Wiley, New York, 1985.

Marlal, G., and M. Bousquet. *Satellite Communications Systems*. Wiley, New York, 1986.

Rainger, Peter, D. N. Gregory, R. V. Harvey, and A. Jennings. *Satellite Broadcasting*. Wiley, New York, 1985.

Slater, J. N., and L. A. Trinogga. *Satellite Broadcasting Systems, Planning and Design*. Halsted (John Wiley) Press, New York, 1985.

8.10 CABLE TELEVISION SYSTEMS

8.10.1 Overview

Cable television systems, sometimes referred to as community-antenna television or CATV, use coaxial cable to distribute television, audio, and data signals to homes or other establishments that subscribe to the service. Systems with bidirectional capability can also transmit signals from various points within the cable network to the central originating point (head end).

Cable distribution systems usually use leased space on utility poles owned by a telephone or power distribution company. In areas with underground utilities, cable systems are normally installed underground, either in conduits or buried directly, depending on local building codes and soil conditions.

Evolution of Cable Television. Cable television began in rural areas in the early 1950s as a means of bringing television service to areas with no broadcast stations. Other early "systems" brought television reception into mountainous areas where the terrain blocked signals from individual home antennas. These systems typically had a limited capability of five channels and carried only the three major television networks. During the 1960s, cable moved into areas served by broadcast stations but without a full complement of network channels. Twelve to 20 channels were common and were used almost exclusively to carry broadcast television signals. By the mid-1970s, satellite distribution of pay television programming made cable viable in urban areas where a good selection of over-the-air programming already existed. The systems offered a greater variety of viewing choices, including independent stations from distant cities and pay television.

Systems constructed in the 1980s typically have a capacity of from 50 to 100 channels and include bidirectional transmission capability. Interactive programming, information retrieval, home monitoring, and point-to-point data transmission are possible. Many systems include an institutional network in addition to one or two subscriber networks.

In 1976, cable television served 12 million subscribers (17 percent of all U.S. television households). Of those, 565,000 were also subscribers to pay television services. By the middle of 1983, cable systems had grown to serve nearly 32 million subscribers (38 percent of all households in the United States), with pay services totaling almost 19 million. Projections for 1990 show more than 59 million homes using cable, a 62 percent penetration of all television households in the United States.

Cable television systems are generally of two types. *Subscriber networks* serve primarily residential subscribers. Services carried by subscriber networks are mainly entertainment and information programming. *Industrial networks* serve commercial, educational, and governmental concerns. The majority of channels on these systems carry informational programming or data.

Program Sources.   Television picture and sound program material provided on cable systems may originate from local or distant broadcast stations or may be relayed by microwave or satellite. A number of advertiser-supported networks available only through cable are distributed by satellite, as are pay television services. In addition, programs may be originated by the cable television operator. Such *local origination* (LO) programming may be produced in studio facilities, reproduced using videotape playback equipment, or generated by automated character-generating systems controlled by wire news services and weather instruments, for example. Bidirectional cable also allows "remote" LO, using an upstream channel to transport the signal from the remote site to the head end, where the signal is converted to a downstream channel for distribution to all subscribers.

8.10.2 Channel Capacity

Although the majority of existing systems have an upper frequency limit of 300 MHz, many newer systems are capable of transmitting signals in the range of 5 to 400 MHz. Systems designed for bidirectional transmission divide the spectrum between the two directions. For subscriber networks, 5 to 35 MHz is used for upstream transmission (toward the head end); 50 to 400 MHz carries downstream transmissions (toward the subscriber). For institutional networks, an equal number of channels in each direction is common. Upstream transmission is usually from 5 to 150 MHz, and downstream channels use frequencies from 200 MHz up.

The upper frequency limit of cable systems is largely a function of the hybrid devices used in the amplifiers. At present, 500-MHz amplifiers are available, and even higher-frequency devices are expected in the future. Other components needed for the system, such as passive line splitters, cable, and converters already exist for operation at 500 MHz or higher.

A maximum of 54 television channels can be carried downstream on a 400-MHz system (36 channels for a 300-MHz system). However, FCC regulations prohibit cable systems from using frequencies in the aircraft navigation and communications bands, making several channels unusable. As a result, most 400-MHz systems carry 50 channels (32 channels for 300 MHz systems). When additional capacity is necessary, the system can be constructed with dual cable, doubling the number of channels.

Channel Assignments. Cable television systems use VHF frequencies for all channels provided to subscribers. Bidirectional systems allow transmissions heading upstream on high frequencies (HF). UHF signals, as a rule, are not used on cable. Some confusion arises in that channel numbers normally associated with UHF channels are used in referring to the additional VHF channels. (See Table 8.2.)

TABLE 8.2 Cable Television Channel Frequencies and Designations

Frequency, MHz	Channel	Designation
5–35	Reverse transmission	Sub-band HF
54–88	2–6	Low-band VHF
120–170	14–22	Mid-band VHF
174–216	7–13	High-band VHF
216–300	23–36	Super-band VHF
300–464	37–64	Hyper-band VHF
468–644	65–94	Ultra-band VHF

Institutional networks generally allocate 20 or 25 channels in the downstream direction and 15 or 20 channels to upstream service. Most institutional networks carry only experimental services, so final channel configurations have not yet evolved. Expectations are that a major portion of the spectrum of institutional networks will be dedicated to digital transmissions.

8.10.3 Elements of a Cable System

System Configuration. Typical cable television systems are comprised of four main elements: a *head end*, the central originating point of all signals carried; a *trunk system*, the main artery carrying signals through a community; a *distribution system*, which is bridged from the trunk system and carries signals into individual neighborhoods for distribution to subscribers; and *subscriber drops*, the individual lines into subscribers' television receivers, fed from taps in the distribution system. (See Fig. 8.19.)

In a subscriber's home, the drop may terminate directly into the television receiver on 12-channel systems, or into a converter where more than 12 channels are provided. Many newer *cable-ready* receivers and videocassette recorders have such a converter integrated into the tuner, with a switch to enable the ad-

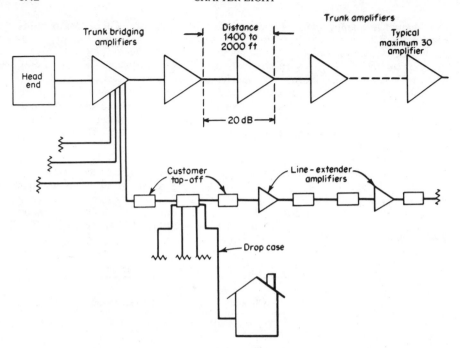

FIGURE 8.19 Typical CATV network. (*Source: D. G. Fink and D. Christiansen, eds.,* Electronic Engineer's Handbook, *2d ed., McGraw-Hill, New York, 1982. In* Television Engineering Handbook, *K. Blair Benson, ed., McGraw-Hill, New York, 1986)*

ditional channel tuning capability. Systems providing pay services may require a descrambler in the subscriber's home to allow the viewer to receive the special channels. Some systems use addressable converters or descramblers, giving the cable operator control over the channels received by subscribers with a computer. Such control enables impulse viewing or pay-per-view pay television service without a technician visiting the home to "install" the special service.

Interactive Service.   While the main purpose of cable television is to deliver a greater variety of high-quality television signals to subscribers, a growing interest is found in interactive communications, which allow subscribers to interact with the program source and to request various types of information, such as videotext. An interactive system can provide monitoring capability for special services such as home security. Additional equipment is required in the subscriber's home for such services. Monitoring requires a home terminal, for example, whereas information retrieval requires a videotext decoder or modem for data transmission.

8.10.4   Head End

The *head end* of a cable system is the origination point for all signals carried on the downstream system. Signal sources include off-the-air stations, satellite services, and terrestrial microwave relays. Programming may also originate at the head end. All signals are processed and then combined for transmission via the cable system. In bidirectional systems the head end also serves as the collection (and turnaround) point for signals originated within the subscriber and institutional networks.

The major elements of the head end are the *antenna system, signal-processing equipment, pilot-carrier generators, combining networks*, and equipment for any bidirectional and interactive services. Figure 8.20 diagrams a typical head end.

A cable television antenna system includes a tower and antennas for reception of local and distant stations. The ideal antenna site is in an area of low ambient electrical noise, where it is possible to receive the desired television channels with a minimum of interference and at a level sufficient to obtain high-quality signals. For distant signals, tall towers and high-gain, directional receiving antennas may be necessary to achieve sufficient gain of desired signals and to discriminate against unwanted adjacent channel, co-channel, or reflected signals. For weak signals, low-noise preamplifiers can be installed near the pickup antennas. Strong adjacent channels are attenuated with bandpass and bandstop filters.

Satellite earth receiving stations, or television receive only (TVRO) systems, are used by most cable companies. Earth station sites are selected for minimum interference from terrestrial microwave transmissions, which share the 4-GHz spectrum, but an unobstructed line-of-sight path to the desired satellites is also necessary. Most earth stations use parabolic receiving antennas, 13 to 23 ft in diameter, a low-noise preamplifier at the focal point of the antenna, and a waveguide to transmit the signal to receivers.

Signal Processing. Several types of signal processing are performed at the head end. Processing includes:

1. Regulation of the signal-to-noise ratio at the highest practical value
2. Automatic control of the output level of the signal to a close tolerance
3. Reduction of the aural carrier level of television signals to avoid interference with adjacent cable channels
4. Suppression of undesired out-of-band signals

Processing is also used to convert received signals to different channel frequencies for cable use. For example, UHF broadcast channels are changed to VHF frequencies. Such processing of off-air signals is accomplished with heterodyne processors or demodulator-modulator pairs. Signals from satellites are processed with a receiver, and they are then placed on a vacant channel by a modulator. Similarly, locally originated signals are converted to cable channels with modulators.

Pilot-carrier generators provide precise reference levels for automatic level control in trunk-system amplifiers. Generally, two reference pilots are used, one near each end of the cable spectrum. Combining networks group the signals from individual processors and modulators into a single output for connection to the cable network.

In bidirectional systems, a computer system is located at the head end. The configuration of the computer varies with the type of service to be offered and can range from a small microprocessor and display terminal to multiprocessor minicomputers with many peripherals. Such computers control the flow of data to and from terminals located within the cable television network. See Fig. 8.21 for a typical bidirectional head end.

Interactive services require one or more data receivers and transmitters at the head end. Polling of home terminals and data collection are controlled by the computer system. Where cable networks are used for point-to-point data trans

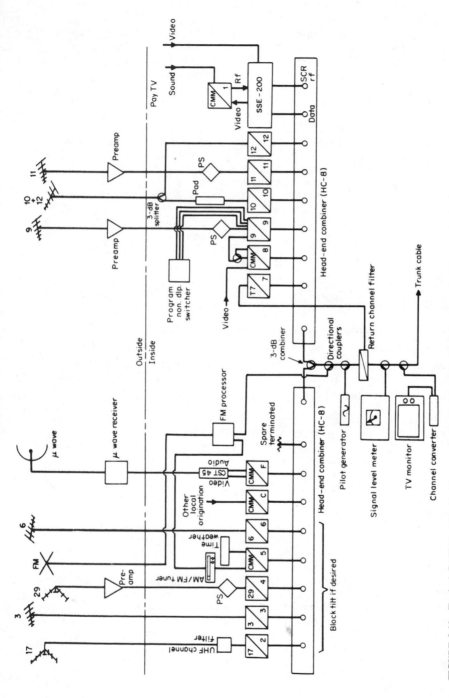

FIGURE 8.20 Typical CATV head end. (*Source: D. G. Fink and D. Christiansen, ed., Electronic Engineer's Handbook, 2d ed., McGraw-Hill, New York, 1982. In Television Engineering Handbook, K. Blair Benson, ed., McGraw-Hill, New York, 1986*)

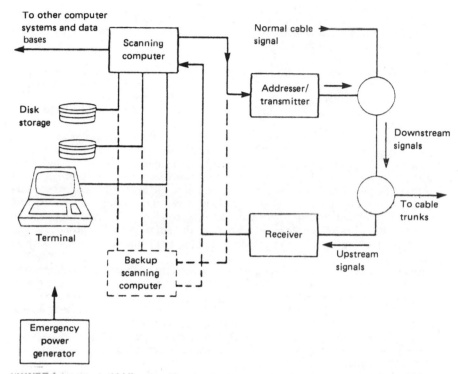

FIGURE 8.21 Typical bidirectional head end. *(Source: T. F. Baldwin and D. S. McVoy,* Cable Communications, *1983. Reprinted by permission of Prentice-Hall, Inc., Englewood Cliffs, NJ. In* Television Engineering Handbook, *K. Blair Benson, ed., McGraw-Hill, New York, 1986)*

mission, modems are required at each end location, with RF-turnaround converters used to redirect incoming upstream signals back downstream.

Modern cable television systems use computerized switching systems to program one or more video channels from multiple-program sources. In addition, computer-controlled alphanumeric character generators are used for development of automated news, weather, stock market, and other information channels.

Institutional networks on cable systems require switching, processing, and turnaround equipment at the head end. Video, data, and audio signals that originate within the network must be routed back out over either the institutional or the subscriber network. One method of accomplishing this is to demodulate the signals to the base band, route them through a switching network, and then remodulate them onto the desired network. Another method is to convert the signals to a common intermediate frequency, route them through an RF switching network, and then up-convert them to the desired frequency with heterodyne processors.

In larger systems, different areas of the network may be tied together with a central head end using supertrunks or multichannel microwave. Supertrunks or microwave systems are also used where the pickup point for distant over-the-air stations is located away from the central head end of the system. In this method, called the *hub system*, supertrunks are high-quality, low-loss trunks, which often

use FM transmission of video signals or feed-forward amplifier techniques to reduce distortion. Multichannel microwave transmission systems may use either
amplitude or frequency modulation.

Heterodyne Processors. The *heterodyne processor* (Fig. 8.22) is the most common type of head-end processor. The processor converts the incoming signal to
an intermediate frequency, where the aural and visual carriers are separated. The
signals are independently amplified and filtered with automatic level control. The
signals are then recombined and heterodyned to the desired output channel.

Prior to cable delivery systems, television license assignments avoided placing
two stations on adjacent channels. As a result, many television receivers were
not designed to discriminate well between adjacent channels. When used with cable television, the receiver experiences interference between the aural carrier of
one channel and the visual carrier of the adjacent channel. Channel processors
alleviate this problem by changing the ratio between the aural and visual carriers
of each channel on the cable. The visual-to-aural carrier ratio is typically 15 to 16
dB on cable television systems. At one time, the aural carrier of broadcast stations was only 10 dB lower than the visual carrier, but some stations now operate
with the aural signal 20 dB below visual.

Channel conversion allows a signal received on one channel to be changed to
a different channel for optimized transmission on the cable. Processors are generally modular, and with appropriate input and output modules any input channel
can be translated to any other channel of the cable spectrum. Conversion is usually necessary for local broadcast stations carried on cable networks. If the local
station were carried on its original carrier frequency, direct pickup of the station
off the air within the subscriber's receiver would create interference with the signal from the cable system.

The visual-signal intermediate-frequency (IF) passband of a typical heterodyne processor is shown in Fig. 8.23. The heterodyne processor is designed to
reproduce the received signal with a minimum of differential phase and amplitude
distortion. This is best accomplished with a flat passband. Note that the visual
carrier is positioned within the flatter portion of the passband response curve.
Television demodulators place the visual carrier at a point 6 dB down on the re-

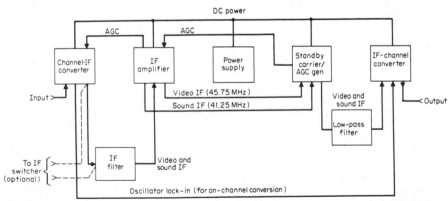

FIGURE 8.22 Block diagram of a typical heterodyne-type head-end processor. *(Courtesy Jerold
Manufacturing Co., Model CHC. In* Television Engineering Handbook, *K. Blair Benson, ed.,
McGraw-Hill, New York, 1986)*

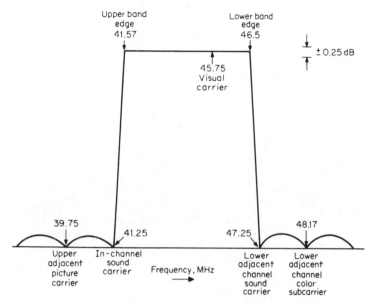

FIGURE 8.23 Typical idealized video-IF response curve of heterodyne processor. Note that the visual carrier is located on the flat-loop portion of the curve in order to provide improved phase response compared to television receivers. *(Source: D. G. Fink and D. Christiansen, ed.,* Electronic Engineer's Handbook, *2d ed., McGraw-Hill, New York, 1982. In* Television Engineering Handbook, *K. Blair Benson, ed., McGraw-Hill, New York, 1986)*

sponse curve. For this reason, the heterodyne processor has better differential phase characteristics than a modulator-demodulator pair.

An integral circuit of heterodyne processors is a standby carrier generator. If the incoming signal fails for any reason, the standby signal is switched in to maintain a constant visual carrier level, particularly on cable systems that use a television channel as a pilot carrier for trunk automatic level control. Such standby carriers may include provisions for modulation from a local source for emergency messages.

Demodulator-Modulator Pairs. Demodulator-modulator pairs are commonly used to convert satellite or microwave signals to cable channels. The demodulator is essentially a high-quality television receiver with base-band video and audio outputs. The modulator is a low-power television transmitter. This approach to channel processing provides increased selectivity, better level control, and more flexibility in switching of input and output signal paths as compared to other types of processors.

In the demodulator (Fig. 8.24), an input amplifier, local oscillator, and mixer down-convert the input signal to an intermediate frequency. Crystal control or phase-lock loops maintain stable frequencies for the conversion. In the IF amplifier section, the aural and visual signals are separated for detection. In comparing the IF response curve of a demodulator (Fig. 8.25) with that of the heterodyne processor (Fig. 8.23), we find that the visual carrier is located on the edge of the passband. The 4.5-MHz intercarrier sound is taken off before video detection,

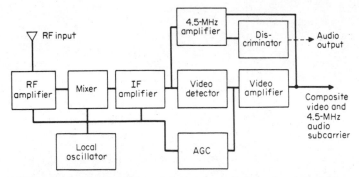

FIGURE 8.24 Block diagram of demodulator portion of a demodulator-modulator pair. *(Source: D. G. Fink and D. Christiansen, ed.,* Electronic Engineer's Handbook, *2d ed., McGraw-Hill, New York, 1982. In* Television Engineering Handbook, *K. Blair Benson, ed., McGraw-Hill, New York, 1986)*

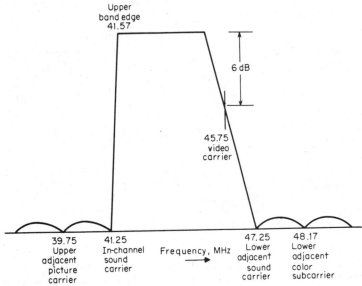

FIGURE 8.25 Idealized video-IF response curve of a demodulator. The video carrier is located 6 dB below the maximum response. *(Source: D. G. Fink and D. Christiansen, ed.,* Electronic Engineer's Handbook, *2d ed., McGraw-Hill, New York, 1982. In* Television Engineering Handbook, *K. Blair Benson, ed., McGraw-Hill, New York, 1986)*

amplified, and limited to remove video components. The sound is often maintained on a 4.5-MHz audio subcarrier for remodulation.

Demodulators are designed to minimize phase and amplitude distortion with linear detector circuits. However, quadrature distortion is inherent in systems that use envelope detectors with vestigial sideband signals. Such distortion can be corrected with video processing circuitry.

In the modulator (Fig. 8.26), a composite input signal is applied to a separation section, where video is separated from the sound subcarrier. The video is amplified, processed for dc restoration, and mixed with a carrier oscillator to an IF frequency. An amplifier boosts the signal to the required power level, and a vestigial sideband filter is used to trim most of the lower sideband for adjacent-channel operation.

Audio, if derived from the composite input, is separated as a 4.5-MHz signal from the visual information, filtered, amplified, and limited before being mixed with the visual carrier. If audio is provided separately to the modulator, an FM modulation circuit generates the 4.5-MHz aural subcarrier to be combined with the visual carrier. Once the two are combined, they are mixed upward to the desired output channel frequency and applied to a bandpass filter to remove spurious products of the upconversion before being applied to the cable system.

Single-Channel Amplifiers. The third type of channel processor is the *single-channel* or *strip amplifier*. These units are designed to amplify one channel and include bandpass and bandstop filters to reject signals from other channels. In simplest form, the unit consists of an amplifier, filter, and power supply, but more elaborate designs may include automatic gain control (AGC) and independent control of the aural and visual carrier levels. (See Fig. 8.27.)

Single-channel amplifiers are useful where the desired signal levels are relatively high and undesired signals are low or absent. Selectivity is low compared with other processor types, and the units generally lack the independent control of carrier levels, the ratio between carriers, independent AGC of the carriers, or limiting for the aural carrier. They cannot be used for channel conversion, and they present difficulties in adjacent-channel applications because of their limited selectivity.

Supplementary Service Equipment. Most cable television systems have the ability to distribute other signals besides television. They may offer additional services to subscribers, such as FM broadcast radio, data transmission, interactive signaling, and specialized audio and video transmissions.

FM radio service on cable systems is commonly provided in the standard FM broadcast band from 88 to 108 MHz. Heterodyne processors are usually used to place an FM radio signal into the cable system spectrum at a frequency where it will not interfere with direct pickup of over-the-air signals. Cable systems have used an FM radio channel to simulcast a stereo audio signal in conjunction with some television programming. Such a processor is diagrammed in Fig. 8.28.

Data transmission on a cable television network requires modems designed for the particular bandwidth, frequency, and data transmission speed required. Many cable systems dedicate a portion of the spectrum on a subscriber network or install a separate institutional network for data transmission. Generally these systems are bidirectional to allow interactive data transmission and signaling. Modems for cable system applications are available for data speeds to 1.544M bits/s. Although data transmissions use discrete frequencies for each point-to-point link, various means to multiplex data channels are under development.

8.10.5 Head-End Signal Distribution

Trunk System. The *trunk system* (Fig. 8.29) is designed for bulk transmission of multiple channels. The trunk may connect numerous distribution points and may

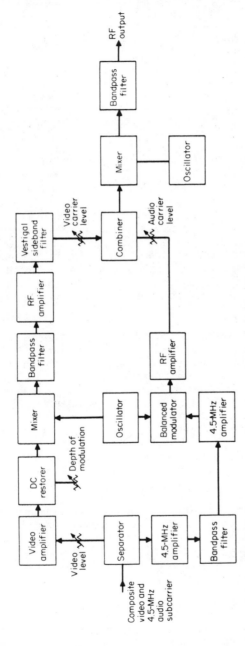

FIGURE 8.26 Modulator block diagram. *(Source: D. G. Fink and D. Christiansen, ed., Electronic Engineer's Handbook, 2d ed., McGraw-Hill, New York, 1982. In Television Engineering Handbook, K. Blair Benson, ed., McGraw-Hill, New York, 1986)*

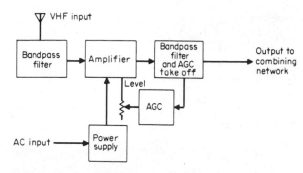

FIGURE 8.27   Block diagram of single-channel amplifier (strip amplifier) with AGC. *(Source: D. G. Fink and D. Christiansen, ed., Electronic Engineer's Handbook, 2d ed., McGraw-Hill, New York, 1982. In Television Engineering Handbook, K. Blair Benson, ed., McGraw-Hill, New York, 1986)*

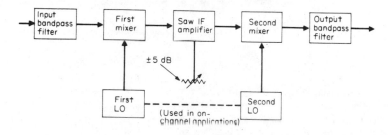

FIGURE 8.28   FM radio heterodyne processor block diagram. *(Source: K. Blair Benson, ed., Television Engineering Handbook, McGraw-Hill, New York, 1986)*

interconnect widely spaced sections of a cable system or even more than one cable system. A trunk system spanning long distances without intermediate distribution is often called a *super trunk.*

The coaxial cable used in trunk systems is made with a solid aluminum shield, a gas-injected polyethylene, low-loss dielectric insulator, and a copper or copper-clad aluminum center conductor. Underground cable has an additional outer polyethylene jacket. Under the jacket is a flooding compound to seal small punctures that might occur during or after installation. Trunk cables are usually ¼ to 1 in. in diameter and typically have a loss of about 1dB per 100 ft at 400 MHz.

For long-distance trunking with minimum distortion, frequency-modulation (FM) techniques are preferred. While FM requires a larger portion of the available bandwidth than amplitude modulation, the distortion in the delivered signal is considerably lower. Thus, FM techniques permit longer trunk lengths.

Amplification.   Amplifiers in the cable system are used to overcome losses in the coaxial cable and other devices of the system. From the output of a given amplifier, through the span of coaxial cable (and an equalizer to linearize the cable losses), and to the output of the next amplifier, unity gain is required so that the same signal level is maintained on all channels at the output of each amplifier unit.

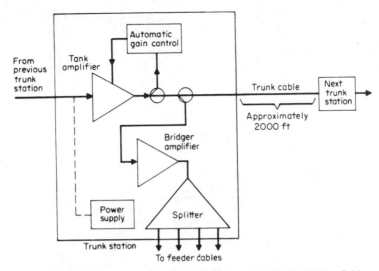

FIGURE 8.29 Trunk system. *(Source: T. F. Baldwin and D. S. McVoy,* Cable Communications, *1983. Reprinted by permission of Prentice-Hall, Inc., Englewood Cliffs, NJ. In* Television Engineering Handbook, *K. Blair Benson, ed., McGraw-Hill, New York, 1986)*

Standard cable system design places repeater amplifiers from 1400 to 2000 ft apart, depending on the diameter of the cable used. This represents an electrical loss of about 20 dB at the highest frequency carried. Systems with amplifier cascades to 64 units are possible, depending on the number of channels carried, the performance specifications chosen, the modulation scheme used, and the distortion-correction techniques used. Table 8.3 tabulates representative figures of distortion versus the number of amplifiers in cascade.

TABLE 8.3 Distortion versus Number of Amplifiers Cascaded in a Typical CATV System

No. of amplifiers in cascade	Cross modulation, dB	Second order, dB	S/N ratio, dB
1	−96	−86	60
2	−90	−81.5	57
4	−84	−77	54
8	−78	−72.5	51
16	−72	−68	48
32	−66	−63.5	45
64	−60	−59	42
128	−54	−54.5	39

*Dashed line indicates lower limit of acceptable system performance.
Source: K. Blair Benson, ed., *Television Engineering Handbook*, McGraw-Hill, New York, 1986.

Bridging Amplifiers. Signals from a trunk system are fed to the distribution system and eventually to subscriber drops. A wide-band directional coupler extracts a portion of the signal from the trunk for use in the bridger amplifier. The bridger acts as a buffer, isolating the distribution system from the trunk, while providing the signal level required to drive distribution lines.

8.10.6 Subscriber-Area Distribution

In Fig. 8.30, four distribution lines are fed from a bridging station. Distribution lines are routed through the subscriber area with amplifiers and tap-off devices to meet the subscriber density of the area. The cable for distribution lines is identical to that used for trunks, except that its diameter is commonly ½ or ⅝ in. with a resulting increase in losses, typically to 1.5 dB per 100 ft at 400 MHz.

Line-Extender Amplifiers. As the signal proceeds on a distribution line, attenuation of the cable and insertion losses of tap devices reduce its level to a point where *line-extender amplifiers* may be required. The gain of line extenders is relative high (25 to 30 dB), so generally no more than two are cascaded, because the high level of operation creates a significant amount of cross-modulation and intermodulation distortion. These amplifiers do not use automatic level-control circuitry to compensate for variations in cable attenuation caused by temperature changes. By limiting the system to no more than two line extenders per distribution leg, the variations are not usually sufficient to affect picture quality.

Multitaps. Two types of tap-off devices are commonly used. The *directional-coupler multitap* provides for connections from two to eight subscribers per tap. Most manufacturers offer a series of taps, each given a value based on the amount of signal taken from the distribution line. (See Fig. 8.31.) Individual *pres-*

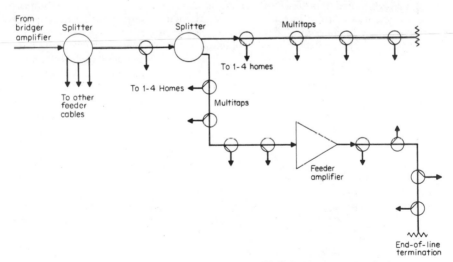

FIGURE 8.30 Distribution and feeder system. (*Source: T. F. Baldwin and D. S. McVoy,* Cable Communications, *1983. Reprinted by permission of Prentice-Hall, Inc., Englewood Cliffs, NJ. In* Television Engineering Handbook, *K. Blair Benson, ed., McGraw-Hill, New York, 1986)*

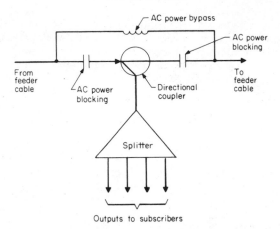

FIGURE 8.31 Multitap block diagram. *(Source: K. Blair Benson,* Television Engineering Handbook, *McGraw-Hill, New York, 1986)*

sure taps fasten directly to the distribution cable, but find limited use today. Pressure taps tend to create signal reflections, because their construction and installation onto the cable simulate a distortion in the center conductor-to-shield distance.

A tap removes an appropriate amount of energy from the distribution cable through a directional coupler, then splits the energy into several paths. Each path proceeds through a drop into a subscriber's home. The value of the taps is selected to produce an adequate signal level for a good signal-to-noise ratio at the subscriber's receiver, but not so high as to create intermodulation in the receiver. A signal-level range of 0–10 dB mV at the subscriber's receiver is generally more than sufficient. Each tap introduces some attenuation into the distribution system, related to the amount of signal tapped off for subscribers. By extracting the signal through a directional coupler, a high attenuation from the tap outlets back into the cable reduces the interference that originates in a subscriber drop cable from reaching subscribers served by other multitaps. The directional characteristics of the taps also prevent reflected signals from traveling along the distribution system and causing ghosting on subscribers' receivers.

The splitter portion of a multitap is a hybrid design that introduces a substantial mount of isolation from reflections or interference coming from a subscriber drop both to the distribution system and to other subscribers connected to the same tap.

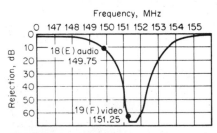

FIGURE 8.32 Notch-filter trap method for rejection of a premium service channel, in this case channel 19 (F), to nonauthorized subscribers. The audio of the adjacent authorized channel, 18 (E), is attenuated only to 10 dB. *(Courtesy Vitek Corp. In* Television Engineering Handbook, *K. Blair Benson, ed., McGraw-Hill, New York, 1986)*

Addressable Subscriber Taps. The tap is the final service point immediately prior to a subscriber's location, so it becomes a convenient point where control of channels authorized to individuals can be applied. An addressable tap, controlled from the system head end, can be used instead of the standard multitap to enable access to basic cable services or to individual channels. Inside the addressable tap, a small receiver with address-recognition circuitry controls a diode switch, notch filters (see Fig. 8.32), jamming oscillators, and scrambling devices. When activated, the switch and the other components can enable or disable a channel, a group of channels, or the entire spectrum.

Control of service tiers and channels can also be effected through the use of filters connected to the output of the multitap. These can take the form of notch filters that are selected to remove a particular channel or a group of channels. Positive filters are used to remove an interfering carrier that would otherwise cause a particular channel to be presented in a scrambled mode.

8.10.7 Trunk and Distribution Amplifiers

Because the trunk is the main artery of a cable television system, trunk amplifiers must provide a minimum of degradation to the system. A typical amplifier station is of modular design with plug-in units for trunk amplification, automatic level control, bridging, and signal splitting to feed distribution cables. The modules receive power from the coaxial cable. A power supply in each amplifier housing taps power from the cable to supply dc voltage to operate the other modules. The power is inserted onto the coaxial cable at points where a connection to the local power utility is convenient. A step-down supply is used to insert 30 V ac or 60 V ac onto the coaxial cable through a passive inductive network.

Figure 8.33 shows a block diagram of the typical *trunk amplifier* with provisions for automatic level control and distribution of gain. The first stage provides low-noise operation, while remaining stages produce gain with low cross-modulation. Modern trunk amplifiers use hybrid circuits, combining integrated circuits and discrete components in a single package. Such hybrid packages are available with a variety of gain and performance characteristics.

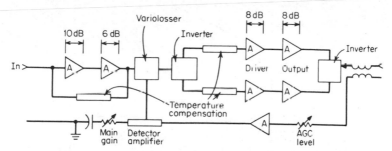

FIGURE 8.33 Functional block diagram of a trunk-AGC amplifier and its gain distribution. *A* indicates an amplifier. (*Source: D. G. Fink and D. Christiansen, eds.,* Electronic Engineer's Handbook, *2d ed., McGraw-Hill, New York, 1982. In* Television Engineering Handbook, *K. Blair Benson, ed., McGraw-Hill, New York, 1986)*

Level Control. Because attenuation of coaxial cable changes with ambient temperature, trunk amplifiers include automatic level-control (ALC) circuits to compensate for the changes and to maintain the output level at a predetermined standard. Automatic control depends on the use of *pilot carriers*. One is at the lower end of the spectrum, while the other is at the upper limit. Often, special pilot carriers are placed on the system at the head end. Other system designs use video channels as pilots. ALC circuits sample the trunk amplifier output at the pilot frequencies and feed a control voltage back to diode attenuators at an intermediate point in the amplifier circuit. The upper pilot adjusts the gain of the amplifier, while the lower pilot controls the slope of an equalizer. This method is effective in maintaining the entire spectrum at a constant output level.

Feed-Forward Distortion Correction. A second design of trunk amplifier uses *feed-forward distortion correction* and permits carriage of a larger number of channels for equivalent gain than the standard trunk amplifier. Feed-forward amplifiers are considerably more expensive than standard amplifiers and used primarily for supertrunks and unusually long trunk lines.

Trunk stations include bridger modules where a connection from the trunk to the distribution system is needed. Bridger amplifiers tap a portion of signal energy from the trunk through a directional coupler and boost the signals to feed multiple distribution lines. Bridgers are the only amplifiers in the system that actually boost signals rather than just recover cable losses.

Line extenders, like trunk stations, are modular, with plug-in amplifiers and a power supply. Because line extenders are located in sections of the system where greater attenuation exists in the distribution cable, taps, and drop cable, gain from the line extender is greater than from trunk amplifiers. Higher levels may be used, at the risk of introducing significant cross-modulation and intermodulation distortion, because only one additional amplifier, at most, will be used prior to a subscriber's connection. Under these circumstances, distortion products should not become apparent on subscribers' receivers.

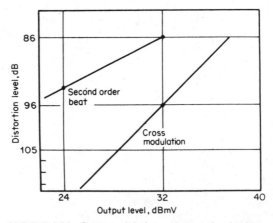

FIGURE 8.34 Cross-modulation and second-order beat distortion versus output level of a trunk amplifier. *(Source: D. G. Fink and D. Christiansen, eds.,* Electronic Engineer's Handbook, *2d ed., McGraw-Hill, New York, 1982. In* Television Engineering Handbook, *K. Blair Benson, ed., McGraw-Hill, New York, 1986)*

Cross-Modulation. The maximum output level permissible in cable television amplifiers depends primarily on cross-modulation of the picture signals. Cross-modulation is most likely in the output stage of an amplifier, where levels are high. The cross-modulation distortion products at the output of a typical amplifier (such as Fig. 8.33) are 96 dB below the desired signal at an output level of +32 dB mV. In Fig. 8.33, gains of the third and fourth stages are each 8 dB. With an amplifier output level of +32 dB mV, the output of the third stage would be +24 dB mV, that of the second stage, 16 dB mV.

Figure 8.34 expresses the relationship between cross-modulation distortion and amplifier output level. A 1-dB decrease in output level corresponds to a 2-dB decrease in cross-modulation distortion. Then the output cross-modulation distortion of the third stage would be 112 dB below the desired signal. Logically, the distortion introduced by the first and second stages of the amplifier is even less and is insignificant.

Cross-modulation increases by 6 dB each time the number of amplifiers in cascade doubles. The distortion also increases by 6 dB each time the number of channels on the system is doubled. Cable subscribers find a cross-modulation level of −60 dB objectionable. Therefore, based on a cross-modulation component of −96 dB relative to normal visual carrier levels in typical trunk amplifiers, 64 such amplifiers can be cascaded before cross-modulation becomes objectionable.

Frequency Response. The most important design consideration of cable amplifiers is frequency response. A response that is flat within ±0.25 dB from 40 MHz to 450 MHz is required of an amplifier carrying 50 or more 6-MHz television channels to permit a cascade of 20 or more amplifiers. Circuit designs must pay special attention to high-frequency parameters of the transistors and associated components as well as circuit layout and packaging techniques.

The circuit in Fig. 8.35 represents an amplifier designed for flat response. The feedback network (C_1, R_1, L_1) introduces sufficient negative feedback to maintain a nearly constant output over a wide frequency range. Collector transformer T_2 is bifilar-wound on a ferrite core and presents a constant 75-Ω impedance over the entire frequency range. Splitting transformers T_1 and

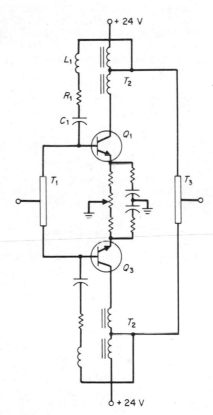

FIGURE 8.35 High-performance, push-pull, wide-band amplifier stage for a trunk amplifier. *(Source: D. G. Fink and D. Christiansen, eds., Electronic Engineer's Handbook, 2d ed., McGraw-Hill, New York, 1982. In Television Engineering Handbook, K. Blair Benson, ed., McGraw-Hill, New York, 1986)*

T_3, of similar construction, have an additional function of introducing 180° phase shift for the push-pull transistor pair Q_1 and Q_2, while maintaining 75-Ω input and output impedances.

Second-Order Distortion.   When cable television systems carried only 12 channels, channels were spaced to avoid the second harmonic of any channel from falling within any other channel. The expansion of systems to carry channels covering more than one octave of spectrum causes second-order beat distortion characteristics of amplifiers to become important. Push-pull amplifier circuits, as shown in Fig. 8.35, provide the most successful method of reducing amplifier second-order distortion. Combining several of these circuit building blocks into an amplifier effectively cancels second-order distortion if the 180° phase relationship is maintained over the full amplifier bandwidth. The gain of both halves of the amplifier must be equal and the individual transistors must be optimized for maximum second-order linearity.

Second-order distortion increases by approximately 1.5 dB for each 1 dB in amplifier output level. Referring to Table 8.3, each amplifier introduces a distortion level of approximately −86 dB relative to the visual carriers. Because second-order beats greater than −55 dB become objectionable, approximately 64 amplifiers can be cascaded before that level is attained.

Noise Figure.   The typical trunk amplifier hybrid module has a noise figure of approximately 7 dB. Equalizers, band-splitting filters, and other components preceding the module add to the overall station noise figure, producing a total of approximately 10 dB.

A single amplifier produces a signal-to-noise ratio of approximately 60 dB for a typical signal input level. Subjective tests show that a signal-to-noise ratio of approximately 43 dB is acceptable to most viewers. Again, based on Table 8.3, noise increases 3 dB as the cascade of amplifiers is doubled. Therefore, a cascade of 32 amplifiers, resulting in a signal-to-noise ratio of 45 dB, is a commonly used cable television system design specification.

Hum Modulation.   Because power for the amplifiers is carried on coaxial cable, some hum modulation of the RF signals occurs. This modulation results from power-supply filtering and saturation of the inductors used to bypass ac current around active and passive devices within the system. A hum modulation specification of −40 dB relative to visual carriers has been found to be acceptable.

8.10.8 Subscriber-Premises Equipment

The output of the tap device feeds a 75-Ω coaxial drop cable into the home. The cable, generally about ¼ in. in diameter, is constructed with an outer polyethylene jacket, a shield of aluminum foil surrounded by braid, a polyethylene dielectric insulator, and a copper-center conductor. A loss at 400 MHz of 6 dB per 100 ft is typical.

Isolation.   For reasons of safety and signal purity, isolation between the subscriber's television receiver and the distribution cable is required. In the multitap, blocking capacitors prevent the ac current used to power amplifiers from reaching the subscriber drop. As the drop enters the subscriber's home, a grounding connection, required by local regulations, protects the television receiver from power surges, such as those caused by lightning discharges, and protects the subscriber from possible shock by voltages that might otherwise be present on the shield of the drop cable. Finally, at the subscriber's receiver, the signal from the unbalanced 75-Ω coaxial cable is converted to a 300-Ω balanced output before connection to the tuner section of the receiver. This matching transformer provides a further level of isolation to the existence of energy along the sheath of the cable.

Converters and Descramblers. Prior to the manufacture of cable-ready television receivers and VCRs, systems with more than 12 channels required the use of a converter at the subscriber's location. A nontunable block converter, which translates a group of channels to a different portion of the spectrum, is probably the least expensive type of converter. Such devices are used primarily when only a small number of channels are to be added to the system.

Most converters used today are tunable units, with microprocessor control of the frequency-synthesizing circuit referenced to a crystal. The viewer selects a channel (often with a hand-held remote control unit), which is converted to a standard VHF frequency (usually channel 3 or 4). In such cases, the television receiver remains tuned to "channel 3," and all tuning is done with the converter.

Cable-Ready Receivers. Cable-ready tuners avoid the need for a converter by including additional tuning capabilities into the receiving equipment. As the viewer selects the desired channel, digital switching sets the input control lines of a phase-locked-loop frequency synthesizer. When selected, any of the additional channel signals are converted to the normal television IF amplifier frequency band.

Prior to cable-ready receivers, the use of nonstandard channels and converters provided an easy method of limiting access to special channels. Now, some method of scrambling may be used to control access. While the addressable tap and notch filters, mentioned earlier, are effective means of control, other approaches are also in use. A common scrambling method operates by suppressing horizontal and vertical synchronizing pulses during the transmission at the head end. At the same time, a pilot signal that contains information about the missing sync is also transmitted on the cable. In a descrambler, located in the subscriber's home, the pilot information provides the key to correctly re-create the synchronization.

A greater degree of security for scrambled signals can be attained through digital methods and *base-band scrambling*. Numerous concepts have been introduced, including the inversion of lines of video on a random basis, shuffling of lines of video, and conversion of the video into a stream of digital information. These methods, requiring that signals be demodulated into audio and video, are more expensive in terms of equipment costs, but they present possible advantages as well, because remote control of the audio level is possible, as is the incorporation of videotext decoding circuits.

By including an addressable switch into a set-top descrambler (Fig. 8.36), the cable operator can use computer control from the head end to restrict access to channels, depending on the tiers of service a subscriber has chosen. In a bidirectional cable system, pay-per-view impulse viewing is relatively easy to implement as well. A subscriber need only key in an appropriate code to request reception of a special program. The head-end computer, constantly polling all con-

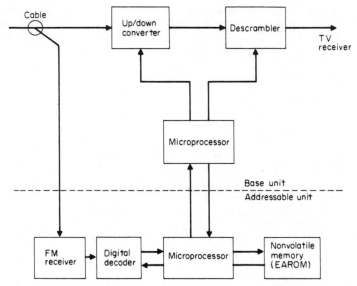

FIGURE 8.36 Addressable-converter block diagram. *(Courtesy Scientific Atlanta. In* Television Engineering Handbook, *K. Blair Benson, ed., McGraw-Hill, New York, 1986)*

verter/descrambler units, detects the request, enables the special channel for the duration of the program request, and subsequently places an entry into the automated billing system for that subscriber.

8.10.9 Bidirectional Cable Television Systems

Spectrum Utilization.   Although a bidirectional cable system could be constructed using separate cables for the upstream and downstream signal paths, the costs are prohibitive. More practical is a single-cable system in which the direction of the signal is related to frequency. Most two-way operations on subscriber networks use the spectrum from 5 to 35 MHz for signals returning to the head end, while signals transmitted for subscriber use are on VHF frequencies above 50 MHz. Using this approach, called *subsplitting*, special circuits must be integrated into all amplifiers in the system. Each amplifier station now has two amplifier sections, one for operation in each direction, with band-splitting and combining filters used to direct signals to the appropriate amplifier modules. Institutional networks may use the same approach, but the frequency of the split is higher, around 150 MHz.

Bidirectional operation introduces a number of new problems into the cable system. Probably the greatest difficulties encountered stem from transmitters external to the system operating on frequencies between 5 and 35 MHz. That includes citizen's band, amateur radio, and shortwave broadcast, as well as some public-service applications. Ingress of signals from these transmitters is difficult to prevent and interferes with reliable upstream transmission. Some of these problems can be reduced by allowing the use of remotely controlled switchers at bridger amplifier locations. In effect, only certain portions of the distribution system can send signals upstream at any given time, which places definite limitations on the effectiveness of bidirectional operation.

CHAPTER 9
TELEVISION RECEPTION

K. Blair Benson
Television Technology Consultant, Norwalk, Connecticut

9.1 ANTENNA SYSTEMS

The antenna system is one of the most important circuit elements in the reception of television signals. The *noise level*, or amount of *snow* in the picture, as well as the clarity of the sound, depends primarily on how well the antenna system is able to capture the signal radiated by the transmitting antenna of a broadcasting

Portions of this chapter were adapted from K. Blair Benson, editor, *Television Engineering Handbook*, McGraw-Hill, New York, 1986.

station and to feed it, with minimum loss or distortion, through a transmission line to the tuner of a receiver.

In urban and residential areas near the television station antennas where strong signals are present, a compact set-top telescoping rod or rabbit ears for very high frequency (VHF) and a single loop for ultrahigh frequency (UHF) usually are quite adequate. The somewhat reduced signal strength in suburban and rural areas generally requires a multielement roof-mounted antenna with either a 300-Ω twin-lead or, to reduce pickup of interference from nearby sources such as automobile ignition, a shielded 75-Ω coaxial transmission line to feed the signal to the receiver.

Fringe areas, where the signal level is substantially lower, usually require a more complicated highly directional antenna, frequently on a tower, to produce an even marginal signal level. The longer transmission line in such installations may dictate the use of an all-channel low-noise preamplifier at the antenna to compensate for the loss in signal level in the line.

9.1.1 Basic Characteristics and Definitions of Terms

Antennas have several characteristics that define their ability to receive energy from a radiated field. These are as follows:

Wavelength is the distance traveled by one cycle of a radiated electric signal. The frequency of the signal is the number of cycles per second. It follows that the frequency f is inversely proportional to the wavelength λ. Both wavelength and frequency are related to the speed of light. Conversion between the two parameters can be made by dividing either into the figure for the speed of light c. The formula is below:

$$\lambda \times f = c$$

In other words, *wavelength* (meters) times *frequency* (MHz) = 300 m. In feet, this is: *wavelength* (feet) times *frequency* (Mhz) = 984 ft.

The velocity of electric signals in air is essentially the same as that of light in free space, which is 2.9983 $\times$ 10^{10} cm/s, or for most calculations, 3 $\times$ 10^{10} cm/s (186,000 mi/s).

Radiation is the emission of coherent modulated electromagnetic waves in free space from a single or a group of radiating antenna elements. Although the radiated energy, by definition, is composed of mutually dependent electric and magnetic field vectors having specific magnitude and direction, conventional practice in television engineering is to measure and specify radiation characteristics in terms of the *electric field*.

Polarization is the angle of the radiated electrical field vector in the direction of maximum radiation. If the plane of the field is parallel to the ground, the signal is *horizontally* polarized; if at right angles to the ground, it is *vertically* polarized. When the receiving antenna is located in the same plane as the transmitting antenna, the received signal strength will be maximum. If the radiated signal is rotated at the operating frequency by electrical means in feeding the transmitting antenna, the radiated signal is *circularly* polarized. Circularly polarized signals produce equal received signal levels with either horizontal or vertical polarized receiving antennas (Fig. 9.1).

Beamwidth, in the plane of the antenna, is the angular width of the directivity pattern where the power level of the received signal is down by 50 per-

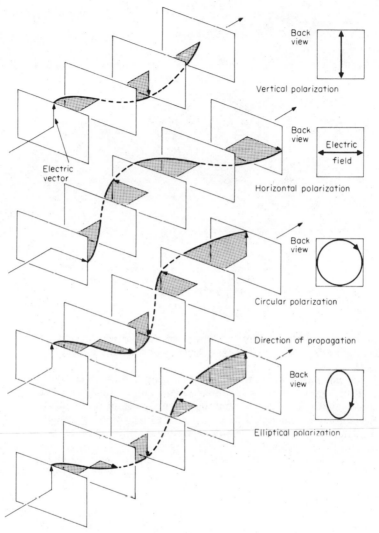

FIGURE 9.1 Polarization of the electric field of the transmitted wave. *[Source: K. Blair Benson (ed.),* Television Engineering Handbook, *McGraw-Hill, New York, 1986]*

cent (3 dB) from the maximum signal in the desired direction of reception. Using Ohm's law ($P = E^2/R$) to convert power to voltage, assuming the same impedance R for the measurements, this is equal to a drop in voltage of 30 percent (Fig. 9.2).

Gain is the signal level produced (or radiated) relative to that of a standard reference dipole. It is used frequently as the *figure of merit*. Gain is closely related to directivity, which in turn is dependent upon the radiation pattern. High values of gain usually are obtained with a reduction in beamwidth.

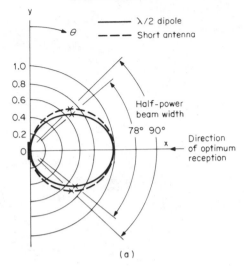

FIGURE 9.2 Normalized E-field pattern of $\Lambda/2$ and short dipole antennas. *[Source: K. Blair Benson (ed.),* Television Engineering Handbook, *McGraw-Hill, New York, 1986]*

Gain can be calculated only for simple antenna configurations. Consequently, it usually is determined by measuring the performance relative to a reference dipole.

Input impedance is the terminating resistance into which a receiving antenna will deliver maximum power. Similar to *gain*, input impedance can be calculated only for very simple formats, and instead is determined by actual measurement.

Radiation resistance is defined in terms of transmission, using Ohm's law, as the radiated power P from an antenna divided by the square of the driving current I at the antenna terminals.

Bandwidth is a general classification of the frequency band over which an antenna is effective. This requires a specification of tolerances on the uniformity of response over not only the effective frequency spectrum but in addition, over that of individual television channels. The tolerances on each channel are important because they can have a significant effect on picture quality and resolution, on color reproduction, and on sound quality. Thus, no single broadband response characteristic is an adequate definition because an antenna's performance depends to a large degree upon the individual channel requirements and gain tolerances and the deviation from flat response over any single channel. This further complicates the antenna-design task because channel width relative to the channel frequency is greatly different between low and high channels.

The problem is readily apparent by a comparison of low VHH and high UHF channel frequencies with the 6-MHz single-channel spectrum width specified in the *FCC Rules and Regulations*. VHF Channel 2 occupies 11 percent of the spectrum up to that channel, while UHF Channel 69 occupies only 0.75 percent of the spectrum up to Channel 69. In other words, a VHF antenna must have a considerably greater broadband characteristic than that of a VHF antenna.

9.1.2 VHF/UHF Receiving Antennas

In strong-signal urban areas, a set-top antenna consisting of rabbit ears for VHF and loops for UHF with a foot or so of 300-Ω twin-lead connecting to the receiver antenna terminals usually is adequate if ghosts due to multipath reception are not present. In suburban and rural areas, a roof-mounted antenna with extra elements to increase the signal level is required. In weak-signal fringe areas, beyond the FCC-specified "Grade B" contour area of 2.5 mV/m covered by the transmitter, in addition to a high outdoor antenna a low-noise signal preamplifier may be necessary at the antenna in order to boost the signal above that of any stray pickup from interference signals such as auto ignition and CBs, and to a level required by the receiver for stable scanning and color synchronization.

Dipole Antenna. The dipole antenna is not only the simplest type suitable for feeding a transmission line but also is the basis for other more complex designs.[2,3] It consists of two in-line rods or wires with a total length equal to a half-wave at the primary band it is intended to cover. In Fig. 9.3, the two quarter-wave elements l are connected at their center to a two-wire transmission line.

Normally a 300-Ω balanced transmission line is connected at the center. However, since the radiation resistance of a half-wave dipole is 73 Ω, use of the usual 300-Ω balanced transmission line will result in a mismatch and some loss in transfer of received signal power. In addition, if the receiving end of the line is not a matching impedance of approximately 73 Ω, there will be reflections up and down the line. These will show up in the received picture as horizontal smear for short lines or ghosts displaced to the right in the case of longer lines. This picture distortion is avoided in receiver design, where the tuner input usually is unbalanced 75-Ω coaxial cable, by the use of a balanced-to-unbalanced (balun) cou-

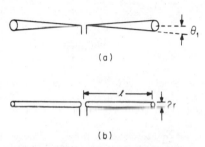

FIGURE 9.3 (*a*) Half-wave dipole antenna. *[Source: K. Blair Benson (ed.),* Television Engineering Handbook, *McGraw-Hill, New York, 1986]* (*b*) Conical dipole. *[Source: K. Blair Benson (ed.),* Television Engineering Handbook, *McGraw-Hill, New York, 1986]* (*c*) Cylindrical dipole. *[Source: K. Blair Benson (ed.),* Television Engineering Handbook, *McGraw-Hill, New York, 1986]*

pler. A balun consists of a compact transformer with closely coupled input and output coils wound on a powdered-iron magnetic core. Early designs used a compact coil of small-gauge parallel wires spaced for a 300-Ω impedance and wound on a small rod (see Fig. 9.4).[4]

Alternatively, a coaxial transmission line can be used to feed the tuner 75-Ω input directly. In this case, a balun is mounted at the antenna to transform the balanced dipole signal to the unbalanced coaxial transmission line and the antenna's balanced impedance of 73 Ω is closely matched to the 75-Ω transmission line. Dissimilar circuits can be matched by a resistive network, but with a resultant loss in signal. A typical case would be between a 300-Ω lead-in from an antenna to a 75-Ω coaxial section of a building distribution system.

The bandwidth of a dipole antenna may be increased by increasing the diameter of the elements by using cones or cylinders rather than wires or rods (see Fig. 9.4*b* and *c*). This also increases the impedance of the antenna so that it more

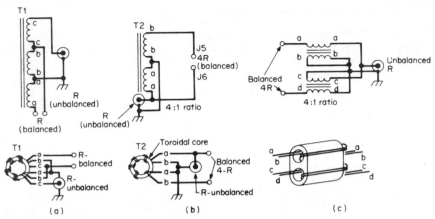

FIGURE 9.4   Balun transformers. (*a*) Balanced to unbalanced, 1/1. (*b*) Balanced to unbalanced, 4/1. (*c*) Balanced to unbalanced on symmetrical balun core, 4/1. *[Source: a and b from* Radio Amateur's Handbook, *American Radio Relay League, Newington, Conn., 1983.* a, b, *and* c *from* Television Engineering Handbook, *K. Blair Benson (ed.), McGraw-Hill, New York, 1986]*

nearly matches the balanced 300-Ω transmission line. The 3-dB, or half-power, beam width, as shown in Fig. 9.2, is 78°.

Short Dipole.   If the total length of the dipole is much less than a half wave, the characteristics become those of a "short dipole" with a substantially lower resistance and high capacity, resulting in an even higher mismatch.

Folded Dipole.   An increase in bandwidth and an increase in impedance may be accomplished also by use of the two-element folded-dipole configuration (see Fig. 9.5). The impedance can be increased by a factor of as much as 10 by using rods of different diameter and varying the spacing. The typical folded dipole with elements of equal diameter has a radiation impedance of 290 Ω, four times that of a standard dipole and closely matching the impedance of a 300-Ω twin lead.

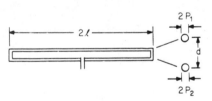

FIGURE 9.5   Folded dipole. *[Source: K. Blair Benson (ed.),* Television Engineering Handbook, *McGraw-Hill, New York, 1986]*

The ¼-wave dipole elements connected to the closely coupled ½-wave single element act as a matching stub between the transmission line and the single-piece half-wave element. This broadbands the folded-dipole antenna to span a frequency range of nearly 2 to 1. For example, the low VHF channels, 2 through 6 (54–88 MHz), can be covered with a single antenna.

V Dipole.   The typical set-top rabbit-ears antenna on VHF Channel 2 is a "short dipole" with an impedance of 3 or 4 Ω and a capacitive reactance of several hundred ohms. Bending the ears into a V in effect tunes the antenna and will affect its response characteristic, as well as the polarization. A change in polarization may be used in city areas to select or reject cross-polarized reflected signals to improve picture quality.

The fan dipole, based upon the V antenna, consists of two or more dipoles connected in parallel at their driving point and spread at the ends, thus giving it a broadband characteristic. It can be optimized for VHF reception by tilting the dipoles forward by 33° to reduce the beam splitting, i.e., improve the gain on the high VHF channels where the length is longer than a half-wave.

A ground plane or flat reflecting sheet placed at a distance of $\frac{1}{16}$ to $\frac{1}{4}$ wave behind a dipole antenna, as shown in Fig. 9.6, can increase the gain in the forward direction by 2. The design is often used for UHF antennas, e.g., a "bow-tie" or cylindrical dipole with reflector. For outdoor applications, a screen or parallel rods may be used to reduce wind resistance. At $\frac{1}{4}$-wave to $\frac{1}{2}$-wave spacing, the antenna impedance is 80 to 125 Ω, respectively.

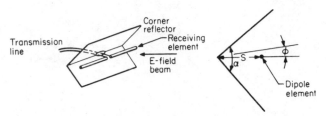

FIGURE 9.6 Corner-reflector antenna. *[Source: K. Blair Benson (ed.), Television Engineering Handbook, McGraw-Hill, New York, 1986]*

The quarter-wave monopole above a ground plane, as shown in Fig. 9.7, is another antenna derived from the elementary dipole. It is supplied with many personal portable television receivers in which the set acts as a ground plane. Although the monopole is intended to receive vertically polarized signals, it usually is moved in its swivel joint to either a horizontal position or to some other angle to a best compromise between direct and reflected signals. The theoretical resistance characteristic of a monopole with an infinite ground plane is 37 Ω.

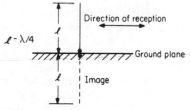

FIGURE 9.7 Vertical monopole above a ground plane. *[Source: K. Blair Benson (ed.), Television Engineering Handbook, McGraw-Hill, New York, 1986]*

The loop antenna set-top configuration—in effect, a form of the folded dipole—is used for UHF reception. Analyzing it as a single-turn magnetic-field pickup loop, the radiation resistance can be calculated and found to vary over a more than 10-to-1 range. The values for a typical 7-in diameter wire loop for several UHF channels are given in Table 9.1.

Multielement arrays can be used to achieve higher gain and directivity.[3] One design that has been popular for single-channel reception has been the Yagi-Uda array (Fig. 9.8). A five-element array is shown in Fig. 9.8. The receiving element is $\frac{1}{2}$ wavelength for the center of the band covered. The single reflector element is slightly longer, and the three director elements slightly shorter, all spaced approximately $\frac{1}{4}$ wavelength from each other. Typically, the bandwidth is only one or two channels. However, this can be increased by

TABLE 9.1   Radiation Resistance for a Single-Conductor Loop Having a Diameter of 17.5 cm (6.9 in) and a Wire Thickness of 0.2 mm (0.8 in)

Channel	Frequency, MHz	Radiation resistance, Ω
14	473	108
42	641	367
69	803	954
83	887	1342

Source: K. Blair Benson (ed.), *Television Engineering Handbook*, McGraw-Hill, New York, 1986.

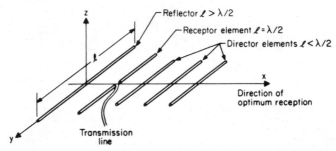

FIGURE 9.8   Yagi-Uda array. *[Source: K. Blair Benson (ed.),* Television Engineering Handbook, *McGraw-Hill, New York, 1986]*

a trade-off of a slightly longer reflector and slightly shorter directors from theoretical for a slight loss in gain. On the other hand, broadbanding a single-channel five-element Yagi to cover all the low-band VHF channels, for example, will reduce the gain figure from 9 dB to between 4 and 6 dB.

Another technique for broadening the bandwidth is to add shorter elements in groups between the directors. For example, in a low-band VHF Yagi, elements can be added for high-band VHF channels. The gain of the Yagi over a standard dipole for up to 15 elements and the beamwidth directivity are listed in Table 9.2.

9.1.3   Direct-Broadcast Satellite (DBS) Service

Satellite Channels.   The two bands allocated for television satellite broadcast are the 4 to 6 GHz (C band) and 12 to 17 GHz (K_u band). The down-link spectrum for C-band satellite systems consists of 24 channels, each with a 40-MHz bandwidth in the 3.7- to 4.2-GHz frequency band. Each channel overlaps its adjacent channels by 20 MHz. Even-numbered channels are horizontally polarized, and the odd-numbered are vertically polarized to give a minimum of 30-dB discrimination between channels. The satellite transponder power is in the order of 20 W. This band is assigned to commercial use, and most of the signals are scrambled, either for privacy or to prevent piracy of entertainment programming.

TABLE 9.2 Typical Characteristics of Single-Channel Yagi-Uda arrays

No. of elements	Gain, dB	Beam width, deg
2	3–4	65
3	6–8	55
4	7–10	50
5	9–11	45
9	12–14	37
15	14–16	30

Source: K. Blair Benson (ed.), *Television Engineering Handbook*, McGraw-Hill, New York, 1986.

Direct-broadcast satellite (DBS) television-program service in the K_u band was established by the 1977 World Administrative Radio Conference (WARC). For down-link transmission, a band between 12.2 and 12.7 GHz was authorized. Channel allocation and technical specifications for the western hemisphere (Region 2 of WARC) were established at the 1983 Western Hemisphere Regional Administrative Radio Conference. Subsequently, the FCC has accepted and approved applications from numerous companies to launch satellites and to provide DBS service for standard broadcast, premium pay-television programing, and high-definition television (HDTV).

Bandwidth for these authorized services ranges from 16 to 27 MHz per channel. Transponder output power required for an acceptable DBS-reception noise level is in the order of 50 W, more than double that used for commercial K_u-band operations. The greater power is necessary because of the smaller parabolic antennas planned for DBS reception. Consequently, regular service has been delayed until the availability of traveling-wave tubes (TWT) capable of providing the necessary high power level reliably in the outer-space environment of a satellite.

DBS Reception. The ground antenna for satellite reception usually consists of a parabolic reflector with a 3- to 5-m diameter for the C band and 1- to 1.5-m for the DBS K_u band. A low-noise amplifier (LNA) with a gain of approximately 40 dB and a noise figure of less than 1.5 dB at 120 K is placed at the antenna feed point in the center of the reflector.

Double-conversion receivers are used to obtain the necessary gain and selectivity. The output of this type of receiver consists of conventionally modulated video and sound carriers on one of the low VHF channels (usually three or four), which feeds the antenna input of a standard VHF/UHF receiver. More recent designs provide decoded *R-G-B* outputs for standard 525-line and future HDTV services.

Phased Arrays. For many reception applications the parabolic antenna is being superseded by a flat multielement antenna known as a *planar array*. Initial use is for commercial and industrial applications. However, the small size of the design makes it an ideal solution to the mounting and space problems associated with parabolic antennas for home use. K_u-band antenna assemblies measuring no more than a foot square have been found to provide acceptable DBS reception in Japan and Europe, and their use may soon spread to North America.

The design principle is based upon that of large phased arrays used by the U.S. Navy with radar systems on aircraft carriers. Rather than collecting the radiated signal by focusing a parabolic reflector, or "dish," on a signal antenna, the electric signal from each antenna is amplified and timed by a filter network so that all signals are in phase at the summing point. A more sophisticated proposed design employs a second smaller antenna to provide an error signal to operate automatic circuits to align the antenna for maximum signal.

Printed circuit board manufacturing techniques are used to etch the antenna elements on a flat panel, along with the PC boards carrying the coupling amplifier and phasing network for each antenna.

DBS Transmission Standards. The FCC has ruled that HDTV service must be compatible with standard 525-line broadcast transmissions. In other words, HDTV signals must be viewable in color after conversion to standard VHF or UHF channels. This requirement has resulted in further delay while agreement is reached on a signal format.

9.2 CABLE DISTRIBUTION

9.2.1 Subscriber Terminal Equipment

The output of the tap device feeds a 75-Ω coaxial cable drop line into the home. Typically, this cable is about ¼ in (0.6 cm) in diameter and is constructed of an outer polyethylene jacket, a shield made of aluminum foil surrounded by braid, a polyethylene insulator, and a copper center conductor. Loss at 400 MHz is typically 6 dB per 100 ft (30 m).

Isolation. Between the subscriber's television receiver and the distribution cable, isolation is provided for the purposes of both safety and signal purity. The multitap utilizes blocking capacitors to prevent amplifier line powering energy from reaching the subscriber drop cable. As the drop cable enters the subscriber's premises, a grounding connection is provided in accordance with local regulations and/or utility requirements. The grounding connection protects the television receiver from power surges, such as those caused by lightning discharges, and protects the subscriber from the possibility of shock due to voltages which might otherwise be present on the shield of the drop cable. The signal from the unbalanced 75-Ω coaxial cable is converted to a 300-Ω balanced output for connection to the subscriber's television set with an unbalanced-to-balanced matching transformer. Where high ambient signal levels exist, excellent balance is required on the 300-Ω side of this transformer. Transformers are available with very good balance plus a form of Faraday shield, which helps to minimize direct pickup of over-the-air signals. The matching transformer also provides one further level of isolation to the carriage of energy along the sheath of the cable.

Converters. In order to allow reception of more than the 12 VHF channels on a standard television set, most CATV systems provide a converter at the subscriber's receiver. In addition, to secure pay television services from unauthorized viewing, signal scrambling is provided on some channels, and descramblers are required at the receiver location.

Most converters in use today are tunable. Incoming signals are first routed through an input bandpass amplifier (sometimes coupled with a buffer amplifier), then converted to a UHF intermediate frequency utilizing a tunable local oscillator and mixer. The signal is then put through a bandpass filter, and next downconverted to a single VHF channel with a fixed-frequency local oscillator and mixer. The output channel is filtered and fed to the subscriber's receiver. Tunable converters are available in many configurations, including models which are located on top of the television set, models with the channel selection switch connected to the converter by a thin cord, and models with channel selection done with a small hand-held remote-control unit which communicates with the converter by infrared light. Tunable converters are used extensively in the cable industry as a means of providing 30 or more channels to subscribers.

The earliest tunable converters utilized switched inductors and capacitors to select channels. Later versions utilized varactors, controlled by a voltage developed by a series of potentiometers selected by the channel switch. The most recent tunable converters utilize frequency synthesizers controlled by microprocessors for the first local oscillator.

The least-expensive tunable converters have a fine-tuning control on one of the local oscillators to allow the subscriber to compensate for converter frequency drift. More sophisticated converters utilize automatic frequency control circuitry to control drift. In these converters, the output signal is compared with the standard output frequency through a discriminator, and a control voltage is generated, which adjusts the frequency of one of the local oscillators through a varactor.

The most recent converters with microprocessor-controlled frequency synthesizers are very frequency-stable and require no additional frequency control, since they are referenced to a crystal.

Scrambling. Descrambling of premium programming is done in many ways. The simplest is the trap. In the positive-trap method, one or more interfering carriers are inserted in the television channel at the head end. Authorized subscribers are given filters which attenuate the interfering signals with minimal distortion of the desired video signal. Though the positive-trap method is very inexpensive, it is relatively easy to defeat, since all that is needed to receive a clear picture is a notch filter of the correct frequency, a device easily constructed by someone with minimal technical knowledge.

The negative-trap method utilizes a sharp-notch filter on the visual carrier frequency of the channel. These filters are installed outside the homes of those subscribers who are not authorized to receive the pay-television channel. This method probably is the most secure security system in use today, since defeating the trap requires access to the CATV plant located outside the home and often on a utility pole. However, control of more than one channel is difficult, since a separate trap is usually required for each channel. In addition, when only a small percentage of subscribers desire the pay television channel, the cost of the negative trap method becomes prohibitive, since a trap must be installed on the drop of every subscriber not authorized to receive that channel.

One common scrambling system, of which there are many versions, operates by suppressing the vertical and horizontal synchronizing signals as the channels are transmitted from the head end. A pilot signal, containing the timing information for the synchronizing signal, is transmitted either on a separate carrier or within the video or audio signals of the television channel. Within the descrambler, which is located within the subscriber's home, the pilot signal is used to add the synchronizing signals back to the video signal. Although this type of scrambling is relatively easy to defeat, it is the predominant method in use to-day. Recent improvements in the technique have made the security of the system greater through the use of random delays between the pilot signal and the desired location of the synchronizing signal, and through transmitting the pilot signal in two or more parts on both the audio and video carriers.

Another method of scrambling, called *baseband scrambling*, utilizes inversion of the polarity of individual lines of the picture on a random basis. The code to instruct the descrambler on which lines to restore to normal polarity is transmit-ted within the video signal of that channel. This type of descrambler requires de-modulation of the television signal to baseband and subsequent remodulation to a channel. As a result, it is considerably more expensive than other methods but is more difficult for unauthorized users to decode. In addition, control of the tele-vision set audio volume level is possible from the converter remote control, and videotext decoders can be more easily incorporated.

In most cases, the descrambler and converter are located in the same housing. Recently, addressable converters have become common in CATV systems. These converters combine the tunable converter and a descrambler with an RF receiver and address-recognition circuitry. Each converter may be addressed and controlled by a computer at the head end. Individual channels may be turned off and on, or the entire unit may be disabled from the head end. In this way a sub-scriber may be authorized for new services or disconnected from them without a CATV service person visiting the subscriber's home.

Addressable converters use either an addressing carrier in the midband or in-corporate the addressing commands within the vertical interval of each video channel. At the head end, a computer system is required to store information on which channels each subscriber is authorized to receive and to continually ad-dress the converters within the network.

Signal security remains a significant problem for the CATV industry, and new technology is emerging in an attempt to provide solutions. Off-premise devices, located outside the subscriber's home or establishment, control which channels are allowed into the home under control from the head end. Since only autho-rized channels are allowed into the subscriber's home, unauthorized viewing is difficult. Two types of off-premise devices are available. The off-premise con-verter performs all the functions of an indoor converter, sending only one chan-nel at a time down the drop to the subscriber. The off-premise addressable tap jams or removes those channels which a subscriber is not authorized to view un-der control of the head end. A converter or television set which can tune mid- and superband CATV channels is needed, but no descrambler is required.

With the off-premise devices, no scrambling is required, since security is pro-vided by preventing nonauthorized signals from entering the subscriber's home. Off-premise equipment is new and for the most part unproven, and significant technical and cost problems remain unsolved.

Manufacturers are presently developing more sophisticated scrambling sys-tems, utilizing many techniques. Digital processing of video and audio, using complicated encryption methods, may emerge in the near future as a solution to the signal security problem.

9.3 RECEIVERS

Television receivers provide black-and-white or color reproduction of pictures and the accompanying monaural or stereophonic sound from signals broadcast through the air or via cable distribution systems. The broadcast channels in the United States are 6 MHz wide for transmission on conventional 525-line standards.[1]

The minimum signal level at which television receivers provide usable pictures and sound, called the *sensitivity level*, generally is in the order of 10 to 20 μV. The maximum level encountered in locations near transmitters may be as high as several hundred millivolts. The Federal Communications Commission has set up two standard signal-level classifications, Grades A and B, for the purpose of licensing television stations and allocating coverage areas. Grade A is to be used in urban areas relatively near the transmitting tower, and Grade B use ranges from suburban to rural and fringe areas a number of miles from the transmitting antenna. The FCC values are expressed in microvolts per meter where meter is the signal wavelength.[5]

The standard transmitter field-strength values for the outer edges of these services for Channels 2 through 69 are listed in Table 9.3. Included for reference in Table 9.3 are the signal levels for what may be considered "city grade" in order to give an indication of the wide range in signal level that a receiver may be required to handle. The actual antenna terminal voltage into a *matched receiver load*, listed in the second column, is calculated from the following equation:

$$e = E \frac{96.68}{\sqrt{f_1 \times f_2}}$$

where e = terminal voltage, μV, 300 Ω
   E = field, μV/m
f_1 and f_2 = band-edge frequencies, MHz

TABLE 9.3 Television Service

Band and channels	Frequency, MHz	City grade		Grade A		Grade B	
		μV/m	μV	μV/m	μV	μV/m	μV
VHF 2–6	54–88 MHz	5,010	7030	2510	3520	224	314
VHF 7–13	174–216 MHz	7,080	3550	3550	1770	631	315
UHF 14–69	470–806 MHz	10,000	1570	5010	787	1580	248
UHF 70–83†	806–890 MHz	10,000	1570	5010	571	1580	180

†Receiver coverage of Channels 70 to 83 has been on a voluntary basis since July 1982. This frequency band was reallocated by the FCC to land mobile use in 1975 with the provision that existing transmitters could continue indefinitely.
Source: K. Blair Benson (ed.), *Television Engineering Handbook*, McGraw-Hill, New York, 1986.

Many sizes and form factors of receivers are manufactured. Portable personal types include pocket-sized or hand-held models with picture sizes of 2 to 4 in (5 to 10 cm) diagonal for monochrome and 5 to 6 in (13 to 15 cm) for color powered by either batteries or ac. The battery power may be either 6 V from four AA cells or up to 12 V from nine C cells. Conventional cathode ray tubes (CRTs) for picture displays in portable sets are rapidly being supplanted by flat CRTs and liquid-crystal displays.

---

1. Pictures of higher resolution that are compatible with 525-line standards and other higher line-number standards are discussed in Sec. 9.3.7.

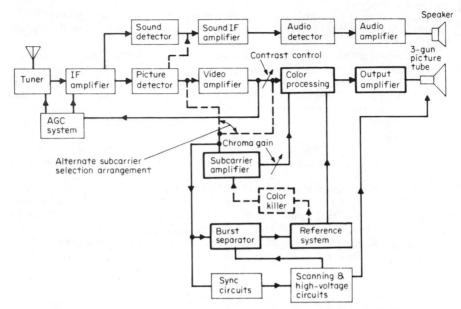

FIGURE 9.9 Fundamental block diagram of a color-receiver with a tri-gun picture-tube display. *[Source: K. Blair Benson (ed.),* Television Engineering Handbook, *McGraw-Hill, New York, 1986]*

Larger screen sizes are available in monochrome where low cost and light weight are prime requirements. However, except where extreme portability is important, the vast majority of television program viewing is in color. The 19-in (48-cm) and 27-in (69-cm) sizes now dominate the market, although the smaller 13-in (33-cm) is popular as a second or semiportable set. In addition, the smaller-size screen in a video monitor configuration generally is used for computer and video game displays.

The television receiver functions may be broken down into several interconnected blocks. With the rapidly increasing use of integrated circuits, the isolation of functions has become more evident in the design and service of receivers. The typical functional configuration of a receiver using a tri-gun picture tube shown in Fig. 9.9 will serve as a guide for the following description of receiver design and operation. The discussions of each major block, in turn, are accompanied with more detailed subblock diagrams.

9.3.1   Tuners

The purpose of the tuner, and the following intermediate amplifier (IF), is to select the desired radio frequency (RF) signals in a 6-MHz channel, to the exclusion of all other signals, available from the antenna or cable system and to amplify the signals to a level adequate for demodulation. Channel selection is accomplished with either mechanically switched and manually tuned circuits, or in *varactor* tuners with electrically switched and controlled circuit components. A mechanical tuner consists of two units, one for the VHF (very high frequency) band from

54 to 88 and 174 to 216 MHz, and the other for the UHF (ultrahigh frequency) band from 470 to 806 MHz. Two separate antenna connections are provided for the VHF and UHF sections of the tuner.[5]

Varactor tuners, on the other hand, have no moving parts or mechanisms and consequently are about a third or a quarter the volume of their mechanical equivalent. Part of this smaller size is the result of combining the VHF and UHF circuits on a single printed circuit board in the same shielded box with a common antenna connection, thus eliminating the need for an outrigger coupling unit.

Selectivity. The tuner bandpass generally is 10 MHz in order to ensure that the picture and sound signals of the full 6-MHz television channel are amplified with no significant imbalance in levels or phase distortion by the skirts of the bandpass filters. This bandpass characteristic usually is provided by three tuned circuits: a single tuned preselector between the antenna input and the RF amplifier stage, a double-tuned interstage network between the RF and mixer stages, and a single-tuned coupling circuit at the mixer output. The first two are frequency-selective to the desired channel by varying either or both the inductance and capacitance. The mixer output is tuned to approximately 44 MHz, the center frequency of the IF channel.

The purpose of the RF selectivity is to reduce all signals that are outside of the selected television channel. For example, the input section of VHF tuners usually contains a high-pass filter and trap section to reject signals lower than Channel 2 (54 MHz) such as standard broadcast, amateur, and citizen's band (CB). In addition, a trap is provided to reduce FM broadcast signal in the 88 to 108 MHz band. A list of the major interference problems is tabulated in Table 9.4 for VHF channels. In Table 9.5 for UHF channels, the formula for calculation of the interfering channels is given in the second column, and the calculation for a receiver tuned to Channel 30 is given in the third column.

VHF Tuner. A block diagram of a typical mechanical tuner is shown in Fig. 9.10. The antenna is coupled to a tunable RF stage through a bandpass filter to reduce spurious interference signals in the intermediate frequency (IF) band, or from FM broadcast stations and CB transmitters. Another bandpass filter is provided in the UHF section for the same purpose. The typical responses of these filters are shown in Fig. 9.11a and b.

TABLE 9.4   Potential VHF-Interference Problems

Desired channel	Interfering signals	Mechanism
5	Ch. 11 picture	2 × ch. 5 osc. − ch. 11 pix = IF
6	Ch. 13 picture	2 × ch. 6 osc. − ch. 13 pix = IF
7 and 8	Ch. 5, FM (98–108 MHz)	Ch. 5 pix + FM = ch. 7 and 8
2–6	Ch. 5, FM (97–99 MHz)	2 × (FM − ch. 5) = IF
7–13	FM (88–108 MHz)	2 × FM = ch. 7–13
6	FM (89–92 MHz)	Ch. 6 pix + FM − ch. 6 osc. = IF
2	6M amateur (52–54 MHz)	2 × ch. 2 pix − 6M = ch. 2
2	CB (27 MHz)	2 × CB = ch. 2
5 and 6	CB (27 MHz)	3 × CB = ch. 5 and 6

Source: K. Blair Benson (ed.), *Television Engineering Handbook*, McGraw-Hill, New York, 1986.

TABLE 9.5 Potential UHF-Interference Problems

Interference type	Interfering channels	Channel 30 example
IF beat	$N \pm 7, \pm 8$	22, 23, 37, 38
Intermodulation	$N \pm 2, \pm 3, \pm 4, \pm 5$	25–28, 32–35
Adjacent channel	$N + 1, - 1$	29, 31
Local oscillator	$N \pm 7 \times$	23, 37
Sound image	$N + \frac{1}{6} (2 \times 41.25)$	44
Picture image	$N + \frac{1}{6} (2 \times 45.75)$	45

Source: K. Blair Benson (ed.), *Television Engineering Handbook*, McGraw-Hill, New York, 1986.

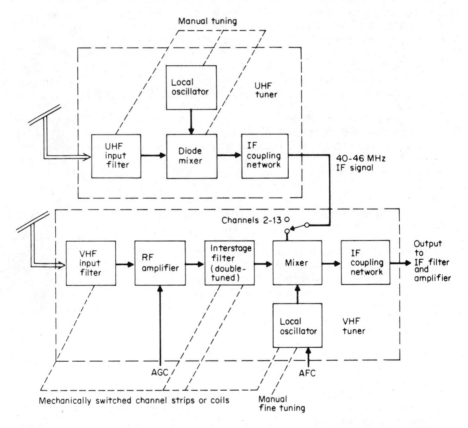

FIGURE 9.10 Typical mechanical-tuner configuration. *[Source: K. Blair Benson (ed.),* Television Engineering Handbook, *McGraw-Hill, New York, 1986]*

The RF stage provides a gain of 15 to 20 dB (10:1 voltage gain) with a bandpass selectivity of about 10 MHz between the −3 dB points on the response curve. The response between these points is relatively flat with a dip of only a decibel or so at the midpoint. Therefore, the response over the narrower 6-MHz television channel bandwidth is essentially flat.

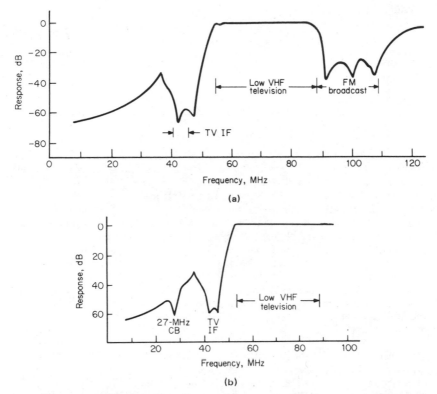

FIGURE 9.11 (a) Response of a tuner input FM bandstop filter. *[Source: K. Blair Benson (ed.),* Television Engineering Handbook, *McGraw-Hill, New York, 1986]* (b) Response of a tuner input with CB and IF traps. *[Source: K. Blair Benson (ed.),* Television Engineering Handbook, *McGraw-Hill, New York, 1986]*

VHF tuners have a rotary shaft that switches a different set of three or four coils or coil taps into the circuit at each VHF channel position (2 to 13). The circuits with these switched coils are the following:

RF input preselection

RF input coupling (single-tuned for monochrome, double-tuned for color)

RF-to-mixer interstage

Local oscillator

In the first switch position (Channel 1), the RF stage is disabled and the mixer stage becomes an intermediate-frequency (IF) amplifier stage, centered on 44 MHz for the UHF tuner.

The mixer stage combines the RF signal with the output of a tunable local oscillator to produce (IF) of 43.75 MHz for the picture-carrier signal and 42.25 for the sound-carrier signal. The local-oscillator signal thus is always 45.75 MHz above that of the selected incoming picture signal. For example, the frequencies for Channel 2 are listed in Table 9.6.

TABLE 9.6   Local-Oscillator and IF Frequencies

	Channel 2, MHz	Channel 6, MHz	Channel 7, MHz
Channel width	54–60	82–88	174–180
Local oscillator	101.00	129.00	221.00
Less picture signal	55.25	83.25	175.25
IF picture signal	45.75	45.75	45.75

These frequencies were chosen to minimize interference from one television receiver into another by always having the local-oscillator signal above the VHF channels. Note that the oscillator frequencies for the low VHF channels (2 to 6) are between Channels 6 (low VHF) and 7 (high VHF), and the oscillator frequencies for the high VHF fall above these channels.

The picture and sound signals of the full 6-MHz television channel are amplified with no significant imbalance in levels or phase distortion by the skirts of the bandpass filters. This bandpass characteristic usually is provided by three tuned circuits: a single-tuned preselector between the antenna input and the RF amplifier stage, a double-tuned interstage network between the RF and mixer stages, and a single-tuned coupling circuit at the mixer output. The first two are frequency-selective to the desired channel by varying either or both the inductance and capacitance. The mixer output is tuned to approximately 44 MHz, the center frequency of the IF channel.

UHF Tuner.   The UHF tuner contains a tunable input circuit to select the desired channel, followed by a diode mixer. As in a VHF tuner, the local oscillator is operated at 45.75 MHz above the selected input-channel signal. The output of the UHF mixer is fed to the mixer of the accompanying VHF tuner, which functions as an IF amplifier. Selection between UHF and VHF is made by applying power to the appropriate tuner RF stage.

Mechanical UHF tuners have a shaft that when rotated moves one set of plates of variable air-dielectric capacitors in three resonant circuits. The first two are a double-tuned preselector in the amplifier-mixer coupling circuit, and the third is the tank circuit of the local oscillator. In order to meet the discrete-selection requirement of the FCC, a mechanical detent on the rotation of the shaft and a channel-selector indicator are provided (Fig. 9.12).

The inductor for each tuned circuit is a rigid metal strip, grounded at one end to the tuner shield and connected at the other end to the fixed plate of a three-section variable capacitor with the rotary plates grounded. The three tuned circuits are separated by two internal shields that divide the tuner box into three compartments.

Tuner-IF Link Coupling.   With mechanically switched tuners, it has been necessary to place the tuner behind the viewer control panel and connect it to the IF section, located on the chassis, with a foot or so length of shielded 50- or 75-Ω coaxial cable. Since the output of the tuner and the input of the IF amplifier are high-impedance-tuned circuits, for maximum signal transfer, it is necessary to couple these to the cable with impedance-matching networks.

Two resonant-circuit arrangements are shown in Fig. 9.13. The low-side capacitive system has a low-pass characteristic that attenuates the local-oscillator

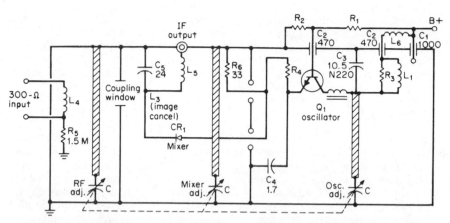

FIGURE 9.12 UHF rotary tuner. *[Source: Courtesy of Philips Consumer Electronics Company. In* Television Engineering Handbook, *K. Blair Benson (ed.), McGraw-Hill, New York, 1986]*

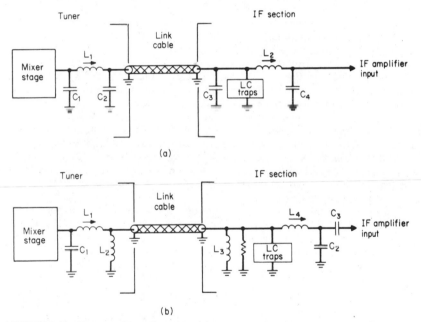

FIGURE 9.13 Tuner-to-IF section link coupling. (*a*) Low-side capacitive. (*b*) Low-side inductive. *[Source: K. Blair Benson (ed.),* Television Engineering Handbook, *McGraw-Hill, New York, 1986]*

and mixer harmonic currents ahead of the IF amplifier. This can be an advantage in controlling local-oscillator radiation and in reducing generation of spurious signals in the IF section. On the other hand, the low-side inductance gives a better termination to the link cable and therefore reduces interstage cable loss.

The necessary bandpass can be obtained either by undercoupled stagger tuning or by overcoupled synchronous tuning (Fig. 9.13).

Varactor Tuner. Tuning is accomplished by the change in capacitance with applied dc voltage to *varicap* diodes. One diode is used in each tuned circuit. Unlike variable air-dielectric capacitors, varicaps have a resistive component in addition to their capacitance that lowers the Q and results in a degraded noise figure. Therefore, varactor UHF tuners usually include an RF amplifier stage, making it functionally similar to a VHF tuner (Fig. 9.14).

The full UHF band can be covered by a single varicap in a tuned circuit since the ratio of highest and lowest frequencies in the UHF bands is less than 2:1 (1.7). However, the ratio of the highest to lowest frequencies in the two VHF bands is over twice (4.07) that of the UHF band. This is beyond the range that can be covered by a tuned circuit using varicaps. The problem is solved by the use of band switching between the low and high VHF channels. This is accomplished rather simply by short-circuiting a part of the tuning coil in each resonant tank circuit to reduce its inductance. The short circuit is provided by a diode that has a low resistance in the forward-biased condition and a low capacitance in the reverse-biased condition. A typical RF and oscillator circuit arrangement is shown in Fig. 9.15. Applying a positive voltage to V_B switches the tuner to high VHF by causing the diodes to conduct and lower the inductance of the tuning circuits.

Tuning Systems. The purpose of the tuning system is to set the tuner, VHF or UHF, to the desired channel and to fine-tune the local oscillator for the video carrier from the mixer to be set at the proper IF frequency of 46.75 MHz. In mechanical tuners, this obviously is an adjustment of the rotary selector switch and the capacitor knob on the switch shaft. In electronically tuned systems, the dc tuning voltage can be supplied from the wiper arm of a potentiometer control connected to a fixed voltage source as shown in Fig. 9.16a.

Alternatively, multiples of this circuit, as shown in Fig. 9.16b, can provide preset fine-tuning for each channel. This arrangement most commonly is found in cable-channel selector boxes supplied with an external cable processor.

In digital-display systems, such as that shown in Fig. 9.16c, the tuning voltage can be read as a digital word from the memory of a keyboard and display station. After conversion from digital to analog voltage, the tuning control voltage is sent to the tuner.

Figure 9.16d shows a popular microprocessor system using a phase-locked loop to compare a medium-frequency square-wave signal from the channel-selector keyboard, corresponding to a specific channel, with a signal divided down by 256 from the local oscillator. The error signal generated by the difference in these two frequencies is filtered and used to correct the tuning voltage being supplied to the tuner.

9.3.2 Intermediate-Frequency (IF) Amplifier Requirements

Carrier Frequencies. The IF picture and sound-carrier frequencies currently standardized for television receivers were chosen with prime consideration of possible degradation of the picture from interfering signals. The picture-carrier frequency is 45.75 MHz, and, with the local oscillator above the received RF signal, the sound-carrier frequency is 41.25 MHz.

The three factors given greatest emphasis in the choice were the following:

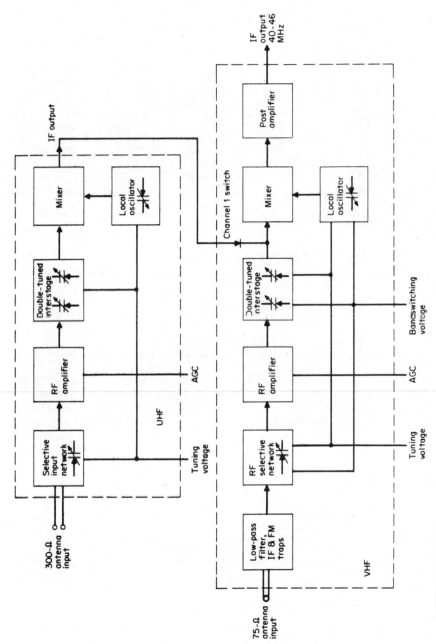

FIGURE 9.14 Typical varactor-tuner block configuration. [*Source: K. Blair Benson (ed.),* Television Engineering Handbook, *McGraw-Hill, New York, 1986*]

9.21

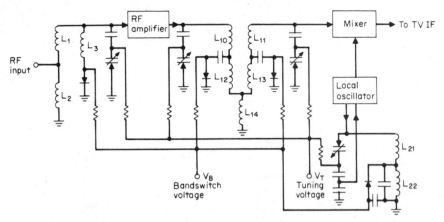

FIGURE 9.15 VHF-tuner band-switching and tuning circuits. $+V_\beta$ = high VHF (active tuning inductors = L_3 in parallel with L_1 and L_2, L_{10}, L_{11}, L_{21}). $-V_\beta$ = low VHF (active tuning inductors = $L_1 + L_2$, L_{12}, $L_{11} + L_{13}$, L_{14}, $L_{21} + L_{22}$). *[Source: K. Blair Benson (ed.), Television Engineering Handbook, McGraw-Hill, New York, 1986]*

1. Interference from other nontelevision services
2. Interference from the fundamental and harmonics of local oscillators in other television receivers
3. Spurious responses from the image signal in the mixer conversion and from harmonics of the IF signals

Analysis of the relationships indicates the soundness of the choice of 45.75 MHz for the IF picture carrier. The important advantages are discussed below:

1. No images from the mixer conversion process fall within the VHF band selected by the tuner except for a negligible interference on the edge of the Channel 7 passband from the image from another receiver tuned to Channel 6.
2. All channels are clear of picture harmonics except that the fourth harmonic of the IF picture carrier falls near the Channel 8 picture carrier. This can cause a noticeable beat pattern in another receiver if the offending receiver has not been designed with adequate shielding.
3. Local-oscillator radiation does not interfere with another receiver on any channel or on any channel image.
4. No UHF signal falls on the image frequency of another station.

On the other hand it should be pointed out that channels for police communication are allocated in the standard IF band. Since these transmitters radiate high power levels, receivers require thorough shielding of the IF amplifier and IF-signal rejection traps in the tuner ahead of the mixer. In locations where severe cases of interference are encountered, the addition of a police-band rejection filter in the antenna input may be necessary.

Gain Characteristics. The output level of the picture and sound carriers from the mixer in the tuner is about 200 μV. The IF section provides amplification to boost this level to about 2 V, which is required for linear operation of the detector or

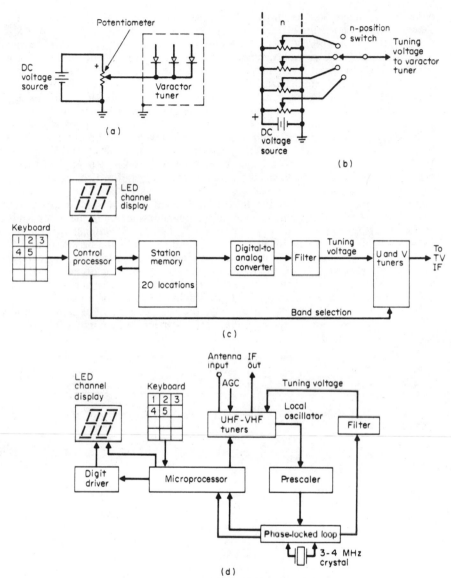

FIGURE 9.16 Varactor-tuning systems. (*a*) Simple potentiometer control varactor-tuning system. (*b*) A multiple of potentiometers provides *n*-channel selection. (*c*) Simplified memory tuning for 20-station selection. (*d*) Microprocessor phase-locked loop. [*Source: K. Blair Benson (ed.), Television Engineering Handbook, McGraw-Hill, New York, 1986*]

demodulator stage. This is an overall gain of 80 dB. The gain distribution in a typical IF amplifier using discrete gain stages is shown in Fig. 9.17a.

Automatic gain control (AGC) in a closed feedback loop is used to prevent overload in the IF, and the mixer stage as well, from strong signals (see Fig.9.17b). Input-signal levels may range from a few microvolts to several hundred microvolts, thus emphasizing the need for AGC. The AGC voltage is applied only to the IF for moderate signal levels so that the low-noise RF amplifier stage in the tuner will operate at maximum gain for relatively weak tuner input signals. A "delay" bias is applied to the tuner gain control to block application of the AGC voltage except at very high antenna signal levels. As the antenna signal level increases, the AGC voltage is applied first to the first and second IF stages. When the input signal reaches about 1 mV, the AGC voltage is applied to the tuner, as well. AGC systems are discussed in greater detail in Sec. 9.3.5.

Response Characteristics. The bandpass of the IF amplifier must be wide enough to cover the picture and sound carriers of the channel selected in the tuner while providing sharp rejection of adjacent channel signals. Specifically the upper adjacent-picture carrier and lower adjacent-sound channel must be attenuated 40 and 50 dB, respectively, to eliminate visible interference patterns in the picture. The sound carrier at 4.5 MHz below the picture carrier must be of adequate level to feed either a separate sound-IF channel or a 4.5-MHz intercarrier sound channel. Furthermore, because in the vestigial-sideband system of transmission the video-carrier lower sidebands are missing, the response characteristic is modified from flat response to attenuate the picture carrier by 50 percent (6 dB).

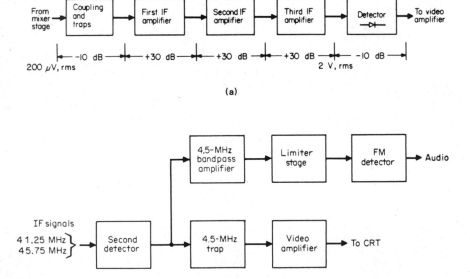

FIGURE 9.17 (*a*) Typical IF-amplifier strip block diagram and gain distribution. *[Source: K. Blair Benson (ed.),* Television Engineering Handbook, *McGraw-Hill, New York, 1986] (b)* Receiver AGC system. *[Source: K. Blair Benson (ed.),* Television Engineering Handbook, *McGraw-Hill, New York, 1986]*

In addition, in color receivers the color subcarrier at 3.58 MHz below the picture carrier must be amplified without time delay relative to the picture carrier or distortion. Ideally this requires the response shown in Fig. 9.18. Notice that the color IF is wider and has greater attenuation of the channel sound carrier in order to reduce the visibility of the 920-kHz beat between color subcarrier and sound carrier.

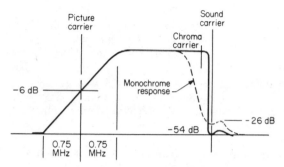

FIGURE 9.18 Ideal IF amplitude response for color and monochrome (dashed line) receivers. *[Source: K. Blair Benson (ed.),* Television Engineering Handbook, *McGraw-Hill, New York, 1986]*

These and other more stringent requirements for color are outlined below and illustrated in Fig. 9.19.

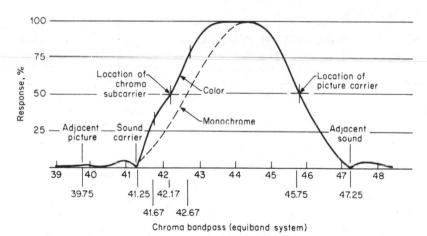

FIGURE 9.19 Overall IF bandwidth for color reception. *[Source: K. Blair Benson (ed.),* Television Engineering Handbook, *McGraw-Hill, New York, 1986]*

1. *IF bandwidth* must be extended on the high-frequency video side to accommodate the color-subcarrier modulation sidebands that extend to 41.25 MHz as shown in Fig. 9.19. The response must be stable and, except in sets with automatic chroma control (ACC), not change with the input-signal level (AGC), in order to maintain a constant level of color saturation.

2. *More accurate tuning* of the received signal must be accomplished in order not to shift the carriers on the tuner-IF passband response. Deviation from their prescribed positions will alter the ratio of luminance to chrominance (saturation). While this is corrected in receivers with automatic fine tuning (AFT) and ACC, it can change the time relationship between color and luminance that is apparent in the color picture as chroma being misplaced horizontally.

3. *Color-subcarrier* presence as a second signal dictates greater freedom from overload of amplifier and detector circuits, which can result in spurious intermodulation signals visible as beat patterns. These cannot be removed by subsequent filtering.

4. *Envelope delay* (time delay) of the narrow-band chroma and wide-band luminance signals must be equalized so that the horizontal position of the two signals match in the color picture. The effect is similar to that in (2) above.

9.3.3 Integrated-Circuit Gain and Filter Block IF Section

The development of integrated-circuit (IC) gain blocks with block filtering indicated the possibility of providing all the required gain and demodulation with one input port and one interstage port for bandpass and trap networks. Unfortunately, the cost of early designs was prohibitive for practical use in the wide range of consumer television receivers. By the mid-1970s, however, acoustic-wave state-of-the-art had progressed to the point where commercially acceptable devices could be manufactured at a price competitive with earlier discrete stage and coupling-filter configurations, and more recently a block filter followed by a gain-block IC amplifier.

Surface Acoustic-Wave (SAW) Filter. A SAW filter can provide the entire passband shape and adjacent-channel attenuation required for a television receiver. A typical amplitude response and group-delay characteristic are shown in Fig. 9.20. The sound-carrier (41.25 MHz) attenuation of the SAW filter has been designed to operate with a synchronous detector; hence the lesser attenuation than in a conventional LC bandpass (see Sec. 9.3.4). The response of an LC discrete stage configuration shows a 60-dB attenuation of the sound carrier, neccessary for suppression of the 920-kHz sound-chroma beat when used with a diode detector. In addition, with SAW technology it possible to make wider adjacent-channel traps, which improves their performance and, in part, makes allowance for the temperature coefficient of the substrate materials used in SAW filters. This drift may be as great as 59 kHZ per 10°C.

The schematic diagram of a SAW filter IF circuit is shown in Fig. 9.21. The filter typically has an insertion loss of 15 to 20 dB and therefore requires a preamplifier to maintain a satisfactory overall receiver signal-to-noise ratio.

The SAW filter consists of a piezoelectric substrate measuring 4 to 8 mm (0.16 to 0.31 in) by 0.4 mm (0.016 in) thick, upon which has been deposited a pattern of two sets of interleaved aluminum fingers. The width of the fingers may be in the order of 50 to 500 mm thick and 10 to 20 μm wide. The pictorial diagram of the layout of a filter is shown in Fig. 9.22.

Although quartz and other materials have been researched for use as SAW

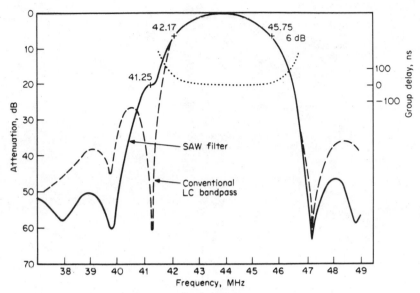

FIGURE 9.20 Surface acoustic wave (SAW) IF filter response. *[Source: K. Blair Benson (ed.), Television Engineering Handbook, McGraw-Hill, New York, 1986]*

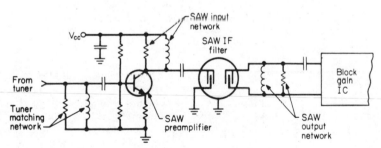

FIGURE 9.21 Schematic of block filter–block gain configuration. *[Source: K. Blair Benson (ed.), Television Engineering Handbook, McGraw-Hill, New York, 1986]*

filter substrates, for television applications lithium niobate and lithium tantalate are typical. When one set is driven by an electric signal voltage, an acoustic wave moves across the surface to the other set of fingers that are connected to the load. The transfer amplitude-frequency response appears as a sin x/x (Fig. 9.23).

A modification to the design in Fig. 9.22 that gives more optimum television bandpass and trap response consists of varying the length of the fingers to form a diamond configuration (Fig. 9.24). This is equivalent to connecting several transducers with slightly different resonant frequencies and bandwidths in parallel. Other modification consists of varying the aperture spacings, distance between the transducers, and the passive coupler-strip patterns in the space between the transducers.

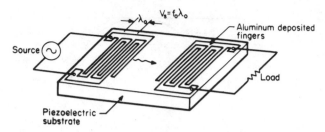

FIGURE 9.22   SAW filter. *[Source: K. Blair Benson (ed.),* Television Engineering Handbook, *McGraw-Hill, New York, 1986]*

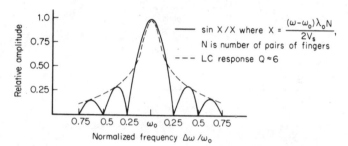

FIGURE 9.23   Response of a uniform interdigitated SAW filter. *[Source:
A. DeVries et al., "Characteristics of Surface-Wave Integratable Filters
(SWIFS),"* IEEE Trans., *vol. BTR-17, no. 1, p. 16. In* Television Engi-
neering Handbook, *K. Blair Benson (ed.), McGraw-Hill, New York, 1986]*

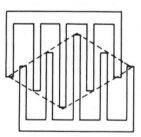

FIGURE   9.24 Apodized   IDT   pattern.
[Source: K. Blair Benson (ed.), Television Engineering Handbook, *McGraw-Hill, New York,
1986]*

Basic Requirements.   Integrated-circuit
gain blocks have the same basic requirements as discrete-stage IF amplifiers (see Fig. 9.25). These are: high
gain, low distortion, and a large linear
gain-control range under all operating
conditions. One typical differential amplifier configuration used in IC gain
blocks that meets these objectives has
a gain of nearly 20 dB and a gain-
control range of 24 dB. A direct-
coupled cascade of three stages yields
an overall gain of 57 dB and a gain-
control range of 64 dB. The gain-
control system internal to this IC begins to gain-reduce the third stage at an
IC input level of 100 μV of IF carrier. With increasing input signal level, the
third-stage gain reduces to 0 dB and then is followed by the second stage to a
similar level, and then followed by the first in the same manner. By this means a
noise figure of 7 dB is held constant over an IF input signal range of 40 dB. The
need for a preamplifier ahead of the SAW IF becomes less important when the IF
amplifier noise figure is maintained constant by this cascaded control system.

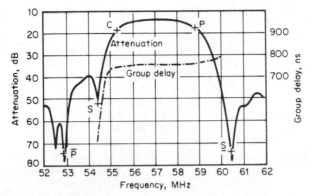

FIGURE 9.25 Frequency response of SAW filter picture-carrier output. *[Source: Yamada and Uematsu, "New Color TV with Composite SAW IF Filter Separating Sound and Picture Signals," IEEE Trans., vol. CE-28, no. 3, p. 193. Copyright 1982 IEEE. In Television Engineering Handbook, K. Blair Benson (ed.), McGraw-Hill, New York, 1986]*

The high gain and small size of an integrated-circuit IF amplifier places greater importance on the PC layout and ground paths if stability is to be achieved under a wide range of operating conditions. These considerations also carry over to the external circuits and components.

9.3.4 Video-Signal Demodulation

The function of the video demodulator is to extract the picture-signal video information that has been placed on the RF carrier as amplitude modulation. The demodulator receives the modulated carrier signal from the IF amplifier that has boosted the peak-to-peak level to 1 or 2 V. The modulation components extend from direct current to 4.5 MHz. The output of the demodulator is fed directly to the video amplifier. There are four types of modulators used in television receivers. These are: the envelope detector, the transistor detector, the synchronous detector, and the feedback balanced diode.

Envelope Detector. Of the several types this is the simplest and the most commonly used. It consists of merely a diode rectifier feeding a parallel load of a resistor and a capacitor (Fig. 9.26). In other words, it is a half-wave rectifier that charges the capacitor to the peak value of the modulation envelope.

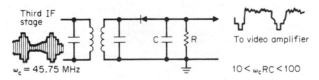

FIGURE 9.26 Envelope detector. *[Source: K. Blair Benson (ed.), Television Engineering Handbook, McGraw-Hill, New York, 1986]*

Because of the large loss in the diode, a high level of IF voltage is required to recover 1 or 2 volts of demodulated video. In addition, unless the circuit is operated at a high signal level, the curvature of the diode impedance curve near cutoff results in undesirable compression of peak-white signals. The requirement for large signal levels and the nonlinearity of detection result in design problems and performance deficiencies.

1. Beat signal products will occur between the color-subcarrier (42.17 MHz), the sound-carrier (41.25 MHz), and high-amplitude components in the video signal. The most serious is a 920-Hz (color to sound) beat and 60-Hz buzz in sound from vertical sync and peak-white video modulation.

2. Distortion of luminance toward black of as much as 10 percent and asymmetric transient response. Called *quadrature distortion*, this characteristic of the vistigial sideband is aggravated by nonlinearity of the diode (see Fig. 9.27).

3. Radiation of the fourth harmonic of the video IF produced by the detection action directly from the chassis to interfere with reception of VHF Channel 8 (180 to 186 MHz).

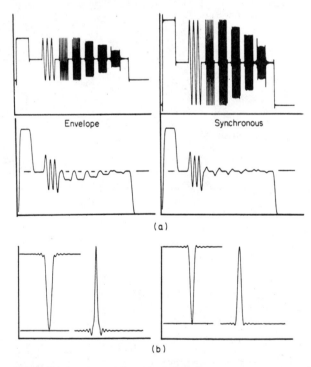

FIGURE 9.27 Comparison of quadrature distortion in envelope and synchronous detectors. (*a*) Axis shift. (*b*) Inverted and normal 2T-pulse response. [*Source: From C. B. Neal and S. Goyal, "Frequency and Amplitude Phase Effects in Television Broadcase Systems," IEEE Trans., vol. CE-23, no. 3, August 1977, p. 241. In K. Blair Benson (ed.),* Television Engineering Handbook, *McGraw-Hill, New York, 1986*]

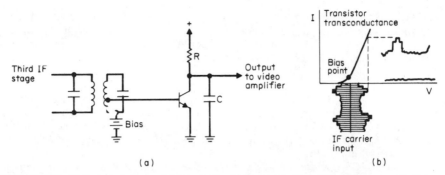

FIGURE 9.28 Transistor detector. (*a*) Schematic. (*b*) Detection characteristic. [*Source: K. Blair Benson (ed.),* Television Engineering Handbook, *McGraw-Hill, New York, 1986*]

Even with these deficiencies, the diode envelope detector has been used in the majority of the monochrome and color television receivers dating back to vacuum-tube designs up to the era of transistors.

Transistor Detector. A transistor biased near collector cutoff and driven with a modulated carrier at an amplitude greater than the bias level (see Fig. 9.28) provides a demodulator that can have a gain of 15 or 20 dB over that of a diode. Consequently, less gain is required in the IF amplifier; in fact, in some receiver designs the third IF stage has been eliminated. Unfortunately, this is offset by the same deficiencies in signal detection as the diode envelope detector.

Synchronous Detector. The third type is basically a balanced rectifier in which the carrier is sampled by an unmodulated carrier at the same frequency as the modulated carrier. The unmodulated reference signal is generated in a separate high-Q limiting circuit that removes the modulation. An alternative system for generating the reference waveform is by means of a local oscillator phase-locked to the IF signal carrier.

The advantages of synchronous demodulation are:

1. Higher gain than a diode detector
2. Low-level input, which considerably reduces beat-signal generation
3. Low-level detection operation, which reduces IF harmonics by more than 20 dB
4. Little or no quadrature distortion (see Fig. 9.27), depending upon the lack of residual phase modulation (purity) of the reference carrier used for detection
5. Circuit easily formatted in an IC

9.3.5 Automatic-Gain Control (AGC)

Each amplifier, mixer, and detector stage in a television receiver has a number of operating conditions that must be met in order to achieve optimum performance, as follows:

1. Input level is greater than the internally generated noise by a factor in excess of the minimum acceptable signal-to-noise ratio.

2. Input level does not overload the amplifier stages, thus causing a bias shift.

3. Bias operates each functional component of the system at its optimum linearity point, that is, lowest third-order product for amplifiers, highest practical second-order for mixers and detectors.

4. Spurious responses in the output are 50 dB below the desired signal.

The function of the automatic-gain control system is to maintain signal levels in these stages at the optimum value over a large range of receiver input signal levels. The control voltage to operate the system usually is derived from the video detector or the first video amplifier stage. A block diagram of an AGC system is shown in Fig. 9.17.

Average AGC operates on the principle of keeping the carrier level constant in the IF. Changes in modulation of the carrier will affect the gain control. It is used only in low-cost receivers.

Peak or sync-clamp AGC compares the video sync-tip level with a fixed dc level. If the sync-tip amplitude exceeds the reference level, a control voltage is applied to the RF and IF stages to their gain and thus restores the sync-tip level to the reference level.

Keyed or gated AGC is similar to sync-clamp AGC. The stage where the comparison of sync-tip and reference signals takes place is activated only during the sync-pulse interval by a horizontal flyback pulse. Since the AGC circuit is insensitive to any spurious noise signals between sync pulses, the noise immunity is considerably improved over the other two systems.

AGC Delay.   For best receiver signal-to-noise ratio, the tuner RF stage is operated at maximum gain for RF signals up to a threshold level of 1 mV. In discrete amplifier chains, the AGC system begins to reduce the gain of the second IF stage proportionately as the RF signal level increases from just above the sensitivity level to the second-stage limit of gain reduction (20 to 25 dB). For increasing signals, the first IF stage gain is reduced. Finally, above the control delay point of 1 mV, the tuner gain is reduced. A plot of the relationships between receiver input RF level and the gain characteristics of the tuner and IF are shown in Fig. 9.29*a* and the noise figure is shown in Fig. 9.29*b*.

System Analysis.   The interconnection of the amplifier stages (RF, mixer, and IF), the detector, and the lowpass-filtered control voltage in effect is a feedback system. The loop gain of the feedback system is nonlinear, increasing with increasing signal level.

First, the loop gain should be large to maintain good regulation of the detector output over a wide range of input signal levels. As the loop gain increases, the stability of the system will decrease and oscillation can occur.

Second, the ability of a receiver to reject impulse noise is inversely proportional to the loop gain. Excessive impulse noise can saturate the detector and reduce the RF-IF gain, thereby causing either a loss in picture contrast or a complete loss of picture. This problem can be alleviated by:

1. Bandwidth limiting the video signal fed to the AGC detector

2. Use of keyed or gated AGC to block out false input signals except during the sync-pulse time

A good compromise between regulation of video level and noise immunity is realized with a loop-gain factor of 20 to 30 dB.

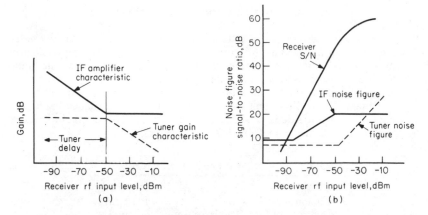

FIGURE 9.29 (*a*) Gain control with input level. (*b*) Noise figure of RF and IF stages with gain control and resultant receiver signal-to-noise ratio. [*Source: K. Blair Benson (ed.), Television Engineering Handbook, McGraw-Hill, New York, 1986*]

The filter network and filter time constants play an important part in the effectiveness of AGC operation. The filter removes the 15,750-Hz horizontal sync pulses and equalizing pulses, the latter in blocks in the 60-Hz vertical-sync interval. The filter time constants must be chosen to eliminate or minimize the following problems:

Airplane flutter is a fluctuation in signal level caused by alternate cancellation and reinforcement of the received signal by reflections from an airplane flying overhead in the path between the transmitting and receiving antennas. The amplitude may vary as much as 2 to 1 at rates from 50 to 150 Hz. If the time constants are too long, especially that of the control voltage to the RF stage, the gain will not change rapidly enough to track the fluctuating level of the signal. The result will be a flutter in contrast and brightness of the picture.

Vertical sync-pulse sag results from the AGC system speed of response being so fast that it will follow changes in sync-pulse energy during the vertical-sync interval. Gain increases during the initial half-width equalizing pulses, then decreases during the slotted vertical-sync pulse, increases again during the final equalizing pulses, and then returns to normal during the end of vertical blanking and the next field of picture. Excessive sag can cause loss in interlace and vertical jitter or bounce. Sag can be reduced by limiting the response of the AGC control loop, or by the use of keyed AGC.

Lock-out of the received signal during channel switching is caused by excessive speed of the AGC system. This can result in as much as a 2:1 decrease in pull-in range for the AGC system. In keyed AGC systems, if the timing of the horizontal gating pulses is incorrect, an excessive or insufficient sync can result at the sync separator and in turn will upset the operation of the AGC loop.

9.3.6 Automatic Frequency Control (AFC)

Also called *automatic fine-tuning (AFT)*, this circuit senses the frequency of the picture carrier in the IF section and sends a correction voltage to the local oscillator in the tuner section if the picture carrier is not on the standard frequency of 45.75 MHz.

Typical AFT systems consist of a frequency discriminator prior to the video detector, a low-pass filter, and a varactor diode controlling the local oscillator. The frequency discriminator in discrete transistor IF systems has been the familiar balanced-diode type used for FM radio receivers with the components adjusted for wide-band operation centering on 45.75 MHz. A small amount of unbalance is designed into the circuit to compensate for the unbalanced sideband components of vestigial sideband signal characteristics. The characteristics of AFC closed loops are shown in Table 9.7.

TABLE 9.7   AFC Closed-Loop Characteristics

Pull-in range	± 750 kHz
Hold-in range	± 1.5 MHz
Frequency error for ±500-kHz offset	< 50 kHz

Source: K. Blair Benson (ed.), *Television Engineering Handbook*, McGraw-Hill, New York, 1986.

In early solid-state designs, the AFT block was a single IC with only two coils as external components. Current designs have included the AFC circuit in the form of a synchronous demodulator on the same IC die as the other functions of the IF section.

9.3.7 Sound-Carrier Separation Systems

Television sound is transmitted as a frequency-modulated signal with a maximum deviation of ±25 kHz (100 percent modulation) capable of providing an audio bandwidth of 50 to 15,000 Hz. The frequency of the sound carrier is 4.5 MHz above the RF picture carrier. The basic system is monaural with dual-channel stereophonic transmission at the option of the broadcaster.

The *intercarrier sound system* passes the IF picture and sound carriers (45.75 and 41.25 MHz, respectively) through a detector (nonlinear stage) to create the intermodulation beat of 4.5 MHz. The intercarrier sound signal is then amplified, limited, and FM-demodulated to recover the audio. Block diagrams of intercarrier sound systems are shown in Figs. 9.30 and 9.31. In a discrete component

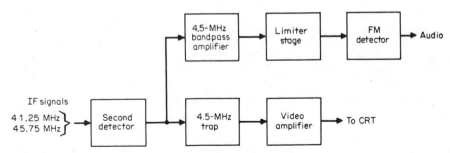

FIGURE 9.30 Typical monochrome intercarrier-sound system. *[Source: K. Blair Benson (ed.), Television Engineering Handbook, McGraw-Hill, New York, 1986]*

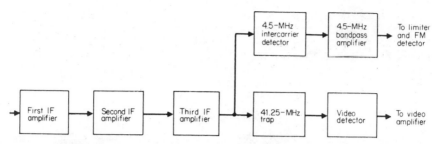

FIGURE 9.31 Intercarrier-sound takeoff ahead of detector. *[Source: K. Blair Benson (ed.), Television Engineering Handbook, McGraw-Hill, New York, 1986]*

format, the intercarrier detector is typically a simple diode detector feeding a 4.5-MHz resonant network. If an IC IF system is used, the sound and picture IF signals are carried all the way to the video detector where one output port of the balanced synchronous demodulator supplies both the 4.5-MHz sound carrier and the composite baseband video.

The coupling network between the intercarrier detector and the sound IF amplifier usually has the form of a half-section high-pass filter which is resonant at 4.5 MHz. This form gives greater attenuation to the video and sync pulses in the frequency range from 4.5 MHz to direct current (carrier), thereby reducing buzz in the recovered audio, especially under the conditions of low picture carrier. An alternative implementation uses a piezoelectric ceramic filter which is designed to have a bandpass characteristic at 4.5 MHz and needs no in-circuit adjustment.

Buzz results when video-related phase-modulated components of the visual carrier (ICPM) are transferred to the sound channel. The generation of incidental carrier phase modulation (ICPM) can be transmitter-related or receiver-related; however, the transfer to the sound channel takes place in the receiver. This can occur in the mixer or the detection circuit. Here a synchronous detector represents very little improvement over an envelope diode detector unless a narrow-band filter is used in the reference channel.[6]

Split-carrier sound processes the IF picture and sound carriers as shown in Fig. 9.32 and described in Sec. 9.3.3.

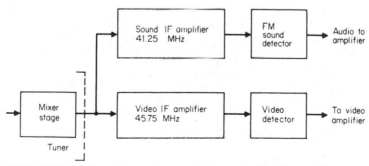

FIGURE 9.32 Split-carrier sound system. *[Source: K. Blair Benson (ed.), Television Engineering Handbook, McGraw-Hill, New York, 1986]*

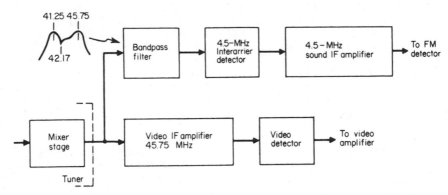

FIGURE 9.33 Quasi-parallel sound system. *[Source: K. Blair Benson (ed.), Television Engineering Handbook, McGraw-Hill, New York, 1986]*

Quasi-parallel sound utilizes a special filter such as the SAW filter of Fig. 9.22 to eliminate the Nyquist slope in the sound-detection channel, thereby eliminating a major source of ICPM generation in the receiver. The block diagram of this system is shown in Fig. 9.33.

Integrated-Circuit Implementation. Nearly all sound channels in present-day television receivers are designed as a one- or two-IC configuration. The single IC contains the functions of sound IF amplifier-limiter, FM detector, volume control, and audio output. Two-chip systems usually have the functions of amplifier-limiter, detector, and volume control in one IC and the audio amplifier and output in the second.

Amplifier-Limiter Sections. These consist of from three to eight direct-coupled, balanced, differential-amplifier stages. Although somewhat similar in design to the amplifier stages in the video-IF IC, the sound-IF stages have no variable gain control (AGC). As signal level increases, the signal in the last stage begins to symmetrically limit (simultaneously clip the positive and negative peaks of the sine wave). Bias networks are stiff so that no shifting of the bias voltage occurs. Increased signal level forces the earlier stages successively into limiting. Limiting action removes much of the amplitude-modulated video and sync components from the sound IF signal. Limiting threshold for an amplifier-limiter section typically occurs at a 4.5-MHz input level of 30 to 100 μV.

FM Detectors. Three types of detector circuits are used in ICs for the demodulation of the FM sound carrier. These are quadrature detector, balanced-peak detector, and the phase-locked loop detector.

The *quadrature detector*, also called gated coincidence detector and analog multiplier, measures the instantaneous phase shift across a reactive circuit as the carrier frequency shifts. At center frequency (zero deviation) the LC phase network gives a 90° phase shift to V_2 compared with V_1. As the carrier deviates, the phase shift changes proportionately to the amount of carrier deviation and direction.[112]

The *balanced peak detector* or differential peak detector, described by

Peterson, utilizes two peak or envelope detectors, a differential amplifier, and a frequency-selective circuit or piezoceramic discriminator. Also shown are the voltages at each detector input and the resultant difference signal.

The differential peak detector operates at a lower voltage level and does not require square-wave switching pulses. Therefore it creates far lower harmonic radiation than the quadrature detector. In some designs a low-pass filter has been placed between the limiter and peak detector to further reduce harmonic radiation and increase AM rejection.

The *phase-locked-loop detector* requires no frequency-selective *LC* network to accomplish demodulation. In this system, the voltage-controlled oscillator (VCO) is phase-locked by the feedback loop into following the deviation of the incoming FM signal. The low-frequency error voltage which forces the VCO to track is indeed the demodulated output.

9.3.8 Video Amplifiers

A range of video signals of 1 to 3 V at the second detector has become standard for many practical reasons, including optimum gain distribution between RF, IF, and video sections and distribution of signal levels so that video detection and sync separation may be effectively performed. The video amplifier gain and output level are designed to drive the picture tube with this input level. Sufficient reserve is provided to allow for low percentage modulation and signal strengths below the AGC threshold.

Picture Controls. A video gain or *contrast control* and a *brightness* or *background control* are provided to allow the viewer to select the contrast ratio and overall brightness level which produce the most pleasing picture for a variety of scene material, transmission characteristics, and ambient lighting conditions. The contrast control usually provides a 4/1 gain change. This is accomplished either by attenuator action between the output of the video stage and the CRT or by changing the ac gain of the video stage by means of an ac-coupled variable resistor in the emitter circuit. The brightness control shifts the dc bias level on the CRT to raise or lower the video signal with respect to the CRT beam cutoff-voltage level. (See Fig. 9.34.)

AC and dc Coupling. For perfect picture transmission and reproduction, it is necessary that all shades of gray are demodulated and reproduced accurately by the display device. This implies that the dc level developed by the video demodulator, in response to the various levels of video carrier, must be carried to the picture tube. Direct coupling or dc restoration is often used, especially in color receivers where color saturation is directly dependent upon luminance level. (See Fig. 9.35.)

Many low-cost monochrome designs utilize only ac coupling with no regard for the dc information. This eases the high-voltage power supply design as well as simplifying the video circuitry. These sets will produce a picture in which the average value of luminance remains nearly constant. For example, a night scene having a level of 15 to 20 IRE units and no peak-white excursions will tend to brighten toward the luminance level of the typical daytime scene (50 IRE units). Likewise a full-raster white scene with few black excursions will tend to darken to the average luminance level. More deluxe monochrome receivers reduce this

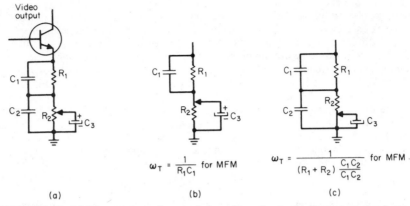

FIGURE 9.34 (*a*) Contrast control network in emitter circuit. (*b*) Equivalent circuit at maximum contrast (maximum gain). (*c*) Minimum contrast. *[Source: K. Blair Benson (ed.), Television Engineering Handbook, McGraw-Hill, New York, 1986]*

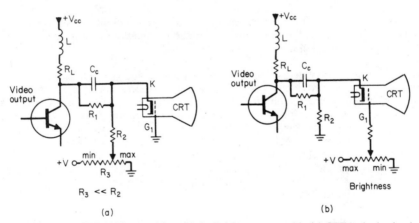

FIGURE 9.35 CRT luminance drive circuit. Brightness control in (*a*) CRT cathode circuit and (*b*) CRT grid circuit. *[Source: K. Blair Benson (ed.), Television Engineering Handbook, McGraw-Hill, New York, 1986]*

condition by use of partial dc coupling in which a high-resistance path exists between the second detector and the CRT. This path usually has a gain of one-half to one-fourth that of the ac-signal path.

The *transient response* of the video amplifier is controlled by its amplitude and phase characteristics. The *low-frequency transient response*, including the effects of dc restoration, if used, is measured in terms of distortion to the vertical blanking pulse. Faithful reproduction requires that the change in amplitude over the pulse duration, usually a decrease from initial value called *sag* or *tilt*, be less than 5 percent. In general, there is no direct and simple relationship between the sag and the lower 3-dB cutoff frequency. However, lowering the 3-dB cutoff frequency will reduce the tilt. (See Fig. 9.36.)

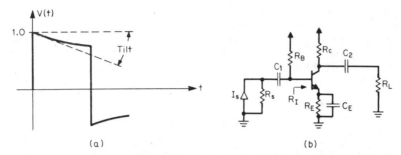

FIGURE 9.36   Video stage low-frequency response. (*a*) Square-wave output showing tilt. (*b*) *RC* time constant circuits in common-emitter stage that affects low-frequency response. *[Source: K. Blair Benson (ed.), Television Engineering Handbook, McGraw-Hill, New York, 1986]*

Emitter bypass circuit:

$$\% \text{ tilt/s} \approx -\frac{100B_0}{(R_1 + R_s')C_E} \tag{9.1}$$

where $R_x' = R_x R_B/(R_s + R_B)$. This equation applies only for small values of tilt, which requires the assumption that the circuit time constants are much longer than the pulse duration.

Low-Frequency Response Requirements.   The effect of inadequate low-frequency response appears in the picture as vertical shading. If the response is so poor as to cause a substantial droop of the top of the vertical blanking pulse, then incomplete blanking of retrace lines may occur.

It is not necessary or desirable to extend the low-frequency response to achieve essentially perfect LF square-wave reproduction. First, the effect of tilt produced by imperfect LF response is modified if dc restoration is employed. Direct-current restorers, particularly the fast-acting variety, substantially reduce tilt, and their effect must be considered in specifying the overall response. Second, extended LF response makes the system more susceptible to instability and low-frequency interference. Current coupling through a common power-supply impedance can produce the low-frequency oscillation known as "motorboating." Motorboating is not usually a problem in television receiver video amplifiers since they seldom employ the number of stages required to produce regenerative feedback, but in multistage amplifiers the tendency toward motorboating is reduced as the LF response is reduced.

A more commonly encountered problem is the effect of airplane flutter and "line bop." Although a fast-acting AGC can substantially reduce the effects of low-frequency amplitude variations produced by airplane reflections, the effect is so annoying visually as to warrant a sacrifice in excellence of LF response to bring about further reduction. A transient in-line voltage amplitude, commonly called a line bop, can also produce an annoying brightness transient which can similarly be reduced through a sacrifice of LF response. Special circuit precautions against line bop include the longest possible power supply time constant, bypassing the picture tube electrodes to the supply instead of ground, and the use of special coupling networks to attenuate the response sharply below the LF cutoff frequency. The overall receiver response is usually an empirically determined compromise.

The *high-frequency transient response* is usually expressed as the amplifier response to an ideal input voltage or current step. This response is described in the following terms and is shown in Fig. 9.37.

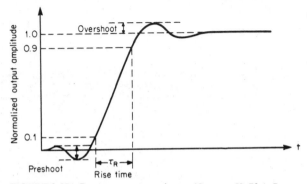

FIGURE 9.37   Response to a step input. *[Source: K. Blair Benson (ed.),* Television Engineering Handbook, *McGraw-Hill, New York, 1986]*

Rise time τ_R is the time required for the output pulse to rise from 10 to 90 percent of its final (steady-state) value.

Overshoot is the amplitude by which the transient rise exceeds its final value, expressed as a percentage of the final value.

Preshoot is the amplitude by which the transient oscillatory output waveform exceeds its initial value.

Smear is an abnormally slow rise as the output wave approaches its final value.

Ringing is an oscillatory approach to the final value.

In practice, rise times of 0.1 to 0.2 μs are typical. Overshoot, smear, and ringing amplitude are usually held to 5 percent of the final value, and ringing is restricted to one complete cycle.

9.3.9   Color-Receiver Luminance Channel

Suppression of chroma subcarrier is necessary to reduce the objectionable dot crawl in and around colored parts of the picture as well as reduce the distortion of luminance levels due to the nonlinear transfer characteristic of CRT electron guns. Traditionally a simple high-Q *LC* trap, centered around the color subcarrier, has been used for rejection, but this necessitates a trade-off between luminance channel bandwidth and the stop band for the chroma sidebands. Luminance channel comb filtering largely avoids this compromise and is one reason why it is becoming widely used in present receivers.

The luminance channel also provides the time delay required to correct the time delay registration with the color difference signals which normally incur delays in the range of from 300 to 1000 ns in their relatively narrow-bandwidth filters.

While delay circuits having substantially flat amplitude and group delay out to the highest baseband frequency of interest can and have been used, this is not necessarily required nor desirable for cost-effective overall design. Since the other links in the chain, i.e., tuner, IF, traps at 4.5 and 3.58 MHz, CRT driver stage, and peaker may all contribute significant linear distortion individually, it is frequently advantageous to allow these distortions to occur and use the *delay block* as an overall group delay and/or amplitude equalizer.

Amplitude Response Shaping (Peaking, Aperture Correction Sharpness Control). Although it is well known that, for "distortionless" transmission, a linear system must possess both uniform amplitude and group delay responses over the frequency band of interest, the limitations of a finite bandwidth lead to noticeably slower rise and fall times rendering edges less sharp or distinct. By intentionally distorting the receiver amplitude response and boosting the relative response to the mid and upper baseband frequencies to varying degrees, both faster rise and fall times can be developed along with enhanced fine detail. If carried too far, however, objectionable *outlining* can occur, especially to those transients in the white direction. Furthermore, the visibility of background noise is increased.

For several reasons, including possible variations in transient response of the transmitted signal, distortion due to multipath, antennas, receiver tolerances, signal-to-noise ratio, and viewer preference, it is difficult to define a fixed response at the receiver that is optimum under all conditions. Therefore it is useful to make the amplitude response variable so it can be controlled to best suit the individual situation. Over the range of adjustment, it is assumed that the overall group delay shall remain reasonably flat across the video band. The exact shape of the amplitude response is directly related to the desired time domain response (height and width of preshoot, overshoot, and ringing) and chroma subcarrier sideband suppression.

Nonlinear Video Compression. Because the peaked signal later operates on the nonlinear CRT gun characteristics, large white preshoots and overshoots can contribute to excessive beam currents, which can cause CRT spot defocusing. To alleviate this, circuits have been developed which compress large excursions of the peaking component in the white direction. For best operation it is desirable that the signals being processed have equal preshoot and overshoot.

Noise Reduction. Low-level, high-frequency noise in the luminance channel can be removed by a technique called *coring*. One coring technique involves nonlinearly operating on the peaking or edge-enhancement signal, discussed earlier in this section. The peaking signal is passed through an amplifier having low or essentially no gain in the midamplitude range. When this modified peaking signal is added to the direct video, the large transitions will be enhanced, but the small ones (noise) will not be, giving the illusion that the picture sharpness has been increased while the noise has been decreased.

9.3.10 Chroma-Subcarrier Processing

Chroma Bandpass. In the equiband chroma system, typical of all consumer receivers, the chroma amplifier must be preceded by a bandpass filter network which complements the chroma sideband response produced by the tuner and IF. Second, frequencies below 3 MHz must be attenuated to reduce not only possible video cross-color disturbances but also crosstalk caused by the lower-frequency *I* channel chroma information.

A third requirement is that the filter have a gentle transition from passband to stop band in order to impart a minimum amount of group delay in the chroma signal which then must be compensated by additional group delay circuitry in the luminance channel. The fourth-order high-pass filter is a practical realization of these requirements.

As described earlier, this stage also serves as the chroma gain control stage. The usual implementation in an IC consists of a differential amplifier having the chroma signal applied to the current source. The gain-control dc voltage is applied to one side of the differential pair to divert the signal current away from the output side.

Burst Separation.   Complete separation of the color synchronizing burst from video requires time gating. The gate requirements are largely determined by the horizontal sync and burst specifications illustrated in Fig. 9.38. It is essential that all video information be excluded. It is also desirable that both the leading and trailing edges of burst be passed so that the complementary phase errors introduced at these points by quadrature distortion average to zero. Widening the gate pulse to minimize the required timing accuracy has negligible effect on the noise performance of the reference system and may be beneficial in the presence of echoes. The ≈ 2μs spacing between trailing edges of burst and horizontal blanking determines the total permissible timing variation. Noise modulation of the gate timing should not permit noise excursions to encroach upon the burst since the resulting cross modulation will have the effect of increasing the noise power delivered to the reference system.

Burst Gating-Signal Generation.   The gate pulse generator must provide both steady-state phase accuracy and reasonable noise immunity. The horizontal flyback pulse has been widely used for burst gating since it is derived from the horizontal-scan oscillator system which meets the noise immunity requirements

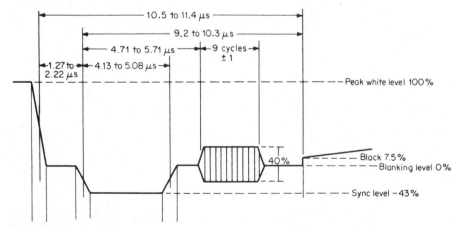

FIGURE 9.38   Horizontal-blanking interval specification. *[Source: K. Blair Benson (ed.), Television Engineering Handbook, McGraw-Hill, New York, 1986]*

and, with appropriate design, can approximate the steady-state requirements. A further improvement in steady-state phase accuracy can be achieved by deriving the gating pulse directly from the trailing edge of the horizontal sync pulse. This technique is utilized in several current chroma system integrated circuits.

The *burst gate* in conventional discrete component circuits has the form of a conventional amplifier which is biased into linear conduction only during the presence of the gating pulse. In IC format, the complete chroma signal is often made available at one input of the automatic phase control (APC) burst-reference phase detector. The gating pulse then enables the phase detector to function only during the presence of burst.

9.3.11 Color-Subcarrier Reference Separation

General Requirements. The color subcarrier reference system converts the synchronizing bursts to a continuous carrier of identical frequency and close phase tolerance. Theoretically, the long-term and repetitive phase inaccuracies should be restricted to the same value, approximately ±5°. Practically, if transmission variations considerably in excess of this value are encountered, and if an operator control of phase ("hue control") is provided, the long-term accuracy need not be so great. Somewhat greater instantaneous inaccuracies can be tolerated in the presence of thermal noise so that an rms phase error specification of 5 to 10° at a signal-to-noise ratio of unity may be regarded as typical.

Reference Systems. Three types of reference synchronization systems have been used. These are automatic phase control (APC) of a voltage-controlled oscillator (VCO), injection lock of a crystal, and ringing of a crystal filter. Best performance can be achieved by the APC loop. In typical applications the *figure of merit* can be made much smaller (better) for the APC loop than for the other systems by making the factor $(1/y) + m$ have a value considerably less than 1, even as small as 0.1. The APC loop system currently is used in all color television receivers. The parts count for each type system, at one time much higher for the APC system, is no longer a consideration, especially in the IC implementation where the oscillator and phase detector are integrated and only the resistors and capacitors of the filter network and oscillator crystal are external.

The *APC circuit* is a phase-actuated feedback system consisting of three functional components: a phase detector, a low-pass filter, and a dc voltage-controlled oscillator (Fig. 9.39). The characteristics of these three units define both the dynamic and static loop characteristics and hence the overall system performance. The following analysis is based on the technique of Richman.

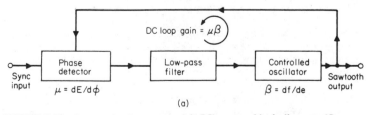

FIGURE 9.39 Automatic phase-control (APC) system block diagram. *[Source: K. Blair Benson (ed.),* Television Engineering Handbook, *McGraw-Hill, New York, 1986]*

The phase detector generates a dc output E whose polarity and amplitude are proportional to the direction and magnitude of the relative phase difference $d\phi$ between the oscillator and synchronizing (burst) signals.

The *voltage-controlled oscillator* (VCO) is an IC implementation which requires only an external crystal and simple phase-shift network. The oscillator can be shifted $\pm45°$ by varying the phase-control voltage. This leads to symmetrical pull-in and hold-in ranges.

9.3.12   Chroma Demodulation

The chroma signal can be considered to be made up of two amplitude-modulated carriers having a quadrature phase relationship. Each of these carriers can be individually recovered by use of a synchronous detector. The reference input to the demodulator is that phase which will demodulate only the I signal. The output contains no Q signal.

Demodulation products of 7.16 MHz in the output current may contribute to an optical interference moiré pattern in the picture. This is related to the line geometry of the shadow mask. The 7.16-MHz output can also result in excessively high line-terminal radiation from the receiver. A first-order low-pass filter with cutoff of 1 to 2 MHz usually provides sufficient attenuation. In extreme cases an *LC* trap may be needed.

Choice of Demodulation Axis.   Over the double sideband region (±500 kHz around the subcarrier frequency) the chrominance signal can be demodulated as pure quadrature AM signals along either the I and Q axis or R-Y and B-Y axis. The latter leads to a simpler matrix for obtaining the color drive signals R, G, and B.

Current practice has moved away from the classic demodulation angles for two main reasons. First, receiver picture tube phosphors have been modified to yield greater light output and can no longer produce the original NTSC primary colors. Currently, the chromaticity coordinates of the primary colors as well as the *RGB* current ratios to produce white balance vary from one CRT manufacturer to another. Second, the white-point setup has, over the years, moved from the cold 9300 K of monochrome tubes to the warmer illuminant C and D65 which produce more vivid colors, more representative of those which can be seen in natural sunlight. Several authors have developed techniques for determining the decoding angles and matrix constants to optimize a given set of chromaticity coordinates, flesh tones, green grass, CRT gamma values, or whatever the design engineer selects.

9.3.13   *RGB* Matrixing and CRT Drive

Since *RGB* primary color signals driving the display are required as the end output, it is necessary to combine or "matrix" the demodulated color-difference signals with the luminance signal. Several circuit configurations can be used to accomplish this.

Color-Difference Drive Matrixing.   In this technique R-Y, G-Y, and B-Y are applied to respective control grids of the CRT while luminance is applied to all three cathodes; the CRT thereby matrixes the primary colors. It has the advantage that gray scale is not a function of linearity matching between the three channels since at any level of gray the color-difference driver stages are at the same dc level. Also, since the luminance driver is common, any dc drift shows up only as a brightness shift. Luminance channel frequency response uniformity is ensured by the common driver.

RGB drive, wherein RGB signals are applied to respective cathodes and G1 is dc biased, requires less drive and has none of the potential color fidelity errors of the former system. *RGB* drive places higher demands on the drive amplifiers for linearity, frequency response, and dc stability, plus requiring a matrixing network in the amplifier chain.

Low-level RGB matrixing and CRT drive are universally used today, especially with the newer CRTs which have unitized guns in which the common G1 and G2 require differential cathode bias adjustments and drive adjustments to give gray-scale tracking. In this technique, *RGB* signals are matrixed at a level of a few volts and then amplified to a higher level (100 to 200 V) suitable for CRT cathode drive.

Direct-current stability, frequency response, and linearity of the three stages, even if somewhat less than ideal, should be reasonably well matched to ensure overall gray-scale tracking. Bias and gain adjustments should be independent in their operation, rather than interdependent, and should minimally affect those characteristics listed above.

Figure 9.40 illustrates a simple example of one of the three CRT drivers. If the amplifier black-level bias voltage equals the black level from the *RGB* decoder, drive adjustment will not change the amplifier black-level output voltage level or affect CRT cutoff. Furthermore, if $R_B \gg R_E$, drive level will be independent of bias setting. Note also that frequency response-determining networks are configured to be unaffected by adjustments.

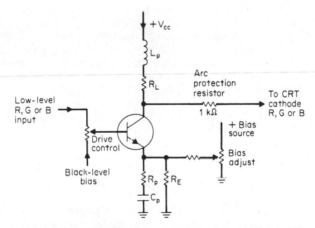

FIGURE 9.40 Simplified *R, G,* and *B* output stage. *[Source: K. Blair Benson (ed.),* Television Engineering Handbook, *McGraw-Hill, New York, 1986]*

Frequently the shunt peaking coil can be made common to all three channels, since differences between the channels are predominantly color-difference signals of relatively narrow bandwidth. Although the frequency responses could be compensated to provide widest possible bandwidth, this is usually not necessary when the frequency response of preceding low-level luminance processing (especially the peaking stage) is factored in. One exception in which output stage bandwidth must be increased to its maximum is in an application, television receiver or video monitor, where direct *RGB* inputs are provided for auxiliary services, such as computers, teletext, and S-VHS wide-band VCRs.

Comb-Filter Processing.   The frequency spectrum of a typical NTSC composite video signal is shown in Fig. 9.41*a*. A filter having amplitude response like teeth of a comb, i.e., 100 percent transmission for desired frequencies of a given channel and substantially zero transmission for the undesired interleaved signal spectrum, can separate chroma and luminance components from the composite signal. Such a filter can be easily made, in principle, by delaying the composite video signal one horizontal scan period (63.555 μs in NTSC-M) and adding or subtracting to the undelayed composite video signal (Fig. 9.41*b* and *c*).

The output of the *sum channel* will have frequencies at f (horizontal), and all integral multiples thereof reinforce in phase, while those interleaved frequencies will be out of phase and will cancel. This can be used as the luminance path. The *difference channel* will have integral frequency multiples cancel while the interleaved ones will reinforce. This channel can serve as the chrominance channel. The filter characteristic and interleaving are shown in Fig. 9.41*c*.

9.3.14   Automatic Circuits

Automatic Chroma Control (ACC).   The relative level of chroma subcarrier in the incoming signal is highly sensitive to transmission path disorders, thereby introducing objectionable variations in saturation. These can be observed between one received signal and another or over a period of time on the same channel unless some adaptive correction is built into the system. The color-burst reference, transmitted at 40 IRE units peak-to-peak, is representative of the same path distortions and is normally used as a reference for automatic gain controlling the chroma channel. A balanced peak detector or synchronous detector, having good noise rejection characteristics, detects the burst level and provides the control signal to the chroma gain-controlled stage.

Color Killer.   Allowing the receiver chroma channel to operate during reception of a monochrome signal will result in unnecessary cross color and colored noise, made worse by the ACC increasing the chroma amplifier gain to the maximum. Most receivers, therefore, cut off the chroma channel transmission when the received burst level goes below approximately 5 to 7 percent. Hysteresis has been used to minimize the flutter or threshold problem with varying signal levels.

Chroma-Limiting.   Burst-referenced ACC systems perform adequately when receiving correctly modulated signals with appropriate burst-to-chroma ratio. Occasionally, however, burst level may not bear a correct relation to its accompanying chroma signal, leading to incorrectly saturated color levels. It has been determined that most viewers are more critical of excessive color levels than insufficient ones. Experience has shown that when peak chroma amplitude exceeds burst by greater than 2/1, limiting of the chroma signal is helpful. This threshold nearly corresponds to the amplitude of a 75 percent modulated color bar chart (2.2/1). At this level, negligible distortion is introduced into correctly modulated

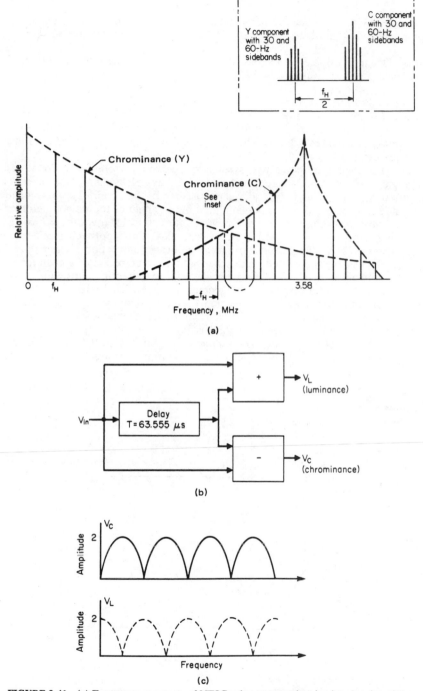

FIGURE 9.41 (a) Frequency spectrum of NTSC color system, showing interleaving of signals. (b) Simplified 1H-delay comb filter block diagram. (c) Chrominance V_C and luminance V_L outputs of comb. [*Source: K. Blair Benson (ed.),* Television Engineering Handbook, *McGraw-Hill, New York, 1986*]

signals. Only those which are nonstandard are affected significantly. The output of the peak chroma detector is also sent to the chroma gain-controlled stage.

Tint. One major objective in color television receiver design is to minimize the incidence of flesh-tone reproduction with incorrect hue. Automatic hue-correcting systems can be categorized into two classes.

Static flesh-tone correction is achieved by selecting the chroma demodulating angles and gain ratios to desensitize the resultant color-difference vector in the flesh-tone region ($+I$ axis). The demodulation parameters remain fixed, but the effective Q axis gain is reduced. This has the disadvantage of distorting hues in all four quadrants.

Dynamic flesh-tone corrective systems can adaptively confine correction to the region within several degrees of the positive I axis, leaving all other hues relatively unaffected. This is typically accomplished by detecting the phase of the incoming chroma signal and modulating the phase angle of the demodulator reference signal to result in an effective phase shift of 10 to 15° toward the I axis for a chroma vector which lies within 30° of the I axis. This approach produces no amplitude change in the chroma. In fact, for chroma saturation greater than 70 percent the system is defeated on the theory that the color is not a flesh tone.

A simplification in circuitry can be achieved if the effective correction area is increased to the entire positive-I 180° sector. A conventional phase detector can be utilized and the maximum correction of approximately 20° will occur for chroma signals having phase of ±45° from the I axis. Signals with phase greater or less than 45° will have increasingly lower correction values (Fig. 9.42).

Vertical-interval reference (VIR) signal, as shown in Fig. 9.42, provides references for black level, luminance, and, in addition to burst, a reference for

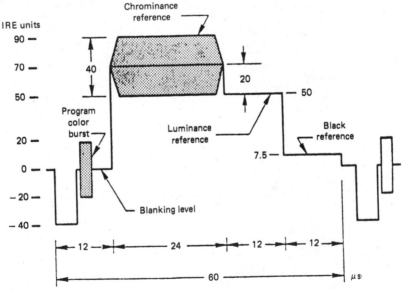

FIGURE 9.42   Vertical-interval reference (VIR) signal. *Note*: The chrominance and the program color burst are in phase. *[Source:* FCC. *In K. Blair Benson (ed.),* Television Engineering Handbook, *McGraw-Hill, New York, 1986]*

chroma amplitude and phase. While originally developed to aid broadcasters, it has been employed in television receivers to correct for saturation and hue errors resulting from transmitter or path distortion errors.

9.3.15 Scanning Synchronization

Sync Separator.   The scan-synchronizing signal, consisting of the horizontal and vertical sync pulses, is removed from the composite video signal by means of a sync-separator circuit. The classic approach has been to ac-couple the video signal to an overdriven transistor amplifier (Fig. 9.43), which is biased off for signal levels smaller than V_{CO} and saturates for signal levels greater than V_{sat}, a range of approximately 0.1 V.

Vertical Integration and Oscillator.   Vertical synchronizing information can be recovered from the output of the sync separator by the technique of integration. The classic two-section *RC* integrator provides a smooth, ramp waveform which corresponds to the vertical-sync block as shown in Fig. 9.44. The ramp is then sent to a relaxation or blocking oscillator, which is operating with a period

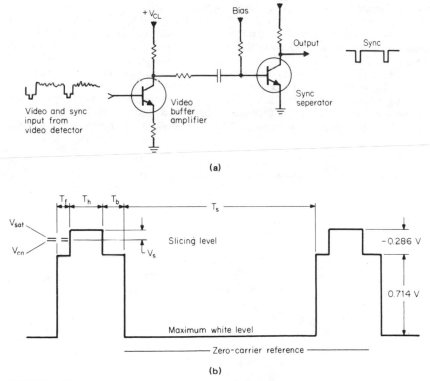

FIGURE 9.43 Sync separator. (*a*) Typical circuit. (*b*) Sync waveform. [*Source: K. Blair Benson (ed.),* Television Engineering Handbook, *McGraw-Hill, New York, 1986*]

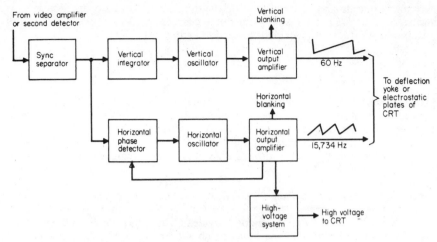

FIGURE 9.44 Picture-scanning section. *[Source: K. Blair Benson (ed.),* Television Engineering Handbook, *McGraw-Hill, New York, 1986]*

slightly longer than the vertical frame period. The upper part of the ramp will trigger the oscillator into conduction to achieve vertical synchronization. The oscillator will then reset and wait for the threshold level of the next ramp. The *vertical-hold* potentiometer controls the free-running frequency or period of the vertical oscillator.

One modification to the integrator design is to reduce the integration (speed up the time constants) and provide for some differentiation of the waveform prior to applying it to the vertical oscillator (Fig. 9.45). Although degrading the noise performance of the system, this technique provides a more certain and repeatable trigger level than the full integrator, thereby leading to an improvement in inter-

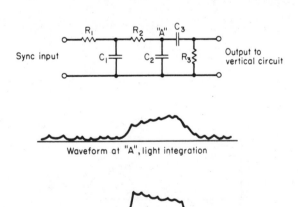

FIGURE 9.45 Modified integrator followed by differentiator. *[Source: K. Blair Benson (ed.),* Television Engineering Handbook, *McGraw-Hill, New York, 1986]*

lace over a larger portion of the hold-in range. A second benefit is to provide a more stable vertical lock when receiving signals which have distorted vertical-sync waveforms. These can be generated by video-cassette recorders in non-standard playback modes, such as fast/slow forward, still, and reverse. Prere-corded tapes having antipiracy nonstandard sync waveforms also contribute to the problem.

Vertical-Countdown Synchronizing Systems. A vertical-scan system using digital logic elements can be based upon the frequency relationship of 525/2 which exists between horizontal and vertical scan. Such a system can be considered to be phase-locked for both horizontal scan and vertical scan which will result in improved noise immunity and picture stability. Vertical sync is derived from a pulse train having a frequency of twice the horizontal rate. The sync is therefore precisely timed for both even and odd fields, resulting in excellent interlace. This system needs no hold control.

The block diagram of one design is shown in Fig. 9.46. The 31.5-kHz clock input is converted to horizontal drive pulses by a divide-by-2 flip-flop. A two-mode counter, set for 525 counts for standard interlaced signals and 541 for noninterlaced signals, produces the vertical output pulse. Noninterlaced signals are produced by a variety of VCR cameras, games, and picture generators used by television servicepeople. The choice of 541 allows the counter to continue until the arrival of vertical sync from the composite video waveform. This actual vertical sync then resets the counter.

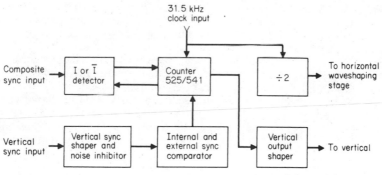

FIGURE 9.46 Vertical-countdown sync processor, using a $2f_H$ clock. *[Source: K. Blair Benson (ed.), Television Engineering Handbook, McGraw-Hill, New York, 1986]*

Other systems make use of clock frequencies of 16 and 32 times the horizontal frequency. Low-cost crystal or ceramic resonator-controlled oscillators can be built to operate at these frequencies. As in the first system, dual-mode operation is necessary in order to handle standard interlaced and nonstandard sync waveforms. Exact 525/2 countdown is used with interlaced signals. For non-interlaced signals, the systems usually operate in a free-running mode with injection of the video-derived sync pulse causing lock.

A critical characteristic, and probably the most complex portion of any countdown system, is the circuitry which properly adjusts for nonstandard sync waveforms. These can be simple 525 noninterlaced fields, distorted vertical-sync blocks, blocks having no horizontal serrations, fields with excessive or insuffi-cient lines, and a combination of the above. Home VCRs operating in the "tricks" modes can present unique sync waveforms.

Impulse Noise Suppression.   The simplest-type suppression which will improve the basic circuit in a noise environment of human origin consists of a parallel re-sistor and capacitor in series with the sync charging capacitor. Variations of this simple circuit which include diode clamps and switches have been developed to speed up the noise performance while not permitting excessive tilt in the sync output during the vertical block.

More complex solutions to the noise problem include the *noise canceler* and *noise gate*. The canceler monitors the video signal between the second detector and the sync separator. A noise spike, which exceeds the sync or pedestal level, is inverted and added to the video after the isolation resistor R. This action can-cels that part of the noise pulse which would otherwise produce an output from the sync separator.

For proper operation the canceler circuit must track the sync-tip amplitude from strong RF signal to fringe level. The noise gate, in a similar manner, recog-nizes a noise pulse of large amplitude and prevents the sync separator from con-ducting either by applying reverse bias or by opening a transistor switch in the emitter circuit.

Horizontal Automatic-Phase Control.   Horizontal scan synchronization is accom-plished by means of an APC loop, with theory and characteristics similar to those used in the chroma reference system. Input signals to the horizontal APC loop are sync pulses from the sync separator and horizontal flyback pulses which are integrated to form a sawtooth waveform. The phase detector compares the phase (time coincidence) of these two waveforms and sends a dc-coupled low-pass-filtered error signal to the voltage-controlled oscillator, to cause the frequency to shift in the direction of minimal phase error between the two input signals.

The recovery of the APC loop to a step transient input involves the param-eters of the natural resonant frequency, as well as the amount of overshoot permitted as the correct phase has been reached. Both of these characteristics can be evaluated by use of a jitter generator which creates a time base error between alternate fields. The result-ant sync error and system dynamics can be seen in the picture display of the receiver shown in Fig. 9.47. This type disturbance occurs when the two or three playback heads of a consumer helical-scan VCR switch tracks. It will also occur when the re-ceiver is operated from a signal which does not have horizontal slices in the vertical sync block.

Phase detector circuitry in the form of classic double-diode bridge detectors, in either the balanced or unbalanced configuration, is still found in low-cost receivers. Inte-grated circuits commonly utilize a gated differential amplifier (Fig. 9.48)

FIGURE 9.47   Pattern on receiver screen dis-playing vertical line with the even and odd fields offset by n μs. [*Source: K. Blair Benson (ed.), Television Engineering Handbook, McGraw-Hill, New York, 1986*]

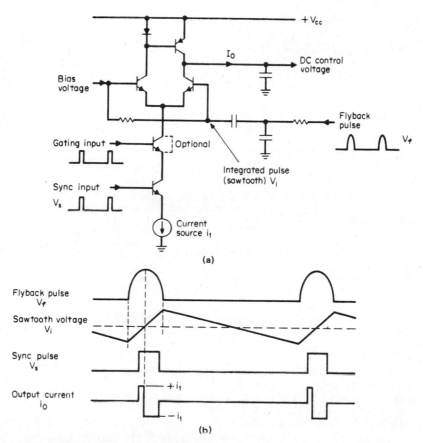

FIGURE 9.48   Gated phase detector for IC format. (*a*) Circuit diagram. (*b*) Pulse-timing waveforms. [*Source: K. Blair Benson (ed.),* Television Engineering Handbook, *McGraw-Hill, New York, 1986]*

in which the common feed from the current source is driven by the sync pulse, and the sawtooth waveform derived from the flyback pulse is applied to one side of the differential pair.

9.3.16   Vertical Scanning

Class-B Amplifiers.   Class-B vertical circuits consist of an audio amplifier with current feedback. This approach maintains linearity without the need for an adjustable linearity control. Yoke-impedance changes caused by temperature changes will not also affect the yoke current; thus the thermistor is not required. The current-sensing resistor must, of course, be temperature stable. An amplifier which uses a single *npn* and a single *pnp* in the form of a complementary output stage is sketched in Fig. 9.49. Quasi-complementary,

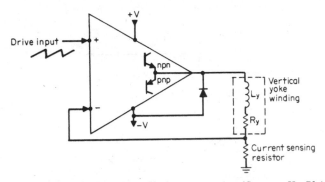

FIGURE 9.49 Class-B vertical-output stage. *[Source: K. Blair Benson (ed.),* Television Engineering Handbook, *McGraw-Hill, New York, 1986]*

Darlington outputs and other common audio-output stage configurations can be used.

Establishing proper dc bias for the output stages is quite critical. Too little quiescent current will result in crossover distortion which will put a faint horizontal white line in the center of the picture even though the distortion may not be detectable on an oscilloscope presentation of the yoke current waveform. Too much quiescent current results in excessive power dissipation in the output transistors.

Retrace Power Reduction.   The voltage required to accomplish retrace results in a substantial portion of the power dissipation in the output devices. The supply voltage to the amplifier and corresponding power dissipation can be reduced by using a "retrace-switch" or "flyback-generator" circuit to provide additional supply voltage during retrace. One version of a retrace switch is given in Fig. 9.50. During trace time the capacitor is charged to a voltage near the supply voltage. As retrace begins, the voltage across the yoke goes positive, thus forcing Q_R into saturation. This places the cathode of C_R at the supply potential and the anode at a level of 1.5 to 2 times the supply, depending upon the values of R_R and C_R.

Integrated-Circuit Vertical Systems.   These systems are becoming commonplace in color television chassis. In one implementation the IC contains the vertical-sync processor, oscillator, and driver stage. The power output stage consists of external discrete components. A second implementation requires drive-pulse input and contains the driver, Class-B output stage, and bias and protection circuits as well as retrace switch. This circuit can drive repetitive yoke current peaks of 1.5 A from a 25-V supply.

9.3.17 Horizontal Scanning

Basic Functions.   The horizontal scan system has two primary functions. It provides a modified sawtooth-shaped current to the horizontal-yoke coils to cause the electron beam to travel horizontally across the face of the CRT. It also provides drive to the high-voltage or *flyback* transformer to create the voltage

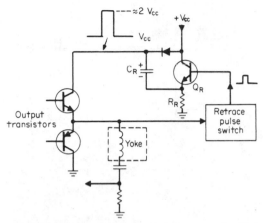

FIGURE 9.50 Retrace switch. *[Source: K. Blair Benson (ed.), Television Engineering Handbook, McGraw-Hill, New York, 1986]*

needed for the CRT anode. Frequently low-voltage supplies are also derived from the flyback transformer. The major components of the horizontal-scan section consist of a driver stage, horizontal output device (either bipolar transistor or SCR), yoke-current damper diode, retrace capacitor, yoke coil, and flyback transformer (see Fig. 9.51).

Scan-Circuit Operation. During the scan or retrace interval, the deflection yoke may be considered a pure inductance with a dc voltage impressed across it. This

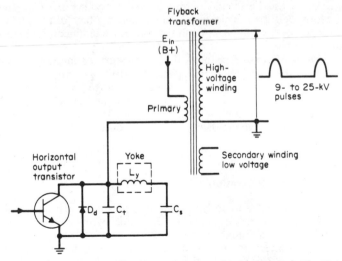

FIGURE 9.51 Simplified horizontal-scan circuit. *[Source: K. Blair Benson (ed.), Television Engineering Handbook, McGraw-Hill, New York, 1986]*

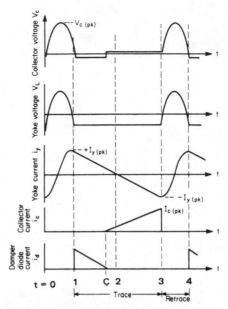

FIGURE 9.52 Horizontal-scan waveform timing diagrams. *[Source: K. Blair Benson (ed.), Television Engineering Handbook, McGraw-Hill, New York, 1986]*

creates a sawtooth waveform of current (see Fig. 9.52). This current flows through the damper diode during the first half of scan. It then reverses direction and flows through the horizontal output transistor collector. This sawtooth-current waveform deflects the electron beam across the face of the picture tube. A similarly shaped current flows through the primary winding of the high-voltage output transformer.

At the beginning of the retrace interval, the transformer and yoke inductances transfer energy to the retrace-tuning capacitor and the accompanying stray capacitances, thereby causing a half sine wave of voltage to be generated. This high-energy pulse appears on the transistor collector and is stepped up, via the flyback transformer, to become the high voltage for the picture-tube anode. Finally, at the end of the cycle the damper diode conducts, and another horizontal scan is started.

9.3.18 High-Voltage Generation

High voltage in the range 8 to 16 kV is required to supply the anode of monochrome picture tubes. Color tubes have anode requirements in the range 20 to 30 kV and focus voltage requirements of 3 to 12 kV. The horizontal flyback transformer is the common element in the current methods of generating high voltage. The three variations of this design which are most popularly used are the flyback with half-wave rectifier, flyback driving a voltage multiplier, and the voltage-adding integrated flyback.

A simplified horizontal-scan circuit is shown in Fig. 9.51. The voltage at the top of the high-voltage (HV) winding consists of a series of pulses delivered during retrace from the stored energy in the yoke field. The yoke voltage pulses are then multiplied by the turns ratio of the HV winding to primary winding. The peak voltage across the primary during retrace is given by

$$V_{p(\text{pk})} = E_{\text{in}}0.8\left(1.79 + 1.57\,\frac{T_{\text{trace}}}{T_{\text{retrace}}}\right)$$

where E_{in} = supply voltage (B^+)
$\quad\;\; 0.8$ = accounts for pulse shape factor with third harmonic tuning
$\quad T_{\text{trace}}$ = trace period, $\approx$ 52.0 μs
$\quad T_{\text{retrace}}$ = retrace period, $\approx$ 11.5 μs

Flyback with Half-Wave Rectifier. The most common HV supply for small-screen monochrome and color television receivers uses a direct half-wave rectifier circuit. The pulses at the top of the HV winding are rectified by the single diode, or composite of several diodes in series. The charge is then stored in the capacitance of the anode region of the picture tube. Large-voltage step-up is required from the primary to the HV winding. This results in a large value of leakage inductance for the HV winding which decreases its efficiency as a step-up transformer.

Harmonic tuning of the HV winding improves the efficiency by making the total inductance and the distributed capacitance of the winding plus the CRT anode capacitance resonate at an odd harmonic of the flyback pulse frequency. For the single-rectifier circuit, usually the third harmonic resonance is most easily implemented by proper choice of winding configuration which results in appropriate leakage inductance and distributed capacitance values. This will result in HV pulse waveforms, which will give an improvement in the HV supply regulation (internal impedance) as well as a reduction in the amplitude of ringing in the current and voltage waveforms at the start of scan.

Voltage-Multiplier Circuits. A voltage-multiplier circuit consists of a combination of diodes and capacitors connected in such a way that the dc output voltage is greater than the peak amplitude of the input pulse.

The *tripler* has been the most popular HV system for color television receivers requiring anode voltages of 25 to 30 kV in the 1970 period. The considerably reduced pulse voltage required from the HV winding has resulted in a flyback transformer with a more tightly coupled HV winding.

Integrated Flybacks. Most current medium- and large-screen color receivers utilize an integrated flyback transformer in which the HV winding is segmented into three or four parallel-wound sections. These sections are series-connected with a diode between adjacent segments. These diodes are physically mounted as part of the HV section. The transformer is then encapsulated in high-voltage polyester or epoxy.

Two HV-winding construction configurations are being used. One, the *layer* or *solenoid-wound* type, has very tight coupling to the primary and operates well with no deliberate harmonic tuning. Each winding (layer) must be designed to have balanced voltage and capacitance with respect to the primary. The second, *bobbin* or *segmented-winding* design, has high leakage inductance and usually requires tuning to an odd harmonic, e.g., the ninth. Regulation of this construction is not quite as good as the solenoid-wound primary winding at a horizontal-frequency rate. The +12-V, +24-V, +25-V, and −27-V supplies are *scan-rectified*. The +185-V, overvoltage sensing, focus voltage, and 25-kV anode voltage are derived by *retrace-rectified* supplies. The CRT filament is used directly in its ac mode.

Flyback-generated supplies provide a convenient means for isolation between different ground systems as needed for iso-hot chassis.

9.3.19 Power Supplies

Most present-day receivers use the flyback pulse from the horizontal transformer as the source of power for the various dc voltages required. Using the pulse waveform at a duty cycle of 10 or 15 percent, by proper winding direction and grounding of either end of the winding, several different voltage souces can be created.

"Scan rectified" supplies are operated at a duty cycle of approximately 80 percent and are thus better able to furnish higher current loads. Also, the diodes used in such supplies must be capable of blocking voltages that are nine to ten times larger than the level they are producing. Diodes having fast-recovery characteristics are used to keep the power dissipation at a minimum during the turn-off interval due to the presence of this high reverse voltage.

A typical receiver system containing the various auxiliary power supplies derived from flyback transformer windings is shown in Fig. 9.53. Transistor Q 452 switches the primary winding at a horizontal-frequency rate. The +12-V, +24-V, +-25-V, and −27-V supplies are *scan-rectified*. The +185-V, overvoltage sens-

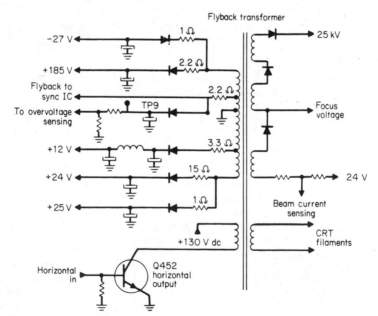

FIGURE 9.53 Auxiliary power sources derived from the horizontal-output transformer. *[Source: Courtesy of Philips Consumer Electronics Co. In K. Blair Benson (ed.),* Television Engineering Handbook, *McGraw-Hill, New York, 1986]*

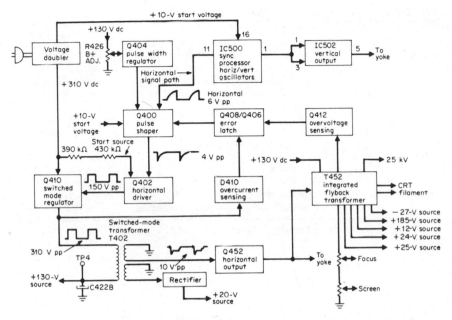

FIGURE 9.54 Color receiver power supply using a switched-mode regulator system. *[Source: Courtesy of Philips Consumer Electronics Co. In K. Blair Benson (ed.),* Television Engineering Handbook, *McGraw-Hill, New York, 1986]*

ing, focus voltage, and 25-kV anode voltage are derived by *retrace-rectified* supplies. The CRT filament is used directly in its ac mode.

Flyback-generated supplies provide a convenient means for isolation between different ground systems as needed for iso-hot chassis. Figure 9.54 shows the block diagram of a television receiver power supply system.[7,8]

AC-DC Supplies for Portable Receivers. Small-screen portable monochrome receivers operate from either the ac power line or a nominal 12-V battery supply. The main voltage in the chassis (+11 V) is supplied either directly from the battery or from a four-diode bridge rectifier and series-pass transistor regulator. The power transformer gives cold-chassis isolation from the power line.

9.3.20 Large-Screen Projection Systems

Two systems of large-screen television employing diffraction principles have been devised. These are the Eidophor system and the General Electric large-screen system. Both systems employ a conventional light source, such as a high-intensity xenon lamp, and modulate the incident light on the screen to form the raster. Of historical interest is the Scophony system, also utilizing diffraction effects. Refractive and reflective systems of conventional types are also briefly described (Table 9.8).

Eidophor Light Modulator. Figure 9.55 is a diagram of one color element of an Eidophor projector. The rays from the lamp (1) pass through a system of

TABLE 9.8   Types of Display Systems

Type	Characteristics	Vendors
1. Light matrix	Nonportable, utilizes light-generating cells, light bulbs or CRTs, very heavy; usually spaced-in building design. Displays from very small to very large.	Mitsubishis, Diamond Vision, JumboTron by Sony
2. Light-valve projection systems	Large-screen video projection system. Portable. Up to 30′ × 60′ screens and HDTV with Eidophor system.	Gretag's Eidophor, General Electric, Soldern
3. Schmidt system	CRT and optical projector for very small displays.	Barco, Sony, Panasonic, and Advent
4. Laser projection	Experimental, high-power requirements, major safety concerns.	Cavendish, VISULUX

Source: Courtesy Intertec Publishing Company, Overland Park, Kansas.

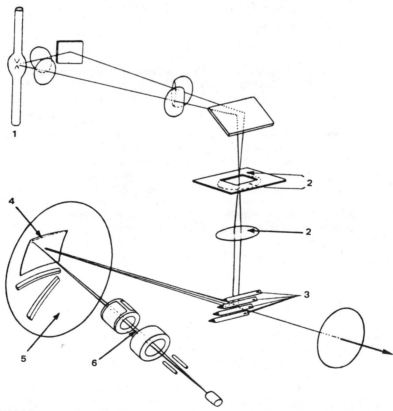

FIGURE 9.55   Schematic diagram of Eidophor projector. *[Source: Courtesy of Eidophor, Ltd., Regensdorf, Zurich, Switzerland. In K. Blair Benson (ed.),* Television Engineering Handbook, *McGraw-Hill, New York, 1986]*

lenses (2) and mirror bars (3) and an oil layer (4) coated on a spherical mirror surface (5). The bars and lenses are aligned so that no light is passed when the liquid surface is undeformed. When the control layer is deformed, part of the light is diffracted and passes between the mirror bars, and a bright spot is then visible on the screen, the brightness dependent on the depth of the deformation. The deformations are produced by bombarding the layer with an electron beam. The electric charge on the thin oil film deforms the surface, which returns to normal as the charge decays. The electron beam scans the picture area of the liquid at constant intensity and variable speed. The variable speed is produced by an alternating voltage of constant frequency and variable amplitude superimposed on the line sweep voltage. The frequency of this alternating voltage determines the size of the picture raster elements; its amplitude determines their relative brightnesses.

General Electric Large-Screen System. A deformable liquid surface modified by an electron gun produces diffraction patterns which are imaged by means of an external light source and Schlieren projection lens upon a screen. See Fig. 9.56. Color filters, input slots, and output bars position the color signals to create three simultaneous and superimposed red, green, and blue images from the same electron beam.

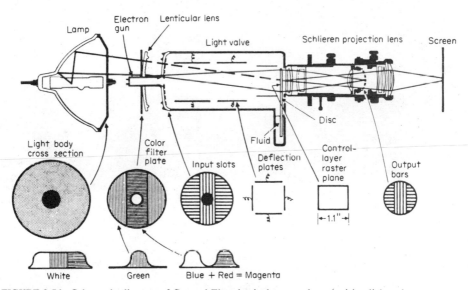

FIGURE 9.56 Schematic diagram of General Electric single-gun color television light-valve assembly. *[Source: K. Blair Benson (ed.),* Television Engineering Handbook, *McGraw-Hill, New York, 1986]*

REFERENCES

1. H. Jasik. "Properties of Antennas and Fundamentals of Antennas," *Antenna Engineering Handbook*. McGraw-Hill, New York, 1961, Chaps. 1 and 2.

2. R. S. Elliot. *Antenna Theory and Design*. Prentice-Hall, Englewood Cliffs, N.J., 1981, p. 64.

3. Y. T. Lo. "TV Receiving Antennas," in H. Jasik (ed.), *Antenna Engineering Handbook*. McGraw-Hill, New York, 1961, pp. 24–25.

4. N. Grossner. *Transformers for Electronic Circuits*, 2d ed. McGraw-Hill, New York, 1983, pp. 344–358.

5. FCC Regulations, 47 CFR, 15.65, Washington, D.C.

6. P. Fockens and C. G. Eilers. "Intercarrier Buzz Phenomena Analysis and Cures," *IEEE Trans. Consumer Electronics*, vol. CE-27, no. 3, August 1981, p. 381.

7. *IEEE Guide for Surge Withstand Capability, (SWC) Tests*, ANSI C37.90a–1974/IEEE Std. 472–1974, IEEE, New York, 1974.

8. *Television Receivers and Video Products UL 1410*, Sec. 71, Underwriters Laboratories, Inc., New York, 1981.

CHAPTER 10
AUDIO EQUIPMENT: STUDIO AND RECORDING

Jerry Whitaker
Editoral Director, Broadcast Engineering *Magazine*

E. Stanley Busby, Jr.
Ampex Corporation (retired), Redwood City, California

Portions of this chapter were adapted from: (1) Tim Schneckloth, Marketing Communications Manager, Shure Brothers, "Microphones: From the Inside Out," *Broadcast Engineering* magazine (September 1986). (2) John F. Phelan, Marketing Manager of Professional Products, Shure Brothers, "Selecting Broadcast Microphones," *Broadcast Engineering* magazine (September 1985). (3) Richard Cabot, Vice President and Principal Engineer, Audio Precision, "Active Balanced Inputs and Outputs," *Broadcast Engineering* magazine (July 1988). (4) Brad Dick, technical editor, "New Approaches to Audio Console Design," *Broadcast Engineering* magazine (July 1988).

10.1 INTRODUCTION

When you say the word *audio* to most people, two things come to mind: microphones and loudspeakers. Microphones and loudspeakers, the two most visible portions of any audio system, are just a small part of the overall chain of equipment that makes audio work for a radio or TV station, production facility, or public address/intercommunication system.

The selection of equipment for a professional studio environment is a critical part of facility planning. Proper planning can make the difference between a production center that can handle whatever jobs may arise and one that is an obstacle to getting projects done. Every part of the audio chain is important. There are no insignificant links.

Key aspects of designing and operating a studio facility include:

• Source origination equipment, including microphones, tape recorders, and disk-based systems

• Audio control and mixing equipment

• Distribution and buffer amplifiers

• Patch bay systems

• Audio equalization and special-effects equipment

10.2 SOURCE ORIGINATION EQUIPMENT

Every electronic system has a starting point, and in the case of audio, that point is usually a microphone. Advancements in recording technology have, however, given operators a multitude of options with regard to programming. The source origination equipment available for a well-equipped audio studio ranges from traditional microphones to recorders of various types to disk-based recording systems. Technology has given audio programmers a whole new menu of options.

10.2.1 Microphone Types

Microphones are transducers—nothing more, nothing less. No matter how large or small, elaborate or simple, expensive or economical a microphone might be, it has only one basic function: to convert acoustical energy to electric energy.

With this fundamental point clearly established, you might wonder why microphones exist in such a mind-boggling array of sizes, shapes, frequency-response characteristics, and element types. The answer is simple. Although the basic function of all microphones is the same, they have to work in many different applications and under various conditions.

Choosing the right microphone for a particular application might seem easy, but it is a decision that deserves considerable thought. Just as no two production sessions are alike, the microphone requirements are varied.

Microphone manufacturers offer a selection of units to match almost any application. If you have a good working knowledge of various microphone designs, choosing the right microphone for the job becomes a much simpler task. The education process begins with a look at some of the microphones commonly used today.

Microphones can be roughly divided into two basic classes: those intended for professional applications, such as broadcast stations and recording studios, and those designed for industrial applications, such as radio communications equipment and special-purpose audio pickup, where faithful audio reproduction is not the primary concern.

10.2.2 Professional Microphones

A wide variety of microphones are available for professional applications. In general, three basic classes of microphones are used by professionals: dynamic, condenser, and ribbon microphones. Dynamic microphones are the most widely used, with condenser microphones steadily gaining ground because of their unique qualities that are well suited to broadcast and other on-camera applications. Ribbon microphones, though once very important, have more or less faded from the scene.

Apart from the basic microphone elements, professional users have a number of product classes from which to choose:

- Lavalier
- Hand-held
- Shotgun
- Wireless
- Parabolic
- Line-level

Each has its place in the professional sound recorder's repertoire.

Dynamic Microphones. The real workhorses of the professional world are dynamic microphones. They can be built for ruggedness and reliability, they are not particularly temperamental, and they do not require a power impedance-conversion circuit (as condenser types do). Dynamic microphones also provide good sound quality, although they cannot provide the detail and precision of the finest, most expensive condenser microphones.

The basic mechanical design of a dynamic microphone is shown in Fig. 10.1.

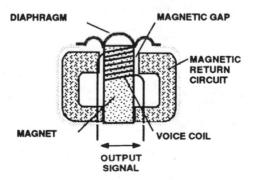

FIGURE 10.1 Basic construction of a dynamic microphone. *(Courtesy* Yamaha Sound Reinforcement Handbook)

The dynamic transducer consists of a lightweight, bobbinless coil (the voice coil) attached to a diaphragm and suspended in the gap between the poles of a permanent magnet. When sound waves cause the diaphragm to move, the coil produces an output voltage. The cartridge is shock-mounted in the case, which normally includes a foam filter screen for protection from dirt and reduction of breath "pop" noise.

The impedance of a dynamic microphone is mostly resistive, and as a result, the output is not substantially affected by electric loading. These microphones are capable of wide dynamic range with low noise and minimal distortion. Dynamic microphones can be readily designed for unidirectional or omnidirectional response patterns. Although dynamic microphones can be made quite small and lightweight, their performance tends to suffer as their size is decreased. For this reason, for the most part condenser-type transducers have taken over the lavalier microphone market.

Condenser Microphones. These microphones have been steadily gaining in popularity over the past decade, and for good reason. They can be made small and lightweight, and they provide clean, precise, and accurate sound reproduction. For these reasons, condenser microphones are often used for applications in which unobtrusiveness and high-quality sound pickup are a must.

In the world of microphones—just as with everything else—there are trade-offs. To get the advantages of a condenser microphone, you have to put up with a few disadvantages. These include the necessity of powering an impedance-converting circuit, high sensitivity to wind, a dynamic range limited by the impedance converter, and the need for shielding against RF and electromagnetic interference. Condenser microphones are also more sensitive to physical shock and humidity than dynamic microphones.

The transducer element of a condenser microphone consists of a lightweight metal or metallized-plastic diaphragm located near a metal backplate. This forms a capacitor, as shown in Fig. 10.2. When sound waves strike the diaphragm, they cause a change in the spacing between the diaphragm and backplate, varying the capacitance. With the application of a dc bias, the capacitance variations are translated to variations in electric voltage. The extremely small capacitance gives condenser microphones a high impedance, making it necessary to add an active circuit that converts the element impedance to a lower value and makes the signal usable in a real-world studio environment.

A polarizing voltage between 9 and 48 V is applied to the diaphragm by an external power supply, often incorporated directly into the audio mixer console.

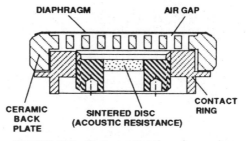

FIGURE 10.2 Basic construction of a condenser microphone. (*Courtesy* Yamaha Sound Reinforcement Handbook)

Because the diaphragm of a condenser microphone is not loaded down with the mass of a voice coil, it can respond quickly and accurately to incident sound. Condensers generally have excellent audio characteristics, making them a good choice for professional audio applications.

A related technique is used in electret microphones (more accurately called *electret-biased condenser microphones*). In this case, the condenser's dc bias is supplied by an electret material, rather than a battery or power supply. The electret material, generally a fluorocarbon polymer, can be a part of the diaphragm or the backplate. Its electrostatic charge lasts indefinitely.

Electrets still require an internal amplifier, however. It is normally housed within the microphone case, powered by either an internal battery or phantom power from an external source. The amplifier is used to buffer the high-impedance condenser capsule from low-impedance loads.

Electrets are increasingly being found in professional applications. They can be made small and lightweight, opening the possibility for unique close-microphoning techniques. Electrets are available for both professional recording and laboratory applications.

Ribbon Microphones. These microphones are not nearly as common as condenser and dynamic microphones, largely because of their reputation for fragility and their tendency to be large. They do, however, provide good frequency response, excellent sound quality, and low handling noise.

The construction of a typical ribbon microphone is shown in Fig. 10.3. Basically, the ribbon microphone produces an output signal in the same way as dynamic microphones do: a conductor moves in a magnetic field and induces a voltage. Instead of a voice coil, however, a ribbon microphone has a thin strip (or ribbon) of aluminum foil suspended between two poles of a permanent magnet. The ribbon, which may be as thin as 0.0001 in, acts as a one-turn coil and also serves as the diaphragm.

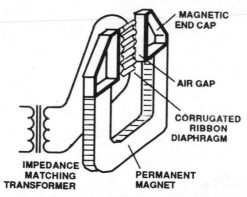

FIGURE 10.3 Basic construction of a ribbon microphone. (*Courtesy* Yamaha Sound Reinforcement Handbook)

The output voltage of the corrugated ribbon diaphragm is very small, and the ribbon impedance is very low. All ribbon microphones, therefore, incorporate a built-in transformer to boost the signal voltage and to isolate the ribbon impedance from the load.

Ribbon microphones provide excellent sonic characteristics, good transient response, and low self-noise. While more delicate than either dynamic or condenser microphones, ribbon microphones still find applications in recording studios, where they are prized as vocal microphones.

10.2.3 Commercial and Industrial Microphones

Microphones intended for applications outside of broadcast stations and recording studios are built to a different set of operational requirements than those for the professional audio industry. The primary requirements of microphones intended for commercial and industrial users are ruggedness and intelligibility under adverse conditions. For the most part, sonic purity does not count.

Microphones of this class include hand-held noise-canceling units for mobile communications and industrial plant operations, handset and headset boom microphones for intercom and order wire circuits, and public address microphones. Each user group has its own particular set of needs, met through variations on a few basic designs.

Carbon Microphones.   Carbon microphones are low-cost, rugged, and reliable, and they produce a high output signal. They are used by the millions in telephone handsets. However, their output signal is inconsistent, distorted, and useful only over a limited frequency range.

The carbon microphone consists of a metal diaphragm, a fixed backplate, and carbon granules sandwiched in between (see Fig. 10.4). When sound waves move the diaphragm, they vary the pressure on the carbon granules. The change in pressure in turn causes the resistance between the diaphragm and backplate to vary. The transducer does not actually generate a voltage; rather, it modulates an externally supplied current.

The output of the carbon button is stepped up by a transformer, as shown in Fig. 10.4. The transformer also serves to isolate the low-impedance button element from the load and to block direct current from the preamplifier input.

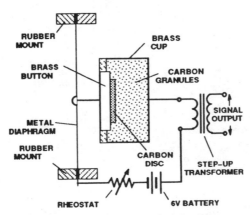

FIGURE 10.4 Basic construction of a carbon microphone. (*Courtesy* Yamaha Sound Reinforcement Handbook)

Piezoelectric Microphones. Also known as ceramic microphones, piezoelectric microphones provide high output and fairly broad frequency response, especially at low frequencies. They are generally inexpensive and reliable, and they are used in communication microphones and some sound-measurement devices. The main disadvantages of piezoelectric microphones are high output impedance, which makes them susceptible to electric noise, and sensitivity to mechanical vibration.

Figure 10.5 shows the basic construction of a piezoelectric microphone. The equivalent circuit is that of a capacitor in series with a voltage generator. When it is mechanically stressed, the element generates a voltage across its opposing surfaces. One end of the piezoelectric element is attached to the center of a diaphragm (usually made of aluminum foil), and the other end of the element is clamped to the microphone frame. When sound energy strikes the diaphragm, the energy vibrates the crystal, deforming it slightly. The crystal, in response, generates a voltage proportional to this "bending," which is an electrical representation of the received sound.

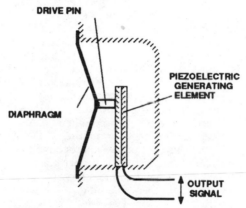

FIGURE 10.5   Basic construction of a piezoelectric microphone. *(Courtesy* Yamaha Sound Reinforcement Handbook)

Controlled Magnetic Microphones.   These microphones consist of a strip of magnetically permeable material suspended in a coil of wire, with one end placed between the poles of a permanent magnet. The center of the diaphragm is attached to the suspended end of the armature.

Movement of the armature in the magnet's gap induces a voltage in the surrounding coil. The voltage is proportional to the armature swing and constitutes the output signal. Electrically, the transducer is equivalent to a voltage generator in series with a resistor and an inductor. Controlled magnetic microphones are also known as magnetic, variable-reluctance, moving-armature, or balanced-armature devices, depending on the manufacturer.

Controlled magnetic microphones are commonly used for communications and paging because of their sensitivity, ruggedness, and dependability.

10.2.4 Selecting Microphone Types

The hand-held microphone, probably the most popular type, is available in many shapes and sizes. Manufactured in both directional and nondirectional versions, the hand-held microphone provides wide frequency response, low handling noise, and a wide choice of characteristic "sounds." Because adequate space is available, a shock-mount system is incorporated in most professional hand-held microphones. Just holding a microphone or dragging the cable across a floor can cause low-frequency noise to be transmitted to the cartridge. A shock-mount system minimizes such noise.

The lavaliere microphone is also in high demand today. Its small size and wide frequency response offer professional users what appears to be the best of all worlds. Small size translates to minimum visual distraction on camera or before an audience. Wide frequency response ensures good audio quality. There are other points to consider, however, before a lavaliere microphone is chosen for a particular application.

The smallest lavaliere microphones available are omnidirectional. This makes the performer's job easier because staying on microphone is less of a problem. However, extraneous noise from the surrounding area can result in generally poor pickup. The omnidirectional lavaliere microphone can pick up unwanted sounds just as easily as it captures the performer's voice. In an indoor, controlled environment, this is usually not a problem. However, if you take the same lavaliere microphone outside, the ambient noise can make the audio track unusable.

Directional lavaliere microphones are available, but they, too, have performance trade-offs. The most obvious is size. To make a lavaliere microphone directional, a back entry must be added to the housing so that sound can reach the back of the microphone. This means a larger housing for the microphone capsule. Although not as large as a hand-held microphone, a unidirectional lavaliere microphone is noticeably larger than its omnidirectional counterpart.

To keep the size under control, shock-mounting of the directional capsule is usually kept to a minimum. This results in a microphone that exhibits more handling noise than a comparable omnidirectional microphone.

Windscreens for lavaliere microphones are a must on any outdoor shoot. Even a soft breeze can cause the audio track to sound as if it were recorded in a wind tunnel. The culprit is turbulence, caused by wind hitting the grille or case of the microphone. The sharper the edges, the greater the turbulence. A good windscreen helps to break up the flow of air around the microphone and to reduce turbulence.

Windscreens work best when fitted loosely around the grille of the microphone. A windscreen that has been jammed down on a microphone only closes off part of the normal acoustic path from the sound source to the diaphragm. The end result is attenuated high-frequency response and reduced wind protection.

Directional Patterns. Every microphone has a directional or polar pattern. That is, each microphone responds in a specific way to sounds arriving from different directions. The polar pattern simply describes graphically the directionality of a given microphone. Although many different polar patterns are possible, the most common are omnidirectional, bidirectional, and varieties of unidirectional.

An omnidirectional microphone responds uniformly to sound arriving from any direction. A unidirectional microphone is most sensitive to sounds arriving at the front of the microphone; it is less sensitive to sounds coming from other directions. Bidirectional microphones are most sensitive to sounds on either side of the pickup element. The pickup pattern for a bidirectional microphone is a figure-eight.

The main disadvantages of an omnidirectional microphone compared with a unidirectional one are a susceptibility to feedback (especially in sound-reinforcement applications) and a lack of rejection of background sounds and noise.

Omnidirectional microphones are, however, widely used in broadcast and other professional applications for several reasons. It is difficult to make a unidirectional microphone very small. As a result, most small lavaliere and other on-camera microphones are omnidirectional. This does not create much of a problem, however, because most broadcast environments in which lavaliere microphones are employed can be controlled to minimize background noise and other extraneous sounds.

Many of the microphones used by radio and TV reporters are omnidirectional because it is often desirable to pick up some ambient sounds in the field to add to the broadcast's sense of liveness. Another plus is the omnidirectional microphone's ability to pick up room ambience and natural reverberation to alleviate the aural flatness a unidirectional microphone might produce.

There are many applications, however, in which a unidirectional pickup pattern is a must. The most obvious is sound reinforcement, where the need to avoid feedback is a primary concern. By preventing the pickup of extraneous sounds, a unidirectional microphone will make available more overall system gain.

The most common form of unidirectional microphone is the cardioid type, whose polar pattern is graphically depicted as heart-shaped (see Fig. 10.6). The area at the bottom of the heart corresponds to the on-axis front of the microphone, the area of greatest sensitivity. The notch at the top of the heart corresponds to the back of the microphone, the area of least receptivity (referred to as the *null*). By contrast, the omnidirectional microphone's polar pattern is (ideally) depicted as a circle.

Special Pattern Microphones.   Other varieties of unidirectional microphones include the supercardioid, hypercardioid, and ultradirectional types such as the shotgun microphone. The supercardioid polar pattern is shown in Fig. 10.7. The nulls of the microphone are at the sides of the pattern, rather than the rear. Note

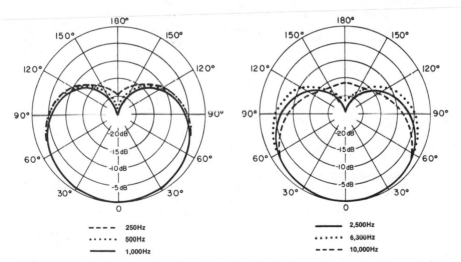

FIGURE 10.6   Typical cardioid polar pattern, plotted at six discrete frequencies. The graphs plot the output in decibels as a function of the microphone's angle relative to the sound source. (Source: Courtesy *Broadcast Engineering* magazine)

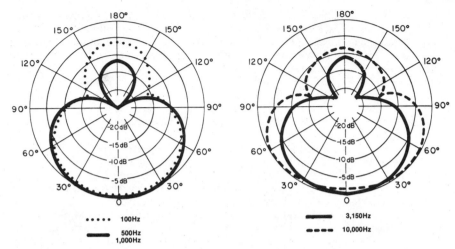

FIGURE 10.7 Typical supercardioid polar pattern, plotted at four discrete frequencies. Note the nulls that appear at the sides of the microphone, rather than at the back. (Source: Courtesy *Broadcast Engineering* magazine)

in Fig. 10.7 the variation in response for the four frequencies shown. Hyper-cardioid microphones offer greater rejection of off-axis sound, but sometimes at the expense of frequency response and/or polar pattern smoothness.

The shotgun microphone is a special type of ultradirectional design. The shotgun offers an extremely tight polar response pattern. Although shotgun microphones seem to reach farther than a normal cardioid microphone, the truth is that they simply reject background noise better. The shotgun microphone can pick out sounds at a distance, while ignoring ambient noise around a shoot. Because the shotgun microphone has a narrow working angle, the microphone must be kept on axis while in use. Generally speaking, the longer the microphone, the more directional a shotgun microphone will be.

New shotgun microphone designs feature a combination of cartridge tuning and interference tube tuning to achieve a smooth frequency response and polar pattern. Compromises in the tuning factors dictate what the microphone will sound like and how it will behave.

Frequency Response.   The frequency response of a microphone describes the sensitivity of the device as a function of frequency. The *range of response* defines the highest and lowest frequencies that the microphone can successfully reproduce. The shape of a microphone's response curve indicates how it responds within this range. (Figure 10.8 shows a typical microphone frequency-response plot.)

Frequency response is often affected by the polar pattern of the microphone. That is, the frequency response may vary depending on the direction of the sound source and the distance from the sound source to the microphone. In addition, some microphones have built-in equalization controls that can alter their frequency response.

The best frequency response for a given microphone depends on the application and the personal taste of the user. There is no such thing as an ideal frequency response. For precise, accurate, lifelike studio recording, a microphone

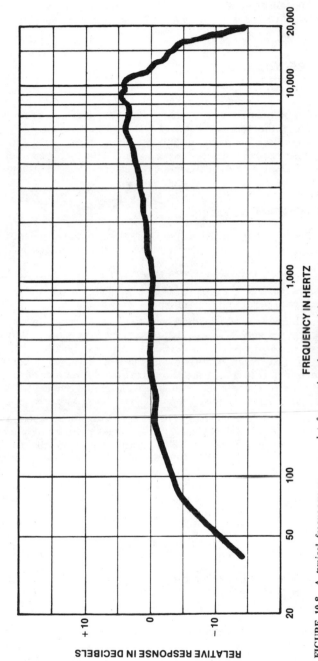

FIGURE 10.8 A typical frequency-response plot for a microphone used in stage and general sound-reinforcement work. Note the high-end presence peak and low-end roll-off. (Source: Courtesy *Broadcast Engineering* magazine)

10.11

with flat response is usually desirable. For most sound-reinforcement applications and other vocal-oriented uses, however, a presence boost in the upper midrange and roll-off on the low end add clarity and intelligibility.

10.3 AUDIO RECORDING

This section describes analog and digital techniques for magnetic-tape recording of audio signals on both audio-only recorders and video-tape recorders. The latter is included here, rather than in Chap. 7, because of the similarity of both systems. However, there are significant differences between operating and maintenance techniques for audio and video equipment. Therefore, these details are covered in Chap. 7 along with the descriptions of each type of recording system.

10.3.1 Longitudinal Analog Recording

This type of analog recording is used in audio only, reel-to-reel systems. The prototype of magnetic recording, audio recording is characterized by:

1. Thick magnetic-tape coatings compared to video and digital recorders.
2. A wide range of track configurations, from a single-track on ¼-in tape to 24 tracks (one channel per track) on 2-in tape.
3. A wide range of tape speeds, usually related to each other by a factor of 2, for example, 30, 15, 7.5, and 3.75 in/s. There are a few very slow reel-to-reel recorders (down to 15/16 in/s), called *logging recorders*, mostly used by radio broadcasters to make a legal record of what they (or their competitors) broadcast. The usual idea is to put 24 h of broadcasting on one reel of tape. Most studio masters are recorded at 15 or 30 in/s, with relatively wide track widths and therefore good signal-to-noise ratios.
4. Generally, three separate head stacks are used, an erase-head stack, a record-head stack, and a playback-head stack. Erase stacks often employ two separate gaps to ensure complete erasure. The record heads usually have relatively wide gaps, on the order of 0.25 to 0.5 mil. While this dimension allows the record heads to be used for synchronous playback, as will be shown later, it is sufficiently large to be an advantage when thickly coated tape is used. Reproduce heads have gaps of 100 to 200 μin, to minimize gap loss at the upper end of the machine's frequency response.

10.3.2 Mechanical Maintenance

A goodly number of mechanical adjustments are usually provided on professional recorders. Some are very important, and misadjustment can ruin valuable recordings.

Tape-Tension Control.   In some older recorders, the tape tension during play, rewind, and fast forward is determined by switched resistors in series with the ac reel motors. This results in a constant-torque tape winding profile, which is not as desirable as a constant-tension one. In a constant-torque system, the variations in tension are a function of the tape pack diameter. Since the inner diameter is fixed by standard, the outer diameter determines the ratio of maximum to minimum tension. Therefore, avoid very large reels on this type of machine. However, most modern transports employ servo-controlled reel motors, producing constant tape tension regardless of the pack diameter.

Typically, the tension is sensed by two tension arms over which the tape passes. Each arm is pulled one way by a spring and deflected the other way by the tension of the tape. The position of the arm is measured (by various means) and provides an input to the two servo systems, so that the torque produced by the reel motors tends to keep the position of the arm, and therefore the tension, constant.

The spring tension is usually adjustable and is the primary determinant of tape tension. See Fig. 10.9. There is often a piston in an air-filled cylinder (with a leak where the spark plug of an automobile engine would be) attached to the tension arm to provide resistive damping to the servo system to prevent oscillation. Sometimes the leak is adjustable. The main potential problem is clogging of the leak by debris.

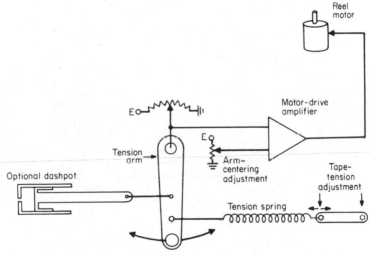

FIGURE 10.9 Reel-servo arrangement. *(Source: K. Blair Benson, ed.,* Audio Engineering Handbook, *McGraw-Hill, New York, 1988)*

In both the above cases, if the tape-tension winding is too small, not all the air will be squeezed out between the layers of tape in the time of one rotation of the reel, and a loose pack will result, which, in the worst case, causes the tape to fold back on itself, usually damaging the tape. If the tape tension is too high, head wear is increased, tape may be permanently stretched, and pressures on the hubs can fracture plastic reel hubs.

Tape-tension transients, stemming from the onset of play, rewind, and fast forward, are a major cause of tape foldback on loose packs, especially large reels.

The thinner the tape is, the greater the problem. To test for a loose pack, pick up a recently rewound tape (always handle tape reels by the hub) and pull on the end of the tape. If the tape pack rotates, it is too loose. Many recorders have a special winding speed, called *spooling speed*, at about 60 to 120 in/s. Using this feature before storing a tape is highly recommended.

Bearings.   Various rotating elements can disturb the otherwise uniform motion of tape across the heads. If any heads are out of round, the disturbance occurs at either a once-around rate, if the element is simply off-center, or at some multiple of that rate. One of the most critical elements is the scrape flutter idler, located on some machines between the record and reproduce heads. Most flutter meters have a demodulator output so that the spectrum for the analysis of wow and flutter on an oscilloscope or on a spectrum analyzer. If there is a predominant periodic frequency in the demodulated output, the formula below yields the diameter of the offending part:

$$\text{Diameter} = \frac{\text{tape speed}}{\pi f}$$

where f = frequency in hertz
       π = 3.14159

The diameter and tape speed are in the same units.

Pinch-Roller Pressure.   Disconnect or disable the capstan motor. Remove reels. Connect a length of tape to a spring scale. Place the tape between the roller and the capstan, and hit the play button to cause the pinch-roller solenoid to operate. While holding the capstan still, measure the force needed to slide the tape past the capstan. The machine's instruction manual should specify a range of acceptable forces. Generally, too little force can cause tape slippage. Too much force accelerates the wear on capstan shaft bearings.

Brakes.   The better professional machines stop the tape by servoing the reels to a stop, and the mechanical reel brakes are used only in case of power failure. The usual brake assembly is asymmetric, with a weak braking force in the winding-on direction and a stronger force in the winding-off direction. This is to ensure that the winding-on reel never decelerates faster than the winding-off reel, which would form a tape loop and spill tape. See Fig. 10.10. The recommended tape tensions and braking forces are very carefully selected during design, so that the tape can be stopped anywhere along the reel without either throwing a loop or cinching the tape pack. Usually there are two adjustable springs on each reel brake, one for each direction of rotation.

Head Azimuth.   Good high-frequency response depends on the record and reproduce head gaps having the same angle with respect to tape motion. The standard for fixed-head audio recorders is 90°. For machines which allow the replacement of individual head stacks in the head assembly, many have provision for azimuth adjustment, while others employ a precision-milled base which was machined after the gap was optically aligned by using a microscope before machining. Alignment tapes have one section with a continuous short-wavelength recording. The reproduce head is adjusted by first adjusting the angle for maximum output. Then the alignment tape is removed, another unrecorded tape is wound on, and the

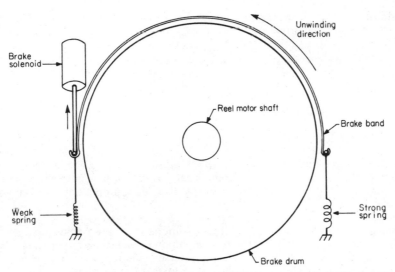

FIGURE 10.10   Brake operation. *(Source: K. Blair Benson, ed.,* Audio Engineering Handbook, *McGraw-Hill, New York, 1988)*

record head is adjusted. Apply a high frequency near the upper range of the machine's capability, and adjust the record-head azimuth for maximum playback response.

Rake Angle.   In most reel-to-reel machines, the tape path is wrapped over the nose of the head for a small angle, about 10° to 15°. The mounting base for some heads allows the head to be pivoted on its vertical axis until the gap is centered in the wrap angle. The usual adjustment method is to vary the angle in one direction until signal is lost, then in the other direction until the signal is lost, and finally to leave the adjustment halfway between. The alternative is to observe the head gap through a piece of uncoated tape, using a suitably mounted microscope.

Cleanliness Rules

1. No smoking.
2. Frequently clean everything the tape touches, especially where the tape is edge-guided. Most damage and debris are generated at the edge of the tape. Most tape debris is abrasive.

   Use a head-cleaning fluid recommended by the manufacturer. Some recorders employ a plastic-coated large-diameter capstan and no pinch roller, while others use a rubber pinch roller which holds the tape against a steel capstan shaft. Clean plastic and rubber parts with alcohol rather than head cleaner. The plastic-coated capstan in particular is supposed to have a "tacky" feel. A coating of dust can lower the coefficient of friction, permitting tape slippage.
3. Frequently demagnetize the heads. A magnetized record head behaves just as if direct current were flowing in the windings. A recording made under this

condition will be noisier than normal and will exhibit elevated levels of even-order harmonic distortion.

4. When the machine will be left overnight, wind off any tape that is threaded on it, and put the tape in its box. If you have a cleaning service, cover the machine with a dust cover.

5. For tape-storage conditions, see Chapter 15.

10.3.3 Electrical Maintenance

A number of adjustments need periodic attention, mostly to account for head wear. If a head is replaced or a different type or brand of tape is selected, some adjustment may be necessary to maintain low distortion, proper output level, and flat frequency response.

Reproduce Equalization. Typically two, sometimes three, adjustments are needed, usually replicated for each speed at which the machine can operate. The main equalizer (usually abbreviated EQ) establishes the reproduce frequency response except for correction of the head-gap loss at the very upper edge of the passband, which may be provided with an adjustment often called *peaking.* Typically, the head inductance is resonated, with a capacitor or with the capacitance of the head cable, at a frequency well above the highest frequency of interest. The resonant circuit is damped by an adjustable resistor so the resonant rise in response can be made equal to the loss. Some playback electronic formats offer a "shelf" adjustment which allows a small but independent control over the upper range of frequencies. The shelf control and the main EQ usually interact.

Reproduce equalization (EQ) is adjusted during playback by using an alignment tape. The mechanical reproduce-head azimuth adjustment should be done first, because high-frequency response is critically dependent on the gap azimuth. The wider the recorded track, the more critical the adjustment. The smaller the gap length, the more critical the adjustment. The alignment tape will have a section recorded at a high frequency for making the azimuth adjustment. Another section of the tape has various frequencies recorded at levels such that a standard reproducer should play back flat. Adjust the reproduce EQ controls until the response most closely approaches flatness. Make a record of deviations from flatness for use later when you are adjusting the record circuitry. It is better to use a good-quality ac voltmeter than to depend on the volume unit (VU) meter on the machine. Allocate time to tweak all the controls for a 24-channel audio machine, for each of four speeds; to do them *right* can take a few days.

Reproduce Level. There will be a section of the alignment tape on which a low-frequency tone at a standard level of magnetization is recorded. The reproduce gain control is then adjusted to provide whatever output level is standard in your installation.

This is a good time to check the input to the noise-reduction circuit if one is installed in the machine.

Reproduce Crosstalk. Some recorders have adjustable circuitry to minimize adjacent-channel playback crosstalk. Prepare a test tape on which one track is recorded while the adjacent one is unbiased; then in another section the first track is unrecorded but the second one is. Start with bulk-erased tape. Adjust for minimum playback signal on the unbiased channel.

Record System Adjustment. With the reproduce system standardized, the record system can be adjusted. The idea is not to obtain the flattest playback response, but to most nearly obtain the same response as reproducing the alignment tape. The order of adjustment is as follows:

1. *Record-head azimuth adjustment:* Use a high frequency, and adjust for maximum playback amplitude. If the head is a combination record and play, this step may be skipped.
2. *Record bias current:* The bias current influences record sensitivity, harmonic distortion, and frequency response, especially at short wavelengths. When the record-head gap length is about the same as the coating thickness, there is a bias current that minimizes distortion, at least for long wavelengths.

The procedure is to record a moderately high frequency, one producing a wavelength of about 1.5 mil. While monitoring the playback, increase the bias current from zero until the response maximizes; then further increase the current until the response falls a prescribed number of decibels, typically 3 to 5 dB. The instruction manual should specify both the frequency to use and the amount of overbias. See Fig. 10.11. Those particles deepest into the thickness of the coating suffer from insufficient bias, but contribute little during playback because of be-

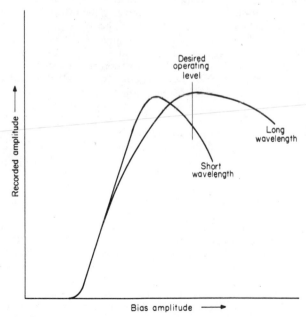

FIGURE 10.11 Influence of bias amplitude. *(Source: K. Blair Benson, ed.,* Audio Engineering Handbook, *McGraw-Hill, New York, 1988)*

ing far from the surface. The short wavelengths, for which only the surface particles contribute, are overbiased; for the most audible frequencies, the majority of those particles which contribute to playback are optimally biased.

Large multichannel recorders have individual bias current adjustments for each channel. These can be used to optimize the bias drive to each head. At the source of the bias current there is a *master* bias current adjustment which affects all channels together. This is used to adjust for different brands and types of tape.

The advantages gained from overbiasing are minimal for longitudinal analog audio recording on typical video recorders for several reasons: The tape is too thin; the gap lengths are too short because the heads, to maintain lip sync, must serve for both record and playback; and for quadruplex recorders, the magnetic-particle orientation is optimized for video, not audio.

Record Equalization.   Since the bias adjustment method left the high frequencies overbiased, a record equalizer is used to boost high frequencies to overcome this loss. There is usually an EQ adjustment for each tape speed. To avoid tape saturation during the recording of high frequencies for adjustment purposes, a reduced record level should be used. For example, -5, -10, -15, and -20 dB are appropriate for 30, 15, 7.5, and 3.75 in/s, respectively. Adjust the record EQ controls to approach as closely as possible the response obtained with the alignment tape. This will minimize interchange problems. *Only* if *every* recording will be played back on the *same* recorder should you consider adjusting for flattest overall response.

Erase Amplitude.   One would think that the more erase current, the better; but, in fact, there is an optimum erase current above which there is a slight tendency to rerecord the erased signal downstream from the erase gap. To properly adjust the erase amplitude requires a good spectrum analyzer. Record a 400- to 1000-Hz signal at about 10 dB above normal. Rewind the tape, disconnect the record head, and enter the record mode. Using a spectrum analyzer in a narrowband mode to minimize noise, observe the playback signal and adjust the erase amplitude for minimum residual signal. Attenuation of approximately 70 dB should be possible.

Erase-Record On-and-Off Timing.   In the process of electronic editing on tape, it is typical to begin erasing and then, just a little later, when the erased tape has arrived at the record head, to ramp up the record bias and the record signal. If the timing is right, the result can be a smooth cross-fade between old and new. See Fig. 10.12. At the outgoing splice, the reverse occurs, producing a smooth transition between new and old. The time between the erasure point and the record point is a function of the tape speed, distance between erase- and record-head gaps, rate of erasure commencement and decay, and rate of ramp-up and decay of bias and record. The adjustments for these delays should be made only after the erase amplitude and bias current have been set, because they influence the rise and fall of erasure and recording, an important component of the total time.

Record Crosstalk Adjustment.   Some recorders, notably 1-in analog video recorders, have an adjustable circuit which injects a small fraction of one channel's signal into the adjacent track's record head, with reversed phase so as to cause cancellation. Arrange to monitor the confidence playback head by using a high-gain oscilloscope or a spectrum analyzer. Place both channels in record, but supply an input signal of about 1000 Hz to channel 1 only. Tune for minimum signal in chan-

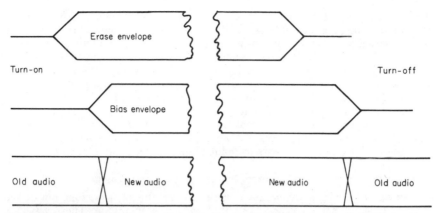

FIGURE 10.12   Erase and bias on-off timing. (*Source: K. Blair Benson, ed.,* Audio Engineering Handbook, *McGraw-Hill, New York, 1988)*

nel 2's playback. Then record a signal on channel 2 only and listen to channel 1's playback. It is essential that both channels have bias current in both cases.

Record Predistortion.   Analog video recorders in particular are often equipped with an adjustable distortion generator which "stretches" both positive and negative signal peaks. This tends to make up for the peak compression normally suffered during recording of high-level signals that approach tape saturation. The maximum level that the tape can accommodate is not affected by the adjustment. The sensory effect of too much predistortion is far worse than that of too little predistortion. A spectrum analyzer is required to isolate the third harmonic of the test tone, so that it can be minimized.

10.3.4 Longitudinal Cassette and Cartridge Recorders

The fundamentals of recording on tape apply to both cassette and cartridge machines. The cartridge recorder is often found in radio stations, where it is used for playing commercials. It uses a single spool with a continuous loop of tape. During use, the tape is pulled out of the open-faced pack at the inside turn, passed over the heads, then wound onto the outside of the pack. Since every turn of tape is moving in respect to its neighboring layers, the tape is heavily lubricated.

This type of cartridge has the advantage that it can be allowed to run forward after use and then stopped at the beginning, ready for reuse with no need for a rewind cycle.

Cassette recorders are found in a multitude of applications. For practical and marketing reasons, they use a smaller tape width (0.15 in) and a slow play speed (1⅞ in/s). The format allows four tracks on the tape, two in each direction. In inexpensive transports, the single record/play head is fixed. When the end of a recording is reached, the cassette is removed, flipped over, and restarted. More expensive units have a movable head, or two fixed heads. There are two capstans and pinch rollers and an end-of-tape sensor which causes the direction to reverse. Cassette recorders are stereo recorders, with a single erase gap for each pair of tracks. Prerecorded cassette tapes are copied by recording all four tracks at once at a very high speed, from a 0.5-in four-track master. The master is generally a continuous tape loop played back on a special bin-loop machine, in which tape falls loosely into a rectangular bin whose sides are barely one tape width apart. Tape is pulled from the bottom of the bin.

The typical cassette recorder has few adjustments, usually confined to a single bias current adjustment. Holes in the plastic case indicate to a probe in the recorder which of three bias levels to select automatically from taps on a resistive voltage divider in the bias driver circuit. The narrow tracks and small head gaps make a noisy recording when normal energy tape is used. The level of tape hiss makes it marginal for consumers and unusable for commercial applications. Improvements made in recent years to reduce noise include the following:

1. Boost the highs more when recording on normal energy tape.

2. Use one of the noise-reduction schemes. These boost, as a function of the record signal itself, those portions that have low amplitude. These schemes are discussed in more detail in Sec. 10.3.5.

3. Use high-energy tape. Recorders automatically reduce the high-frequency boost when recording on high-energy tapes. Prerecorded tapes which use high-energy tapes, however, use the high-end boost intended for normal tape. Much of the market for these tapes is for use in very small battery-powered portable reproducers which have no equalization switch.

At least one cassette recorder brand offers a scheme for automatically setting record bias current. A unit having separate record and play heads is required, so that simultaneous playback and recording is possible. Two tones are recorded, one high and one low. During bias setting, the bias is varied until a desired ratio between the playback tones is obtained.

The cassette also has two holes covered by plastic knock-outs, one hole for each direction. Removing a tab will inhibit recording in that direction. The cassette also contains a felt spring-loaded pad to ensure good head-to-tape contact.

10.3.5   Dynamic Noise Reduction

The basic idea in all the noise-reduction schemes for analog tape is to amplify weak passages during recording and attenuate them during playback, which attenuates the playback noise in the process. How much to amplify and attenuate is based on the amplitude of the signal itself. Six systems are outlined below. They are described in greater detail in Ref. 4. These schemes are often used for recording and for transmission over noisy channels and are sometimes used in broadcasting in the expectation that the receiver can process the reception in the same way as it would a tape playback.

Dolby A.   The first of a series of four designs, Dolby A separates the audio spectrum into four bands. The amplification applied to each band is less than 2-to-1; that is, for a 6-dB decrease in energy in any one band, there will be a 3-dB or less increase in circuit gain. The overall noise reduction is about 10 dB.

Dolby B.   Widely employed in consumer recorders, Dolby B is like a Dolby A with only two bands; the upper one slides up and down the spectrum depending on the amplitude and spectral content of the record signal. The intent is to boost only those frequencies higher than those which contain the bulk of the energy.

This avoids boosting a signal which is already large, thereby avoiding tape saturation.

Dolby C. This has been described as two Dolby B's in series. Two Dolby B sliding-band-type filters are used, but the levels at which they begin to compress are staggered. The result is a modest increase in the compression ratio and a maximum noise reduction of 20 dB.

Dolby SR (Spectral Recording). Using two low-frequency bands in series and three in the high frequency ranges, Dolby SR achieves a maximum of 30-dB noise reduction.

In an effort to reduce artifacts of the circuit's action to inaudibility, all the Dolby systems do not affect the signal if it is greater than about 20 dB below the standard record level. Figure 10.13 illustrates the input-output relationship in these systems. These systems are said to be *level-sensitive*, so it is extremely important for Dolby circuitry to receive accurately calibrated input levels.

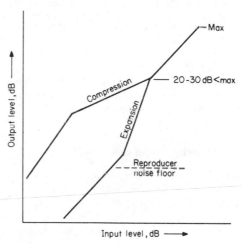

FIGURE 10.13   Input-output relation in a Dolby system. *(Source: K. Blair Benson, ed.,* Audio Engineering Handbook, *McGraw-Hill, New York, 1988)*

Telcon. Telefunken of West Germany developed a noise-reduction circuit called Telcon C4. One of its two versions can be used with Dolby A. Figure 10.14 shows how the degree of compression and expansion is uniform at all levels. Such a system will amplify frequency-response variations to the same degree as the compression ratio; i.e., if there is a 10 percent variation in the frequency response of the recorder, a 2-to-1 compression ratio will double it to 20 percent, even at normal record level, at which it is easier to hear this kind of artifact. Therefore it is important to keep frequency-response variations to within narrow limits.

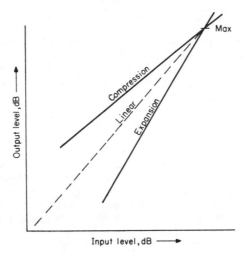

FIGURE 10.14 Input-output relation in the Telcon C4 system. *(Source: K. Blair Benson, ed., Audio Engineering Handbook, McGraw-Hill, New York, 1988)*

dBx. This noise-reduction system does not divide the spectrum, but treats all frequencies alike. It has a 2-to-1 compression ratio over the entire input-output amplitude range. The dBx system is specified by the FCC for TV stereo sound.

10.3.6 Longitudinal Digital Audio Recording

The first digital audio recordings were made on unmodified instrumentation recorders at a sampling rate of 50 kHz. Just as is true today, each sample of the audio waveform was a 16-bit two's-complement number and was a linear representation of the signal (no compression was used). Later, modified helical-scan video recorders were used, at a sampling rate of 44.1 kHz, which is related to the vertical-scanning rate of monochrome television. Compact disks (CDs) use 44.1 kHz. Some recordings are made at 44.056 kHz, which is a rate suitable for video recorders locked to National Television System Committee (NTSC) color television vertical scanning. Digital television recorders use 48 kHz, which can be derived from the horizontal-scanning rate of either NTSC or the two European 625-line systems. Still later, a rotary-head audio-recording format was devised (this is discussed in Sec. 10.3.7).

Digital recording offers the opportunity to build protection against tape dropouts. Additional data, generated from the digital numbers being recorded, are either appended to the digital data on the same track (Sony, Studor, Matsushita) or recorded on separate tracks of a multitrack recorder (Mitsubishi). Although the formatting of analog recordings is only a matter of track width and tape speed, the distribution of the digital data and of the error detection and correction overhead data is quite complex and must be known in detail if tape interchange is to be achieved.

In general, with 2 bytes (16 bits) of overhead, you can locate 2 bytes in error, correct 2 bytes in error if you already know where they are, or locate 1 byte in error and correct it. With 4 bytes, you can detect and correct 2 damaged bytes, and so on. Data is recorded in blocks of 30 to 40 bytes, but before recording data is scattered along the tape or across the tape or both, so that a long dropout or the mess left by a punch-in edit will appear during playback as a large number of widely scattered single-byte errors, not close enough together to overload the correction system. The longitudinal formats permit razor-blade splices with good results, even though the power of the error detection and correction is briefly diminished for a few hundred milliseconds at the splice point.

Digital error detection and correction schemes exhibit these characteristics:

1. Up to some error rate around 1 bit in 1000 wrong, they work perfectly, correcting all errors.

2. At an error rate above some threshold, the scheme fails to correct errors but reports the failure to the correction circuitry, so that the damaged sample can be replaced with an estimate of its value. This technique is called *error concealment*. It works better for video than for audio and is not applicable at all to instrumentation recording, where the nature of the input signal cannot be predicted.

3. At a still higher error rate, approaching random noise, the scheme will occasionally lie to you, reporting that all is well when it is not, and uncorrected errors can sneak through. The sound level of these can range from unnoticeable to that of a 0.38 magnum firing through the loudspeaker cone.

4. Data is scrambled before recording such that when the scheme fails as in characteristic 2 above, bad samples are usually preceded and followed by good ones. The estimate to replace the bad (and uncorrectable) sample is formed by adding the preceding and following samples and dividing by 2. The system works well in an audio environment because the predominant energy in the audio spectrum is concentrated toward the low-frequency end.

5. If things are really bad and the signal cannot be corrected or concealed, the output is muted.

Many digital recorders, both audio and video, are equipped to reproduce a digital-signal stream well in advance of the time at which the record head addresses the same location on tape. Therefore a read-modify-write mode can be implemented in which data is retrieved, processed, mixed with or otherwise combined with other data (either from another track or from the outside world), and then processed and rerecorded at the same spot where it would have been had it not been modified. While this is a valuable feature, it does require a total commitment to each edit, because data is replaced at each edit and old data is always lost. It should be used with caution. The duration of the cross-fade between old and new and new and old is typically programmable and ranges between 5 and 100 ms.

DASH (Digital Audio Stationary Head) Format.   Sony and other manufacturers offer stereo recording at 7.5 in/s with eight tracks, 15 in/s with two tracks per channel, or 30 in/s with one track per channel for 24 channels. (See Ref. 1.) The Mitsubishi format offers 32 channels using 40 tracks, on 1-in tape arranged in four groups of eight channels for the data of each channel plus two tracks in each group for the error detection and correction information. (See Ref. 2.)

Reel-to-reel tapes having a thin coating are available for the above two formats from major tape manufacturers.

The S-DAT (*s*tationary *d*igital *a*udio on *t*ape) format is a longitudinal format of

many skinny tracks. Thin-film magnetoresistive technology is used for the reproduce heads. The iron-particle tape is housed in a cassette. Sampling rates of 48,
44.1, and 32 kHz are supported, and the tape speed varies directly with the sampling rate. One operational mode, available only at the 32-kHz sampling rate,
uses a nonlinear 12-bit code to extend the recording time. It should be adequate
for recording speech. (See Ref. 3 for more detail.)

Digital recorders have reproduce equalizer adjustments and record current adjustments, but these in no way affect either frequency response or distortion.
They do affect the error rate. Most recorders have some indication, on a control
panel or on a plug-in card edge, of the level of activity of error-correction circuitry and the error-concealment rate. If the correction rate is high or any con-
cealments are seen, copy that tape and discard it. The threshold between good
error correction and inadequate correction is a very sharp one.

10.3.7 Rotary-Head Recording of Digital Audio

Very early digital recordings were made on laboratory-modified helical-scan
video recorders. Later adapters were manufactured and sold which packaged
bursts of digital audio samples between television sync pulses so they would be
acceptable to a video recorder. Professional equipment uses a ¾-in cassette-
based format and offers time-base-corrected crystal-accurate output. Others were
designed for ½-in home video recorders and used 14-bit quantization. Even later
developments include the following:

R-DAT Format. R-DAT (*r*otary *d*igital *a*udio on *t*ape) is a format designed just for
digital audio. Using a tiny cassette with metal-particle tape just 3.81 mm wide and a
maximum of 70 m long with only about a 90° wrap angle around the helical scanner,
it forms a compact, easily threaded assembly. See Fig. 10.15. The planar packing
density of this format is high, partly because alternate heads on the scanning
asembly have their azimuth angles tilted by 20° in opposite directions. This allows

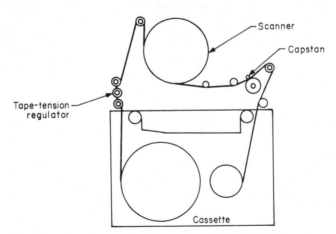

FIGURE 10.15 Schematic layout of an R-DAT cassette system.
(Source: K. Blair Benson, ed., Audio Engineering Handbook,
McGraw-Hill, New York, 1988)

recording tracks which have no guard band at all or even a negative guard band. (See Ref. 5.)

As of this writing, consumer R-DAT machines *record* at 48 and 32 kHz, but will play back only at 44.1 kHz, the sampling frequency most used on prerecorded material and for CDs. This was done to protect the manufacturers of CDs from digitally dubbed piracy. Even so, the consumer machines are not now available in the United States, being vehemently opposed by record manufacturers. Professional machines are available, cost several thousand dollars, and *will* record at 44.1 kHz, but not if the digital input stream contains the infamous "copy protect" bit. Most engineers in the audio business know someone who knows which diode to clip out to defeat this inhibition. Consumer units may be legally imported, but that means a trip to Japan to buy one.

8-mm Video Format. A rotary video/audio format, in which digital audio is optional, is the 8-mm video format. The wrap angle around the scanner is more than 180°, allowing for continuous recording of video and recording of PCM audio during the overlap time (the time that two heads are in contact with the tape). The audio signal is sampled at twice the horizontal-scanning rate, or about 31.5 kHz. Eight bits per sample are used, in a nonlinear representation which tends to make noise at signal amplitudes at which the ear is least sensitive to it. Both *analog* companding, which has well-known attack and decay time limitations, and *digital* companding, which is instantaneous, are used. Small signals are quantized to 10-bit accuracy. (See Ref. 6.)

D1 Component Video Format. Manufactured by Sony and Bosch, this recording method writes four channels of digital video and audio data using the video heads, with the digital audio data centered at the middle of the tape. There are virtually no audio adjustments since the video heads do all the recording. Auxiliary tracks include control track, time code, and voice-grade comment but are not discussed here. The first level of digital error protection is applied independently of video and includes optimal scrambling of the input data to allow alternate-sample error concealment, should it become necessary. The second level of error correction and detection is common to the video and audio record channels. The audio data is written by the video heads as short bursts at the video rate. Each burst is separated from the rest by a short period of zero data, to allow punch-in editing of any channel independent of the others. The data from each channel is recorded twice except for a brief period at the beginning and end of an edit, when only one copy is recorded. This permits a cross-fade to be made between old and new at the edit boundaries during playback.

The format will accommodate up to 20 bits per sample, and each burst of recording contains other ancillary data, including

1. Whether preemphasis was employed and which kind
2. How many bits per sample were used
3. Two edit flags
4. Whether this is a long burst or a short burst
5. Data to reconstruct the block boundaries of the AES/EBU digital audio-transmission format

The tape is in a cassette, with small, medium, and large sizes available. The tape is made of cobalt-doped ferric oxide.

D2 Composite Video Format.   Offered by Sony (DVR-10) and Ampex (VPR-300 and ACR-225), D2 records audio samples in much the same way as the D1 format except that the bursts of audio are recorded at the edges of the tape. There is no overlap of old and new data at an edit point, but the edit point is marked by a flag so the reproducer can do a quick fade at the edit point if desired. The tape is iron-particle, and the D2 *uses the same cassette* as the D1 machine. Users are cautioned to beware of tape type mixups.

Both the above machines offer the AES/EBU serial input-output format. Each input and output handles two channels or one stereo pair. The same type of XLR connector is used for both analog and digital input-output, so read the labels. The D2 machines also offer a byte-wide digital interface, suitable for flat ribbon cable over short distances.

Both formats offer the ability to reproduce at a variable speed, over a range of ±15 percent or so. Audio, unlike video, cannot afford to throw away or iterate a whole chunk of data, so variable-rate playback in these machines involves slowing down or speeding up the capstan *and* the scanner so that no tracks are missed. This results in a nonstandard sampling rate at the digital interface. The usual application is to compress a slightly overlong commercial into a standard length. Equipment is available which can restore the resulting pitch shift. When digitally dubbing at a nonstandard playback rate, you must convert the nonstandard sampling rate to 48 kHz. Both Sony and Studer make such converters. It is best to choose just one nonstandard speed to use throughout the playback. Sample-rate converters do not work well when the input frequency is *changing*.

References

1. W. J. van Gestel, H. G. de Haan, and T. G. J. A. Martens. "Digital Magnetic-Tape Recording and Reproduction", in *Audio Engineering Handbook*, K. Blair Benson, editor. McGraw-Hill, New York, 1988, p. 11.7.2.

2. P. Jeffery Bloom and Guy W. McNally. "Digital Techniques," in *Audio Engineering Handbook*, K. Blair Benson, editor. McGraw-Hill, New York, 1988, p. 4.5.1.

3. W. J. van Gestel, H. G. de Haan, and T. G. J. A. Martens. "Digital Magnetic-Tape Recording and Reproduction," in *Audio Engineering Handbook*, K. Blair Benson, editor. McGraw-Hill, New York, 1988, p. 11.7.4.

4. Ray Dolby, David P. Robinson, and Leslie B. Tyler. "Noise Reduction Systems," in *Audio Engineering Handbook*, K. Blair Benson, editor. McGraw-Hill, New York, 1988, p. 15.8.

5. W. J. van Gestel, H. G. de Haan, and T. G. J. A. Martens. "Digital Magnetic-Tape Recording and Reproduction," in *Audio Engineering Handbook*, K. Blair Benson, editor. McGraw-Hill, New York, 1988, p. 11.7.5.

6. W. J. van Gestel, H. G. de Haan, and T. G. J. A. Martens. "Digital Magnetic-Tape Recording and Reproduction," in *Audio Engineering Handbook*, K. Blair Benson, editor. McGraw-Hill, New York, 1988, p. 11.7.3.

10.4   INTERFACING AUDIO SYSTEMS

System interfacing problems have existed since the first audio engineer tried to wire two pieces of equipment together. Hardware that works flawlessly alone can fail miserably when connected to other hardware.

In years past, audio engineers routinely used transformers to correct for differences in ground potential between equipment and to eliminate electric noise picked up in cabling. These transformers helped to usher in the age of the balanced and floating interface. However, they also added significant cost, weight, and distortion to the equipment.

As the performance of electronic equipment improved, the transformer's shortcomings became more noticeable. The large physical size of transformers also made them inconvenient for use in the ever-shrinking chassis of transistorized equipment. As solid-state technology became more common, designers began to look for ways to eliminate transformers. The result was electronically balanced (active) inputs and outputs (I/O).

The move away from transformers has presented audio engineers with several difficult problems. First, installation practices that provided acceptable performance with transformer-isolated equipment may not be acceptable for active balanced I/O equipment. Second, active balanced I/O circuits are less forgiving of wiring errors and accidental short-circuits than transformers. Some active balanced output stages will self-destruct if their output lines are short-circuited. Third, attention to ground loops is more important with an active balanced I/O system than with a transformer system. Active balanced input circuits can provide excellent noise rejection, but only to a point. Beyond a certain threshold, the noise rejection of an active balanced input circuit deteriorates rapidly.

10.4.1 Balanced and Unbalanced Systems

Nearly all professional audio systems use balanced inputs and outputs because of the noise immunity they provide. Figure 10.16 shows a basic source and load connection. No grounds are present, and both the source and load float. This is the optimum condition for equipment interconnection.

Either the source or the load may be tied to ground with no problems, provided only one ground connection exists. Unbalanced systems are created when each piece of equipment has one of its connections tied to ground, as shown in Fig. 10.17. This condition occurs if the source and the load equipment have unbalanced (single-ended) inputs and outputs. This type of equipment utilizes the chassis ground for one of the audio conductors. This is common practice in consumer audio products. Equipment with unbalanced inputs and outputs is cheaper to build; however, when various pieces of audio gear are tied together, trouble often results. These problems are compounded when the equipment is separated by a significant distance.

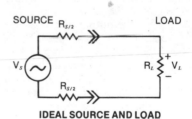

FIGURE 10.16 A basic source and load connection. No grounds are indicated, and both the source and the load float. (Source: Courtesy *Broadcast Engineering* magazine)

As shown in Fig. 10.17, a difference in ground potential causes current flow in the ground wire. This current develops a voltage across the wire resistance. The ground noise voltage adds directly to the signal itself. Because the ground current is usually from leakage in power transformers and line filters, the 60-Hz signal

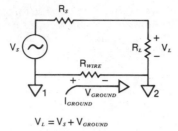

$$V_L = V_s + V_{GROUND}$$

FIGURE 10.17 An unbalanced system in which each piece of equipment has one of its connections tied to ground. (Source: Courtesy *Broadcast Engineering* magazine)

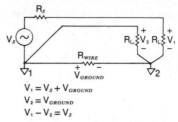

$$V_1 = V_s + V_{GROUND}$$
$$V_2 = V_{GROUND}$$
$$V_1 - V_2 = V_s$$

FIGURE 10.18 Ground loop noise can be canceled by amplifying both the high side and the ground side of the source and subtracting the two signals.

gives rise to hum. Reducing the wire resistance through a heavier ground wire helps reduce the hum, but cannot eliminate it completely.

By amplifying both the high side and the ground side of the source and subtracting the two to obtain a *difference signal*, it is possible to cancel the ground loop noise (see Fig. 10.18). This is the basis of the *differential input*.

Unfortunately, problems can still exist with the unbalanced source to balanced load system shown in Fig. 10.18. The reason centers on the impedance of the unbalanced source. One side of the line will have a slightly lower amplitude because of impedance differences in the output lines.

By creating an output signal that is out of phase with the original, a balanced source can be created to eliminate this error (see Fig. 10.19). As an added benefit, for a given maximum output voltage from the source, the signal voltage is doubled compared to the unbalanced case.

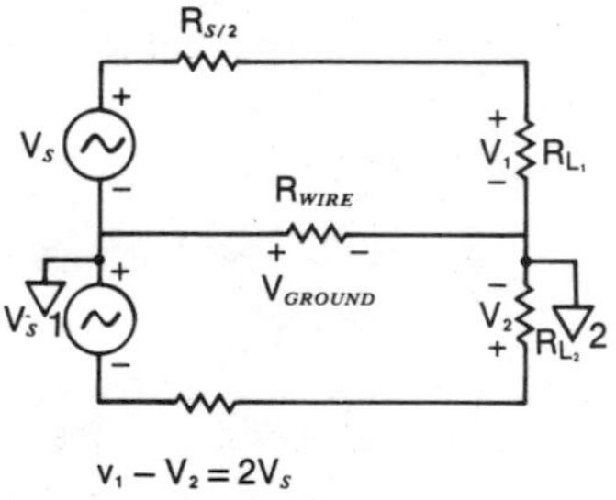

$$v_1 - V_2 = 2V_s$$

FIGURE 10.19 A balanced source configuration where the inherent amplitude error of the system shown in Fig. 10.18 is eliminated. (Source: Courtesy *Broadcast Engineering* magazine)

Common Mode Rejection Ratio. This ratio, abbreviated CMRR, is the measure of how well an input rejects ground noise. The concept is illustrated in Fig. 10.20. The input signal to a differential amplifier is applied between the plus and minus amplifier inputs. The stage will have a certain gain for this signal condition, called the *differential gain*. Because the ground-noise voltage appears on both the plus and minus inputs simultaneously, it is common to both inputs.

The amplifier subtracts the two inputs, giving only the difference between the voltages at the input terminals at the output of the stage. The gain under this condition should be zero, but in practice it is not. The CMRR is the ratio of these two gains (the differential gain and the common-mode gain) in decibels. The larger the number, the better. For example, a 60-dB CMRR means that a ground signal common to the two inputs will have 60 dB less gain than the desired differential signal. If the ground noise is already 40 dB below the desired signal level, the output noise level will be 100 dB below the desired signal level. If, however, the noise is already part of the differential signal, the CMRR will do nothing to improve the signal.

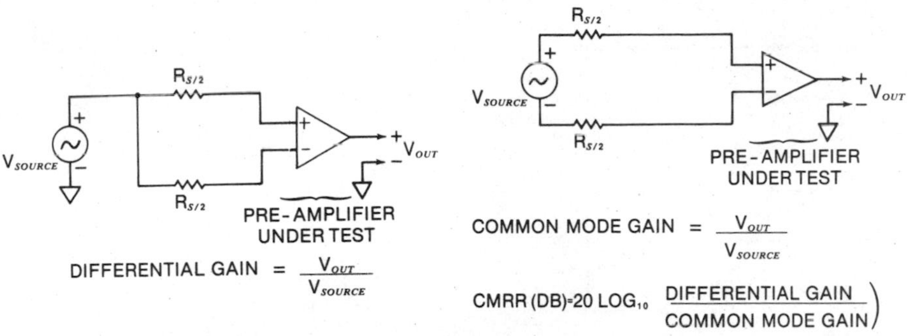

$R_{S/2}$

V_{SOURCE}

$+V_{OUT}$
$-$

$R_{S/2}$

PRE-AMPLIFIER
UNDER TEST

DIFFERENTIAL GAIN $= \dfrac{V_{OUT}}{V_{SOURCE}}$

$R_{S/2}$

V_{SOURCE}

$+V_{OUT}$
$-$

$R_{S/2}$

PRE-AMPLIFIER
UNDER TEST

COMMON MODE GAIN $= \dfrac{V_{OUT}}{V_{SOURCE}}$

CMRR (DB)=20 LOG$_{10}\left(\dfrac{\text{DIFFERENTIAL GAIN}}{\text{COMMON MODE GAIN}}\right)$

FIGURE 10.20 An illustration of the concept of common-mode rejection ratio for an active balanced input circuit. (Source: Courtesy *Broadcast Engineering* magazine)

Although active balanced I/O circuits are not as effective as transformers in rejecting ground noise and radio-frequency interference (RFI), they usually are adequate if they are well designed. Because of their advantages in cost, weight, and low distortion, active balanced I/O circuits are generally the best choice for all but the most demanding environments.

10.4.2   Active Balanced Input Circuits

A wide variety of circuit designs have been devised for active balanced inputs. All have the common goal of providing high CMRR and adequate gain for subsequent stages. All are also built around a few basic principles.

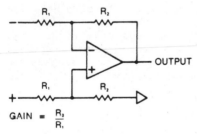

R_1 R_2

OUTPUT

R_1 R_2

GAIN $= \dfrac{R_2}{R_1}$

FIGURE 10.21 The simplest and least expensive active balanced input op-amp circuit. Performance depends on resistor matching and the balance of the source impedance. (Source: Courtesy *Broadcast Engineering* magazine)

Figure 10.21 shows the simplest and least expensive approach using a single operational amplifier (op-amp). For a unity-gain stage, all the resistors are the same value. This circuit presents an input impedance to the line that is different for the two sides. The positive input impedance will be twice the negative input impedance. The CMRR is dependent on the matching of the four resistors and the balance of the source impedance. The noise performance of this circuit, which is usually limited by the resistors, is a trade-off between low loading of the line and low noise.

Another approach (shown in Fig. 10.22) uses a buffering op-amp stage for the positive input. The positive signal is inverted by the op-amp and then added to the negative input of the second inverting amplifier stage. Any common-mode signal on the positive input (which has been inverted) will cancel when it is added to the negative input signal. Both inputs have the same impedance. The matching of resistors limits the CMRR to about 50 dB. With the addition of an adjustment pot, it is possible to achieve an 80-dB CMRR, but component aging will degrade this over time.

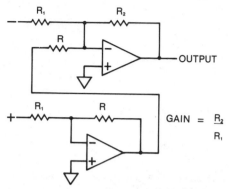

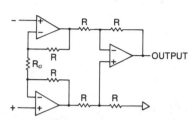

$$\text{GAIN} = \frac{R_2}{R_1}$$

FIGURE 10.22 An active balanced input circuit using two op-amps, one to invert the positive input terminal and the other to buffer the difference signal. Without adjustments, this circuit will provide about 50 dB of CMRR. (Source: Courtesy *Broadcast Engineering* magazine)

FIGURE 10.23 An active balanced input circuit using three op-amps to form an instrumentation-grade circuit. The input signals are buffered and then applied to a differential amplifier. (Source: Courtesy *Broadcast Engineering* magazine)

Adding a pair of buffer amplifiers before the summing stage results in an instrumentation-grade circuit, as shown in Fig. 10.23. The input impedance is increased substantially, and any source-impedance effects are eliminated. Additional noise is introduced by the added op-amp, but the resistor noise usually can be decreased by reducing impedances, causing a net improvement in system noise.

10.4.3   Active Balanced Output Circuits

Early active balanced output circuits used the approach shown in Fig. 10.24. The signal is buffered to provide one phase of the balanced output signal. This signal is then inverted with another op-amp to provide the other phase of the output signal. The outputs are taken through two resistors, each of which is half of the desired source impedance. Because the load is driven from the two outputs, the maximum output voltage is double that of an unbalanced output.

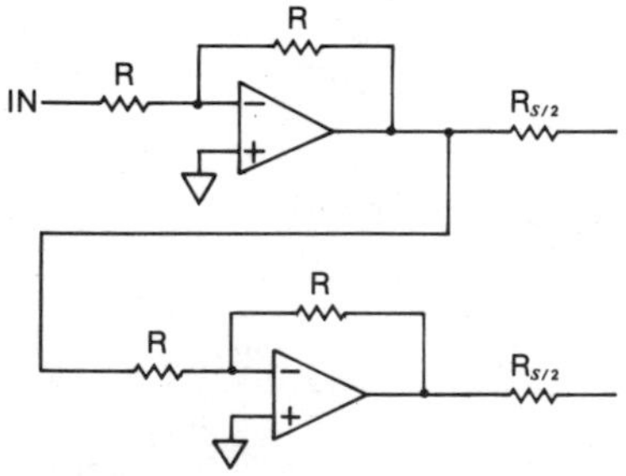

The circuit in Fig. 10.24 works reasonably well if the load is always balanced, but it suffers from two problems when the load is not balanced. If the positive output is short-circuited to ground by an unbalanced load connection, the first op-amp is likely to distort. This produces a distorted signal at the input to the other op-amp. Even if the circuit is arranged so that the second op-amp is grounded by an unbalanced load, the distorted output current probably will show up in the audio output from coupling through grounds or circuit-board traces. Equipment that uses this type of balanced output often provides a second set of output jacks that are wired to only one amplifier for unbalanced applications.

FIGURE 10.24 A basic active balanced output circuit. This configuration works well when driving a single balanced load. (Source: Courtesy *Broadcast Engineering* magazine)

The second problem with the circuit in Fig. 10.24 is that the output does not float. If any voltage difference (such as power-line hum) exists between the local ground and the ground at the device receiving the signal, the difference will appear as an addition to the signal. The only ground-noise rejection will be from the CMRR of the input stage at the receiving end.

The preferred output stage is the electronically balanced and floating design shown in Fig. 10.25. The circuit consists of two op-amps that are cross-coupled with positive and negative feedback. The output of each amplifier is dependent on the input signal and the signal present at the output of the other amplifier. These designs may

FIGURE 10.25 An electronically balanced and floating output circuit. A stage such as this will perform well even when driving unbalanced loads. (Source: Courtesy *Broadcast Engineering* magazine)

have gain or loss depending on the selection of resistor values. The output impedance is set via appropriate selection of resistor values. Some resistance is needed from the output terminal to ground, to keep the output voltage from floating to one of the power supply rails. Care must also be taken to properly compensate the devices, or else stability problems may result.

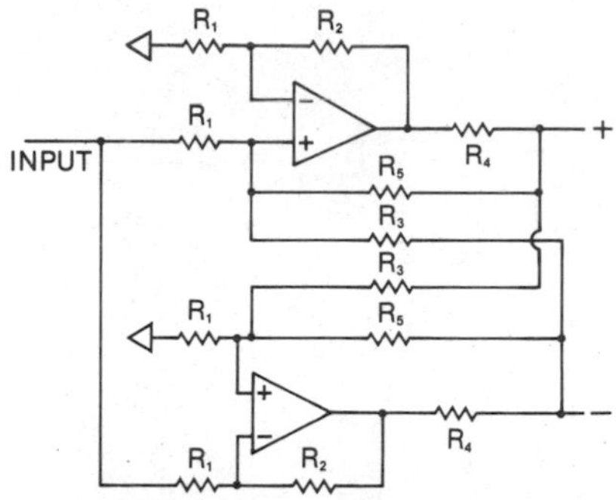

10.4.4 Interconnecting Audio Equipment

Susceptibility to RFI is a common problem with active balanced circuits. Strong radio signals often can be rectified by nonlinearities in the input operational amplifiers or transistors. Although wideband, low-distortion circuits are less prone to this problem, they are not immune to it. Therefore, signals outside the range of the active circuits must be filtered before they are inadvertently amplified or demodulated.

To reduce this problem, most manufacturers add small series resistors and capacitors to ground at the input terminals. Inductors also may be added, but they are susceptible to external magnetic fields. If package shielding is inadequate, the inductors may pick up as much garbage as they are supposed to filter out.

Successful equipment interfacing is a function of how well the I/O circuits were designed and the care taken by the equipment installer. Mixing of balanced and unbalanced inputs and outputs can be done if care is taken in planning where signals go and how wiring is to be done. Remember that the ground potential of one device may not be the same as the ground potential of another.

Wiring Guidelines. When it comes to designing and building a broadcast or production studio, theory and practice often go in divergent directions. A system design may look elegant on paper, yet turn out to be a nightmare when all the hardware is installed and turned on. Strict adherence to some basic wiring conventions will help prevent equipment interfacing problems.

The microphone is the starting point for most production work. Microphones are considered to be floating sources. Use only two-conductor shielded cable for all microphone wiring. Connect the microphone shield only to the microphone

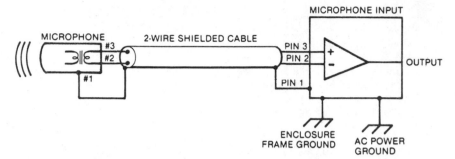

FIGURE 10.26 Interconnection arrangement for a floating source such as a microphone. Note that the microphone shield is bonded to the microphone case. It also should be bonded to the microphone preamplifier chassis. (Source: Courtesy *Broadcast Engineering* magazine)

case and input chassis (see Fig. 10.26). If a grounded-shell XLR connector is used, connect the shell pin to the ground (shield) wire. The shell ground provides complete RFI shielding at the point where two cables join. This allows users to serially connect several microphone cables and to maintain complete shielding. Be certain, however, that the XLR shells are not allowed to come in contact with any metallic structures or fixtures.

For audio equipment other than microphones, use only two-conductor shielded cable and tie the shield of each cable to the signal ground point at the respective source equipment chassis. Do not connect the other end of the shield. Dress the cable so that the shield connects to ground at only one point. Do not use the shield of an audio cable to provide safety (ac) grounding under any circumstances. Instead, build a separate ground system to tie the equipment together.

Equipment Grounding. It is a common misconception that the only way for an audio system to be free from hum and buzz is to secure a good earth ground. While this will certainly help, it is only one part of the picture. The term *grounding* when applied to audio often refers loosely to the interconnection of the individual chassis with the earth. It is better, however, to view grounding as an entire system. One ground system mistake can seriously impair the performance of an otherwise perfect arrangement.

Grounding schemes can range from simple to complex, but any system serves three primary purposes:

1. It provides for operator safety.
2. It protects electronic equipment from damage caused by transient disturbances.
3. It diverts stray radio-frequency energy away from sensitive audio equipment.

Most engineers view grounding mainly as a method to protect equipment from damage or malfunction. However, the most important element is operator safety. The 120-Vac line current that powers most audio equipment can be dangerous—even deadly—if improperly handled. Grounding of equipment and structures provides protection against wiring errors or faults that could endanger human life.

Many different approaches can be taken to designing a studio facility ground system. But some conventions should always be followed to ensure a low-resistance (and low-inductance) ground that will perform as required. Proper grounding is important whether or not the studio is located in an RF field.

Figure 10.27 shows the recommended grounding arrangement for a typical broadcast facility. The configuration is called a *star system* because of the way the various pieces of equipment are connected in a radial pattern around the central ground point. For larger systems, a "star of stars" system is used, as illustrated in Fig. 10.28.

Distribution Amplifiers.   It is often necessary in an audio system to connect a single source to several different loads. While this can be accomplished simply by tying the inputs in parallel, it is a poor solution. The parallel arrangement is cheap but leaves the audio system vulnerable to problems caused by changing levels, ground loops, and wiring faults. With a parallel hookup, short-circuiting one of the input lines will result in a complete loss of signal to all the other inputs.

Distribution amplifier (DA) systems have been developed to solve these problems. DAs are available in stand-alone and modular configurations. Stand-alone DAs are appropriate if you have only a small number of sources to distribute. Modular DAs are better suited to multisource distribution. Generally, if more than five channels must be distributed and they can all be fed conveniently from the same physical location, the modular approach is better.

DAs are available with either fixed or adjustable output levels. If all the devices being fed require approximately the same input level, then a fixed-output DA may be adequate. If, however, the expected levels vary, it might be worth the extra cost of individual output adjustments. Most DAs use active balanced inputs and outputs. The output stage may use one of two basic approaches: a resistive divider network or individual driver amplifiers. Figure 10.29 shows the resistive-divider concept, while Fig. 10.30 shows the individual amplifier arrangement. Each approach has benefits for users. Some engineers like the redundancy provided by having an amplifier for each output. Such a design prevents a single component failure from crashing the entire system. The cost for a multiple-output DA system, however, is higher than for a comparable resistive-divider circuit.

Audio Patch Bays.   Designing patch bays into an audio system can provide a number of benefits to the users. Without patch bays, interconnections between the devices in a system are fixed, permanent, and difficult to access. If troubleshooting or setup adjustments are required, it is usually necessary to access the connections on the equipment itself. With patch bays, if access to equipment inputs or outputs is required, it is readily available.

Most patch bays are configured with two horizontal rows of vertically aligned three-conductor tip, ring, sleeve (TRS) jacks. The jacks on the top row are connected to the equipment outputs or sources. The jacks on the bottom row are connected to the inputs or loads.

If the signal path changes frequently, patch cords are used to complete the circuit between the desired jacks in the patch bay. When the system's circuit path is relatively permanent, switching jacks can be used. These jacks incorporate TRS terminals which are an integral part of the contacts that mate with the patch cord plug. When a pair of jacks is *normaled* (connected) to each other, the signal is routed through the pair. When a patch cord is inserted in either of the two normaled jacks, the circuit path is broken and the signal is routed to the patch cord.

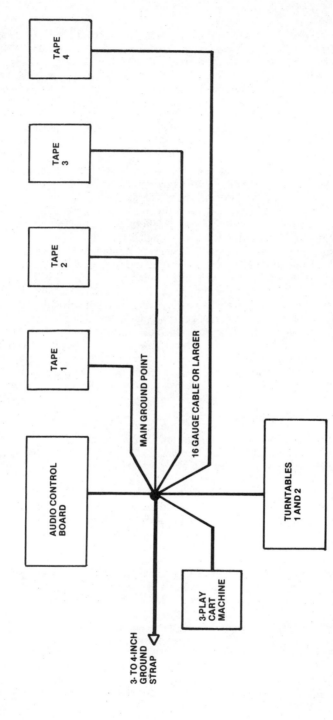

FIGURE 10.27 A typical grounding arrangement for a broadcast studio facility. The *main station ground point* is the reference from which all grounding is done at the facility. (Source: Courtesy *Broadcast Engineering* magazine)

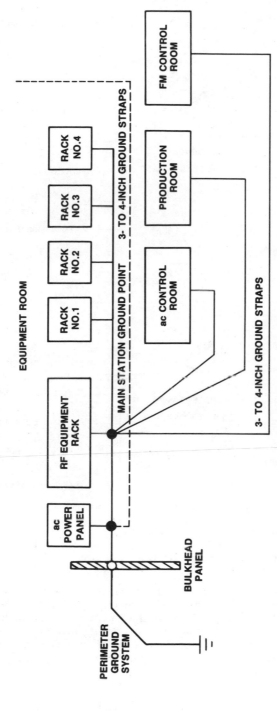

FIGURE 10.28 A typical grounding arrangement for individual equipment rooms at a broadcast facility. The ground strap from the station ground point establishes a *local ground point* in each room, to which all source and control equipment is bonded. (Source: Courtesy *Broadcast Engineering* magazine)

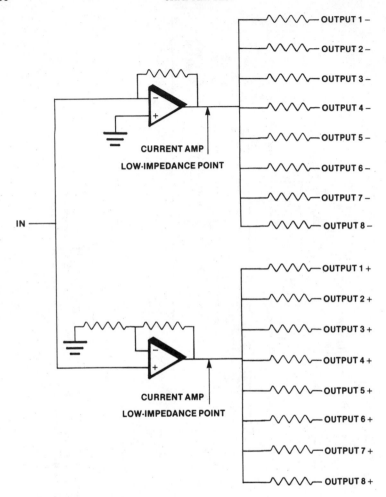

FIGURE 10.29 A distribution amplifier output stage using resistive dividers. This system relies on a single high-current driver stage to feed all outputs. (Source: Courtesy *Broadcast Engineering* magazine)

10.5 AUDIO CONTROL EQUIPMENT

Audio quality has never been more important than it is today for professional users. For radio stations, audio is their only product. The advent of stereo TV has led many television stations to make major improvements in their audio systems. Independent production facilities are likewise finding that customers and consumers expect a high-quality audio product. Some of the most significant changes in audio development can be found in mixing consoles and processing equipment.

10.5.1 Audio Mixing Consoles

The mixing console is the hub of any audio facility's program work. Audio *boards*, as they are called, come in basically two varieties: on-air and production

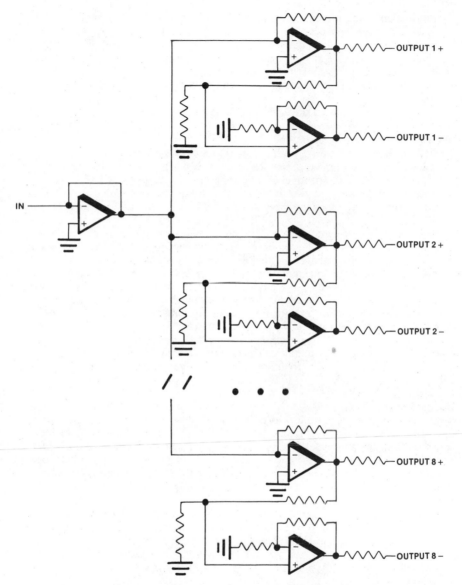

FIGURE 10.30 A distribution amplifier output stage using individual driver amplifiers. With this arrangement, the failure of one driver will not cause the entire system to fail. (Source: Courtesy *Broadcast Engineering* magazine)

boards. A console designed for use as an on-air board for a radio or TV station is characterized by relatively simple, straightforward design. The board will have limited equalization capabilities and few, if any, special-effects features. Machine control, status feedback, and monitoring provisions are paramount. The function of an on-air console is simply to get program material on the air, not to do anything fancy with it.

Production boards for broadcast or other applications are another breed. In the production room, the emphasis is on versatility and features. Production consoles usually have provisions for equalization on some or all input channels, special-effects send/return, and multiple-track tape recorder operation. A production board will take longer than an on-air board to set up, but it gives the user extensive capabilities to meet the creative needs of the facility. As the creative needs grow, so do the requirements for a complex and versatile audio production board.

An important aspect of any audio console design is the human element. A board that can perform every imaginable function is of limited value if it cannot be easily understood by operators.

Many manufacturers offer both production and on-air versions of their audio consoles. Some use common modules in each design and add extra features to tailor the console to the production environment. These options include equalizer, filter, compressor, studio monitor, effects send/return, and slate/talkback modules. The facility planner simply selects the types of module required for the installation.

Expandable configurations have become standard among many console manufacturers, allowing installation of a custom-designed system that meets a facility's exact requirements. Ultimately, the expansion limits are set by the mainframe size. Careful consideration must be given to selection of a mainframe that will meet future needs and lend itself to facility expansion. Mainframes are commonly available for 10 to 34 input positions, in a variety of steps. Large production consoles used for a major postproduction facility or network television production center are commonly available in 48 to 64 input positions.

The functions of the console are dictated by the individual input, output, monitor, and special-purpose modules. The channel elements vary in function and form from one manufacturer to another, but some generalizations apply to most units.

Microphone/Line Input Module. The input channel module for an on-air console is relatively simple compared to its production facility counterpart. Figure 10.31 shows the layout of a typical microphone/line input module. The module includes on/off pushbutton switches that control audio signal flow and operate a machine interface circuit. The interface may consist of a contact closure, low-voltage logic signal, or open-collector transistor output. Lamps in the buttons can be used to indicate signal flow status or external machine status in the case of a line amplifier module.

A cue button allows the audio source to be monitored without disturbing the attenuator setting. A number of options are available in the selection of an attenuator. User preference and field experience are the best guides. A pan pot provides the means to balance a stereo signal source or to position a mono signal in the desired stereo image, a feature typically used for announce or interview microphones.

Output routing switches on the input module allow selection of the program, audition, or utility buses. Channel input selectors permit two or more inputs to be used. Modules that have provisions for external machine control usually include logic that allows output commands to follow the input selector switches. Front panel or internal gain trim controls are sometimes provided to allow convenient adjustment of input levels.

Figure 10.32 shows a typical input module for production room applications. Although the configuration is different from that shown in Fig. 10.31, the module functions basically the same, with a couple of additional features. A multi-

FIGURE 10.31 The layout of a typical microphone or line-level input module for an on-air audio console. (Source: Courtesy *Broadcast Engineering* magazine)

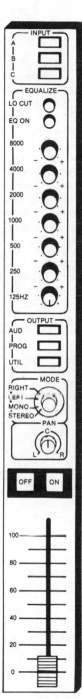

FIGURE 10.32 The layout of a microphone or line-level input module for a broadcast production console. (Source: Courtesy *Broadcast Engineering* magazine)

band equalizer has been added to allow the operator to tailor the sound to the production requirements. In this example, a seven-band equalizer is utilized. Other designs offer two- or three-band equalizers and separate shelving or notch filters. An equalization in/out switch allows the production setup to be bypassed when not in use. An input-mode control enables the operator to select stereo, mono, left, or right signals for mixing. Modules can also be equipped with provisions for cue, solo, effects send/return, multitrack assignment, and external machine control.

Output Module. The output module for an on-air or production console is fairly straightforward. A master gain control may be provided for each of the three common buses: program, audition, and utility. Individual gain-trim posts may also be available on the modules for precise adjustment of channel balance and output level.

Because the output of a mixing console usually must be distributed to several pieces of equipment external to the board, an internal distribution amplifier is sometimes provided. Complex distribution requirements, however, dictate the use of an external DA.

Monitor Module. The ability to monitor signals from various sources is of prime importance in both on-air and production situations. Figure 10.33 shows a control-room monitor module that offers the features desired in an on-air operating environment. Volume controls are provided for the headphones and the control-room monitor, cue and talkback speakers. A bank of selector switches allows the user to hear program, audition, or utility channels as well as any of several external signal sources.

The headphone feed may follow the monitor speaker signal or be independently selectable. Provisions for switching the headphone input from stereo to left, right, or mono mix may also be useful to the operator.

Meter Display Options. A number of methods are available to monitor program levels on an audio console. These range from the classic analog VU meter to solid-state designs of various types. Selection of the metering is usually determined by the personal preferences of the users. Regardless of the type of metering chosen, two basic display characteristics are used: VU ballistics and peak program meter (PPM) ballistics. The VU meter is more familiar to operators in the United States while the PPM is more familiar to users in Europe.

The VU meter provides an easy-to-read display of average program energy. It does not, however, provide accurate data on the presence of short-duration program peaks. The PPM is an increasingly popular method of monitoring audio levels for professional applications. The dynamic characteristics of a PPM are radically different from those of a VU meter. The PPM is designed to follow and display the peak energy of the audio waveform. Probably the best method of monitoring the output level of an audio console is to use both PPM and VU meters. With this arrangement, the operator can regularly observe the VU meter for program level setting and use the PPM to check for high-level peaks that are not displayed by the VU meter. The increased use of solid-state bar graph displays for program level metering permits combined or switchable characteristics. A simpler solution involves the use of a standard VU meter and a peak indicator lamp or LED built into the meter that is set to trip at a given reference point to display short-duration program peaks.

10.5.2 Automating Audio Consoles

Through digital control techniques, consoles now can provide more features in a smaller amount of space. Even more important is the enhanced operational control available through *assignable controls*. To better understand how assignable controls work, you must first understand how a large console is typically used.

Console operations can be broken down into several basic tasks: input selection, equalization adjustments, processing adjustments, output selection, level setting, and monitoring. Historically, each of these tasks has required the use of separate knobs, switches, and faders. However, digital technology has opened the door to a different approach.

A console with assignable controls allows one or more sets of controls to be assigned to many functions. With digital storage of commands and perhaps some digital processing, a fewer number of knobs can perform a wide variety of tasks. The advantages are reduced console size and easier operation.

Although large consoles are impressive-looking, realistically an engineer can accurately set the controls of only one channel or module at a time. A recent study showed that on a conventional console equipped with 40 modules, an engineer uses only 2.5 percent of the controls at one time. The philosophy of assignable controls takes advantage of this fact. Figure 10.34 shows an equalization module that can be assigned to any input module on its parent console.

The disadvantage of assignability is that it makes visual feedback a sequential accessing routine through the use of shared lights and displays, robbing the audio engineer of much vital, immediately available information.

The degree to which a console design implements assignability is a matter of balance. A successful design incorporates the correct trade-offs for all parameters, including which functions are assignable and which have dedicated controls and displays.

As assignability functions are incorpo-

FIGURE 10.33 The layout of a typical monitor speaker and headphone control module for either an on-air or a production-room audio console. (*Broadcast Engineering*)

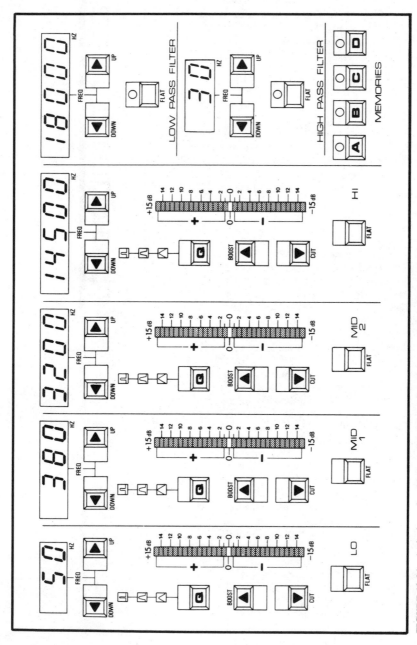

FIGURE 10.34 A single assignable equalization module for an audio console. This design features four 16-frequency parametric EQ sections, plus sweep high- and low-pass filters. (Source: Courtesy *Broadcast Engineering* magazine)

rated into a console design, it becomes easier to take the next step and provide programmable features. Automated or programmed actions can be performed by a console operating under the control of a computer. Such an approach allows the operator to set up the equalization controls, presets, input channel switching, and even level settings; store them on a computer disk; and recall the entire setup with the touch of a button.

10.5.3 Remote Audio Processing

Engineers usually think of consoles in terms of the front panel or control surface with an array of knobs, switches, lights, and displays. Historically, each of these devices was linked mechanically to the circuitry located beneath the console. The knob was simply an extension of the potentiometer or switch to which it was connected. Advances in voltage-controlled amplifiers and components and in digital technology have made it possible to physically separate the knobs and buttons from the circuit elements they control. Now the knobs and switches can communicate electronically, rather than by more direct mechanical means. The result is an audio control surface with no audio passing through the control knobs or switches.

This approach is relatively new to audio engineers, but not to video engineers. Video switchers and similar devices seldom route the video signal through the control panel. Instead, the video remains within a rack-mounted assembly that is directed by a control surface located elsewhere.

Remote processing techniques offer many advantages to console designers. First is a significant reduction in the size of the actual work surface in the studio. Second is that the design lends itself to complete or partial automation by a supervisory computer. Such a system also opens the possibility for parallel-control operation in which more than one operator can have access to the system.

The changing demands of audio production are driving console manufacturers to devise new ways of mixing and modifying audio. In the recording business, an important shift is occurring in the origin of recorded sound. A large portion of the sound recorded in studios today does not originate in the acoustical domain. Instead, these sounds come from electronic instruments, which can place different requirements from acoustical instruments on the mixing system.

Audio consoles must also meet new performance and operational demands. Recording and audio-for-video work today requires more inputs, additional control over each audio source with equalization and audio processing, and sometimes automation. As facilities search for ways to produce a unique product, automation will provide users with the tools they need.

CHAPTER 11
AM AND FM TRANSMISSION SYSTEMS

Jerry Whitaker
Editorial Director, Broadcast Engineering *Magazine*

11.1 INTRODUCTION

Radio broadcasting has been around for a long time. Amplitude modulation (AM) was the first modulation system that permitted voice communications to take

---

Portions of this chapter were adapted from: (1) Donald Markley and James Carpenter, "Broadcast Transmission Technology," in *Audio Engineering Handbook*, edited by Blair K. Benson, McGraw-Hill, New York, 1988, Chap. 5. (2) Robert Perelman and Thomas Sullivan, Andrew Corporation, "Selecting Flexible Coaxial Cable," *Broadcast Engineering* magazine (May 1988). (3) Dominic Bordonaro, WFAA-AM and WFTQ-FM, Worcester, MA, "Reducing IPM in AM Transmitters," *Broadcast Engineering* magazine (May 1988).

place. In this scheme, the magnitude of the carrier wave is varied in accordance with the amplitude and frequency of the applied audio input. The magnitude of the incoming signal determines the magnitude of the carrier wave, while the frequency of the modulating signal determines the rate at which the carrier wave is varied. This simple modulation system was predominant through the 1920s and 1930s. Frequency modulation (FM) came into regular broadcast service during the 1940s.

These two basic approaches to modulating a carrier have served the broadcast industry well for many decades. While the basic modulating schemes still exist today, numerous enhancements have been made, including stereo operation and subcarrier programming. New technology has given radio broadcasters new ways to serve listeners.

Technology has also changed the rules by which the AM and FM broadcasting systems developed. AM radio, as a technical system, offered limited audio fidelity but provided design engineers with a system that allowed uncomplicated transmitters and simple, inexpensive receivers. FM radio, on the other hand, offered excellent audio fidelity but required a complex and unstable transmitter (in the early days) and complex, expensive receivers. It is, therefore, no wonder that AM radio flourished and FM remained relatively static for at least 20 years.

Advancements in transmitter and receiver technology during the 1960s changed the picture dramatically. Designers found ways of producing FM transmission equipment that was reliable and capable of high power output, for a reasonable price. Receiver manufacturers then began producing compact, high-quality receivers at prices consumers could afford. These developments began a significant change in listening habits that is still unfolding today. As late as the mid-1970s, AM radio stations as a whole commanded the majority of all listeners in the United States. By 1980, however, the tables had turned. FM is now king, and AM is struggling to keep up with its former stepchild.

11.1.1 Radio Transmission Standards

AM radio stations operate on 10-kHz channels spaced evenly from 540 to 1600 kHz. Various classes of stations have been established by the Federal Communications Commission (FCC) and agencies in other countries to allocate the available spectrum to given regions and communities. In the United States, the basic classes are clear, regional, and local. Current practice uses the CCIR (international) designations as class A, B, and C, respectively. Operating power levels range from 50 kW for a clear channel station to as little as 250 W for a local station.

FM radio stations operate on 200-kHz channels spaced evenly from 88.1 to 107.9 MHz. In the United States, channels below 92.1 MHz are reserved for non-commercial, educational stations. The FCC has established three classifications for FM stations operating east of the Mississippi River and four classifications for stations west of the Mississippi. Power levels range from a high of 100 kW *effective radiated power* (ERP) to 3 kW or less for a lower classification.

AM Stereo. Stations choosing to do so may operate in stereo using one of two incompatible formats. Known in the broadcast industry as the *Kahn system* and *C-QUAM system*,[1] both formats are fully compatible with existing monophonic

---

1. C-QUAM is a registered trademark of Motorola, Inc.

radios. In order to receive a stereo AM broadcast, consumers must purchase a new stereo radio. Depending on the design of the radio, however, the consumer may or may not be able to receive in stereo two stereo AM stations in the same area transmitting on different systems. The receiver, instead, usually will receive one system in stereo and the other station (using the competing system) in mono.

This unusual situation is the result of a decision by the FCC to let the "marketplace" decide which system, Kahn or C-QUAM, would be the de-facto broadcast standard. Unfortunately, the marketplace has not, as of this writing, completely decided which way to go. This confusion has slowed the introduction of stereo AM broadcasting and the purchase by consumers of new, high-fidelity stereo AM receivers.

The two systems have fundamental differences. The Kahn system independently modulates the sideband signals of the AM carrier, one for the right channel and the other for the left channel, to provide the stereo effect. The C-QUAM system transmits the stereo *sum signal* (in other words, the monophonic signal) in the usual manner and places the stereo *difference signal* on a phase-modulate subchannel. Decoder circuits in the receiver reconstruct the stereo signals. (Chapter 12 provides a technical description of both the Kahn and C-QUAM systems.)

FM Stereo.   Stereo broadcasting is used almost universally in FM radio today. Introduced in the mid-1960s, stereo has contributed in large part to the success of FM radio. The left and right sum (monophonic) information is transmitted as a standard frequency-modulated signal. Filters restrict this *main channel* signal to a maximum of about 17 kHz. A pilot signal is transmitted at low amplitude at 19 kHz to enable decoding at the receiver. The left and right difference signal is transmitted as an amplitude-modulated subcarrier that frequency modulates the main FM carrier. The center frequency of the subcarrier is 38 kHz. Decoder circuits in the FM receiver matrix the sum and difference signals to reproduce the left and right audio channels. (See Chapter 12 for a detailed discussion of FM stereo.)

11.2   AMPLITUDE MODULATION SYSTEMS

In the simplest form of amplitude modulation, an analog carrier signal is controlled by an analog modulating signal. The desired result is an RF waveform whose amplitude is varied by the magnitude of the audio signal, and at a rate equal to the frequency of the audio signal. The resulting waveform consists of a carrier wave plus two additional signals: an upper-sideband signal, which is equal in frequency to the carrier plus the frequency of the modulating signal, and a lower-sideband signal, which is equal in frequency to the carrier minus the frequency of the modulating signal. The bandwidth of an AM transmission, therefore, is determined by the modulating frequency. The bandwidth required for full fidelity reproduction in a receiver is equal to twice the applied modulating frequency.

The magnitude of the upper sideband and lower sideband will not normally exceed 50 percent of the carrier amplitude during modulation. This results in an upper-sideband power of one-fourth of the carrier power. The same power exists in the lower sideband. As a result, up to one-half of the actual carrier power appears additionally in the sum of the sidebands of the modulated signal. A repre-

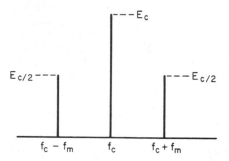

FIGURE 11.1 Frequency-domain representation of an amplitude modulated signal at 100 percent modulation (E_c = carrier power, F_c = frequency of the carrier, and F_m = frequency of the modulating signal). *(Source: K. Blair Benson,* Audio Engineering Handbook, *McGraw-Hill, New York, 1988)*

sentation of the AM carrier and its sidebands is shown in Fig. 11.1. The actual occupied bandwidth, assuming pure sinusoidal modulating signals and no distortion during the modulation process, is equal to twice the frequency of the modulating signal.

Full (100 percent) modulation occurs when the peak value of the modulated envelope is twice the value of the unmodulated carrier, and the minimum value of the envelope is 0. The envelope of a modulated AM signal in the time domain is shown in Fig. 11.2.

When modulation exceeds 100 percent on the negative swing of the carrier, spurious signals are emitted. It is possible to modulate an AM carrier asymmetrically—that is, to restrict modulation in the negative direction to 100 percent but to allow modulation in the positive direction to exceed 100 percent without a significant loss of fidelity. In fact, many modulating signals normally exhibit asymmetry, most notably speech waveforms.

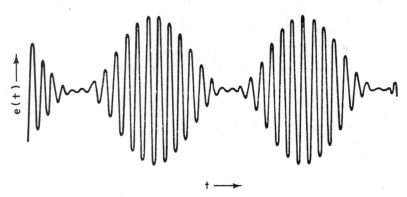

FIGURE 11.2 Time-domain representation of an amplitude-modulated signal. Modulation at 100 percent is defined as the point at which the peak of the waveform reaches twice the carrier level and the minimum point of the waveform is 0. *(Source: K. Blair Benson,* Audio Engineering Handbook, *McGraw-Hill, New York, 1988)*

11.2.1 High-Level AM Modulation

High-level anode modulation is the oldest and simplest way of generating a high-power AM signal. In this system, the modulating signal is amplified and combined with the dc supply source to the anode of the final RF amplifier stage. The RF amplifier is normally operated class C. The final stage of the modulator usually consists of a pair of tubes operating class B. A basic modulator of this type is shown in Fig. 11.3.

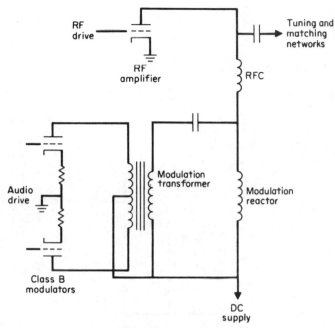

FIGURE 11.3 Simplified diagram of a high-level amplitude-modulated amplifier. *(Source: K. Blair Benson, Audio Engineering Handbook, McGraw-Hill, New York, 1988)*

The RF signal is normally generated in a low-level transistorized oscillator. It is then amplified by one or more solid-state or vacuum-tube stages to provide final RF drive at the appropriate frequency to the grid of the final class C amplifier. The audio input is applied to an intermediate power amplifier (usually solid state) and used to drive two class B (or class AB) push-pull output stages. The final amplifiers provide the necessary modulating power to drive the final RF stage. For 100 percent modulation, this modulating power is 50 percent of the actual carrier power.

The modulation transformer shown in Fig. 11.3 does not usually carry the dc supply current for the final RF amplifier. The modulation reactor and capacitor shown provide a means to combine the audio signal voltage from the modulator with the dc supply to the final RF amplifier. This arrangement eliminates the necessity of having dc current flow through the secondary of the modulation transformer, which would result in magnetic losses and saturation effects. In some newer transmitter designs, the modulation reactor has been eliminated from the system, thanks to improvements in transformer technology.

The RF amplifier normally operates class C with grid current drawn during positive peaks of the cycle. Typical stage efficiency is 75 to 83 percent.

This type of system was popular in AM broadcasting for many years, primarily because of its simplicity. The primary drawback was low overall system efficiency. The class B modulator tubes cannot operate with greater than 50 percent efficiency. Still, with inexpensive electricity, this was not considered to be a significant problem. As energy costs increased, however, more efficient methods of generating high-power AM signals were developed. Increased efficiency normally came at the expense of added technical complexity.

11.2.2 Pulse-Width Modulation

Pulse-width (also known as pulse-duration) modulation is one of the most popular systems developed for modern vacuum-tube AM transmitters. Figure 11.4 shows a scheme for pulse-width modulation identified as PDM (as patented by the Harris Corporation). The PDM system works by utilizing a square wave switching system, illustrated in Fig. 11.5.

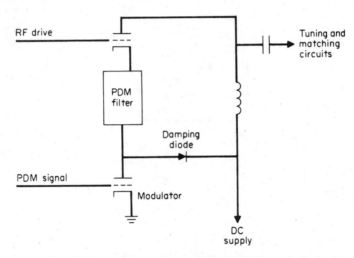

FIGURE 11.4 The pulse-duration modulation (PDM) method of pulse-width modulation patented by the Harris Corporation. *(Source: K. Blair Benson,* Audio Engineering Handbook, *McGraw-Hill, New York, 1988)*

The PDM process begins with a signal generator (see Fig. 11.6). A 75-kHz sine wave is produced by an oscillator and used to drive a square wave generator, resulting in a simple 75-kHz square wave. The square wave is then integrated, resulting in a triangular waveform, which is mixed with the input audio in a summing circuit. The result is a triangular waveform that effectively rides on the incoming audio (as shown in Fig. 11.5). This composite signal is then applied to a threshold amplifier, which functions as a switch that is turned on whenever the value of the input signal exceeds a certain limit. The result is a string of pulses in which the width of pulse is proportional to the period of time the triangular waveform exceeds the threshold. The pulse output is applied to an amplifier to obtain the necessary power to drive subsequent stages. A filter eliminates whatever transients exist after the switching process is complete.

The PDM scheme is, in effect, a digital modulation system with the audio information being sampled at a 75-kHz rate. The width of the pulses contain all the audio information. The pulse-width-modulated signal is applied to a *switch* or *modulator tube*. The tube is simply turned on, to a fully saturated state, or off in accordance with the instantaneous value of the pulse. When the pulse goes pos-

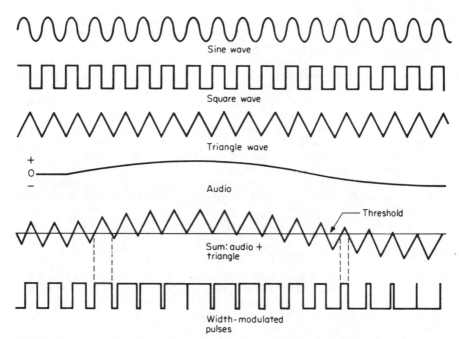

FIGURE 11.5 The principal waveforms of the PDM system. *(Source: K. Blair Benson,* Audio Engineering Handbook, *McGraw-Hill, New York, 1988)*

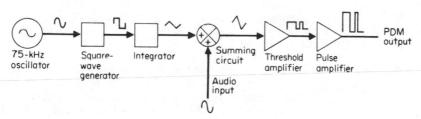

FIGURE 11.6 Block diagram of a PDM waveform generator. *(Source: K. Blair Benson,* Audio Engineering Handbook, *McGraw-Hill, New York, 1988)*

itive, the modulator tube is turned on and the voltage across the tube drops to a minimum. When the pulse returns to its minimum value, the modulator tube turns off.

This PDM signal becomes the power supply to the final RF amplifier tube. When the modulator is switched on, the final amplifier will have current flow and RF will be generated. When the switch or modulator tube goes off, the final amplifier current will cease. This system, in effect, causes the final amplifier to operate in a highly efficient class D switching mode. A dc offset voltage to the summing amplifier of Fig. 11.6 is used to set the carrier (no modulation) level of the transmitter.

A high degree of third-harmonic energy will exist at the output of the final amplifier because of the switching-mode operation. This energy is eliminated by a third-harmonic trap. The result is a stable amplifier system that normally operates in excess of 90 percent efficiency. The power consumed by the modulator and its driver is usually a fraction of a full class B amplifier stage.

The *damping diode* shown in Fig. 11.4 is used to prevent potentially damaging transient overvoltages during the switching process. When the switching tube turns off the supply current during a period when the final amplifier is conducting, the high current through the inductors contained in the PDM filters could cause a large transient voltage to be generated. The energy in the PDM filter is returned to the power supply by the damping diode. If no alternative route were established, the energy would return by arcing through the modulator tube itself.

The pulse-width-modulation system makes it possible to completely eliminate audio frequency transformers in the transmitter. The result is wide frequency response and low distortion. It should be noted that variations on this amplifier and modulation scheme have been used by other manufacturers for both standard broadcast and short-wave service.

11.2.3 Power Grid Vacuum Tubes

Power grid vacuum tubes have been the mainstay of AM transmitters since the beginning of radio. The need for power tubes capable of higher output and greater reliability is being met with new processes and materials. Users today are asking for systems that incorporate solid-state components in low- and medium-power stages and vacuum-tube components in high-power stages of broadcast equipment. Each technology has its place, and each has its strengths and weaknesses.

Triodes. Power triodes are commonly used in AM transmitters today. Most tubes are cylindrically symmetrical. The filament or cathode structure, the grid, and the anode are all cylindrical in shape and are mounted with the axis of each cylinder along the centerline of the tube. Triodes have three internal elements: the cathode, control grid, and plate.

Tetrode. The tetrode is a four-element tube with two grids. The control grid serves the same purpose as the grid in a triode, while a second (screen) grid with the same number of vertical elements (bars) as the control grid is mounted between the control grid and the anode. The grid bars of the screen grid are mounted directly behind the control grid bars, as observed from the cathode surface, and serve as a shield or screen between the input circuit and the output circuit of the tetrode. The principal advantages of a tetrode over a triode include lower internal plate-to-grid capacitance and lower driving power requirements (in other words, higher gain).

Inside a Vacuum Tube. Each type of power grid tube is unique insofar as its operating characteristics are concerned. The basic physical construction, however, is common to most devices. An example will help illustrate what is inside a vacuum tube.

The plate of a tetrode power tube resembles a copper cup with the upper half of a plate contact ring welded to the mouth and cooling fins silver soldered or welded to the outside of the assembly (see Fig. 11.7). The lower half of the anode contact ring is bonded to the base ceramic spacer. At the time of assembly, the two halves of the ring are welded together to form a complete unit.

The screen grid consists of a number of vertical supports fastened to a metal

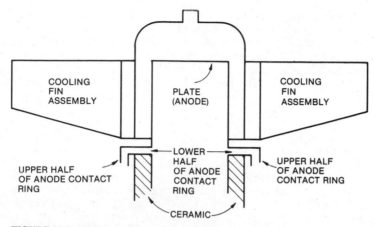

FIGURE 11.7   A cutaway view of the anode structure of a tetrode RF power amplifier tube of the type used in AM and FM transmitters. (Source: Courtesy *Broadcast Engineering* magazine)

base cone. The lower end of the cone is bonded to the screen contact ring, as shown in Fig. 11.8. The control grid and cathode assembly are also cylindrical and concentric. The control grid is built in a manner similar to the screen grid but slightly smaller in height and diameter.

The thoriated-tungsten filament (or cathode) commonly used in broadcast power tubes is created in a high-temperature atmosphere that forms a deep layer of ditungsten carbide on the surface of the elements. The end of a tube's useful

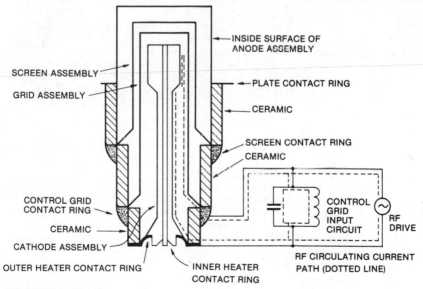

FIGURE 11.8   The internal arrangement of the anode, screen, control grid, and cathode assemblies of a tetrode power tube. Also shown is a simplified representation of the grid input circuit. (Source: Courtesy *Broadcast Engineering* magazine)

life (for this type of filament) occurs when most of the carbon has evaporated or has combined with residual gas, depleting the carbide surface layer.

The actual construction of the filament assembly can have an effect on the performance of the tube, and the performance of the transmission system as a whole. Filaments built in a basket-weave mesh arrangement offer lower distortion in critical high-power AM modulation circuits. The mesh filament is also more rugged and resistant to physical shock than simpler geometries.

Neutralization. An RF power amplifier must be properly neutralized to provide acceptable performance in broadcast applications. The means to accomplish this end can vary considerably from one transmitter design to another.

An RF amplifier is neutralized when two conditions are met. First, the interelectrode capacitance between the input and output circuits must be canceled out. Second, the inductance of the screen grid and cathode assemblies must be completely canceled. The cancellation of these common forms of coupling between the input and output circuits of intermediate-power amplifier (IPA) and PA tubes prevents self-oscillation of the system.

For operation at AM frequencies, neutralization typically employs a capacitance bridge circuit to balance out the RF feedback caused by residual plate-to-grid capacitance.

11.2.4 Solid-State Transmitters

Solid-state transmitters make use of various schemes employing pulse-width-modulation techniques. One system is shown in simplified form in Fig. 11.9a. This method of generating a modulated RF signal employs a solid-state switch to swing back and forth between two voltage levels at the carrier frequency. The result is a square-wave signal that is filtered to eliminate all components except the fundamental frequency itself. A push-pull version of the circuit is shown in Fig. 11.9b. These examples would be adequate if you were interested only in generating a simple sinusoid at the carrier frequency.

Figure 11.10 illustrates a class D switching system utilizing bipolar transistors that permits the generation of a modulated carrier. Modern solid-state transmitters use field-effect-transistor (FET) power devices as RF amplifiers. Basically, the dc supply to the RF amplifier stages is switched on and off by an electronic switch in series with a filter. Operating in the class D mode results in a composite signal similar to that generated by the vacuum-tube class D amplifier in the PDM system (at much higher power). The individual solid-state amplifiers are typically combined through a toroidal filter. The result is a group of low-powered amplifiers operating in parallel and combined to generate the required energy.

Transmitters up to 50 kW have been constructed using this design philosophy. They have proved to be reliable and efficient. The parallel design provides users with automatic redundancy and has ushered in the era of "graceful degradation" failures. In almost all solid-state transmitters, the failure of a single power device or stage will reduce the overall power output or modulation capability of the system but will not take the transmitter off the air. This feature allows repair of defective modules or components at times convenient to the service engineer. The negative effects on system performance are usually negligible.

The audio performance of current technology solid-state transmitters is usually better than vacuum-tube designs of comparable power levels. Frequency re-

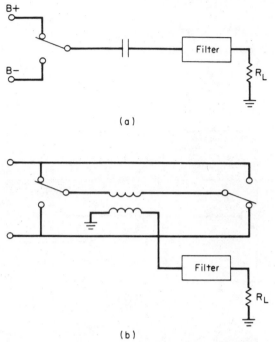

FIGURE 11.9 The basic concept behind class D radio frequency amplifiers for AM service. (*a*) A single-ended system. (*b*) A push-pull system. *(Source: K. Blair Benson, Audio Engineering Handbook, McGraw-Hill, New York, 1988)*

sponse is typically within ±1 dB from 50 Hz to 10 kHz. Distortion is usually less than 1 percent at 95 percent modulation.

11.3 FREQUENCY MODULATION SYSTEMS

Frequency modulation is a technique whereby the phase angle or phase shift of a carrier is varied by an applied modulating signal. The *magnitude* of frequency change of the carrier is a direct function of the *magnitude* of the modulating signal. The *rate* at which the frequency of the carrier is changed is a direct function of the *frequency* of the modulating signal. In FM modulation, multiple pairs of sidebands are produced. The actual number of sidebands that make up the modulated wave is determined by the *modulation index* (MI) of the system. The modulation index is a function of the frequency deviation of the system and the applied modulating signal:

$$\text{MI} = \frac{\text{Frequency deviation}}{\text{Modulating frequency}} \qquad (11.1)$$

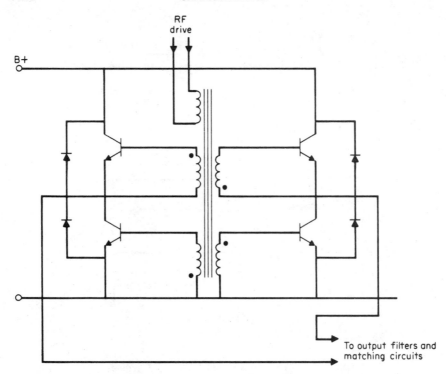

FIGURE 11.10 Bipolar transistor RF stage operating in a class D switching mode for AM service. *(Source: K. Blair Benson, Audio Engineering Handbook, McGraw-Hill, New York, 1988)*

The higher the MI, the more sidebands produced. It follows that the higher the modulating frequency for a given deviation, the fewer number of sidebands produced, but the greater their spacing.

To determine the frequency spectrum of a transmitted FM waveform, it is necessary to compute a Fourier series or Fourier expansion to show the actual signal components involved. This work is difficult for a waveform of this type, as the integrals that must be performed in the Fourier expansion or Fourier series are not easily solved. The actual result is that the integral produces a particular class of solution that is identified as the *Bessel function*.

Supporting mathematics will show that an FM signal using the modulation indices that occur in a broadcast system will have a multitude of sidebands. From the purist point of view, *all* sidebands would have to be transmitted, received, and demodulated in order to accurately reconstruct the modulating signal. In practice, however, the 200-kHz channel bandwidth permitted FM broadcast stations is sufficient to reconstruct the modulating signal with no discernible loss in fidelity.

The deviation permitted an FM broadcast service is determined by the FCC and similar organizations in other countries. For commercial FM broadcasting, 100 percent modulation is defined as ±75 kHz.

The power in an FM system is constant throughout the modulation process. While the output power is increased in the amplitude modulation system by the modulation process, the FM system simply distributes the power throughout the various frequency components that are produced by modulation. While being modulated, a wideband FM system does not have a high amount of energy present in the carrier. Most of the energy will be found in the sum of the side-bands.

The constant-amplitude characteristic of FM greatly assists in capitalizing on the low-noise advantage of FM reception. Upon being received and amplified, the FM signal normally is clipped to eliminate all amplitude variations beyond a certain threshold. This removes all noise picked up by the receiver as a result of man-made or atmospheric signals. It is not possible for these random noise sources to change the frequency of the desired signal; they can only affect its amplitude. The use of hard limiting in the receiver will strip off such interference.

11.3.1 Preemphasis and Deemphasis

The FM broadcast-reception system offers significantly better noise rejection characteristics than AM radio. However, FM noise rejection is more favorable at low modulating frequencies than at high frequencies because of the reduction in the number of sidebands at higher frequencies. To offset this problem, the audio input to FM transmitters is *preemphasized* to increase the amplitude of higher-frequency signal components in normal program material. FM receivers utilize complementary *deemphasis* to produce flat overall system frequency response.

FM broadcasting uses a 75-μs preemphasis curve, meaning that the time con-stant of the resistance-inductance (RL) or resistance-capacitance (RC) circuit used to provide the boost of high frequencies is 75 μs. Other values of pre-emphasis have been used in limited amounts in other FM broadcast systems.

11.3.2 Modulation Circuits

Early FM transmitters used *reactance modulators* that operated at a low fre-quency. The output of the modulator was then multiplied to reach the desired output frequency. This approach was acceptable for monaural FM transmission but not for modern stereo systems or other applications that utilize subcarriers on the FM broadcast signal. Modern FM systems all utilize what is referred to as *direct modulation*. That is, the frequency modulation occurs in a modulated os-cillator that operates on a center frequency equal to the desired transmitter out-put frequency. In stereo broadcast systems, a composite FM signal is applied to the FM modulator. The basic parameters of this composite signal are shown in Fig. 11.11.

Various techniques have been developed to generate the direct-FM signal. One of the most popular uses a variable-capacity diode as the reactive element in

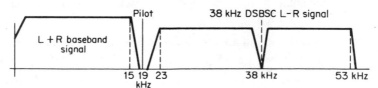

FIGURE 11.11 Composite stereo FM signal. (*Source: K. Blair Benson,* Audio Engineering Handbook, *McGraw-Hill, New York, 1988*)

the oscillator. The modulating signal is applied to the diode, which causes the capacitance of the device to vary as a function of the magnitude of the modulating signal. Variations in the capacitance cause the frequency of the oscillator to vary. Again, the magnitude of the frequency shift is proportional to the amplitude of the modulating signal, and the rate of frequency shift is equal to the frequency of the modulating signal.

The direct-FM modulator is one element of an FM transmitter exciter, which generates the composite FM waveform. A block diagram of a complete FM exciter is shown in Fig. 11.12. Audio inputs of various types (stereo left and right signals, plus subcarrier programming, if used) are buffered, filtered, and preemphasized before being summed to feed the modulated oscillator. It should be noted that the oscillator is not normally coupled directly to a crystal but is a free-running oscillator adjusted as near as practical to the carrier frequency of the transmitter. The final operating frequency is carefully maintained by an automatic frequency control system employing a phase-locked loop (PLL) tied to a reference crystal oscillator or frequency synthesizer. The operation of the PLL is identical to that used for a TV transmitter exciter (see Sec. 8.3.5).

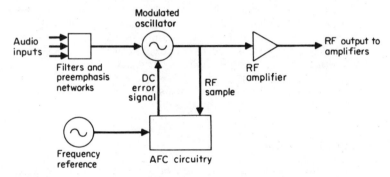

FIGURE 11.12 Block diagram of an FM exciter. (*Source: K. Blair Benson,* Audio Engineering Handbook, *McGraw-Hill, New York, 1988*)

A solid-state class C amplifier follows the modulated oscillator and raises the operating power of the FM signal from 20 to 30 W. One or more subsequent amplifiers in the transmitter raise the signal power to several hundred watts for application to the final power amplifier (PA) stage. Nearly all current high-power FM transmitters utilize solid-state amplifiers up to the final RF stage, which is generally a vacuum tube for operating powers of 5 kW and above. All stages operate in the class C mode. In contrast to AM systems, FM power amplifiers can operate class C because no information is lost from the frequency-modulated signal because of amplitude changes. As mentioned previously, FM is a constant-power system.

Auxiliary Services.   Modern FM broadcast stations are capable of not only broadcasting stereo programming, but one or more subsidiary channels as well. These signals, referred to by the FCC as *Subsidiary Communications Authorization* (SCA) services, are used for the transmission of stock market data, background music, control signals, and other information not normally part of the station's main programming. These services do not provide the same range of coverage or audio fidelity as the main stereo program; however, they perform a public service and can represent a valuable source of income for the broadcaster.

SCA systems provide efficient use of the available spectrum. The most common subcarrier frequency is 67 kHz, although higher subcarrier frequencies have been utilized. Stations that operate subcarrier systems are permitted by the FCC to exceed (by a small amount) the maximum 75-kHz deviation limit, under certain conditions. The subcarriers utilize low modulation levels, and the energy produced is maintained essentially within the 200-kHz bandwidth limitation of FM channel radiation.

11.3.3 FM Power Amplifiers

Nearly all high-power FM transmitters manufactured today employ cavity designs. The ¼-wavelength cavity is the most common. The design is simple and straightforward. A number of variations can be found in different transmitters, but the underlying theory of operation is the same.

The goal of any cavity amplifier is to simulate a resonant tank circuit at the operating frequency and provide a means to couple the energy in the cavity to the transmission line. Because of the operating frequencies involved (88 to 108 MHz), the elements of the "tank" take on unfamiliar forms.

A typical ¼-wave cavity is shown in Fig. 11.13. The plate of the tube connects directly to the inner section (tube) of the plate-blocking capacitor. The blocking capacitor can be formed in one of several ways. In at least one design, it is made by wrapping the outside surface of the inner tube conductor with multiple layers

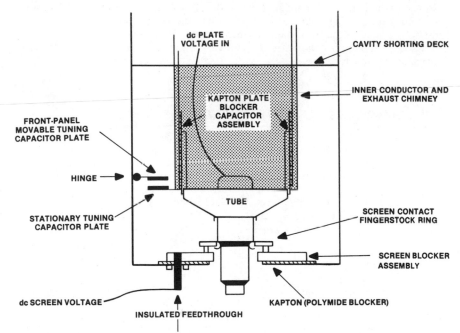

FIGURE 11.13 The layout of a common type of ¼-wave PA cavity for FM broadcast service. (Source: Courtesy *Broadcast Engineering* magazine)

of 8-in-wide and 0.005-in-thick polymide (Kapton) film. The exhaust chimney–inner conductor forms the other element of the blocking capacitor. The cavity walls form the outer conductor of the ¼-wave transmission line circuit. The dc plate voltage is applied to the PA tube by a cable routed inside the exhaust chimney and inner tube conductor.

In the design shown in Fig. 11.13, the screen-contact fingerstock ring mounts on a metal plate that is insulated from the grounded-cavity deck by a Kapton blocker. This hardware makes up the screen-blocker assembly. The dc screen voltage feeds to the fingerstock ring from underneath the cavity deck through an insulated feedthrough.

Some transmitters that employ the ¼-wave cavity design use a grounded-screen configuration in which the screen contact fingerstock ring is connected directly to the grounded cavity deck. The PA cathode then operates at below ground potential (in other words, at a negative voltage), establishing the required screen voltage for the tube.

The cavity design shown in Fig. 11.13 is set up to be slightly shorter than a full ¼-wavelength at the operating frequency. This makes the load inductive and resonates the tube's output capacity. Thus, the physically foreshortened shorted transmission line is resonated and electrically lengthened to ¼-wavelength.

Tuning the Cavity.   Coarse tuning of the cavity is accomplished by adjusting the cavity length. The top of the cavity (the cavity shorting deck) is fastened by screws or clamps and can be raised or lowered to set the length of the assembly for the particular operating frequency.

Fine-tuning is accomplished by a variable-capacity plate-tuning control built into the cavity. In the example, one plate of this capacitor, the stationary plate, is fastened to the inner conductor just above the plate-blocking capacitor. The movable tuning plate is fastened to the cavity box, the outer conductor, and is mechanically linked to the front-panel tuning control. This capacity shunts the inner conductor to the outer conductor and can vary the electrical length and resonant frequency of the cavity.

11.4   TRANSMISSION SYSTEM HARDWARE

The mechanical and electrical characteristics of the hardware located after the transmitter are critical to proper operation of an AM or FM station. Mechanical considerations determine the ability of the transmission line, antenna-tuning unit (if used), and antenna to withstand temperature extremes, lightning, rain, ice, and wind. Electrical considerations directly relate to overall transmission system efficiency, reliability, and coverage area.

11.4.1   Transmission Line

A number of different types of coaxial cable are available to users. The type chosen depends on the power level required, length of the run from the transmitter to the antenna, and the installation method preferred.

Rigid coaxial cables, constructed of heavy-wall, copper tubes with Teflon or ceramic spacers, provide electrical performance approaching an ideal transmission line. They are, however, expensive to purchase and install. Single-piece corrugated copper outer conductor cables, on the other hand, have the combined characteristics that make them a popular choice for many AM and FM applications.

Cable Basics.   A coaxial cable is a transmission line in which concentric center and outer conductors are separated by a dielectric material. When current flows along the center conductor, it establishes an electric field. The electric flux density and the electric field intensity are determined by the dielectric constant of the dielectric material. The dielectric material becomes polarized with positive charges on one side and negative charges on the opposite side. Therefore, the dielectric acts as a capacitor with a given capacitance per unit length of the coaxial line.

Properties of this field establish a given inductance per unit length of the coaxial line. The transmission line also exhibits a given series resistance in ohms per unit length. If the transmission line resistance is negligible and the line is terminated properly, the following simple formula can be used to determine the characteristic impedance of the cable:

$$Z_o = \sqrt{\frac{L}{C}} \qquad\qquad (11.2)$$

where L = inductance in henries per foot
      C = capacitance in farads per foot

Coaxial cables typically are manufactured with 50- or 75-Ω characteristic impedances. Other characteristic impedances are possible by changing the spacing between the center and outer conductors.

Electrical Considerations.   VSWR, attenuation, and power-handling capability are key electrical factors in the application of coaxial cable. High VSWR can cause power loss, voltage breakdown, and thermal degradation of the line. High attenuation means less power delivered to the antenna, higher power consumption at the transmitter, and increased heating of the transmission line itself. A system suffering from inadequate power-handling capability can experience thermal breakdown and, perhaps, catastrophic failure.

11.4.2 AM Antenna Systems

The primary purpose of an AM antenna is to radiate efficiently the power supplied to it by the transmitter. A simple antenna, consisting of a single vertical tower that radiates equally in all directions, can do this job quite well. The antenna system is also required at many stations to concentrate the radiated power in desired directions and minimize radiation in the direction of other stations sharing the same or adjacent frequencies. To achieve such directionality may require a complicated antenna that incorporates a number of individual towers and matching networks.

Vertical polarization of the transmitted signal is used for AM stations because of its superior groundwave propagation and because of the simple antenna designs that it affords. The FCC and licensing authorities in other countries have established classifications of AM stations with specified power levels and hours of operation. It is common for AM stations to reduce their transmitter power at sunset and return to full power at sunrise. This method of operation is based on the propagation characteristics of AM band frequencies. AM signals propagate better at nighttime than during the day. In theory, the day-night operating powers are designed to provide AM stations the same coverage area around the clock. Theory rarely translates into practice insofar as coverage is concerned, however, because of the increased interference that all AM stations suffer at nighttime.

The tower you see at any AM radio station transmitter site is only half of the antenna system. The second element is a buried ground system. Current on a tower does not simply disappear; rather, it returns to earth through the capacitance between the earth and the tower. Ground losses are greatly reduced if the tower has a radial copper ground system. A typical single tower ground system is made up of 120 radial ground wires each 140 electrical degrees long (at the operating frequency) equally spaced out from the tower base. This is often augmented with an additional 120 interspersed radials 50 ft long.

AM Antenna Tuning Unit (ATU). Unlike an FM transmission system, you can't just connect the output of an AM transmitter to an antenna. An antenna tuning unit is required at the tower base (in the case of a nondirectional tower) or at each tower base (in the case of a directional antenna system). The function of an ATU is to match the tower load to the transmission line from the transmitter. The networks used for matching impedances are typically T or L configurations. Pi networks have shunt elements at the input and output and a central series element. They can be made electrically identical to any T network; however, T networks are usually easier to adjust.

Figure 11.14 shows a typical ATU built around a T network. The static drain choke is used to prevent the buildup of dangerous static electricity on a series-fed (nongrounded) tower. The choke exhibits high reactance at the operating frequency of the station but presents a dc path to ground for static energy on the tower. The ammeter shown in the diagram provides an indication of the power being radiated by the antenna.

Directional AM Antennas. When a nondirectional antenna with a given power does not radiate enough energy to serve the station's primary service area or radiates too much energy toward other radio stations on the same or nearby fre-

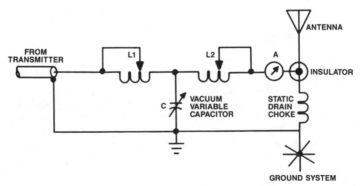

FIGURE 11.14 Basic antenna tuning unit (ATU) for an AM broadcast tower. In a directional antenna system, an ATU is used at the base of each tower.

quencies, it is necessary to employ a directional antenna system. Rules set out by the FCC and regulatory agencies in other countries specify the protection requirements to be provided by various classes of stations, for both daytime and nighttime hours. These limits tend to define the shape and size of the most desirable antenna pattern.

A directional antenna functions by carefully controlling the amplitude and phase of the RF currents fed to each tower in the system. The directional pattern is a function of the number and spacing of the towers (vertical radiators) and the relative phase and magnitude of their currents. The number of towers in a directional AM array can range from two to six or even more in a very complex system. One tower is usually defined as the *reference tower*. The amplitude and phase of the other towers are measured relative to this reference.

A complex network of power splitting, phasing, and antenna coupling elements is required to make a directional system work. Figure 11.15 shows a block diagram of a basic two-tower array. A power divider network controls the relative current amplitude in each tower. A phasing network provides control of the phase of each tower current, relative to the reference tower. Matching networks at the base of each tower couple the transmission line impedance to the characteristic impedance of the radiating towers.

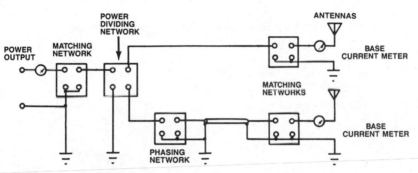

FIGURE 11.15 Block diagram of an AM directional antenna feeder system for a two-tower array.

In practice, the system shown in Fig. 11.15 would not consist of individual elements. Instead, the matching network, power dividing network, and phasing network would all probably be combined into a single unit, referred to as the *phasor*.

11.4.3 FM Antenna Systems

The propagation characteristics of FM radio are much different than for AM. There is essentially no difference between day and night FM propagation. FM stations have relatively uniform day and night service areas with the same operating power.

A wide variety of antennas is available for use in the FM broadcast band. Although these antennas differ from one manufacturer to another, generalizations can be made that apply to most designs.

Circular polarization (CP) of the transmitted signal is generally employed for FM broadcasting. CP results from equal electrical fields in the vertical and horizontal planes of radiation that are out of phase by 90° and rotating a full 360° in one wavelength of the operating frequency. The rotation can be clockwise or counterclockwise, depending on the antenna design. This continuously rotating field gives CP good signal penetration capabilities because it can be received efficiently by an antenna of any random orientation.

Antenna Gain.   FM broadcasters are able to increase the effective radiated power of their transmitters by installing antennas that are said to exhibit "gain." This gain is the result of narrowing the beamwidth of the radiated signal to concentrate energy toward the station's primary coverage area. The actual amount of energy being radiated is the same with a unity-gain antenna or a high-gain antenna, but the useful energy (known in the industry as the ERP) can be increased significantly.

The ideal combination of antenna gain and transmitter power for a particular installation requires a careful examination of the antenna height, location of the primary service area, and economic restraints placed on the station. A high-power transmitter is much more expensive than a high-gain antenna. As shown in Table 11.1, a variety of pairings can be made to achieve the same ERP.

TABLE 11.1   Various Combinations of Transmitter Power and Antenna Gain to Achieve 100-kW ERP

No. bays	Antenna gain	Transmitter power* (kilowatts)
3	1.5888	66.3
4	2.1332	49.3
5	2.7154	38.8
6	3.3028	31.8
7	3.8935	27.0
8	4.4872	23.5
10	5.6800	18.5
12	6.8781	15.3

*Transmitter power based on 95 percent line efficiency.
Source: Courtesy of *Broadcast Engineering* magazine.

For greatest efficiency, electrical *beam tilt* can be designed into a high-gain antenna. A standard antenna radiates more than half of its energy above the horizon. This energy is lost for practical purposes. Electrical beam tilt, caused by delaying the RF current to the lower elements of the antenna, can be used to provide more useful power to the service area.

Directional FM antennas can also be used to increase field strength over the primary service area. The directional characteristics are created by the placement of reflecting and parasitic elements.

Pattern optimization is another method to maximize radiation to the intended service area. The characteristics of the transmitting antenna are affected, sometimes greatly, by the presence of the supporting tower, if side-mounted, or by nearby tall obstructions (such as another transmitting tower) if top-mounted. Antenna manufacturers use various methods to tune out pattern distortions. These generally involve changing the orientation of the radiators with respect to the tower and adding parasitic elements.

Antenna Designs. There are several basic classes of FM transmitting antennas available to stations today:

- Ring stub and twisted ring
- Shunt- and series-fed slanted dipole
- Multiarm short helix

While each design is unique, all have some things in common:

1. The antennas are designed for side mounting to a steel tower or pole.
2. The radiating elements are shunted across a common rigid coaxial transmission line.
3. The elements are placed along the rigid line every one wavelength.
4. Antennas with one to seven bays are fed from the bottom of the coaxial transmission line.
5. Antennas with more than seven bays are fed from the center of the array to provide more predictable performance in the field.
6. The antennas generally include a means of tuning out reactances after the antennas have been installed, through the adjustment of variable capacitive or inductive elements at the feed point.

Figure 11.16 shows a shunt-fed slanted dipole antenna that consists of two half-wave dipoles offset 90°. The two sets of dipoles are rotated 22.5° (from their normal plane) and are delta-matched to provide a 50-Ω impedance at the radiator input flange. The lengths of all four dipole arms may be matched to resonance by mechanical adjustment of the end fittings. Shunt feeding (when properly adjusted) provides equal currents in all four arms.

Wideband *panel antennas* are a fourth common type of antenna used for FM broadcasting. Panel designs share some of the characteristics listed previously but are intended primarily for specialized installations in which two or more stations will use the antenna simultaneously. Panel antennas are larger and more complex than other FM antennas but offer the possibility for shared tower space among several stations and custom coverage patterns that would be difficult or even impossible with more common designs. Stations using wideband panel antennas at a *community transmission site* feed their transmitters into an RF combiner that sums the individual outputs and applies the signal to a single transmitting antenna.

11.5 RADIO TRANSMITTER PERFORMANCE AND MAINTENANCE

Assessing the performance of an AM or FM transmitter has been greatly simplified by recent advancements in RF system design. Most transmitters can be

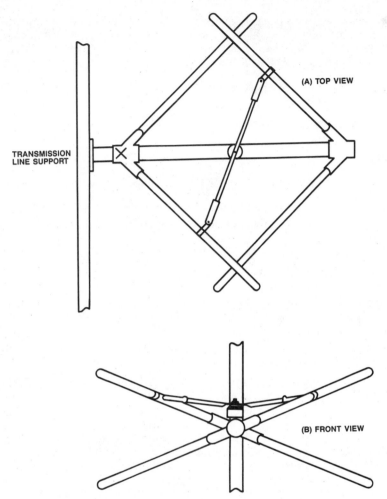

FIGURE 11.16 Mechanical configuration of one bay of a circularly polarized FM transmitting antenna (Jampro JSCP series). (*a*) Top view of the antenna. (*b*) Antenna as viewed from the front.

checked by measuring the audio performance of the overall system. If the audio tests indicate good performance, chances are good that all the RF parameters also meet specifications.

Both AM and FM broadcasting have certain limitations that prevent them from ever being a completely transparent medium. Top on the list for FM is multipath distortion. In many locations, some degree of multipath is unavoidable. Running a close second is the practical audio bandwidth limitation of the FM stereo multiplex system. The theoretical limit is 19 kHz; however, real-world filter designs result in a high-end passband of between 15 and 17 kHz.

For AM radio, the primary limitation is restricted bandwidth. Modern transmitters are capable of faithfully transmitting wideband audio, but most receivers are not capable of receiving it. The problem is not just poor-quality AM radios but rather the high-noise environment of the AM band. To reduce adjacent channel interference and atmospheric and artificial noise, receiver designers have restricted the IF and audio bandwidth of their products. The result is a typical AM receiver in which the audio response falls off rapidly beyond 3 kHz.

11.5.1 Key System Measurements

The procedures for measuring AM and FM transmitter performance vary widely depending on the type of equipment being used. Some generalizations, however, can be made with respect to equipment performance measurements that apply to most systems:

Frequency response is the actual deviation from a constant amplitude across a given span of frequencies. Research has shown that frequency-response accuracy over the audio passband does make an audible difference. Researchers exploring subtle differences in audio amplifier design have found that errors as small as 0.2 dB can be heard. Therefore, flat response (strict adherence to the 75-μs preemphasis curve) is an important goal in FM broadcasting. Typical performance targets for an FM station are ±1 dB, 50 Hz to 15 kHz. Typical targets for an AM station are ±1 dB, 50 Hz to 10 kHz.

Total harmonic distortion (THD) is the creation by a nonlinear device of spurious signals harmonically related to the applied audio waveform. Research has shown that although THD levels greater than 1 percent are detectable during sine-wave testing, listeners will tolerate somewhat higher levels of THD on musical material. It must be noted that depending upon the setup of the station's audio processor, distortion targets of 1 percent or less may be impossible to meet with the processor on-line. Because of the highly competitive nature of broadcasting today, audio processors cannot always be adjusted to provide for the purest reproduction of the incoming signal.

Audio processing aside, most FM transmitters will yield THD readings below 1 percent at frequencies up to 7.5 kHz. AM transmitters can typically produce distortion figures below 1.5 percent at frequencies up to 5 kHz at 95 percent modulation.

The THD test is sensitive to the noise floor of the system under test. If the system has a signal-to-noise ratio of 60 dB, the distortion analyzer's best possible reading will be greater than 0.1 percent (60 dB = 0.001 = 0.1 percent).

Intermodulation distortion (IMD) is the creation by a nonlinear device of spurious signals not harmonically related to the audio waveform. These distortion components are sum-and-difference (beat notes) mixing products that research has shown are more objectional to listeners than even-harmonic distortion products. The IMD measurement is relatively impervious to the noise floor of the system under test. IMD performance targets for AM and FM transmitters are the same as the THD targets mentioned previously.

Signal-to-noise ratio (S/N) is the amplitude difference, expressed in decibels, between a reference level audio signal and the system's residual noise and hum. In many cases, system noise is the most difficult parameter to bring under control. An FM performance target of −60 dB per stereo channel reflects state-of-the-art, exciter-transmitter performance. Most AM transmitters are capable of −56 dB S/N for monaural operation, and −50 dB per channel for stereo operation.

Separation is a specialized definition for signal crosstalk between the left and

right channels of a stereo system. The separation test is performed by feeding a
test tone into one channel while measuring leakage into the other channel (whose
input is terminated with a 600-Ω wirewound resistor, or other appropriate value).
Typical performance targets for an FM station are −30 dB from 50 to 400 Hz, and
−35 dB from 400 Hz to 15 kHz. Targets for a stereo AM station are −20 dB from
50 to 400 Hz and −25 dB from 400 Hz to 10 kHz.

11.5.2 Synchronous AM in FM Systems

The output spectrum on an FM transmitter contains many sideband frequency
components, theoretically an infinite number. The practical considerations of
transmitter design and frequency allocation make it necessary to restrict the
bandwidth of all FM broadcast signals. Bandwidth restriction brings with it the
undesirable side-effects of phase shifts through the transmission chain, the generation of synchronous AM components, and distortion in the demodulated output of a receiver.

In most medium- and high-power FM transmitters, the primary out-of-band
filtering is performed in the output cavity of the final stage. Previous stages in the
transmitter (the exciter and IPA) are designed to be broadband, or at least more
broadband than the PA.

As the bandwidth of an FM transmission system is reduced, synchronous amplitude modulation increases for a given carrier deviation (modulation). Synchronous AM is generated as tuned circuits with finite bandwidth are swept by the
frequency of modulation. The amount of synchronous AM generated is dependent on tuning, which determines (to a large extent) the bandwidth of the system.
Figure 11.17 illustrates how synchronous AM is generated through modulation of
the FM carrier.

11.5.3 Incidental Phase Modulation

Before the advent of AM stereo, few AM broadcast engineers understood the
term *incidental phase modulation* (IPM), let alone concerned themselves with its
effect on AM transmitters. IPM is defined as phase modulation produced by an
AM transmitter as a result of amplitude modulation. In other words, as an AM
transmitter develops the amplitude-modulated signal, it also produces a phase-
modulated, or PM, version of the audio as well.

In theory, IPM is of little consequence for monophonic AM because FM and
PM do not affect the carrier amplitude. An envelope detector will, in theory at
least, ignore IPM. This is true to a point; however, stations attempting to broadcast AM stereo must pay close attention to IPM. Any undesired carrier phase
modulation, such as IPM, will have an effect on stereo performance. Although
the effects of IPM on stereo signals vary from one system to another, optimum
performance is realized when IPM is minimized.

Causes of IPM.   As a general rule, because IPM is a direct result of the modulation process, it can be generated in any stage that is influenced by modulation.
The most common cause of IPM in plate-modulated and pulse-modulated transmitters is improper neutralization of the final RF amplifier. Adjusting the transmitter for minimum IPM is an accurate way of achieving proper neutralization.

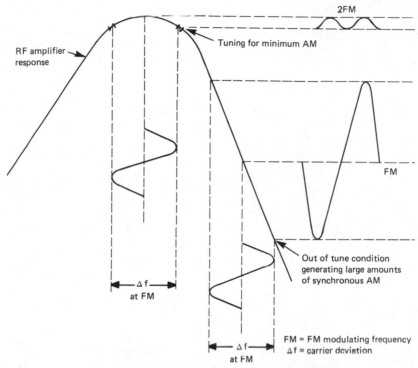

FIGURE 11.17 Synchronous AM is generated by one or more narrow-band stages in an FM transmitter. The amount of AM produced is a function of the flatness of the composite RF system. If one or more stages is out of tune, as illustrated, significant amounts of synchronous AM will be produced. Note that at the point of minimum synchronous AM (and therefore correct tuning), the demodulated output of an AM detector will double in frequency. (Source: Courtesy *Broadcast Engineering* magazine)

The reverse is not always true, however, because some neutralization methods will not necessarily result in the lowest amount of IPM.

Improper tuning of the IPA stage is the second most common cause of IPM. As modulation changes the driver loading to the PA grid, the driver output also may change. The circuits that feed the driver stage usually are isolated enough from the PA that they do not produce IPM. An exception can be found when the power supply for the IPA is influenced by modulation. Such a problem could be caused by a loss of capacitance in the high-voltage power supply itself.

BIBLIOGRAPHY

Crutchfield, E. B. *National Association of Broadcasters Engineering Handbook*, 7th ed. Sec. 2.4, AM Broadcast Antenna Systems, Part 1 (by Carl Smith); Sec. 2.4, AM Broadcast Antenna Systems, Part 2 (by Edward Edison); and Sec. 2.5, Antennas for FM Broadcasting (by Peter Onnigian). National Association of Broadcasters, Washington, D.C.

CHAPTER 12
AM AND FM RECEPTION

Jerry Whitaker
Editorial Director, Broadcast Engineering *Magazine*

12.1 INTRODUCTION

The development of radio transmission and reception must be considered one of the major technical achievements of the twentieth century. The impact of voice broadcasts to the public, whether by commercial stations or government-run organizations, has expanded the horizons of everyday citizens in virtually every country on earth. It is hard to overestimate the power of radio broadcasting.

As discussed in Chap. 11, technology has reshaped the transmission side of AM and FM broadcasting. Profound changes have also occurred in receiver technology. Up until 1960, radio broadcasting was basically a stationary medium. The receivers of that time were physically large and heavy, and required 120 V ac power to drive them. The so-called portable radios of the day relied on bulky batteries that offered only a limited amount of listening time. Automobile radios incorporated vibrator-choppers to simulate ac current. All the receivers available for commercial use during the 1940s and 1950s used vacuum tubes exclusively.

The first technical breakthrough for radio broadcasting was the transistor, available commercially at reasonable prices during the early 1960s. The transistor

Portions of this chapter were adapted from Ulrich L. Rohde and T. T. N. Bucher, *Communications Receivers, Principles and Design*, McGraw-Hill Publishing Company, New York, 1988.

brought with it greatly reduced physical size and weight, and even more importantly, it eliminated the necessity of ac line current to power the radio. The first truly portable AM radios began to appear during the early 1960s, with AM-FM versions following by the middle of the decade.

Many of the early receiver designs were marginal from a performance standpoint. The really good receivers were still built around vacuum tubes. As designers learned more about transistors, and as better transistors became available, tube-based receivers began to disappear. By 1970, *transistorized* radios, as they were called, commanded the consumer market.

The integrated circuit (IC) was the second technical breakthrough in consumer receiver design. This advance, more than anything else, made high-quality portable radios possible. It also accelerated the change in listening habits from AM to FM. IC-based receivers allowed designers to put more sophisticated circuitry into a smaller package, permitting consumers to enjoy the benefits of FM broadcasting without the penalties of the more complicated receiver required for FM stereo.

The move to smaller, lighter, more power-efficient radios has led to fundamental changes in the way radios are built and serviced. In the days of vacuum-tube and transistor-based receivers, the designer would build a radio out of individual stages that interconnected to provide a working unit. The stages for a typical radio included:

- RF amplifier
- Mixer
- Local oscillator
- Intermediate frequency (IF) amplifier
- Detector and audio preamplifier

Today, however, large-scale integration (LSI) or even very large scale integration (VLSI) techniques have permitted virtually all the active circuits of an AM-FM radio to be placed on *one single* IC. Advanced circuitry has also permitted radio designers to incorporate all-electronic tuning, eliminating troublesome and sometimes expensive mechanical components. Electronically tuned radios (ETRs) have made features such as "station scan" and "station seek" possible. Some attempts were made to incorporate scan and seek features in mechanically tuned radios, but the results were never very successful.

The result of LSI-based receiver design has been twofold. First, radios based on advanced chip technology are much easier to build and are, therefore, usually less expensive to consumers. Second, such radios are not serviceable. Most consumers today would not bother to have a broken radio repaired. They would simply buy a new one and throw the old one away.

Still, however, it is important to know what makes a radio work. Although radios being built with LSI and VLSI technology do not lend themselves to stage-by-stage troubleshooting as earlier radios did, it is important to know how each part of the system functions to make a working unit. Regardless of the sophistication of a VLSI-based receiver, the basic principles of operation are the same as a radio built of discrete stages.

12.1.1 Radio Reception Fundamentals

A *simplex* transmitter-receiver system is the simplest type of communication system possible. A single source transmits to a single receiver. When the receiver can respond to a transmission from the source and relay voice or other information back to the transmission site, a *duplex* arrangement is realized. A two-way business radio system operating on different transmit and receive frequencies employs a duplex arrangement. AM and FM broadcast systems utilize a *simplex star* configuration in which one transmitter feeds many receivers. For greatest efficiency, the receivers in a simplex star should be made as simple as possible, concentrating technical complexity at the transmit site. Radio broadcasting today follows this basic rule.

AM and FM stations transmit at high power levels to facilitate simpler radio designs. AM stations in the United States can operate at up to 50 kW, while FM stations can operate at up to 100 kW. This essentially eliminates the need for a sensitive antenna at the receiver. Stereo AM and stereo FM systems are, likewise, designed to permit the more complex operations to be performed in the encoding stage at the transmitter, rather than in the decoding circuits of the receiver.

Superheterodyne Receiver. Virtually every AM-FM radio manufactured today incorporates the superheterodyne method of reception. This system makes use of the heterodyne principle of mixing an incoming signal with a signal generated by a local oscillator (LO), as illustrated in Fig. 12.1. The LO is offset by a fixed intermediate frequency (IF) from the desired signal. Because the mixer (a nonlinear device) generates a difference frequency that is identical if the desired signal is either above or below the LO frequency (and also a number of other spurious responses), filtering is required prior to the mixer to suppress the undesired signal. The frequency of the undesired signal is referred to as an *image frequency*

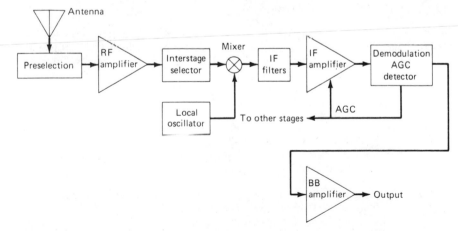

FIGURE 12.1 Simplified block diagram of a superheterodyne receiver. Depending on the sophistication of the radio, some of the stages shown may be combined into a single circuit. *(Source: Ulrich L. Rohde and T. T. N. Bucher, Communications Receivers: Principles and Design, McGraw-Hill, New York, 1988)*

and is separated from the desired signal frequency by a difference equal to twice the IF.

Channel filtering is accomplished by one or more fixed-frequency filters in the IF. This is a decided advantage when the receiver must cover a wide frequency band, because it is more difficult to maintain constant bandwidth in a tunable filter than in a fixed one. An IF frequency of 455 kHz is used for AM receivers, and 10.7 MHz for FM receivers.

As shown in Fig. 12.1, the input signal is fed from the antenna to a preselector filter and RF amplifier. The input circuit matches the antenna to the first amplifying device to achieve the best sensitivity. It also provides the necessary selectivity to reduce the possibility of overload in the first amplifier stage caused by strong, undesired signals. Because sufficient selectivity must be provided to eliminate the image and other spurious signals prior to the mixer, preselection filtering may be broken into two or more parts with intervening amplifiers (in premium receiver designs) to minimize the effects of filter losses on the receiver *noise figure* (NF). The noise figure is a comparison in decibels (dB) of the noise power generated by a receiver with the noise power generated in an equivalent resistor.

The LO provides a strong stable signal at the proper frequency to the mixer for conversion of the desired signal to the IF. (The mixer stage may also be called the *first detector, converter*, or *frequency changer*.) The operating frequency of the LO may be controlled by a variable capacitor or varactor diode operating in a phase-locked-loop (PLL) system. Frequency synthesizer PLL designs are the basis for ETRs.

The output of the mixer is applied to the IF amplifier, which raises the desired signal to a suitable power level for the demodulator. The demodulator derives from the IF signal the modulated baseband waveform, which may be amplified by a baseband amplifier before being applied to a decoder (for stereo broadcasts) or audio preamplifier (for monophonic broadcasts).

12.1.2 Radio Wave Propagation

Radio waves are electromagnetic energy that can be attenuated, reflected, refracted, and scattered by changes in the media through which they propagate. Radio waves in free space have electric and magnetic field components that are mutually perpendicular and lie in a plane transverse to the direction of propagation. Radio waves travel at a velocity of about 300,000 km/s.

The received signal field is accompanied by noise generated as the result of natural sources (electrical activities in the atmosphere and space) and/or artificial sources (other radio transmissions and machinery of various types). In addition, the receiver itself is a source of noise. Electrical noise limits the range and performance of radio communications by requiring a sufficient signal strength at the receiver to overcome the undesired noise.

AM Band Propagation. AM broadcasting (530 to 1610 kHz) falls in the middle of the medium-frequency (MF) band, which extends from 300 kH to 3 MHz. During daylight hours, ground wave propagation is predominate, and the effects of atmospheric noise are minimal. The receiver NF has little effect on overall signal quality unless the antenna system is very inefficient. At night, however, sky wave propagation is significant, permitting reception of signals hundreds or even thousands of miles away. Unfortunately, atmospheric noise is also greatest at night.

Fading at the receiver occurs in portions of a station's coverage area where both the ground wave and sky wave are comparable in strength. Fading can become quite deep during periods when the two waves are nearly equal. The ground wave will always reach the receiver through the most direct route. The sky wave will arrive later in time, out of phase with the ground wave, resulting in attenuation of the composite signal. Fading can also occur as a result of two or more sky waves with different numbers of reflections combining at the receiver.

When fading is caused by two or more waves that interfere as a result of having traveled over paths of different lengths, different frequencies within the transmitted spectrum can be attenuated to various extents. This phenomenon is known as *selective fading*. It can result in severe distortion of the signal.

Noise-free AM reception is particularly difficult at night because of the tremendous amount of noise present on the AM dial caused by sky wave interference from distant stations on the same or adjacent frequencies. The only solution to this problem is to increase the signal strength of the desired station at the antenna of the receiver. This essentially means that the effective service area of an AM station will be restricted during nighttime hours to locations in which a strong local ground wave signal is available.

FM Band Propagation.   FM broadcasting (88 to 108 MHz) occupies a portion of the very-high-frequency (VHF) band, which extends from 30 to 300 MHz. For greatest reliability, most VHF communication relies on a line-of-sight path from the transmitter to the receivers. For this reason, FM transmitting antennas are located on high towers, tall buildings or mountaintop sites. The wavelength at FM frequencies is sufficiently short that FM can penetrate automobiles and steel-frame buildings with a minimum of loss. FM is not affected by nighttime sky wave interference. The propagation properties of FM radio are essentially the same from day to night.

FM is, however, affected by reflections of the transmitted signal that can mix at the receiver and cause fading or distortion. The phenomenon is known as *multipath* (see Fig. 12.2). Reflections can be caused by mountains, steel-frame buildings, vehicles, and other objects. The interference patterns set up as a result of multipath cause the signal strength to vary from one location to another in an apparently random manner. It is often possible, for example, to improve reception simply moving the antenna 2 ft or so (about ¼ wavelength at FM frequencies).

The noise level at VHF is low compared to the MF band. Artificial noise can produce impulsive interference, but the use of hard limiting in receivers will eliminate most amplitude-related noise. Interference from other FM stations can, however, produce noise at the receiver that usually cannot be stripped off by limiting.

12.1.3   Radio Receiver Characteristics

Receiver design varies significantly from one model to another and from one manufacturer to another. Each company has its own approach to system design and may specify the performance of its products using unique procedures. It is difficult, therefore, to directly compare performance specifications without some

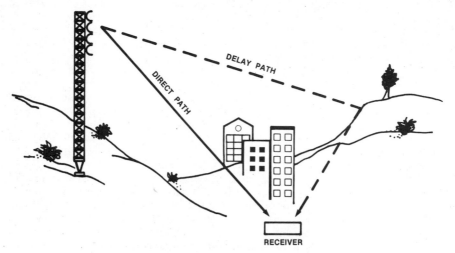

FIGURE 12.2 The causes of multipath in FM broadcasting. Distortion of the demodulated FM signal is related to both the direct-reflected signal ratio and the delay time of the reflected signal. Distortion increases as the signal ratio approaches 1:1, and as the secondary path delay rises.

knowledge of the test procedures. However, some generalizations can be made that apply to most radios in the field today.

Noise and RF Gain.   Receivers are required to intercept and decode a wide range of signal levels. The extent to which a received signal can be useful to the consumer is dictated by the noise levels picked up by the antenna on the desired and adjacent frequencies and by the noise generated within the receiver itself.

The design of any radio includes a gain distribution plan, which specifies the gain required of each stage in the system. Given sufficient system gain and ignoring atmospheric (and other natural) noise, the weakest signal that may be demodulated satisfactorily is limited by the receiver's internal noise floor. The noise figure (NF, expressed in decibels) of a receiver compares the total receiver noise with the noise that would be present if the receiver generated no noise. (This ratio is called the noise factor F.) Receiver designers have little control over atmospheric and artificial noise picked up by the antenna. They do, however, have control over internal noise sources.

Passive devices, such as resistors and coils, generate noise as a result of the continuous thermal motion of free electrons. This type of noise is generally referred to as *thermal noise*. Semiconductor devices, however, usually represent the greatest source of noise in a receiver. Transistors, diodes, and ICs (as did their vacuum-tube predecessors) produce two primary types of noise:

- *Shot noise* results from fluctuations in carrier flow through semiconductors and can result in wide-band noise similar to thermal noise.
- *Flicker noise*, also called the *flicker effect*, is inversely proportional to frequency. It results from imperfect contact between two materials, somewhat similar to *contact noise* in the adjustment of a variable resistor. The most com-

mon source is from the flow of currents in transistors and other semiconductors.

Selectivity.   The ability of a receiver to separate a given signal on one frequency from those on all other frequencies is known as *selectivity*. At least two characteristics must be considered in establishing the required selectivity of a receiver: intermodulation and dynamic range. The selective circuits must be sharp enough to suppress interference from adjacent channels and spurious signals, but they must also be broad enough to pass the highest sideband frequencies of the desired transmission with low amplitude and phase distortion. A typical IF selectivity curve for an AM receiver (not designed for AM stereo reception) is shown in Fig. 12.3.

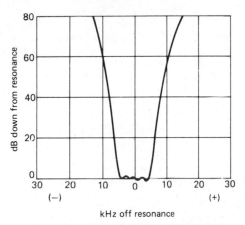

FIGURE 12.3   An example of an IF selectivity curve for an AM broadcast band receiver. *(Source: Ulrich L. Rohde and T. T. N. Bucher,* Communications Receivers: Principles and Design, *McGraw-Hill, New York, 1988)*

Intermodulation.   Intermodulation (IM) performance is one of the most important attributes of a high-quality receiver. No matter how sensitive a receiver may be, if it has poor immunity to strong interfering signals, it will be of little use. IM produces sum and difference frequency products of many orders that manifest themselves as interference. Odd-order products are generally more troublesome than even-order products and are, therefore, given more consideration by designers.

   Cross-modulation, another form of IM, occurs in an AM radio when a strong modulated signal amplitude-modulates a weak signal through the inherent nonlinearities of the receiver. Cross-modulation requires a much higher interfering signal level than is typically necessary for the generation of IM products.

Dynamic Range.   The ratio in decibels between the strongest and the weakest signals that a receiver can handle with acceptable noise or distortion is referred to as the *dynamic range*. This strict definition, however, must be considered in light of the real-world conditions under which a receiver must operate. In the normal signal environment of a broadcast receiver, the desired signal may have a range of typical values; however, this range occurs within a group of other signals that range from very weak to very strong. The selectivity of the receiver front end will have a significant effect on the *usable dynamic range* of a radio in the field. Very strong signals on frequencies nearby the desired signal can degrade performance substantially because of the nonlinearity of the active devices necessary to provide amplification and frequency conversion. Odd-order IM products can limit the dynamic range of a receiver significantly through the generation of a distorted and/or noisy signal, or through *desensitization* of the receiver front end (reduced gain of one or more active devices).

A better way to specify dynamic range is the ratio of strong out-of-band signals to the level of the weakest acceptable desired signal. The level of the strong signal must be sufficient to cause the weak signal to become unacceptable.

If the foregoing discussion of dynamic range seems vague, it is because several characteristics are encompassed by the term, not just one. A receiver is a complex device with many active stages providing differing degrees of selectivity and response to strong, off-frequency signals.

Because AM and FM receivers must be capable of handling widely varying input signal levels, some form of automatic gain control (AGC) is required. The AGC system of a receiver provides the means to sense signal strength and adjust the gain of the RF or IF amplifiers. The demands placed on the AGC circuit of a receiver are particularly tough when applied to auto radios. The incoming signal may change by tens of decibels over periods ranging from fractions of a second to several minutes. The level may also change significantly when the receiver is switched to a different frequency. The principal AGC characteristics of importance are the input-output gain curve and the attack and decay times.

12.2 ELEMENTS OF RECEIVER DESIGN

The design of an AM or FM receiver is a complicated process involving a huge number of technical, economic, and perceptual characteristics. It is not enough for a receiver to perform well—it must also meet a particular price point and respond to the needs (or wants) of the consumer. Take, for example, the small, portable ("Walkman"-type) AM-FM radios available today. The consumer wants a radio that is small and light enough to clip on a belt or slip into a shirt pocket. The radio does not need to be feature-rich, but it must perform well without any external antenna in a wide variety of working environments. The price point of the radio to consumers must be below $20.

These specifications place enormous demands on the manufacturer. It is far easier to build a large, heavy, feature-rich radio that uses an outdoor antenna, is powered from the ac line, and can be priced upwards of $400 than it is to build the radio just described. The design of any AM-FM radio for use by consumers is an exercise in compromise.

Many radios today have been reduced to just a handful of LSI chips, or even a single VLSI chip. This dramatic move toward miniaturization has eliminated the traditional stages that technicians are familiar with. Still, each stage exists in one form or another in virtually every radio produced today. The circuits may be hidden on a slab of silicon, but they are there just the same.

12.2.1 Antenna Systems

The antenna must be designed to efficiently capture the transmitted signal and meet the physical limitations of the receiver. This is easier said than done. The ideal antenna for an AM radio would be a tall vertical wire or tower placed over a buried ground system. The FM equivalent would be a high-gain Yagi antenna mounted 10 to 30 ft above the ground. Such antennas are completely impractical for all but a few specialized, fixed installations. For most applications, especially mobile or portable radios, compromises in antenna design must be made.

Antenna efficiency, impedance, bandwidth, and reception pattern are all a function of the antenna dimensions relative to wavelength. A ferrite loop antenna is typically used for fixed and portable AM radios, while a short (less than 3 ft) telescoping vertical antenna is normally used for automobile AM radios. FM antennas usually consist of simply a length of wire (about 18 in) that is tacked up on a nearby wall or run along a baseboard for fixed-location receivers, or a short telescoping antenna for portable radios. Automobile FM radios use the AM auto antenna discussed previously. Shirt-pocket-sized radios simply rely on a length of wire run within the case of the unit.

None of the antennas just discussed are particularly effective in converting radio waves into electric signals. The constraints of space and cost have led manufacturers to sacrifice antenna efficiency for a compact, convenient product. Supplemental external antennas are used only in fixed applications requiring exceptional performance.

Antenna Coupling Network. Whatever form the antenna takes, it must be coupled efficiently to the first RF stage of the receiver. The coupling network must exhibit low loss and adequate bandwidth to cover the band selected (AM or FM). Until recently, it was customary practice to couple the antenna to a tuned circuit connected to the input RF amplifier. Figure 12.4 illustrates several common coupling arrangements. The examples differ mainly in the details of coupling. With ETR systems gaining wide acceptance, mechanical tuning is becoming obsolete. Voltage-tuned capacitors (varactors) are increasingly being used in radios today.

Whip Antenna. Nearly all automobile and some portable radios use whip antennas of about 2 to 3 ft in length for both AM and FM reception. Seldom does the mounting surface resemble a plane. The problem of coupling a whip optimally to the first active circuit is a difficult one. The antenna represents a complex set of impedance, capacitance, reactance, and resistive components that vary with the frequency and the surrounding physical structures. Still, given the ability to switch coupling circuitry between AM and FM bands, coupling to the RF amplifier can be satisfactory for reasonable efficiency and bandwidth.

Loop Antenna. The loop antenna has been used in portable AM receivers for years. Its response differs from the monopole in that when the face of the loop is vertical, it responds to the magnetic field rather than the electric field. Instead of being omnidirectional in azimuth (like a whip), the loop responds to the cosine of the angle between its face and the direction of the desired transmission. This

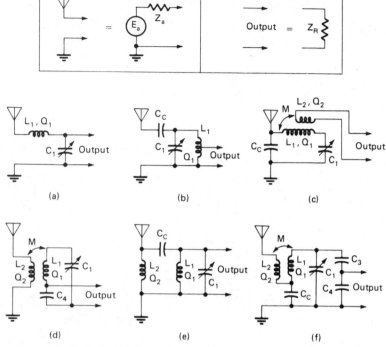

FIGURE 12.4   Typical circuits used for coupling an antenna to a tuned resonant circuit. (*Source: Ulrich L. Rohde and T. T. N. Bucher,* Communications Receivers: Principles and Design, *McGraw-Hill, New York, 1988)*

yields the familiar *figure-eight* pattern, which makes the loop useful for direction finding by providing a sharp null for waves arriving perpendicular to the face.

Loops used for AM broadcast reception incorporate a high-permeability (ferrite) core to facilitate reduced size. Such a loop may be tuned by a capacitance and connected directly to the input device of the receiver. Coupling is usually simple, as illustrated in Fig. 12.5. If the loop has an inductance lower than that required for proper input impedance, it may be connected in series with an additional inductance for tuning (as shown). If the loop impedance is too high, the receiver input may be tapped down on the circuit.

12.2.2   Filter Types

Filters are perhaps the most basic element of any receiver. If you are old enough to remember "crystal radio" hobbyist sets popular some years ago, you will recall that these childhood toys consisted of little more than a tuned circuit (filter) and detector. Although these sets have pretty much been retired to museums, the principle still holds true.

Filter design has come a long way within the past 10 to 20 years. Filters that once were used only in military applications because of their high cost are now produced by the tens of thousands. With reduced costs have come better designs.

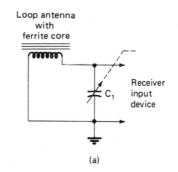

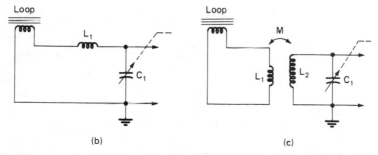

FIGURE 12.5 Examples of coupling circuits used for AM broadcast reception using a loop antenna. (*Source: Ulrich L. Rohde and T. T. N. Bucher*, Communications Receivers: Principles and Design, *McGraw-Hill, New York, 1988*)

Radios today include filters that designers of consumer products could have only dreamed about 20 years ago.

Conventional filter design may be implemented using a number of different types of resonators. The principal available technologies include:

- LC filter
- Electrical resonator
- Quartz crystal filter
- Monolithic quartz filter
- Ceramic filter

The classical approach to radio filtering involved cascading single- or dual-resonator filters separated by amplifier stages. Overall selectivity was provided by this combination of one- or two-pole filters. This approach, however, had two distinct disadvantages: The circuits were difficult to align properly, and the system was susceptible to IM and overload (even in the early IF stages) from out-of-band signals. The classic approach did have some advantages, though. First, limiting from strong impulse noise would occur in early stages where the broad bandwidth would reduce the noise energy more than after complete selectivity had been achieved. Second, such designs were relatively inexpensive to manufacture.

Modern radios use multiresonator filters inserted as early as possible in the amplification chain to reduce nonlinear distortion, simplify alignment, and permit easy attainment of a variety of selectivity patterns. The simple single- or dual-resonator pairs are now used primarily for impedance matching between stages or to reduce noise between broadband cascaded amplifiers.

LC Filter.   Inductor-capacitor resonators are limited to Q values on the order of a few hundred for reasonable sizes. In most cases, designers must be satisfied with rather low Q. The size of the filter depends on the center frequency. Two separate LC filters can easily cover the AM and FM broadcast bands. Skirt selectivity depends on the number of resonators used. Ultimate filter rejection can be made higher than 100 dB with careful design. Filter loss depends on the percentage bandwidth required and the resonator Q. It can be as high as 1 dB per resonator at narrow bandwidths.

This type of filter does not generally suffer from nonlinearities. Frequency stability is limited by the individual components and cannot be expected to achieve much better than 0.1 percent of center frequency under extremes of temperature and aging. Except for front ends that require broad bandwidth filters, *LC* filters have been largely superseded in modern radios by new filter technologies.

Electric Resonator.   At VHF frequencies and above, the construction of inductors for use in LC resonant circuits becomes more difficult. The *helical resonator* is an effective alternative in some applications.

The helical resonator looks like a shielded coil, as shown in Fig. 12.6, but it acts like a resonant transmission line section. High Q can be achieved in reasonable sizes for upper VHF and UHF frequencies (Fig. 12.7), but not for FM frequencies (88 to 108 MHz). For this reason, helical resonators find little, if any, application in consumer radios.

Coupling in and out of the resonator may be achieved by a tap on the coil, a loop in the cavity near the grounded end (the high magnetic field area), or a probe near the ungrounded end (the high electric field area). The length of the coil is somewhat less than the predicted open-circuit quarter-wave line because of the end capacity to the shield. A separate adjustable screw or vane may be inserted near the open end of the coil to provide tuning of the resonator.

Quartz Crystal Filter.   While several piezoelectric materials have been used for filter resonators, quartz crystal units have proved to be the most satisfactory. Filters are available in frequencies that can easily cover the AM broadcast band, and the FM band IF frequency (10.7 MHz). Standard filter response shapes are available to simplify the receiver design process and keep costs down. Ultimate filter rejection can be in excess of 100 dB. Input and output impedances are determined by input and output matching net-

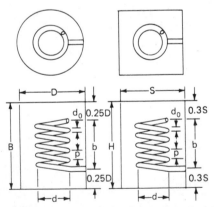

FIGURE 12.6   Basic physical construction of a helical resonator. Round or square types may be built. The diameter D (or side S) is determined by the desired unloaded Q (approximate dimensions are shown in Fig. 12.7). *(Source: J. R. Fisk, "Helical Resonator Design Techniques," QST, July 1976. Reprinted with permission.)*

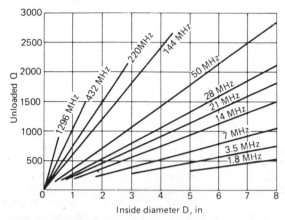

FIGURE 12.7 Helical resonator unloaded Q versus shield diameter D for frequencies from 1.8 MHz to 1.3 GHz. Because of the large dimensions required for operation in the FM broadcast band, helical resonators are generally not used in consumer radios. *(Source: J. R. Fisk, "Helical Resonator Design Techniques," QST, July 1976. Reprinted with permission.)*

works in the filters. Insertion loss varies from about 1 to 10 dB, depending on filter bandwidth and complexity. Frequency accuracy and temperature stability can be maintained to tight tolerances.

Monolithic Quartz Filter. A monolithic quartz filter is made up of a number of resonators constructed on a single quartz substrate. The principal energy of each resonator is confined primarily to the region between the plated electrodes, with a small amount of energy escaping to provide coupling. Usually these filters are limited to about four resonators. Monolithic quartz filters are available for operation through the FM band with characteristics generally similar to discrete quartz resonator filters, except that the bandwidth is limited to several tenths of a percent. Monolithic quartz filters are also smaller and lighter than discrete resonator filters.

Ceramic Filter. Piezoelectric ceramics can be used to achieve some of the characteristics of a quartz filter, but at a lower cost. Such filters are comparable in size to monolithic quartz filters but are available over a limited center frequency range (100 to 700 kHz). This limits ceramic filters to IF applications for AM reception (455 kHz). The cutoff rate, stability, and accuracy of a ceramic filter are not as good as quartz but are adequate for many applications. Single- and double-resonator structures are available. Multiple-resonator filters use electrical coupling between sections.

12.2.3 RF Amplifier and AGC

Depending on the receiver design, an amplifier stage may be used to increase the voltage of the signal received by the antenna (which may be 1 μV or less) to a level sufficient to drive the mixer. Because of the wide range of signals to which a receiver must respond, the input device must be capable of wide dynamic range. It must also be as linear as possible to minimize the generation of IM products from strong signals at the input. It follows that the number of strong signals should be minimized by restricting receiver bandwidth at as low a gain level as possible. Thus, gain should be low prior to the most narrow bandwidth stage in the receiver (the IF).

At AM frequencies, it is common practice to avoid RF amplification and use the mixer as the input device of the receiver. Bandwidth restriction is handled by filters in the first IF amplifier stage.

When the desired signal is relatively strong, the RF amplifier may raise it to a level that will cause distortion in later stages. Automatic gain control (AGC) is, therefore, provided for most RF amplifiers. AGC circuits are basically low-frequency feedback systems. They are necessary in a receiver to maintain a relatively constant output level when the input signal changes frequently. The successful design of an AGC circuit that will perform satisfactorily under all expected signal conditions is a major challenge.

Circuit Designs. A wide variety of transistors are available for use in RF amplifier stages, including *pnp* or *npn* bipolar, or FET configurations. A FET may be classified as junction FET (JFET), metallic-oxide semiconductor FET (MOSFET), and gallium-arsenide (GaAs) FET. These devices differ mainly in the manufacturing process. Bipolar transistors may be used in several amplifying configurations, as shown in Table 12.1. Modern devices have a gain-bandwidth product of from 1 to 6 GHz. The basic amplifier configurations of a JFET are shown in Table 12.2. They are analogous to the amplifier configurations for bipolar transistors.

The MOSFET transistor has an insulation layer between the gate and the source-drain channel and, therefore, has an extremely high impedance at direct current. Several thousand megohms have been measured. JFETS have somewhat better NFs than MOSFETS; otherwise there is little difference between parameters of the two devices.

GaAs FETs are designed for applications above 1 GHz. The carrier mobility of GaAs is much higher than that of silicon. For the same geometry, a significantly higher cutoff frequency is possible with a GaAs FET.

Gain Control. The large dynamic range of signals that must be handled by most receivers requires gain adjustment to prevent overload or IM. A simple method of gain control would involve the use of a variable attenuator between the input and the first active stage. Such an attenuator would decrease the signal level, but it would also reduce the S/N of the receiver for all but the weakest acceptable signal. Most listeners are willing to tolerate an S/N of 20 dB for weak signals, but they expect an S/N of 40 dB or more for strong, local signals.

Gain control, therefore, is generally distributed over a number of stages. The gain in later IF stages is reduced first, and gain in earlier (RF and first IF) stages is reduced only for signal levels sufficiently high to assure the desired S/N. Variable gain amplifiers are controlled electronically. When attenuators are used in a receiver, they are usually operated electrically by either variable voltages for continuous attenuators or by electric switches (relays or diodes) for fixed or stepped attenuators. This switching action may be accomplished automatically or manually, depending on the design of the receiver. Some radios sold today have

TABLE 12.1 Basic Amplifier Configurations and Operating Characteristics of Bipolar Transistors

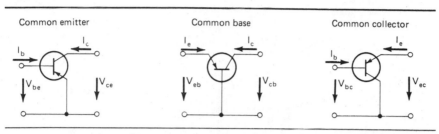

Characteristics of basic configurations			
	Common emitter	Common base	Common collector
Input impedance Z_1	Medium	Low	High
	Z_{1e}	$Z_{1b} \approx \dfrac{Z_{1e}}{h_{fe}}$	$Z_{1c} \approx h_{fe} R_L$
Output impedance Z_2	High	Very high	Low
	Z_{2e}	$Z_{2b} \approx Z_{2e} h_{fe}$	$Z_{2c} \approx \dfrac{Z_{1e} + R_g}{h_{fe}}$
Small-signal current gain	High	< 1	High
	h_{fe}	$h_{fb} \approx \dfrac{h_{fe}}{h_{fe} + 1}$	$\gamma \approx h_{fe} + 1$
Voltage gain	High	High	< 1
Power gain	Very high	High	Medium
Cutoff frequency	Low	High	Low
	$f_{h_{fe}}$	$f_{h_{fb}} \approx h_{fe} f_{h_{fc}}$	$f_{h_{fc}} \approx f_{h_{fe}}$

Source: Ulrich L. Rohde and T. T. N. Bucher, *Communications Receivers: Principles and Design,* McGraw-Hill, New York, 1988.

"local-distant" controls that switch an attenuator or RF amplifier stage in or out of the circuit. Other radios perform this switching automatically by monitoring the AGC signal.

The simplest method of gain control is to design one or more amplifier stages to change gain in response to a control voltage. One common configuration uses a differential pair of common-emitter amplifiers whose emitters are supplied through a separate common-emitter stage. The gain-control voltage is applied to the base of the latter stage, while the signal is applied to one (or both, if balanced) of the bases of the differential pair. This approach has been implemented in a variety of linear integrated circuits. Figure 12.8 shows a schematic diagram of one such IC, the CA3002 (RCA). The gain-control voltage curves are shown in Fig. 12.9.

A PIN diode attenuator can also provide the low-distortion gain control that is especially important prior to the mixer. Figure 12.10 shows such a circuit and its control elements. The control curve has approximately linear decibel variation over most of its 60-dB range. The pi-type attenuator circuit provides a good

TABLE 12.2 Basic Amplifier Configurations and Operating Characteristics of FET Transistors

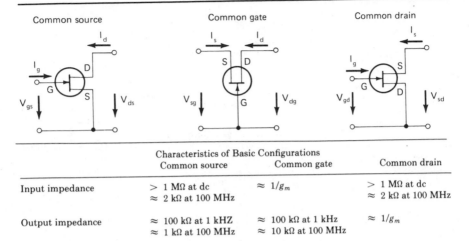

| | Characteristics of Basic Configurations | | |
	Common source	Common gate	Common drain
Input impedance	> 1 MΩ at dc ≈ 2 kΩ at 100 MHz	$\approx 1/g_m$	> 1 MΩ at dc ≈ 2 kΩ at 100 MHz
Output impedance	≈ 100 kΩ at 1 kHZ ≈ 1 kΩ at 100 MHz	≈ 100 kΩ at 1 kHz ≈ 10 kΩ at 100 MHz	$\approx 1/g_m$
Small-signal current gain	> 1000	≈ 0.99	> 1000
Voltage gain	> 10	> 10	< 1.0
Power gain	≈ 20 dB	≈ 14 dB	≈ 10 dB
Cutoff frequency	$g_m/2\pi C_{gs}$	$g_m/2\pi C_{ds}$	$g_m/2\pi C_{gd}$

Source: Ulrich L. Rohde and T. T. N. Bucher, *Communications Receivers: Principles and Design*, McGraw-Hill, New York, 1988.

match between terminations over the control range. The minimum useful frequency for a PIN diode attenuator varies inversely with the minority carrier lifetime. For available diodes, the low end of the HF band is near this limit.

The AGC Loop. The design objective of the AGC loop is to provide a substantially constant signal level to the demodulator despite changes in the input signal level. Figure 12.11 shows a simplified block diagram of a receiver with AGC. The input voltage to the RF amplifier stage may range between 1 μV and 1 V. The envelope of this voltage is detected at the input to the detector stage. This signal is processed to produce the control voltages necessary for the variable-gain device (or devices) in the IF amplifier chain. This control voltage may be increased in level by an amplifier stage, as shown in Fig. 12.12.

As the AGC tries to maintain constant output voltage with a varying input voltage, the issue of attack and decay times comes into play. The AGC system has a finite delay in its response to a change at the input of the receiver. In practice, it is not desirable for the AGC to have too fast a reaction time. In such a case, any static pulse, ignition noise, or other impulsive interference with a fast rise time would be detected by the AGC and desensitize the receiver for a given "hold time." The loop must be fast enough, though, to respond to signal level variations caused by terrain, artificial structures, and multipath.

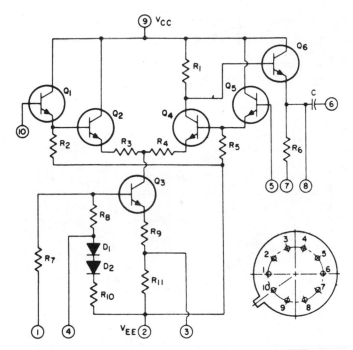

FIGURE 12.8 A gain-controlled RF amplifier integrated circuit (RCA CA3002) suitable for use in AM-FM receivers. *(Courtesy of RCA Corp., Solid-State Division)*

12.2.4 Mixer

In the mixer circuit, the RF and LO signals are acted upon by the nonlinear properties of a device (or devices) to produce a third frequency, the IF. At AM frequencies, receivers are usually built without RF preamplifiers; the antenna is fed directly to the mixer stage. In this frequency range, artificial and atmospheric noise is usually greater than the receiver NF. At FM frequencies, an RF amplifier is typically used. The mixer is located in the signal chain prior to the narrow filtering of the first IF.

Ideally, the mixer should accept the RF and LO inputs and produce an output having only one frequency (sum or difference), with signal modulation precisely transferred to this IF. In actual practice, however, mixers produce the desired IF but also many undesired components that must be filtered. Any device with nonlinear transfer characteristics can act as a mixer. For the purposes of this discussion, two classes will be discussed:

- Passive mixers, which use diodes as the mixing elements
- Active mixers, which employ gain devices (e.g., bipolar transistors or FETs)

Passive Mixer. Passive mixers have been built using germanium and silicon diodes. The development of *hot carrier diodes*, however, has resulted in a significant improvement in passive mixers.

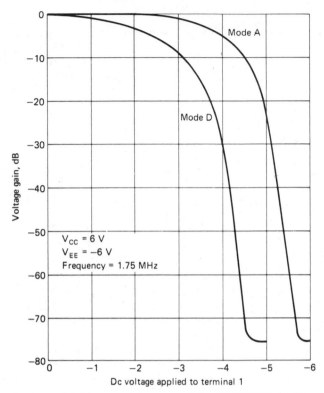

FIGURE 12.9 Gain-control curves of the CA3002 RF amplifier shown in Fig. 12.8. The "Mode A" and "Mode D" traces refer to different circuit configurations for the CA3002. *(Courtesy of RCA Corp., Solid-State Division)*

A single diode can be used to build a mixer. The performance of such a circuit is poor, though, because the RF and LO frequencies (as well as their harmonics and other odd and even mixing products) all appear at the output. As a result, a large number of spurious components are produced that are difficult to remove. Moreover, there is no isolation of the LO and its harmonics from the input circuit, necessitating the use of an RF amplifier to prevent oscillator radiation from the antenna.

A better approach can be found in the double-balanced mixer, shown in Fig. 12.13. This circuit, with its balanced diodes and transformers, cancels even harmonics of both RF and LO frequencies and provides isolation among the various ports. For optimum performance, careful matching of the diodes and transformers is necessary. The manufacturing process for hot carrier diodes has provided the tight tolerances that make them substantially better than other diode types available for mixer applications. Passive mixers can be described as low-, medium- or high-level mixers, depending on the diodes used and the number of diodes in the ring.

Active Mixer. The simplest type of active mixer uses a FET or bipolar transistor with the LO and RF signals applied to the gate-source or base-emitter junction.

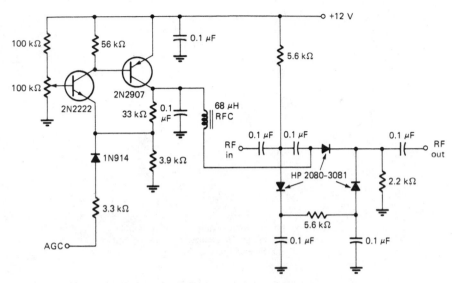

FIGURE 12.10 Schematic diagram of a PIN diode attenuator. *(Source: Ulrich L. Rohde and T. T. N. Bucher,* Communications Receivers: Principles and Design, *McGraw-Hill, New York, 1988)*

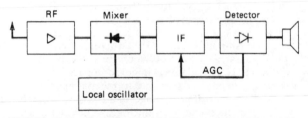

FIGURE 12.11 Simplified block diagram of a receiver with AGC. *(Source: Ulrich L. Rohde and T. T. N. Bucher,* Communications Receivers: Principles and Design, *McGraw-Hill, New York, 1988)*

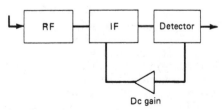

FIGURE 12.12 Block diagram of a receiver with amplified AGC. *(Source: Ulrich L. Rohde and T. T. N. Bucher,* Communications Receivers: Principles and Design, *McGraw-Hill, New York, 1988)*

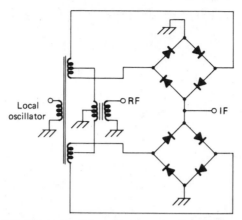

FIGURE 12.13 Schematic diagram of a double-balanced passive mixer. *(Reprinted with permission from* Ham Radio Magazine.*)*

This unbalanced mixer has the same drawbacks as the simple diode mixer and is not used for high-performance receivers.

An improved configuration uses a dual-gate FET or cascode bipolar circuit with the LO and RF signals applied to different gates (bases). The balanced transistor arrangement of Fig. 12.8 can also be used as a mixer with the LO applied to the base of Q_3 and the signal applied to the bases of Q_1 and/or Q_5.

Active mixers can be implemented using a wide variety of devices and configurations, depending on the specifications and cost structure of the receiver. Figure 12.14 shows a push-pull balanced FET mixer. The circuit uses two dual-gate FETs in a push-pull arrangement between the RF input (applied to the first gates) and the IF output. The oscillator is injected in parallel on the second gates.

Active mixers have gain and are sensitive to mismatch conditions. If operated at high levels, the collector or drain voltage can become so high that the base-collector or gate-drain junction can open during a cycle and cause severe distortion. Control of the RF input by filtering out-of-band signals and AGC are important considerations for active mixer designs. Advantages of the active mixer include lower LO drive requirements and the possible elimination of an RF preamplifier stage.

12.2.5 Local Oscillator

Most modern high-performance receivers utilize a frequency-synthesized local oscillator to generate all individual frequencies needed over the required band(s). Synthesized oscillators generally use varactor diodes for the required variable-tuning capability. This arrangement offers a stable frequency source that is easily tunable in precise steps and is easily adapted to preprogrammed stations. The days of station presets based on moving electromechanical elements are, thankfully, gone for good. The LO must meet three basic requirements:

- Spectral purity to simplify IF filtering requirements
- Frequency agility to facilitate rapid station selection

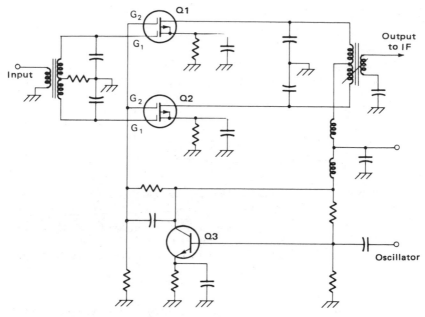

FIGURE 12.14   Schematic diagram of a push-pull dual-gate FET balanced mixer. *(Reprinted with permission from* Ham Radio Magazine.*)*

• Frequency adjustment accuracy to match the center carrier frequencies of AM and FM broadcast stations

  Synthesizers may categorized into two basic classes:

• *Direct*, in which the LO output is derived from the product of multiple mixing and filtering

• *Indirect*, in which the LO output is derived from a phase-locked loop (PLL) that samples the direct output to reduce spurious signals

PLL Synthesizer.   Most AM-FM receivers incorporating a frequency synthesized LO use a single-loop digital PLL of the type shown in Fig. 12.15. When describing frequency synthesizers mathematically, a linearized model is generally used. Because most effects occurring in the phase detector are highly nonlinear, however, only the so-called piecewise linear treatment allows adequate approximation. The PLL is nonlinear because the phase detector is nonlinear. However, it can be accurately approximated by a linear model when the loop is in lock.

  Assume that the voltage-controlled oscillator (VCO) of Fig. 12.15 is tunable over a range of 88 to 108 MHz. The output is divided to the reference frequency in a programmable divider stage whose output is fed to one of the inputs of the phase-frequency detector and compared with the reference frequency (fed to the other input). The loop filter at the output of the phase detector suppresses the reference frequency components, while also serving as an integrator. The dc control voltage output of the loop filter pulls the VCO until the divided frequency and phase equal those of the reference. A fixed division of the frequency standard

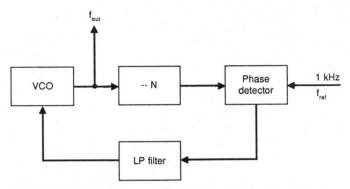

FIGURE 12.15 Block diagram of a single-loop frequency synthesizer.

oscillator (not shown in Fig. 12.15) produces the reference frequency of appropriate step size. The operating range of the PLL is determined by the maximum operating frequency of the programmable divider, its division range ratio, and the tuning range of the VCO.

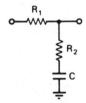

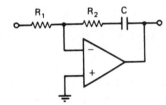

FIGURE 12.16 Schematic diagram of a PLL passive RC filter. *(Source: Ulrich L. Rohde,* Digital PLL Frequency Synthesizers, *Prentice-Hall, Englewood Cliffs, NJ, 1983. Reprinted with permission.)*

FIGURE 12.17 Schematic diagram of an active filter for a second-order PLL. *(Source: Ulrich L. Rohde,* Digital PLL Frequency Synthesizers, *Prentice-Hall, Englewood Cliffs, NJ, 1983. Reprinted with permission.)*

There are various choices of loop filter types and response. Because the VCO by itself is an integrator, a simple RC filter following the phase detector can be used. If the gain of the passive loop is too low to provide adequate drift stability of the output phase (especially if a high division ratio is used), an active amplifier may be used as an integrator. In most frequency synthesizers, an active filter-integrator is preferred to a passive one. Figure 12.16 shows a passive RC filter for the second-order loop typically used in PLL synthesizers. Figure 12.17 shows an active filter for the second-order loop.

Frequency Divider. Frequency dividers are commonly built using transistor-transistor logic (TTL), complementary MOS (CMOS), and low-power emitter-coupled logic (ECL) IC technologies. Dividers come in two common categories: synchronous counters and asynchronous counters. The frequency range of the CMOS, depending on the process, is limited to 10 to 30 MHz. TTL operates successfully up to 100 MHz in a ripple counter configuration. In a synchronous counter configuration, TTL is limited to 30 MHz.

Frequency extension is possible through the use of an ECL *prescaler*, available in variable-ratio and fixed-ratio configurations. The term *prescaling* is generally used in the sense of a predivider that is nonsynchronous with the rest of the chain. Fixed-ratio prescalers are used as ripple counters preceding a synchronous counter. A single-loop synthesizer loses resolution by the amount of prescaling.

Figure 12.18 shows a block diagram of the MC12012 (Motorola) variable-ratio dual-modulus prescaler. Through external programming, this ECL divider can be made to divide in various ratios. With proper clocking, the device can be considered a synchronous counter. With such a system (at the present state of the art), it is possible to increase the maximum operating frequency to about 400 MHz without losing resolution.

Phase-Frequency Detector. The simplest form of phase detector is the double-balanced mixer shown in Fig. 12.19. This circuit provides very low dc output and so a postamplifier is required. Because the signal from a digital PLL synthesizer is typically taken from a digital output of the divider chain, a digital phase detector is usually preferable. The minimum requirement is an exclusive OR gate for the digital signals being compared.

In actual implementation, frequency detectors typically use edge-triggered J-K master-slave flip-flops, as shown in Fig. 12.20. The reference signal frequency R is applied to pin 6 and the divided-down VCO output V is applied to pin 9. The logic states are shown representing the condition when the reference frequency is greater than the divided VCO, U, and when the divided VCO is greater than the reference D.

Still better performance can be obtained from a tristate phase-frequency comparator, which requires a *charge-pump-type* integrator at the output. Modern tristate circuits avoid the dead zone at zero phase difference. Figure 12.21 shows one such phase detector, the MC145156. A type of *antibacklash* circuit is incorporated to avoid loop instabilities. At zero correction a charge-pump would not supply any correcting voltage to the VCO. Because gain is determined by the amount of energy supplied by the charge-pump, zero correction would result in zero gain. In practice the result is that gain can drop 20 dB and heavy phase jitter can occur. Correction of this situation requires an increase in leakage with consequent reduction in reference suppression, or the use of an antibacklash circuit, as shown in the figure. For a perfect tristate phase-frequency comparator, reference frequency suppression would be infinite. Typical values of real circuits fall between 40 and 60 dB. This type of detector is available in chips using CMOS, TTL, or ECL.

Optimizing the Loop. A PLL system must be matched to its intended application. Optimization is aimed at achieving minimum noise, fastest acquisition, maximum reference frequency suppression, and a minimum number of parts. Figure 12.22 shows the major portions of a frequency synthesizer for VHF operation built around the MC145152 (Motorola) LSI frequency synthesizer chip, an external SP8690 divide-by-10/11 counter and SP1671B divide-by-2 counter, the loop filter and amplifiers. (The VCO is not shown in the schematic.) The MC145152 has two divider chains, one for the reference divider and one for the programmable divider. The fifth-order loop includes an active integrator and an additional second-order low-pass filter. Because of the additional low-pass filtering, the reference frequency suppression is substantially higher than could be provided by a simple filter. Reference suppression increases at harmonics of the reference frequency.

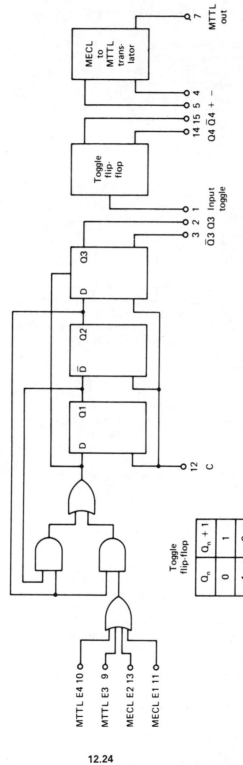

FIGURE 12.18 Block diagram of a divide-by-10/11 dual-modulus prescaler IC, the Motorola MC12012. *(Courtesy of Motorola, Inc.)*

12.24

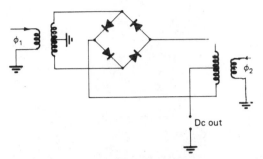

FIGURE 12.19 Schematic diagram of a double-balanced mixer used as a PLL phase detector. *(Source: Ulrich L. Rohde and T. T. N. Bucher,* Communications Receivers: Principles and Design, *McGraw-Hill, New York, 1988)*

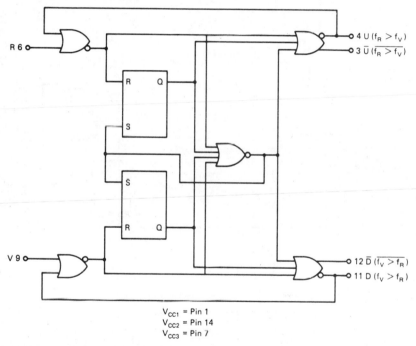

V_{CC1} = Pin 1
V_{CC2} = Pin 14
V_{CC3} = Pin 7

FIGURE 12.20 Block diagram of the MC12040 (Motorola) phase-frequency comparator. *(Courtesy of Motorola, Inc.)*

Because the LSI chips used in programmable counter applications have only a limited number of pins available, a programming technique must be used to keep the number of connections required for this function to a minimum. Some modern frequency divider chips have been optimized for use with a microprocessor bus input. The IC shown in Fig. 12.22, however, works very well with an external

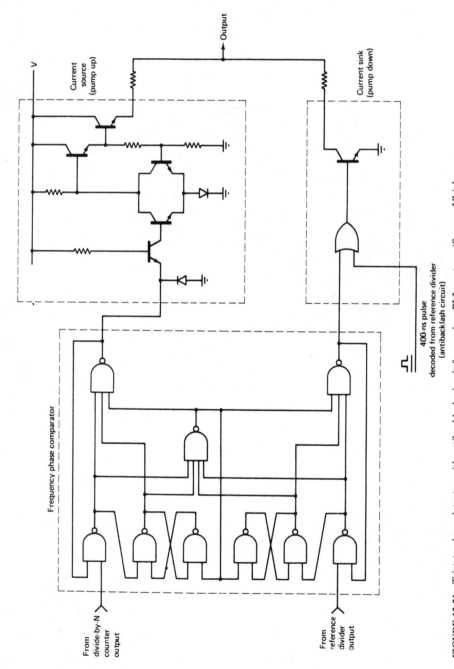

FIGURE 12.21 Tristate phase detector with antibacklash circuit for use in a PLL system. *(Source: Ulrich L. Rohde, Digital PLL Frequency Synthesizers, Prentice-Hall, Englewood Cliffs, NJ, 1983. Reprinted with permission.)*

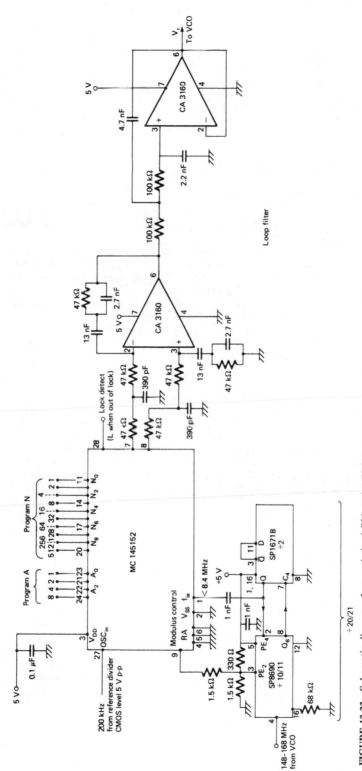

FIGURE 12.22 Schematic diagram of an optimized fifth-order PLL system incorporating a programmable divider chain. (*Source: Ulrich L. Rohde and T. T. N. Bucher, Communications Receivers: Principles and Design, McGraw-Hill, New York, 1988*)

12.27

ROM. The frequency is controlled by the loop dividers designated N and A in the figure.

Variable-Frequency Oscillator. The LO in the receiver must be capable of being turned over a specified frequency range, offset from the desired operating band(s) by the IF. Prior to the advent of the varactor diode and good switching diodes, it was customary to tune an oscillator mechanically using a variable capacitor with an air dielectric, or in some cases by moving a powdered iron core inside a coil to make a variable inductor. Automobile radios typically used the variable-inductor method of tuning the AM broadcast band. Figure 12.23 shows the classic VFO circuits commonly used in receivers. Different configurations are used in different applications, depending on the range of tuning and whether the tuning elements are completely independent or have a common element (such as the rotor of a tuning capacitor).

Newer receivers control the oscillator frequency by electronic rather than mechanical means. Tuning is accomplished by a voltage-sensitive capacitor (varactor diode). Oscillators that are tuned by varying the input voltage are referred to as *voltage-controlled oscillators* (VCOs).

The capacitance versus voltage curves of a varactor diode depend on the physical composition of the diode junction. Maximum values range from a few hundred picofarads, and useful capacitance ratios range from about 5 to 15. Figure 12.24 shows three typical tuning circuits incorporating varactor diodes. In all cases the voltage is applied through a large value resistor.

Diode Switching. Because diodes have a low resistance when biased in one direction and a very high resistance when biased in the other, they may be used to switch RF circuits. A sufficiently large bias voltage may be applied to keep the diode *on* when it is carrying RF currents, or *off* when it is subjected to RF voltages. It is important that, in the forward-biased condition, the diode add as little resistance as possible to the circuit and that it be capable of handling the maximum RF current plus the bias current. When the diode is reverse-biased, the breakdown voltage must be higher than the combined bias and RF peak voltage in the circuit. Almost any type of diode can perform switching, but at high frequencies, PIN diodes are especially useful. Figure 12.25 shows three examples of diode switching in RF circuits.

The advantage of electronic tuning using varactor diodes is only fully realized when band selection also takes place electronically. Diode switches are preferable to mechanical switches because of their high reliability. Diode switches eliminate the need for a mechanical link between front panel controls and the tuned circuits to be switched.

Crystal-Controlled Oscillator. Piezoelectric quartz crystals are the basis for most PLL reference oscillators. Quartz crystals have resonances that are much more stable than the LC circuits discussed so far and also have very high Q. Consequently, quartz crystal resonators are typically used for high-stability fixed-frequency oscillators. A piezoelectric material is one that develops a voltage when it is under a mechanical strain or is placed under strain by an applied voltage. A physical piece of such material, depending upon its shape, can have a number of mechanical resonances. By appropriate shaping and location of the electrodes, one or another resonant mode of vibration can be favored, so that the resonance may be excited by an external voltage.

The crystal exhibits at its frequency of oscillation the equivalent electric

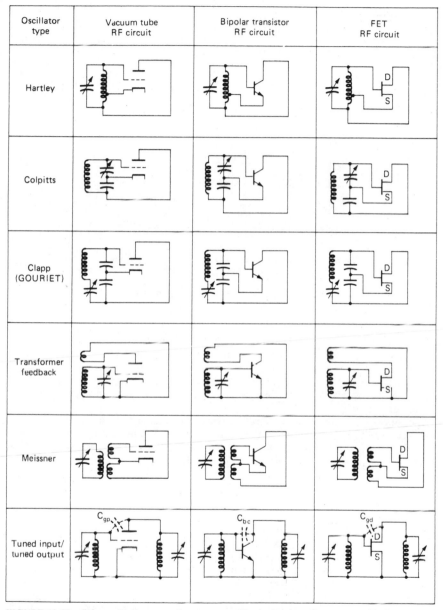

Oscillator type	Vacuum tube RF circuit	Bipolar transistor RF circuit	FET RF circuit
Hartley			
Colpitts			
Clapp (GOURIET)			
Transformer feedback			
Meissner			
Tuned input/ tuned output			

FIGURE 12.23 Schematic diagrams of common oscillator circuits using vacuum-tube, transistor, and FET active circuits. DC and biasing circuits are not shown. *(Source: Ulrich L. Rohde and T. T. N. Bucher,* Communications Receivers: Principles and Design, *McGraw-Hill, New York, 1988)*

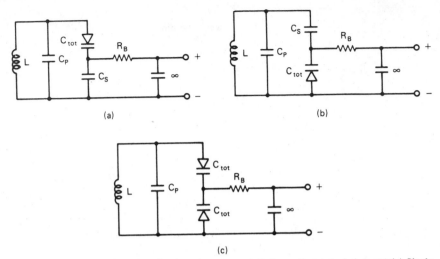

FIGURE 12.24 Typical tuning circuits using varactor diodes as the control element. (*a*) Single diode in the circuit low side. (*b*) Single diode in the circuit high side. (*c*) Two diodes in a series back-to-back arrangement. (*Source: Ulrich L. Rohde,* Digital PLL Frequency Synthesizers, *Prentice-Hall, Englewood Cliffs, NJ, 1983. Reprinted with permission.*)

circuit shown in Fig. 12.26. The series resonant circuit represents the effect of the crystal vibrator, and the shunt capacitance is the result of the coupling plates and of capacitance to surrounding metallic objects (such as the metal case). The resonant circuit represents the particular vibrating mode that is excited. If more than one mode can be excited, a more complex circuit would be required to represent the crystal.

The most common type of circuit using a fundamental (AT) crystal is an *aperiodic oscillator*, which has no selective circuits other than the crystal. Such oscillators, often referred to as *parallel resonant oscillators*, use the familiar Pierce and Clapp configurations (see Fig. 12.27).

12.2.6 AM-FM Demodulation

The function of any receiver is to recover the original information used to modulate the transmitter. This process is referred to as *demodulation*, and the circuits that perform the recovery are called *demodulators*. The term *detector* is also used, and the demodulator in a single-superheterodyne receiver is sometimes called a *second detector*. Today, however, the term *detector* is seldom used in this fashion.

Because of transmission and reception system distortions and noise caused by thermal, atmospheric, and artificial sources, the demodulated signal is—to some extent—a distorted version of the original modulating signal and is corrupted by the addition of noise. It is the obvious goal of the demodulator to minimize these corrupting effects and provide an output signal that is as close to the original modulating waveform as possible.

Analog modulated waves come in a variety of forms, including:

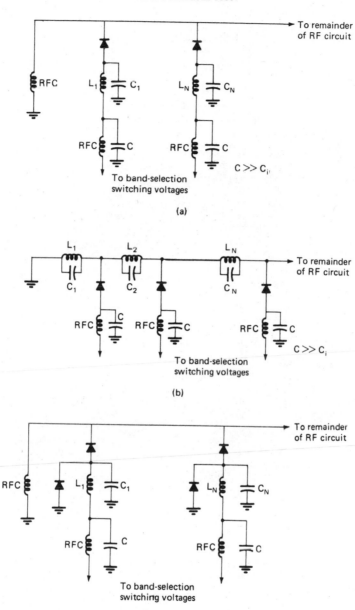

FIGURE 12.25 Typical circuits using diodes for band switching. (*a*) Series diode arrangement. (*b*) Shunt-diode arrangement. (*c*) Use of both series and shunt diodes. (*Source: Ulrich L. Rohde and T. T. N. Bucher,* Communications Receivers: Principles and Design, *McGraw-Hill, New York, 1988*)

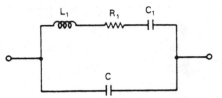

FIGURE 12.26 The equivalent electric circuit of a crystal at resonance (spurious and overtone modes not shown). *(Source: Ulrich L. Rohde,* Digital PLL Frequency Synthesizers, *Prentice-Hall, Englewood Cliffs, NJ, 1983. Reprinted with permission.)*

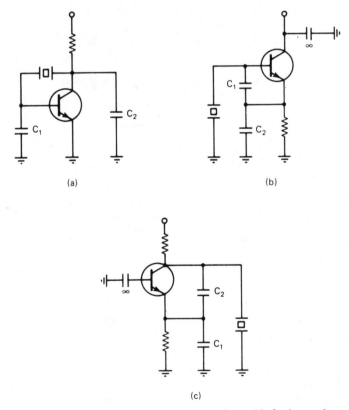

FIGURE 12.27 Common parallel resonant circuits used in fundamental crystal oscillators. (*a*) Pierce circuit. (*b*) Clapp circuit, collector grounded. (*c*) Clapp circuit, base grounded. *(Source: Ulrich L. Rohde,* Digital PLL Frequency Synthesizers, *Prentice-Hall, Englewood Cliffs, NJ, 1983. Reprinted with permission.)*

- Conventional AM
- Double-sideband suppressed carrier (DSSC) AM
- Single-sideband (SSB) AM
- Vestigial-sideband (VSB) AM
- Phase modulation (PM)
- Frequency modulation (FM)

For the purposes of this discussion, we will concentrate on the two forms used for AM-FM broadcast transmission.

AM Demodulation. An AM signal is made up of an RF sinusoid whose envelope varies at a relatively slow rate about an average (carrier) level. Any sort of rectifier circuit will produce an output component at the modulation frequency. Figure 12.28 illustrates two of the simple diode rectifier circuits that may be used, along with idealized waveforms. The average output of the rectifier of Fig. 12.28a is proportional to the carrier plus the signal. The circuit exhibits, however, significant output energy at the RF and its harmonics. A low-pass filter is necessary to eliminate these components. If the selected filter incorporates a sufficiently large capacitor at its input, the effect is to produce a peak rectifier, with the idealized waveform shown in Fig. 12.28b. In this case the demodulated output is increased from the average of a half a sine wave (0.637 peak) to the full peak, and the RF components are substantially reduced. A peak rectifier used in this way is often referred to as an *envelope detector* or demodulator. It is the circuit most frequently used for demodulating AM broadcast signals.

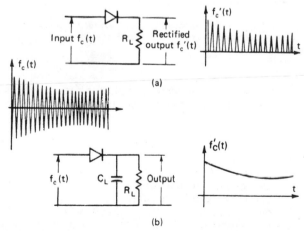

FIGURE 12.28 AM demodulators with idealized waveforms. (*a*) Average demodulator and resulting waveform. (*b*) Envelope detector and resulting waveform. (*Source: M. Schwartz,* Information, Transmission and Noise, *2d ed., McGraw-Hill, New York, 1970. Reprinted by permission of McGraw-Hill Publishing Company.*)

AM signals may also be demodulated by using a coherent or *synchronous demodulator*. This type of demodulator uses a mixer circuit, with an LO signal synchronized in frequency and phase to the carrier of the AM input. Figure 12.29

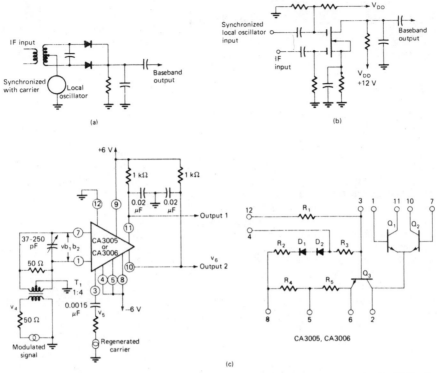

FIGURE 12.29 Three types of synchronous demodulators. (*a*) Diode-based circuit. (*b*) Dual-gate MOSFET-based circuit. (*c*) Bipolar IC- (CA3005/CA3006) based circuit. (*Courtesy of RCA Corp., Solid-State Division*)

illustrates three approaches to synchronous demodulation. The synchronous component may be generated by an oscillator phase-locked to the carrier, as illustrated in Fig. 12.30.

The synchronous demodulator translates the carrier and sidebands to baseband. As long as the LO is locked to the carrier phase, baseband noise results only from the in-phase component of the noise input. Consequently the noise increase and S/N reduction that occur at low levels in the envelope demodulator are absent in the synchronous demodulator. The recovered carrier filtering is narrow band, so that phase lock can be maintained at carrier-to-noise levels below useful modulation output levels. This type of circuit, while better than an envelope demodulator, is not generally used for AM broadcast demodulation because of its complexity. Most stereo AM receivers, however, incorporate synchronous demodulators as part of the decoding circuit.

FM Demodulation. The most common technique for FM demodulation incorporates the use of linear circuits to convert frequency variations to envelope variations, followed by an envelope detector. Another technique used with linear integrated circuits involves the conversion of frequency variations to phase variations that are then applied to a phase demodulator. Still other FM demodu-

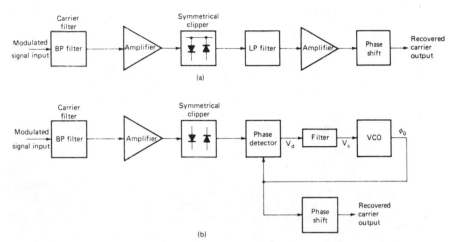

FIGURE 12.30 Two approaches to AM carrier recovery. (*a*) A system based on a filter, clipper and amplifier. (*b*) A system based on a filter, clipper, and PLL. (*Source: Ulrich L. Rohde and T. T. N. Bucher,* Communications Receivers: Principles and Design, *McGraw-Hill, New York, 1988*)

lators employ PLLs and frequency-locked loops (FM feedback circuits), or counter circuits whose output is proportional to the rate of zero crossings of the wave. Frequency demodulators are often referred to as *discriminators* or *frequency detectors*.

Resonant circuits are used in discriminators to provide adequate sensitivity to small-percentage frequency changes. To eliminate the dc components, two circuits may be used, one tuned above and one tuned below the carrier frequency. The outputs are demodulated by envelope demodulators and are then subtracted, eliminating the dc component. Voltage sensitivity is doubled compared to the use of a single circuit. This balanced arrangement also eliminates all even-order distortion so that the first remaining distortion term is third-order. Figure 12.31

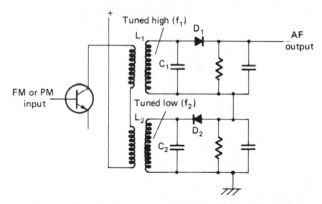

FIGURE 12.31 Schematic diagram of the Travis discriminator. (*Source: S. W. Amos, "FM Detectors,"* Wireless World, **87** *(no. 1540) (January 1981): 77. Reprinted with permission from Electronics and Wireless World.*)

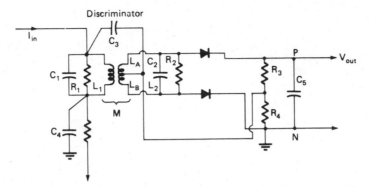

FIGURE 12.32 The Foster-Seeley FM discriminator circuit with tuned primary. *(Source: Ulrich L. Rohde and T. T. N. Bucher,* Communications Receivers: Principles and Design, *McGraw-Hill, New York, 1988)*

shows one implementation of this scheme, known in the United States as the *Travis discriminator*. Because the circuit depends on the different amplitude responses of two circuits, it has sometimes been called an *amplitude discriminator*.

The *Foster-Seeley discriminator*, shown in Fig. 12.32, is a more common approach to FM demodulation. In this circuit, the voltage across the primary is added to the voltage across each of the two halves of the tuned secondary. At resonance the secondary voltage is in quadrature with the primary voltage, but as the frequency changes, so do the phase shifts. The voltages from the upper and lower halves of the secondary add to the primary voltage in opposition. As the frequency rises, the phase shift increases, and as the frequency falls, it decreases. The opposite phase additions cause the resulting amplitudes of the upper and lower voltages to differ, producing the discriminator effect. When the primary circuit is also tuned to the center frequency (which produces much higher demodulation sensitivity), the phase of the primary voltage also varies slightly, as does its amplitude. The proper selection of coupling factor is required to produce optimum sensitivity and linearity of the discriminator. Because of the method of arriving at the amplitude difference in this demodulator, it is sometimes referred to as a *phase discriminator*. The *ratio detector* is a variant of the phase discriminator that has an inherent degree of AM suppression. The circuit tolerates less effective limiting in prior circuits and thus can reduce the cost of the receiver. Figure 12.33 shows the basic concept of the ratio detector. It resembles the Foster-Seeley circuit, except that the diodes are reversed. The combination of R_1, R_2, and C_3 has a time constant that is long compared to the lowest modulating frequency (on the order of 0.1 s for audio modulation). The result is that during modulation, the voltage to the (grounded) center tap across load resistor R_2 is $(E_1 + E_2)/2$, and across R_1 it is $-(E_1 + E_2)/2$. Following the circuit from ground through R_2 and C_2, we see that the voltage at the center tape of the capacitors is $[(E_1 + E_2)/2] - E_2 = (E_1 - E_2)/2$, or half of the value of the Foster-Seeley discriminator.

The long time constant associated with C_3 reduces the required current from the diodes when $E_1 + E_2$ drops and increases it when $E_1 + E_2$ rises. This changes the load on the RF circuit and causes higher drive when the output falls and lower drive when it rises. This tends to further stabilize the voltage $E_1 + E_2$ against incidental AM. The sum voltage can also be used to generate an AGC signal, so

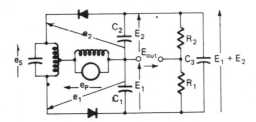

FIGURE 12.33 Schematic diagram of a basic ratio detector circuit. *(Source: Ulrich L. Rohde and T. T. N. Bucher,* Communications Receivers: Principles and Design, *McGraw-Hill, New York, 1988)*

that the prior circuits need not limit. This can be advantageous when a minimum number of circuits is required and the selectivity is distributed.

Figure 12.34 shows an implementation of the ratio detector. The primary is tuned; the tuned secondary is coupled similar to the circuit shown in Fig. 12.33; and the untuned tertiary, when used, is tightly coupled to the primary to provide a lower voltage than appears across the primary. It may be replaced by a tap, isolating capacitor, and RF choke for the same effect. A lower-impedance primary can achieve a comparable performance, except for gain. Use of the tertiary winding allows the primary to be designed for optimum gain from the driving amplifier.

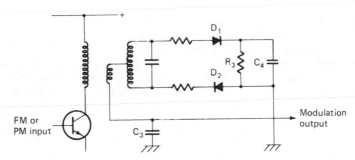

FIGURE 12.34 Circuit implementation of a ratio detector. *(Source: S. W. Amos, "FM Detectors," Wireless World, 87 (no. 1540) (January 1981): 77. Reprinted with permission from Electronics and Wireless World.)*

Amplitude Limiter. Amplitude limiting is essential for FM demodulators using analog circuits. Although solid-state amplifiers tend to limit when the input signal becomes excessive, limiters that make use of this characteristic often limit the envelope asymmetrically. For angle demodulation, symmetrical limiting is desirable. AGC circuits, which can keep the signal output constant over wide ranges of input signals, are unsuitable for limiting because they cannot be designed with sufficiently rapid response to eliminate the envelope variations encountered in angle modulation interference. One or more cascaded limiter stages are required for good FM demodulation.

Almost any amplifier circuit, when sufficiently driven, provides limiting. However, balanced limiting circuits produce better results than those that are not

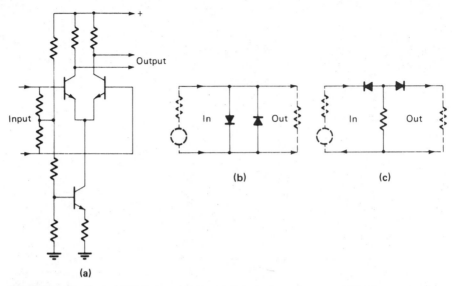

FIGURE 12.35 Typical FM limiter circuits. (*a*) Balanced-transistor amplifier. (*b*) Shunt-diode limiter. (*c*) Series diode limiter. (*Source: Ulrich L. Rohde and T. T. N. Bucher,* Communications Receivers: Principles and Design, *McGraw-Hill, New York, 1988*)

balanced. In general, current cutoff is more effective than current saturation in producing sharp limiting thresholds. Nonetheless, overdriven amplifiers have been used in many FM systems to provide limiting. If the amplifier is operated with low supply voltage and near cutoff, it becomes a more effective limiter. The transistor differential amplifier shown in Fig. 12.35*a* is an excellent limiter when the bias of the emitter load transistor is adjusted to cause cutoff to occur at small base-emitter input levels.

The classic shunt-diode limiter is shown in Fig. 12.35*b*. It is important that the *off* resistance of the diodes be much higher than the driving and load impedances, and the *on* resistance be much lower. Figure 12.35*c* shows the classic series diode limiter. In this example the diodes are normally biased *on*, so that they permit current flow between the driver and the load. As the RF input voltage rises, one diode is cut off; and as it falls, the other is cut off. The effectiveness of limiting is determined by the difference in *off* and *on* resistances of the diode, compared to the driving and load impedances.

12.3 STEREO SYSTEMS

Stereophonic broadcasting has reshaped the radio industry during the last 25 years. Stereo FM, introduced during the early 1960s, gave FM broadcasters a powerful new marketing tool. Receiver manufacturers responded with new radios that offered high-quality performance at an affordable price.

Although few people outside the broadcast industry know it, AM broadcasters have been talking about stereo operation as long as FM engineers. It was not until the early 1980s, however, before AM stereo broadcasting became a reality.

12.3.1 FM Stereo

The system devised for broadcasting stereo audio over FM has served the industry well. Key requirements for the scheme were: (1) compatibility with monophonic receivers that existed at the time the standard was developed, and (2) a robust signal that would not be degraded significantly by multipath. Figure 12.36 shows the composite baseband that modulates the FM carrier for stereophonic broadcasting. The two-channel baseband has a bandwidth of 53 kHz and is made up of:

- A *main channel* $(L + R)$ signal, which consists of the sum of left plus right audio signals (the same signal broadcast by a monaural station). A fully modulated main channel will modulate the FM transmitter to 45 percent when broadcasting stereo programming.
- A *stereophonic subchannel* $(L - R)$, which consists of a double-sideband AM modulated carrier with a 38-kHz center frequency. The modulating signal is equal to the difference of the left and right audio inputs. The subcarrier is suppressed to conserve modulation capability. As a result, the AM sidebands have the same modulation potential as the main channel. A fully modulated subchannel will modulate the FM transmitter to 45 percent when broadcasting stereo programming.
- A 19-kHz subcarrier *pilot*, which is one-half the frequency of the stereophonic subchannel and in phase with it. The pilot supplies the reference signal needed by stereo receivers to reinsert the 38-kHz carrier for demodulation of the double-sideband suppressed carrier transmission. The pilot, in other words, is used to synchronize the decoder circuitry in the receiver to the stereo generator at the transmitter. The frequency tolerance of the pilot is ±2 Hz. The pilot modulates the transmitter 8 to 10 percent.

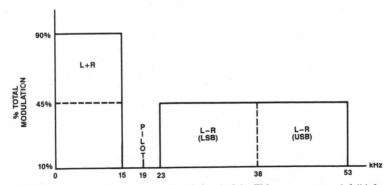

FIGURE 12.36 The composite baseband signal of the FM stereo system. A full left-only or right-only signal will modulate the main $(L + R)$ channel to a maximum of 45%. The stereophonic subchannel is composed of upper sideband (USB) and lower sideband (LSB) components.

Generating the Stereo Signal.   Two basic approaches have been used to generate the stereophonic subchannel: time-division multiplexing (TDM), the switching method; and frequency-division multiplexing (FDM), the matrix method.

A simplified block diagram of the FDM approach is shown in Fig. 12.37. The left and right audio channels first pass through a preemphasis circuit and low-pass filter. They are then supplied to the matrix, which produces *sum* and *difference* components. The audio signals are added to form the $L + R$ main channel signal. The difference signal is fed to a balanced modulator that generates the $L - R$ subchannel. Because a balanced modulator is used, the 38-kHz carrier is suppressed, leaving only the modulated sidebands. The 19-kHz pilot signal is derived by dividing the 38-kHz oscillator by 2. The main channel, stereophonic subchannel and pilot are then combined in the proper (45/45/10 percent) ratio to form the composite baseband.

The TDM method of generating a stereo signal is shown in block diagram form in Fig. 12.38. The $L + R$ and $L - R$ signals are generated by an electronic switch that is toggled at a 38-kHz rate. The switch samples one audio channel and then the other. Considerable harmonic energy is generated in this process, requiring the use of a low-pass filter. When the harmonics are filtered out, the proper composite waveform results. This approach, while simple and stable, may produce unwanted artifacts, most notably reduced stereo separation, because of the filtering requirements.

An improvement to the basic TDM concept is shown in Fig. 12.39. By using a *soft switch* to sample the left and right channels, it is possible to eliminate the low-pass filter and its side-effects. The variable element shown in the figure consists of an electronic attenuator that is capable of swinging between its minimum and maximum attenuation values at a 38-kHz rate. Like the fast-switching TDM system, the $L + R$ and $L - R$ channels are generated in one operation. No filter is required at output of the generator as long as the 38-kHz sine wave is free from harmonics and the variable attenuator has good linearity.

Decoding the Stereo Signal.   All stereo FM receivers include a circuit to convert the multiplexed signal at the FM detector to the left and right audio channels originally transmitted by the station. There are a number of ways to accomplish this task. In practice, however, one type of decoder is most commonly found, built

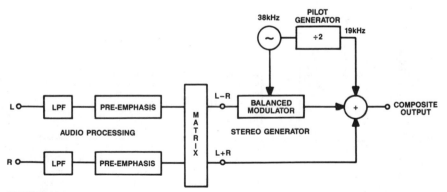

FIGURE 12.37   Functional block diagram of a frequency-division multiplexing (FDM) FM stereo generator.

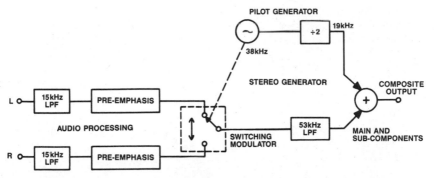

FIGURE 12.38   Functional block diagram of a time-division multiplexing (TDM) FM stereo generator.

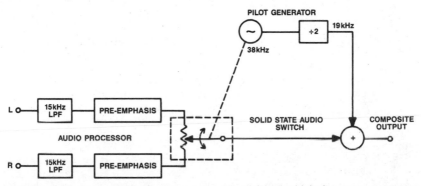

FIGURE 12.39   Functional block diagram of a time-division multiplexing stereo generator using a variable electronic attenuator.

around a PLL integrated circuit (see Fig. 12.40). This system offers both high performance and low cost.

The composite signal from the demodulator is fed to a buffer amplifier and sampled by a PLL within the decoder IC. A voltage controlled oscillator, typically running at 76 kHz (four times the pilot frequency) is locked in phase with the pilot by the error output voltage of the PLL. The oscillator signal is divided by 2, resulting in a square wave at 38 kHz with nearly perfect duty cycle and fast rise and fall times. This signal drives the audio switcher (demultiplexer) to transfer the composite baseband to the left and right audio outputs in synchronization with the station's stereo generator. A deemphasis circuit follows the matrix to complement the signal preemphasis at the FM transmitter.

12.3.2 AM Stereo

AM stereo operation was approved by the FCC in 1981. In what has turned out to be a very controversial decision, the commission introduced a new concept in the process of selecting new broadcast transmission standards: selection by the "marketplace." No single system of the several offered by various manufactur-

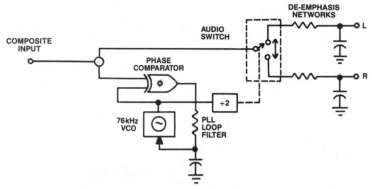

FIGURE 12.40 Block diagram of a stereo decoder using PLL-controlled time-division demultiplexing. In modern receivers, the functions of such a stereo decoder are integrated into one IC.

ers was approved by the FCC. It was, instead, left to broadcasters to decide which system to use. The net effect has been to slow significantly the conversion of AM stations to stereo operation.

The proponents of AM stereo systems developed two distinctly different means of transmitting left and right audio channels on an AM carrier. Both schemes provide good stereo performance and result in no significant increase in interference levels. Both are also compatible with monophonic receivers, a key requirement for success in the marketplace. The two systems are not, however, compatible with each other, requiring different means to decode the signals.

One approach involves independently modulating the two sidebands, with the left channel on one sideband and the right channel on the other sideband (commonly known as the *Kahn* system, named after its inventor, Leonard Kahn). The second approach involves modulating the main AM carrier with the sum ($L + R$) information and the use of quadrature modulation to convey the difference ($L - R$) audio channel (compatible quadrature AM, better known as the *C-QUAM*[1] system, developed by Motorola).

Generating the C-QUAM Signal.   It is not too difficult to convey two signals on a single AM broadcast carrier by transmitting one using the standard AM method and applying the other signal in quadrature on the same carrier. The problem, however, is how to maintain compatibility with monophonic receivers in the field, many of which have unsophisticated detectors and narrow-bandwidth IF chains. The C-QUAM system is a modification of simple quadrature modulation that is designed to maintain monophonic compatibility.

The C-QUAM encoder is shown in Fig. 12.41. As in FM stereo broadcasting, sum and difference signals of the left and right audio inputs are produced. Pure quadrature is generated by taking the $L + R$ and $L - R$ signals and modulating two balanced modulators fed with RF signals that are out of phase by 90° (producing components referred to as I and Q). As shown in the figure, the 90° phase shift is derived by using a *Johnson counter*, which divides an input frequency (4 times the station carrier frequency) by 4 and provides digital signals precisely 90° out of phase for the balanced modulators. The carrier is inserted directly from the

---

1. C-QUAM is a registered trademark of Motorola, Inc.

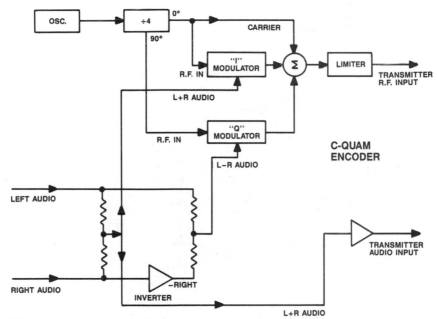

FIGURE 12.41　Block diagram of a C-QUAM AM stereo exciter.

Johnson counter. At the output of the summing network, the result is a pure quadrature AM stereo signal. From there it is passed through a limiter that strips the incompatible AM components from the signal. The output of the limiter is amplified and sent to the transmitter in place of the crystal oscillator.

The left and right audio signals are summed and sent as compatible $L + R$ to the audio input terminals of the transmitter.

Decoding the C-QUAM Signal.　C-QUAM AM stereo is decoded by converting the demodulated AM broadcast waveform (which is already close to a quadrature signal) to pure quadrature and then using a quadrature detector to extract the $L - R$ component (see Fig. 12.42). In order to prepare the received signal for the quadrature demodulator, it has to be converted from the envelope-detector-compatible signal that is broadcast to the original quadrature signal that was not envelope detector compatible. This is accomplished by demodulating the broadcast signal in two ways: with an envelope detector and with an I detector. The two signals are compared, and the resultant error is used to gain-modulate the input of the I and Q demodulators.

When the transmitted signal is $L + R$ (monaural, no stereo), it is pure AM (only I sidebands). In this case the envelope detector and the I demodulator see the same thing. There is no error signal, the inverse modulator does nothing, and the signal passes without change. However, when a left or right only signal is transmitted, both AM and PM are present, and the input signal is shifted in phase to the I demodulator, which loses some of its I amplitude. The envelope detector sees no difference in the AM because of the phase modulation. When the envelope detector and the I demodulator are compared, there is an error signal. The error signal increases the input level to the detector. This makes the input signal

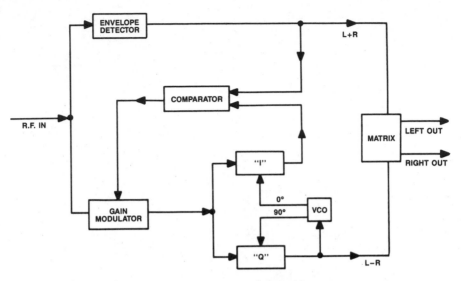

FIGURE 12.42 Block diagram of a C-QUAM AM stereo decoder.

to the I and Q demodulators look like a pure quadrature signal, and the audio output yields the $L - R$ information. The demodulator output is combined with the envelope-detector output in a matrix to reconstruct the left and right audio channels.

Generating the Kahn Signal. The Kahn AM stereo system uses each sideband of the AM envelope to convey stereo information. Figure 12.43 shows a block diagram of the stereo exciter. The left and right audio sources are summed and fed through a constant-amplitude, phase-difference network. The output of this network drives an audio frequency amplifier to produce a signal suitable for modulating a conventional AM transmitter.

The difference between the L and R components drives a companion phase-difference network to assure a quadrature relationship between the output of the sum circuit and the difference circuit. This signal is next applied to a phase modulator, and the resulting phase-modulated wave is frequency translated to the station's carrier frequency. The phase-modulated wave is envelope modulated by the transmitter to produce an independent sideband wave in which the lower sideband carries the left information and the upper sideband carries the right information.

Additional circuitry is provided to ensure proper time delay so that the AM and PM components arrive at the modulated stage of the transmitter coincidentally. A second-order $L - R$ component is also added to the fundamental $L - R$ audio signal supplied to the phase modulator. This component minimizes bandwidth requirements and aids in low-distortion reception.

Decoding the Kahn Signal. Figure 12.44 shows a simplified block diagram of an independent sideband AM stereo receiver. The envelope-detector output contains both the left and right audio channels ($L + R$) by nature of the modulating system. An *inverse modulator circuit*, driven by the output of the receiver IF and

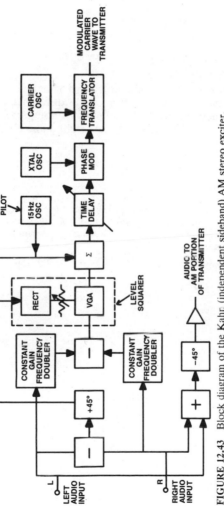

FIGURE 12.43 Block diagram of the Kahn (independent sideband) AM stereo exciter.

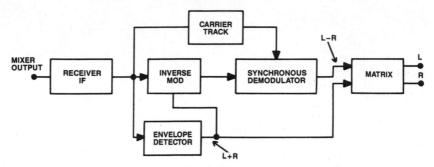

FIGURE 12.44 Simplified block diagram of an AM stereo decoder for the Kahn system.

the envelope detector, feeds a synchronous demodulator. The addition of a *carrier track signal* allows construction of an audio difference ($L - R$) component. When the envelope-detector ($L + R$) and synchronous-detector ($L - R$) signals are combined in a matrix, the original left and right audio channels are recovered.

BIBLIOGRAPHY

1. E. B. Crutchfield. *National Association of Broadcasters Handbook*, 7th ed. Sec. 3.2, AM Stereo Systems—An Overview (by Edmund Williams), and Kahn AM Stereo System (by Leonard Kahn); and Sec. 3.4, Subcarrier Transmissions and Stereophonic Broadcasting (by John Kean).
2. *Introduction to the Motorola C-QUAM AM Stereo System*, Motorola, 1985.

CHAPTER 13
LOUDSPEAKERS AND SOUND SYSTEMS

Jerry Whitaker
Editorial Director, Broadcast Engineering *Magazine*

13.1 INTRODUCTION

In days gone by, engineers seldom thought much about the speakers or power amplifiers used in their facilities. If sound was coming from the speakers, they assumed everything must be working properly. Today, increasing demands are being placed on professional monitoring systems. Radio and TV stations, recording studios, and a wide variety of sound reinforcement users now employ high-

---

Portions of this chapter were adapted from the following references: (1) Cal Perkins, "Inside Power Amplifiers," *Broadcast Engineering* magazine, August 1986. (2) Brad Dick, "Inside Monitor Loudspeakers," *Broadcast Engineering* magazine, June 1987. (3) Richard Cabot, "Dimensions in Equalization," *Broadcast Engineering* magazine, August 1985. (4) Richard Cabot, "Limiters, Compressors and Expanders," *Broadcast Engineering* magazine, August 1986.

quality analog and digital audio equipment that provides sonic quality never before available.

This performance does not come, however, without a fair amount of work and attention to detail in specifying and constructing the sound system, from the audio preamplifier to the speaker. Unfortunately, when it comes to selection of the individual components, users may end up purchasing amplifiers or speakers that are inadequate for the particular application. A common mistake is to purchase an amplifier with insufficient power capability. On the other hand, an amplifier with far more power than the speaker is capable of handling is also a bad buy. Too often, equipment is purchased on the basis of brand loyalty or familiarity. Informed decisions are based on a thorough knowledge of how the system works and its strong points and weak points.

13.2 POWER AMPLIFIERS

Circuit tricks and beyond-the-state-of-the-art data-sheet specifications are of little value if an amplifier will not perform adequately on the job. Products with similar specifications can exhibit surprising real-world performance differences. In most cases, audible differences between two amplifiers with similar specs can be attributed directly to abnormal performance of one (or both) of the products. Because commercial usage may sometimes stress an amplifier beyond its designed operating region, performance outside of the typical conditions can be more important than data-sheet specifications. Low- and high-frequency overload characteristics, intermodulation distortion products, high-frequency stability, and the ability to handle real-world reactive loads are some important criteria by which to evaluate an amplifier.

For a manufacturer, the major costs of a power amplifier include the package, power supply, output devices, and heat exchanger system (heat sink). The actual circuit configuration used may have only a minimal effect on the total cost of the product.

Most power amplifiers have three basic stages of amplification:

- The input stage, usually a differential amplifier.
- The intermediate stage, which level-shifts and amplifies the input-stage signal. Unless the input stage has a current mirror and is well buffered, most of the amplifier gain is provided by the intermediate stage (if the topology is a totally discrete design). Designs incorporating IC front ends have voltage gains of 10,000 or more. Usually the intermediate stage has a gain of from 50 to 5000.
- The output stage, which may or may not have gain. Output stages with gain are generally limited to a value of less than 20 (26 dB).

13.2.1 Input Stage

The input stage of most power amplifiers manufactured today consists of bipolar or FET differential amplifiers. This is true whether the circuit design uses all discrete devices or has an IC front end. The popular 5532 IC has an *npn* bipolar FET input stage, while the TI072 uses an FET input stage. Discrete designs often use cascode-connected bipolar or FET differential amplifiers. A cascode-connected input circuit (Fig. 13.1) provides superior performance to the more common dif-

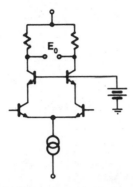

FIGURE 13.1 Cascode bipolar differential amplifier designed for use as the input stage of a power amplifier. *(Source: Cal Perkins, "Inside Power Amplifiers," Broadcast Engineering, August 1986. Reprinted with permission of Intertec Publishing Corp.)*

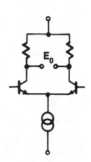

FIGURE 13.2 Standard bipolar differential amplifier. *(Source: Cal Perkins, "Inside Power Amplifiers," Broadcast Engineering, August 1986. Reprinted with permission of Intertec Publishing Corp.)*

ferential amplifier (shown in Fig. 13.2) in the following areas:

- Better high-frequency response
- Increased high-frequency input impedance [the cascode connection eliminates the input transistor *Miller effect* caused by the collector-to-base capacitance (C_{ob})]
- Superior power-supply noise rejection.

Because of their high output impedance, cascode-connected input differential amplifiers are excellent voltage-to-current converters. If the circuit topology takes full advantage of the differential input stage, the amplifier power-supply rejection ratio can easily exceed 120 dB.

13.2.2 Intermediate Stage

The intermediate-stage circuit is used to drive the output stage to its designed power level. Figures 13.3 through 13.6 show four basic approaches to intermediate-amplifier design. Circuit types can be classified into two basic categories: symmetrical and asymmetrical. In a symmetrical scheme, both the positive and the negative signal swings are driven equally. This ensures that the interelement capacities of the output stage and frequency compensation networks do not unbalance the signal, causing second-harmonic distortion. When the output has equal drive, the large-signal frequency response of the

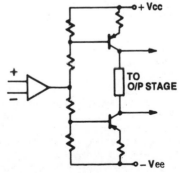

FIGURE 13.3 IC front end with discrete level shift, symmetrical drive. *(Source: Cal Perkins, "Inside Power Amplifiers," Broadcast Engineering, August 1986. Reprinted with permission of Intertec Publishing Corp.)*

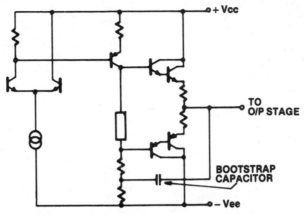

FIGURE 13.4 Asymmetrical discrete design using a boot-strapped collector load in an intermediate (voltage gain) stage. *(Source: Cal Perkins, "Inside Power Amplifiers," Broadcast Engineering, August 1986. Reprinted with permission of Intertec Publishing Corp.)*

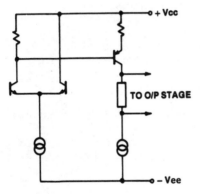

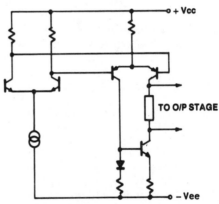

FIGURE 13.5 Asymmetrical discrete design using a current source for the second-stage load. *(Source: Cal Perkins, "Inside Power Amplifiers," Broadcast Engineering, August 1986. Reprinted with permission of Intertec Publishing Corp.)*

FIGURE 13.6 Symmetrical discrete design, also known as *Wilson current mirror*. *(Source: Cal Perkins, "Inside Power Amplifiers," Broadcast Engineering, August 1986. Reprinted with permission of Intertec Publishing Corp.)*

positive and negative halves should also be equal. For this reason, the amplifier should exhibit a symmetrical slew rate (a measure of how fast a signal changes from one instantaneous value to another). Surprisingly, however, amplifiers with symmetrical slew rates are in the minority.

An asymmetrical drive scheme is one in which the positive- and negative-half *pullup* and *pulldown* are not equal. Active current source loads and bootstrapped resistive collector loads, unless carefully designed, will not deliver the same low-distortion performance as a symmetrical drive scheme. This is especially true at high frequencies. There are some 0.002 percent total harmonic distortion (THD) amplifiers on the market that use asymmetrical drive. However, these amplifiers achieve low distortion with more negative feedback than would be used with a symmetrical drive approach.

Slew symmetry, high-frequency stability, and overload recovery are the most visible performance characteristics of an asymmetrical drive scheme. These factors are especially noticeable when the amplifier is driven into its nonlinear region. Small-signal analysis flies out the window when an amplifier is driven into clipping. The differences between asymmetrical and symmetrical drive become especially apparent during overdrive conditions.

13.2.3 Output Stage

Whether symmetrical or asymmetrical in design, the output-stage voltage and current drivers use discrete devices. These components provide voltage amplification and the necessary level shifting to provide the full-load signal swing needed with a unity-gain output stage, such as an emitter-follower (bipolar) or source-follower (MOSFET) circuit. In many designs, most of the distortion produced by the amplifier is generated not by the output stage but by the drive stage. A well-designed driver will measure less than 0.05 percent THD with no feedback applied. Depending on the beta linearity, transconductance, and circuit design, it is possible to get the *output-stage* distortion down to the 0.05 percent level before the feedback loop is closed around the amplifier.

The output configuration of a power amplifier can take on a number of forms. To understand how the stage functions, it is necessary to first know the electrical conditions under which the power devices will operate. Figure 13.7 shows a simplified two-transistor complementary output stage driving a resistive load. It can be shown mathematically that class B efficiency is not a linear function of output power. In a class B amplifier, maximum dissipation occurs at 40.5 percent of full

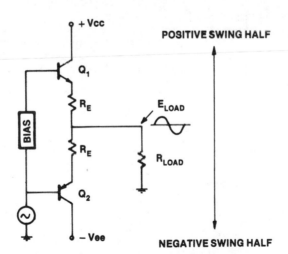

FIGURE 13.7 Simplified bipolar complementary output stage. *(Source: Cal Perkins, "Inside Power Amplifiers,"* Broadcast Engineering, *August 1986. Reprinted with permission of Intertec Publishing Corp.)*

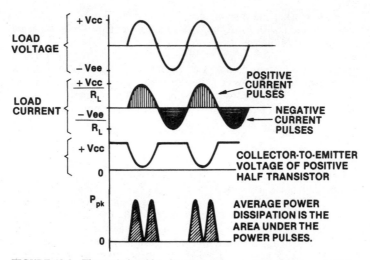

FIGURE 13.8 Time relationship of output-stage current, voltage, and dissipation. *(Source: Cal Perkins, "Inside Power Amplifiers,"* Broadcast Engineering, *August 1986. Reprinted with permission of Intertec Publishing Corp.)*

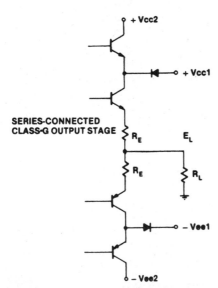

FIGURE 13.9 Typical class G power output circuit design. *(Source: Cal Perkins, "Inside Power Amplifiers,"* Broadcast Engineering, *August 1986. Reprinted with permission of Intertec Publishing Corp.)*

output. Figure 13.8 shows the relationship of the output current, load voltage, voltage across the transistors, and power pulses (which have a frequency that is double the output voltage frequency).

Two classes of amplifiers have been developed to reduce output-stage dissipation. The class G amplifier (shown in Fig. 13.9) consists of a series-connected output stage with diode switching to two different power supply levels. The second type of amplifier, class H (shown in Fig. 13.10), uses switched power supplies in conjunction with a conventional output stage. These designs realize an approximate 4:1 reduction in maximum worst-case dissipation and a reduction in power-transformer size. At full power, the two classes can be a few percentage points more efficient than class B. However, actual circuit implementation may limit the full power dissipation to a value close to that for class B because of the added saturation losses of the additional series transistors.

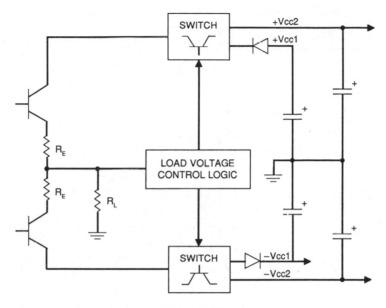

SWITCH POWER SUPPLY METHOD

FIGURE 13.10 Typical class H power amplifier circuit design. *(Source: Cal Perkins, "Inside Power Amplifiers," Broadcast Engineering, August 1986. Reprinted with permission of Intertec Publishing Corp.)*

The preceding examples assume that the amplifier is feeding a resistive load. With a reactive load, the voltage-current relationships are displaced in time (out of phase), and the relationships change somewhat from the resistive case. The worst-case dissipation occurs with a 90° reactive load and can result in a 214 percent increase in output-stage dissipation. Dissipation under this worst-case scenario increases from 40.5 percent of output power to 127 percent of output power. In other words, more power is generated in heat than is delivered to the load.

Safe Operating Area. The safe operating area (SOA) of a power transistor is the single most important parameter in the design of a solid-state amplifier. Fortunately, advances in diffusion technology, masking, and device geometry have enhanced the power-handling capabilities of semiconductor devices. A bipolar transistor exhibits two regions of operation that must be avoided:

- The *dissipation* region, where the voltage-current product remains unchanged over any combination of voltage V and current I. Gradually, as the collector-to-emitter voltage increases, the electric field through the base region causes hot spots to form, and the carriers actually punch a hole by melting the silicon. The result is a dead (shorted) transistor.

- The *secondary breakdown* $(I_{s/b})$ region, where power transistor dissipation varies in a nonlinear inverse relationship with the applied collector-to-emitter voltages when the transistor is forward-biased. The $I_{s/b}$ point on the power transistor's safe operating area (SOA) chart is the *inflection point* at which the

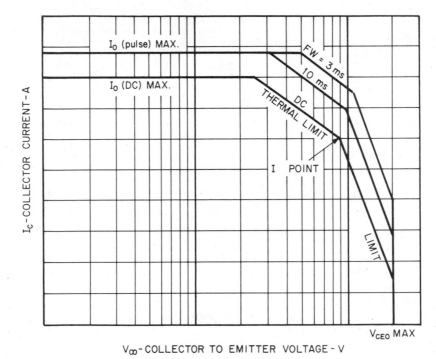

FIGURE 13.11   Typical manufacturer's data sheet for a transistor's safe operating area (SOA). *(Source: Cal Perkins, "Inside Power Amplifiers," Broadcast Engineering, August 1986. Reprinted with permission of Intertec Publishing Corp.)*

secondary breakdown phenomenon occurs. Figure 13.11 shows a power transistor SOA curve.

To put the manufacturer's information into some type of useful format, a family of curves at various operating temperatures must be developed and plotted on a linear graph. This replotting gives a clear picture of what the data sheet indicates, compared with what happens in actual operation. With improvements in semiconductor fabrication processes, output-device SOA is primarily a function of the size of the silicon slab inside the package. Package type, of course, determines the ultimate dissipation because of thermal saturation with temperature rise. A good TO-3 or a two-screw-mounted plastic package will dissipate approximately 350 to 375 W if properly mounted. Figure 13.12 demonstrates the relationships between case size and power dissipation for a TO-3 package.

Circuit Design.   There are nearly as many circuit topologies for power amplification as there are companies producing amplifiers. The most popular types of output-stage designs fall into one or more of the following categories:

- Unity-gain voltage-follower output, such as full complementary/symmetrical or quasi-complementary designs
- Voltage-gain output stage that performs the requisite small-to-large signal level shifting

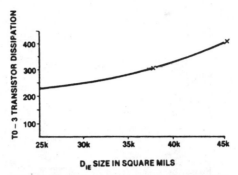

FIGURE 13.12 Relationship between case (die) size and transistor dissipation. *(Source: Cal Perkins, "Inside Power Amplifiers," Broadcast Engineering, August 1986. Reprinted with permission of Intertec Publishing Corp.)*

- Conventional design, chassis-ground referenced power supply
- Floating bridge output with no chassis-ground reference
- Grounded bridge output with a chassis-ground reference (this approach necessitates a floating power supply)

For each of these five categories, the circuit can be implemented with series-connected output devices, parallel-connected output devices, or any combination of series-parallel configurations. Usually the more esoteric approaches are used to design around some inherent component limitation, either real or self-imposed. Figures 13.13 through 13.18 show various examples of power-output stages.

In some specialized cases, the conventional approach simply will not work. A high-voltage amplifier is a good case in point. The breakdown-voltage limitations of most devices necessitate some type of series-connected output stage.

13.2.4 Amplifier Load

The ideal load for almost any power amplifier is a resistive load that exhibits the following properties:

- All the power delivered to the load is dissipated as heat.
- The load resistance does not change with frequency.
- The load resistance does not change with power level.

Maximum voltage (and power) at the load occurs when the voltage across the output transistors is at the minimum value, and the load current is maximum. The amplifier output stage then sees minimum voltage at maximum current. Inductors and capacitors used in speakers are altogether different. Reactive (inductive or capacitive) loads are the nemesis of power amplifiers. Some units will oscillate or even self-destruct when presented with a capacitive load. An inductive load may trigger the amplifier's protective circuits, which will then act in concert with the output stage and power supply to force the load further into an unsafe operating condition.

Figure 13.19 illustrates the voltage-current relationship of a resistive load and

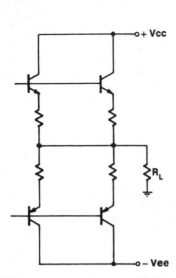

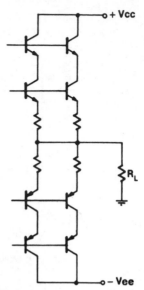

FIGURE 13.13 Power supply ground-referenced conventional, parallel-connected unity-gain output stage. (*Source: Cal Perkins, "Inside Power Amplifiers," Broadcast Engineering, August 1986. Reprinted with permission of Intertec Publishing Corp.*)

FIGURE 13.14 Power supply ground-referenced series-parallel-connected output stage with unity gain. (*Source: Cal Perkins, "Inside Power Amplifiers," Broadcast Engineering, August 1986. Reprinted with permission of Intertec Publishing Corp.*)

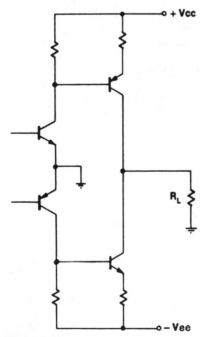

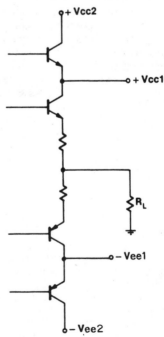

FIGURE 13.15 Power supply ground-referenced output stage with gain. (*Source: Cal Perkins, "Inside Power Amplifiers," Broadcast Engineering, August 1986. Reprinted with permission of Intertec Publishing Corp.*)

FIGURE 13.16 Power supply ground-referenced high-efficiency class G output stage with unity gain. (*Source: Cal Perkins, "Inside Power Amplifiers," Broadcast Engineering, August 1986. Reprinted with permission of Intertec Publishing Corp.*)

a pure reactive load. Loudspeakers present the power amplifier with a load that is a hybrid of *R, L,* and *C.* The minimum impedance of a loudspeaker is determined by the dc resistance of the speaker voice coil, and this determines the maximum peak current into the device. Although there are amplifiers on the market that can synthesize *negative output resistance,* which theoretically cancels the real dc resistance of the loudspeaker voice coil, the voice-coil resistance is present and still dissipates power. The negative output impedance of the amplifier cannot compensate for the lost dissipation, but it can correct drive aberrations. Figure 13.20 shows a typical speaker impedance curve.

A typical loudspeaker mounted in a vented box effectively forms two closely coupled resonant circuits, as shown in Fig. 13.21. Fortunately, the minimum impedance is limited by the

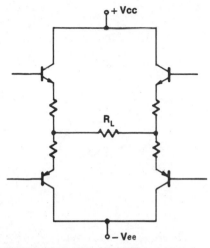

FIGURE 13.17 Floating bridge output stage with unity gain. *(Source: Cal Perkins, "Inside Power Amplifiers,"* Broadcast Engineering, *August 1986. Reprinted with permission of Intertec Publishing Corp.)*

dc resistance of the speaker voice coil, and when going through resonance, the overall impedance increases. Nonetheless, the real speaker load is far more difficult to drive than a pure resistive load. Loudspeakers store mechanical energy as well as electric energy, and when pushed to the limits, the load becomes nonlinear. Actual measurements show that a loudspeaker load angle (phase shift between voltage and current) can vary by as much as 60°.

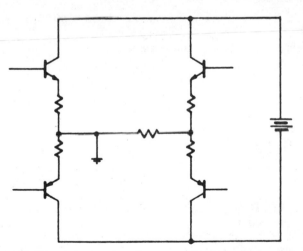

FIGURE 13.18 Floating power supply, ground-referenced bridge with unity gain. *(Source: Cal Perkins, "Inside Power Amplifiers,"* Broadcast Engineering, *August 1986. Reprinted with permission of Intertec Publishing Corp.)*

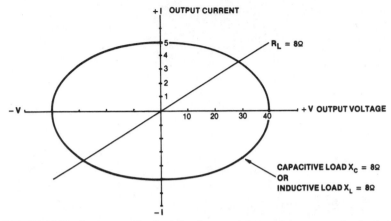

FIGURE 13.19 Power amplifier load line for a resistive and a pure reactive load. *(Source: Cal Perkins, "Inside Power Amplifiers," Broadcast Engineering, August 1986. Reprinted with permission of Intertec Publishing Corp.)*

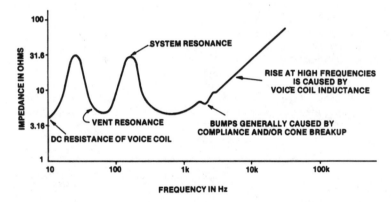

FIGURE 13.20 Typical impedance curve for a speaker mounted in a vented enclosure. *(Source: Cal Perkins, "Inside Power Amplifiers," Broadcast Engineering, August 1986. Reprinted with permission of Intertec Publishing Corp.)*

Mathematical analysis shows that for a load of 60°, the current flowing in both the amplifier and the load is at one-half the peak value when the amplifier output voltage is at zero. When the output is at 0 V, there is one-half the total supply voltage across the output transistors. The transistors see combinations of $\frac{1}{2}I_{pk}$ and E_{pk} or $\frac{1}{2}E_{pk}$ and I_{pk} simultaneously. Designing the output stage to handle a 60° reactive load without the encroachment of protection circuits is at least a minimum requirement for proper amplifier design. Because speakers seldom meet their nominal spec-sheet impedance specifications, it is best to consider the speaker dc resistance as the minimum value.

Prudent design would call for an amplifier intended for a nominal 4-Ω rating for unabated protection circuit operation. This does not imply that the power supply needs to supply steady-state 4-Ω operational power. Nor does it imply that

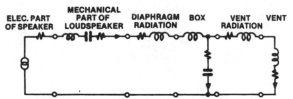

FIGURE 13.21 Equivalent electromechanical circuit for a loudspeaker in a vented box. *(Source: Cal Perkins, "Inside Power Amplifiers," Broadcast Engineering, August 1986. Reprinted with permission of Intertec Publishing Corp.)*

the heat sink must be designed for 4-Ω operation under worst-case reactive dissipation conditions.

13.2.5 Power Supply

Before an amplifier can provide power to a load, it must first get its power from somewhere. In all ac-operated amplifiers, the power supply converts the 50-Hz or 60-Hz 120 V ac source into direct current with a value between ±20 V dc and ±125 V dc, depending on the output devices used. The most widely used power-supply design is a conventional capacitor input filter full-wave bridge with a center-tapped transformer, although other, more sophisticated supplies are being used by manufacturers.

Full-Wave Bridge Supply. The capacitor input filter, full-wave bridge power supply (shown in Fig. 13.22) is the workhorse of power amplifier technology. Although relatively unsophisticated when compared to today's switching technology, the conventional power supply has several redeeming qualities. First, it basically uses only four parts. Second, it does not produce a large amount of RFI or power-line noise. Third, the supply does not require shielding or heat sinking, as a switching-type supply does. The disadvantage, however, is that 50- or 60-Hz

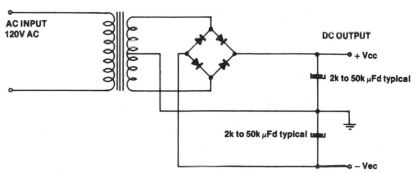

FIGURE 13.22 Conventional capacitor input filter, full-wave bridge power supply. *(Source: Cal Perkins, "Inside Power Amplifiers," Broadcast Engineering, August 1986. Reprinted with permission of Intertec Publishing Corp.)*

power transformers are large and heavy. A transformer used for a dual 250-W amplifier can weigh from 20 to 30 lb.

The diodes conduct only near the peak of the ac waveform. Because the filter capacitors charge to the peak voltage, it is important that the primary ac wiring be sized not for the rms current draw, but for the *peak* current draw. For this reason, contractors often use 00 (double 0) or even 0000 (4-0) gauge wire in their power distribution systems when the total rms current draw would indicate that 4 gauge wire would be sufficient. Remember that the output of the power amplifier is proportional to the square of the voltage, and a 10 percent loss in peak line voltage is reflected back as a 19 percent loss in power. Typical power losses can run as high as 20 to 35 percent if the ac line voltage is not sufficiently regulated.

Power supply regulation is primarily determined by the dc resistance of the power transformer. Once the filter capacitors reach a limiting value, any further increase in capacitor size does not significantly affect the long-term dc regulation.

Because the energy storage capability of the supply is directly proportional to the capacitance and the square of the voltage, a 40 percent increase in voltage will double the energy. However, large capacitors are not necessarily better. The larger the capacitor, the more inductive it becomes. The high-frequency impedance of the power supply may also be increased by using too large a filter capacitor. A better approach is to bypass the large electrolytic capacitor with a smaller film and foil (not metallized film) capacitor to ensure that the high-frequency impedance will remain low.

Phase-Controlled Supply. The phase-controlled backslope power supply (shown in Fig. 13.23) has recently been rediscovered by the audio industry. The phase-controlled supply was originally used for preregulation in large industrial power control systems. Basically, the phase-controlled supply consists of a silicon controlled rectifier (SCR) that turns on (and off) the ac line voltage to the power transformer primary as a function of phase angle.

The narrow conduction angles caused by the SCR switching necessitate that the peak line current be quite high because the total energy is a voltage-current-time product. Because the peak current is quite high, so is the rms current. High

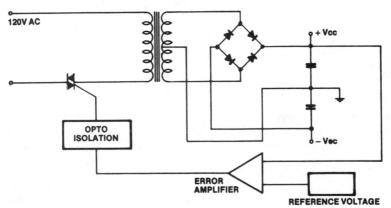

FIGURE 13.23 Phase-controlled power supply with primary regulation. *(Source: Cal Perkins, "Inside Power Amplifiers," Broadcast Engineering, August 1986. Reprinted with permission of Intertec Publishing Corp.)*

rms current means high transformer temperatures because the power dissipated by the transformer is I^2R. A well-designed backslope system draws about 10 percent more line current than a conventional power supply. Perhaps the biggest problems with a phase-controlled supply are the transient overvoltages it can create on the ac line. Because of the low dc resistance of the power transformer, large chunks are effectively taken out of the line every time the SCR switches on. When the SCR turns on, the power transformer appears as an effective short across the power line.

Phase-controlled power supplies can provide excellent regulation and offer protection against excessively high input voltages. With this regulation scheme, the amplifier can be designed to maintain its specified output power with line voltages as low as 108 V ac. Conversely, a power amplifier with a conventional supply will lose 19 percent of its output power if operated at 108 V. Another design benefit from using the phase-angle concept is the convenient control mechanism it provides for protecting the amplifier.

Switching Supply. High-frequency switching power supplies have, so far, found limited application in high-power audio amplifiers. Component cost, reliability, power dissipation requirements, and electrical noise problems have slowed widespread acceptance. As with phase-controlled power supplies, power-line disturbances are difficult to eliminate. In some cases, the power-line problems can seriously degrade amplifier performance.

13.2.6 Maintenance

To ensure amplifier reliability, periodic maintenance is a necessity. If the units are fan-cooled, filters will need to be cleaned. After several thousand hours of operation, the fans may have to be replaced because of failed bearings. More insidious, however, are failures that occur after several years of operation. Experience indicates that the dominant cause of component failure is a dramatic increase in thermal resistance between the output transistor and its heat sink. There are two mechanisms that typically cause output transistors to fail after a period of time.

The first, and most obvious, is the gradual loosening of the mounting screws. These screws clamp the output transistor to the heat sink. The continuous expansion and contraction of the heat sink and the semiconductor during thermal cycles may cause the fasteners to loosen. The second centers around the eventual evaporation of the silicon oil used in the zinc oxide–silicon grease thermal joint compound. After repeated thermal cycles, the silicon oil can be squeezed out of the compound and evaporate. This results in a poor thermal joint that can cause the transistor to run hot, even to the point of destruction. The solution to these two problems is simple: Check the mounting screws and joint compound every two or three years. Tighten the screws and replace the joint compound as needed.

13.3 LOUDSPEAKERS

One of the problems with selecting a high-quality monitor system lies in the difficulty of defining *quality*. Some not-too-scientific descriptions may develop from discussions with other users, and even with knowledgeable authorities. Terms such as "solid bass," "smooth highs," "tight," or "clean" all may be mentioned as monitor system requirements. Trying to incorporate these subjective requirements into a working system is almost impossible. On the other hand, selecting a speaker solely on the basis of frequency response and harmonic distortion is likewise inappropriate. It is difficult to equate either scientific measurements or subjective considerations with how "good" or "bad" a particular speaker system sounds.

13.3.1 Loudspeaker Properties

For the purposes of this examination, the term *speaker* will refer to a single transducer. *Monitor* or *monitor system* will refer to an assembly of speaker(s), enclosures, and, where appropriate, crossovers and amplifiers. In fact, it might be best to consider a monitor system as having at least three major components: source driver (amplifier), transducer (speaker), and mounting assembly (cabinet).

A speaker cannot produce acoustic energy without being driven by an electronic source. It likewise requires an enclosure to properly couple acoustic energy into the listening environment in a controlled manner. Even the most expensive speaker, if set on a shelf without an appropriate enclosure, will perform poorly.

A speaker is really an electromagnetic transducer that converts electric energy into acoustic energy (sounds). There are two principal types of speakers:

- Direct radiator, which uses a vibrating surface (the diaphragm) to couple sound directly into the air. The direct-radiator type is used in most applications.
- Horn radiator, which uses a horn attached near the diaphragm to couple sound into the air. The horn type is often used in large monitor systems that need high volume levels or that cover large areas. Horns also are used with high-frequency speakers.

This discussion will be limited to direct-radiator (diaphragm) speakers. The principal advantages of the direct-radiator speaker are small size, low cost, and satisfactory performance over a comparatively wide frequency range. The disadvantages include low efficiency, narrow directivity pattern at high frequencies, and irregular response curves at high frequencies.

Speaker Construction.   In the simplest of analogies, the speaker is a motor whose motion is directed in a straight line forward and backward. The motion of the cone is dependent upon the current flow through the voice coil, the length of the conductor in the magnetic field, and the strength of the speaker's magnetic field.

Most manufacturers honor the convention that application of a positive voltage to the positive speaker terminal produces forward movement of the cone. If an alternating current is applied, the cone will move forward and backward at the same rate as the current. In the case of a simple sine wave, the applied current will reverse direction from positive to negative or vice versa, passing through 0 V during the transition. Therefore, in one sine wave, the current applied to the speaker will reach one positive peak, one negative peak, and three zero points. The speaker cone follows the current by moving forward and backward and returning to the center (rest) point three times.

The cross section of a typical direct-radiator speaker is shown in Fig. 13.24.

The diaphragm or cone is generally composed of paper or some other stiff material. Speakers use springs in the form of the flexible edge suspensions surrounding the outer edge of the cone, and the spider around the diaphragm. These springs resist the force of the speaker's motor (the voice coil and magnet) and return the cone to center rest position after it is driven forward or backward by an electric signal. The springs provide what is called the *restoring force*.

Speaker Oscillation. A speaker is capable of storing energy. This takes place when the cone is displaced from rest. After the drive signal is removed, the cone tries to return to its natural resting state because of the springs. However, the springs have a limited ability to accelerate the mass of the cone. This means that after the cone is extended, it slowly starts to return to the resting point. However, as the cone begins to move, its velocity increases. The inertia that develops forces the cone to traverse past the resting point, and the cone is again displaced from center. Then the stored energy (springs) again attempts to return the cone to the center resting point.

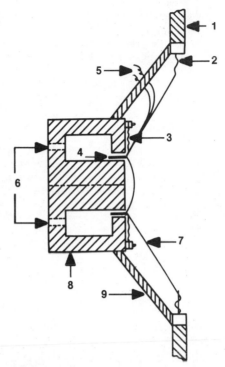

FIGURE 13.24 Cutaway drawing of a direct-radiator speaker. *(Source: Brad Dick, "Inside Monitor Loudspeakers," Broadcast Engineering, June 1987. Reprinted with permission of Intertec Publishing.)*

In a world without mechanical or electric resistance, this oscillation would continue indefinitely. However, the laws of physics apply, and after a period of time, the cone comes to rest. From an acoustical standpoint, this oscillation or ringing must be limited or *damped*. Otherwise, the ringing will become a source of distortion.

One way to dampen the cone motion is to apply a counter EMF (electromagnetic force) to the speaker voice coil by shorting the coil terminals. This causes the coil to generate a counterforce. This force opposes the cone movement and is produced by the motion of the coil through the magnetic field. This braking effect is called *back EMF*.

Proper control of a speaker's movement requires current to put the cone in motion. As already noted, once the cone is in motion, it must be stopped if ringing or oscillation (distortion) is to be prevented. If the amplifier can apply a sufficiently low-impedance path, then a back EMF will be developed, and cone ringing will be limited. This low-impedance path often is referred to as *amplifier damping*, or the ability of a power amplifier to electronically brake excessive diaphragm motion. How well a speaker performs depends, to a degree, on the damping applied by the amplifier.

Amplifiers are usually rated by a *damping factor*. The amplifier damping factor is equal to Z_L divided by Z_o in ohms, where Z_L is the rated load impedance of the amplifier, and Z_o is the actual output impedance of the amplifier. Simply put, the lower the amplifier output impedance, the more like a "dead short" it will appear to a speaker when the output voltage is zero. The damping factor can be degraded drastically by the addition of resistance to the circuit, such as would occur with a long speaker cable run.

13.3.2 Speaker Enclosure

Part of the problem in designing a speaker-enclosure system is that psycho-acoustic factors involved in the reproduction of speech and music are not fully understood. Any four listeners may rate differently four identical speakers mounted in four identical cabinets. The engineer's task is to find some way to equate subjective criteria with objective specifications.

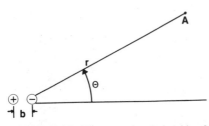

FIGURE 13.25 The acoustic relationship of a speaker and its environment. The device is both directional and frequency-sensitive. *See the text. (Source: Brad Dick, "Inside Monitor Loudspeakers," Broadcast Engineering, June 1987. Reprinted with permission of Intertec Publishing.)*

Why mount the speaker in an enclosure anyway? As already mentioned, the speaker cone moves forward and backward in relation to the applied signal. At low frequencies, the speaker can be represented by a pair of sound sources of equal strength, located near each other and pulsing out of phase. The back of the speaker represents one of these sources, and the front represents the other. Figure 13.25 depicts this situation with a point-source sound (speaker) and a monitoring location A. In mathematical terms, the sound power level at point A, P, can be represented by the following equation:

$$P = \frac{p_o f^2 \, U_o \, b\pi \cos \theta}{rc} \tag{13.1}$$

where U_o = rms strength of each simple source, m^3/s
     b = separation between the sound sources, m
     p_o = density of air, kg/m^3
     r = distance from sources to point A, m
     θ = angle shown in Fig. 13.25
     c = speed of sound, m/s

What emerges, upon examination of the equation, is that for a constant-volume velocity of the speaker diaphragm, the pressure P measured at a distance r is directly proportional to the square of the frequency f, and the cosine of the angle θ, and is inversely proportional to r. In terms of decibels, the sound pressure P increases at the rate of 12 dB for each octave (doubling) in frequency.

This increase is not constant and changes depending upon factors such as operation below or above speaker resonance. Below the first resonance, the increase is +18 dB. Above the speaker's first resonance, the sound pressure P increases by only 6 dB per octave.

In less complex terms, as the cone moves forward, the air immediately in front of the cone is compressed. The tendency of this compressed air is to rush in to fill the rarefaction created by the opposite action behind the cone. It is only at high frequencies, where the speaker assembly is large in relation to the wavelength of the sound, that there is appreciable sound radiation.

Speaker Baffle. Placing the speaker in a large baffle will improve the low-frequency response, because the distance from the front to the back of the speaker is increased greatly. The term *infinite baffle* is often used to describe mounting the speaker in a wall of infinite size, which prevents energy from the front of the speaker from reaching the back of the speaker. Infinite baffles are not practical in the real world, so other designs have been developed to address the problem. Early solutions used boxes or flat planes to shield the speaker. Designers quickly discovered that the boxes resonated at various frequencies, causing even more problems, and that sound leaked around the flat planes.

One of the early successful solutions is shown in Fig. 13.26. The original closed box has been modified by the installation of a carefully designed opening in the front mounting plate. The *vented* or *ported* enclosure (or *bass-reflex monitor*) provided a method of coupling the energy radiated from the back of the speaker with that radiated from the front of the speaker. If the port size is carefully designed, it becomes a second diaphragm driven by the back side of the speaker. At low frequencies, the port is equivalent to a short length of tube with an acoustic reactance and a series acoustic resistance. A carefully designed port can add an octave or more to the system's low-frequency response.

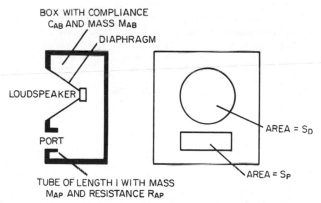

FIGURE 13.26 The simplest form of ported or reflex enclosure uses a hole to create a second diaphragm for increased efficiency and extended low-frequency performance. *(Source: Brad Dick, "Inside Monitor Loudspeakers," Broadcast Engineering, June 1987. Reprinted with permission of Intertec Publishing.)*

One key to the successful design of a ported enclosure is to match the enclosure resonance to that of the speaker alone. The process effectively reverses the phase of the backwave at the port, resulting in a radiated sound that is in phase with the speaker's sound. This design allows the system to produce substantial radiation below the speaker's own free-air resonance frequency. Many other designs have been developed over the years, but this one remains a favorite with manufacturers.

13.3.3 Integrated Monitor System

Up to this point, the discussion has been limited to single-speaker systems. For a number of reasons, high-quality monitors usually rely on several speakers of different characteristics mounted within a single enclosure, referred to as two-way, three-way, or even four-way systems. The advantages include increased acoustic output, wider sound dispersion, and reduced IM distortion.

In any multiple-speaker monitor, some way must be found to divide the output of the amplifier to the individual speakers. A *crossover network* is used to route the low frequencies to the low-frequency speaker and the high frequencies to the high-frequency speaker. Three-way systems rely on three crossovers for signal routing. Crossovers are designed as either high-level or low-level networks. High-level crossovers, located after the amplifier and prior to the individual speakers, are passive devices (see Fig. 13.27). Low-level crossovers may be either passive, as shown in Fig. 13.28, or active, as shown in Fig. 13.29. These crossovers are located prior to the amplifiers. Low-level crossover systems require an amplifier and speaker for each band of frequencies. Low-level crossover systems often are referred to as bi-amplified (two-way systems) or tri-amplified (three-way systems).

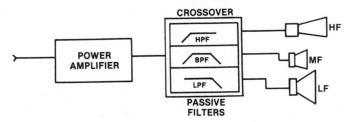

FIGURE 13.27 Block diagram of a passive high-level crossover, sometimes called a *three-way system*. *(Source: Brad Dick, "Inside Monitor Loudspeakers," Broadcast Engineering, June 1987. Reprinted with permission of Intertec Publishing.)*

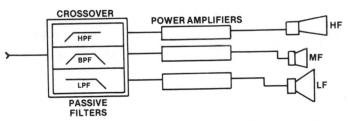

FIGURE 13.28 Block diagram of a passive low-level crossover. *(Source: Brad Dick, "Inside Monitor Loudspeakers," Broadcast Engineering, June 1987. Reprinted with permission of Intertec Publishing.)*

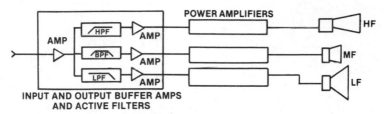

FIGURE 13.29   Block diagram of an active low-level crossover. Loss in the cross-overs is compensated for by gain from the internal amplifiers. *(Source: Brad Dick, "Inside Monitor Loudspeakers," Broadcast Engineering, June 1987. Reprinted with permission of Intertec Publishing.)*

As mentioned previously, adding any resistance between the amplifier output and the speaker affects amplifier damping. Therefore, using a high-level cross-over carries with it a design penalty. Because the crossover is in series with the signal, the amplifier becomes less effective in controlling speaker ringing, possibly causing increased distortion. Another drawback to high-level crossovers centers on cost. Even if only one amplifier is needed per system, one crossover still is required for each monitor. In sound reinforcement applications, this requirement can significantly increase system cost because of the large number of monitors typically used.

System Performance.   A number of important factors must be taken into consideration when multiple-driver systems are selected. The crossover shown in Fig. 13.27 is, by application, passive. Typically, such devices rely on first- or second-order Butterworth designs. Such crossovers produce a relatively slow rolloff of 6 or 12 dB per octave, respectively. One problem with this approach is that out-of-band energy can be delivered to the individual speakers. Active crossovers, on the other hand, often are designed with third-order, 18-dB/octave active cross-overs, or first- or second-order crossovers connected in series. A steeper rolloff characteristic results in less out-of-band energy being delivered to the individual speakers in the monitor.

The possibility of reduced distortion is another important reason to consider multidriver systems. Once the decision is made to use multiple amplifiers, certain operating problems cease to seriously degrade system performance. Take the case of a single amplifier feeding a passive crossover. If the amplifier is driven into distortion by a low-frequency signal, all speakers in the monitor receive the distorted signal, either as a fundamental waveform or as higher-order harmonics. Conversely, in a multiway system, if the low-frequency amplifier distorts, the remaining amplifier(s) will continue to operate normally.

One drawback to a multiamplifier system is cost. The designs shown in Figs. 13.28 and 13.29 require three amplifiers instead of one. Cost factors and design goals must be weighed against possible sonic improvement.

Phase Response.   Modern technology has helped identify speaker performance characteristics that previously remained unknown. Phase response recently has been quantified by several methods, and can now be measured by various techniques. The term refers to the alignment, in the time domain, of sound coming from a speaker. The goal is to ensure that the relationship between the fundamental and the overtones of a complex signal remains unchanged, and that the sound elements arrive at the listening position at the same time.

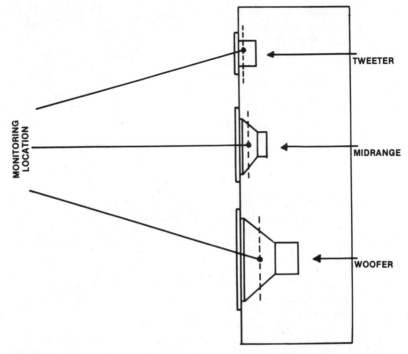

FIGURE 13.30 In a multidriver monitor system, any difference in the acoustical center of the drivers will cause a phase shift or time delay. Whether this represents a problem for most monitoring applications is a point of controversy. *(Source: Brad Dick, "Inside Monitor Loudspeakers,"* Broadcast Engineering, *June 1987. Reprinted with permission of Intertec Publishing.)*

Figure 13.30 shows a typical monitor cabinet with three speakers mounted on the front panel. The acoustical center of each speaker is marked by a dotted line. Notice how the speaker's center lines do not match. According to some experts, this misalignment of centers can cause phase distortion. Some manufacturers address this problem by using a stepped enclosure facing. The technique physically positions the acoustical center of each speaker on the same plane. Phase-delay networks also can be used to compensate for the delay.

Various studies have been undertaken in an effort to identify the audibility of such phenomena. One study showed that phase shifts as low as 15° in midrange frequencies were audible. Other studies indicated that, although many top-grade monitor systems produce delays, these delays are inaudible.

Monitor Environment.   The performance of even the best monitor system is dependent upon the environment in which it is operating. The way the monitor interacts with its environment is complex, and volumes of data have been developed on the interaction between sound sources and rooms. One of the first things to realize is that the closer a monitor is mounted to the walls, ceiling, or floor, the more bass it produces. For example, if a monitor is mounted in the center of a room, the low frequencies are radiated in all directions. If the same speaker is mounted next to a wall, the low-frequency output increases by 3 dB. Moving the speaker into a corner further increases the low-frequency output.

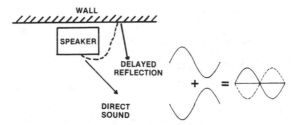

FIGURE 13.31 Placing a monitor near a wall may produce a comb-filter effect as low-frequency energy from the back combines with direct sounds from the front. *(Source: Brad Dick, "Inside Monitor Loudspeakers,"* Broadcast Engineering, *June 1987. Reprinted with permission of Intertec Publishing.)*

A related problem exists when a speaker is mounted next to a wall or ceiling. Low-frequency sounds can reflect from the back wall or ceiling and combine with the sound from the front of the speaker, as shown in Fig. 13.31. This results in phase cancellations that produce a comb-filter effect. Typically, the solution is either to locate the speakers near the monitoring position or to isolate the speaker from the effects of the walls.

The first technique, called *near-field monitoring*, was developed by audio consultant Ed Long. The monitors are located approximately 3 ft apart, at ear level, near the listening position. This allows the operator to hear primarily the direct sound from the speakers. Reflections from other surfaces are minimized, which results in fewer cancellations within the monitoring area. The second technique involves flush-mounting and isolating the monitors within the control room wall. This isolates them from resonances and reflections that the wall might produce, and reduces problems caused by low-frequency reflections. In both cases, the speakers are not physically connected to the wall. To do so may cause the wall or surface to act as a low-frequency radiator.

Room Equalization. Equalization is an often discussed, yet seldom understood technique. Contrary to its name, room equalization is not an attempt to modify the room. Rather, the monitor system is distorted to match the deficiencies of the room. Room equalization is not expensive, but it does require use of the proper test equipment and an engineer trained in equalization techniques. Problems often develop when an installer who doesn't understand the intricacies involved attempts to equalize a room. For instance, installers sometimes try to equalize by ear, rather than using the proper instruments. The result may be that they end up trying to equalize for standing-wave problems or poor room acoustics. Equalizers cannot make an inferior monitor system into a good one, and cannot impart good acoustics to a poorly designed room.

Figure 13.32 shows the frequency response of a speaker mounted next to a wall. The solid line shows the response prior to room equalization and relocation of the speaker. Notice the low-frequency bump at about 150 Hz. Although an equalizer could help smooth out the bump, the irregular peaks and notches in the upper range would be more difficult to correct. The improvement that can result from moving the speaker *and* using an equalizer is shown by the dotted line in the graph. This example points out the occasional need to combine techniques to obtain satisfactory performance.

Keep in mind that a ⅓-octave equalizer is just that; if a system has a problem at any non-⅓-octave frequency, it may be masked by the analyzer. It is not un-

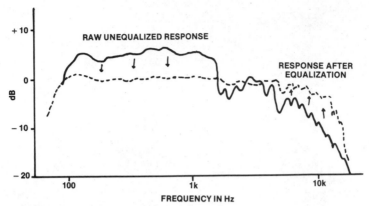

FIGURE 13.32 Example of the use of equalization and monitor relocation to solve a difficult monitor response problem. The solid line represents the original frequency response. The dotted line shows the frequency response after the monitor was moved and room equalization was applied. *(Source: Brad Dick, "Inside Monitor Loudspeakers," Broadcast Engineering, June 1987. Reprinted with permission of Intertec Publishing.)*

common for an inexperienced installer to chase a minor problem and add excessive amounts of equalization in the process. Also, remember that equalizers can add phase shift, especially if large amounts of gain and/or boost are used. The low-pass Butterworth filters typically used in equalizers produce significant phase shift, as shown in Fig. 13.33. In applications where the user is concerned about time alignment of speakers, adding phase shift from an equalizer does not make sense.

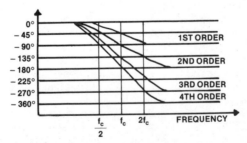

FIGURE 13.33 The phase shift produced by some equalizers can be significant. This graph shows the low-pass filter phase response for various orders of Butterworth filters. *(Source: Brad Dick, "Inside Monitor Loudspeakers," Broadcast Engineering, June 1987. Reprinted with permission of Intertec Publishing.)*

13.4 EQUALIZATION

For more than 25 years, audio systems have used some form of equalizer technology. From the earliest days of the tube-based Lang and Pultec filters to today's multiband graphic and parametric systems, the equalizer has been relied upon to correct and enhance sound. It has also formed the basis for many of the sophisticated automatic audio-processing systems in use today. Perhaps because of this popularity, equalizers are some of the most overused and misunderstood devices in the field of sound.

13.4.1 Types of Equalization

The frequency response curves of three simple equalizer circuits are shown in Figs. 13.34 through 13.36. These curves represent the gain of each filter with respect to frequency. Figure 13.34 shows the response of a *boost* circuit, with the important characteristics labeled. The center frequency is the frequency of maximum boost, so called because it marks the center of the response peak. On each side of the center frequency is a point at which the amplitude is 3 dB lower than the maximum level. The range in frequency from the first point to the second is the *bandwidth*, an indicator of the sharpness of the filter. The smaller the bandwidth, the smaller the range of frequencies that will be affected by the actions of the filter. Because the ear hears changes in frequency on a percentage or octave scale, a filter of a given bandwidth will have substantially greater effect with a low center frequency than with a high center frequency. For example, a 50-Hz-bandwidth filter centered at 10 kHz affects less than 1 percent of an octave and will be virtually inaudible. A 50-Hz-bandwidth filter at 100 Hz, however, will affect almost an entire octave, radically altering the sound.

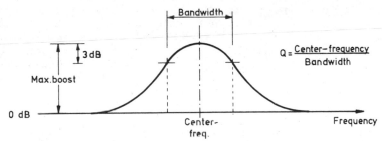

FIGURE 13.34 Basic boost filter response. *(Source: Richard Cabot, "Dimensions in Equalization," Broadcast Engineering, August 1985. Reprinted with permission of Intertec Publishing Corp.)*

The filter *sharpness* can be defined by dividing the center frequency by the bandwidth. The resulting number is the Q of the filter. For a 50-Hz-bandwidth filter at 100 Hz, the Q is 2. For a 50-Hz-bandwidth filter with a center frequency of 10 kHz, the Q is 200. Typical boost filters have Qs between 1 and 10.

Virtually all equalizers give the user control over the amount of boost or cut. The maximum boost on most equalizers ranges from 10 to 15 dB. At small values of boost, the concept of bandwidth or Q becomes hard to define. If the maximum

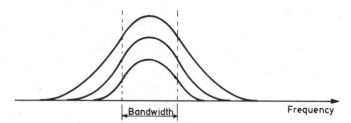

FIGURE 13.35 Constant-Q boost filter response. *(Source: Richard Cabot, "Dimensions in Equalization,"* Broadcast Engineering, *August 1985. Reprinted with permission of Intertec Publishing Corp.)*

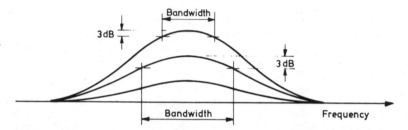

FIGURE 13.36 Constant-shape boost filter response. *(Source: Richard Cabot, "Dimensions in Equalization,"* Broadcast Engineering, *August 1985. Reprinted with permission of Intertec Publishing Corp.)*

boost is 2 dB, how do you define the 3-dB (down) frequencies? Most manufacturers, therefore, only specify the Q at full boost.

The filter response curve shown in Fig. 13.35 maintains constant Q as the gain varies. This causes the frequency range that the filter affects to decrease as the maximum boost is reduced. The Q of the filter shown in Fig. 13.36 decreases as the maximum boost is reduced, although the shape of the filter remains the same. The filter affects all frequencies around the center frequency by the same relative amount as the degree of boost is varied. Both characteristics are used in commercially available equipment, although the manufacturer does not usually specify which type of circuit is used. In practice, there are situations in which one has advantages over the other, but it is difficult to say that one or the other is superior.

So far, we have considered boost circuits. The situation changes, however, when the filter is set to produce a cut or reduction in gain over some frequency band. Figure 13.37 shows the response of a simple filter circuit when the boost/cut control is adjusted for a 10-dB cut (-10 dB). The center frequency of the filter is defined as the frequency of minimum gain, the opposite of a boost filter. The bandwidth, and therefore the Q, is defined by the frequencies at which the gain is 3 dB less than maximum.

Figure 13.38 shows the response curves of a constant-Q filter when adjusted for varying values of boost and cut. The boost and cut curves are not mirror images of each other. This type of filter produces sharp nulls in the response curve at the center frequency. There are applications in which this characteristic is not desirable, and the inverse of the boost curve would be more appropriate. Such a response characteristic is shown in Fig. 13.39. The Q of the filter is not constant

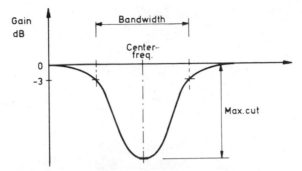

FIGURE 13.37   Basic cut filter response. (*Source: Richard Cabot, "Dimensions in Equalization,"* Broadcast Engineering, *August 1985. Reprinted with permission of Intertec Publishing Corp.*)

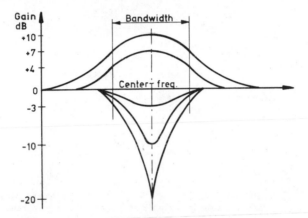

FIGURE 13.38   Constant-Q boost/cut filter response. (*Source: Richard Cabot, "Dimensions in Equalization,"* Broadcast Engineering, *August 1985. Reprinted with permission of Intertec Publishing Corp.*)

in the cut mode. It is much lower than the corresponding boost position. This characteristic is called a *reciprocal peaking filter*. Most graphic equalizers use reciprocal peaking filters, while most parametrics offer constant-Q filters.

Shelving Equalizer.   The need often arises in professional applications to boost or cut all frequencies above or below some selected frequency. A unit that performs this function is called a shelving filter. The frequency-response curve for a low-frequency shelving filter is shown in Fig. 13.40, and the curve for a high-frequency shelving filter is shown in Fig. 13.41. These filters are effective in eliminating or producing frequency rolloff at the extremes of the audio band.

The boost or cut amplitude of a shelving equalizer is defined as the maximum deviation from the nominal (flat) gain of the filter. For a high-frequency shelving equalizer, it is the high-frequency gain minus the low-frequency gain. For a low-frequency shelving equalizer, it is the reverse. The frequency characteristics of a

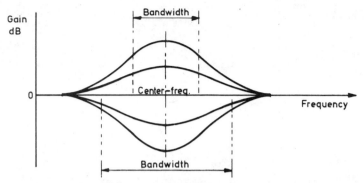

FIGURE 13.39 Reciprocal peaking filter response. (*Source: Richard Cabot, "Dimensions in Equalization,"* Broadcast Engineering, *August 1985. Reprinted with permission of Intertec Publishing Corp.*)

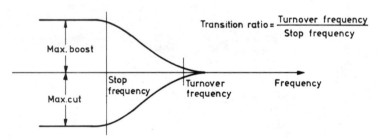

FIGURE 13.40 Low-frequency shelving equalizer response. (*Source: Richard Cabot, "Dimensions in Equalization,"* Broadcast Engineering, *August 1985. Reprinted with permission of Intertec Publishing Corp.*)

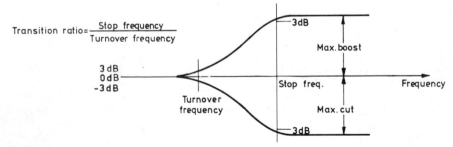

FIGURE 13.41 High-frequency shelving equalizer response. (*Source: Richard Cabot, "Dimensions in Equalization,"* Broadcast Engineering, *August 1985. Reprinted with permission of Intertec Publishing Corp.*)

shelving filter are described by the *turnover frequency*, the *stop frequency*, and the *transition ratio*:

- The turnover frequency is the point at which gain changes from the nominal value by 3 dB. In a shelving equalizer adjusted for a boost, the turnover frequency is the frequency at which the gain is 3 dB above the midband value.
- The stop frequency is the point at which gain stops increasing or decreasing. This is taken as the frequency at which the gain is within 3 dB of maximum or minimum, for boost and cut settings, respectively. When small boost or cut values are selected, these definitions can become unclear. It is common to approximate the shelving equalizer curve with straight lines and use the points where they intersect as the appropriate frequencies, as shown in Fig. 13.42.
- The transition ratio is the ratio of the stop frequency to the turnover frequency. It is analogous to the Q of the peaking filter.

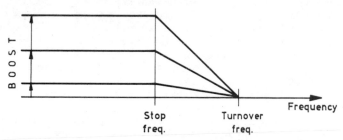

FIGURE 13.42   A straight-line approximation of the shelving equalizer response. *(Source: Richard Cabot, "Dimensions in Equalization," Broadcast Engineering, August 1985. Reprinted with permission of Intertec Publishing Corp.)*

Parametric Equalizer.   The most flexible equalizers on the market today provide independent control over all parameters of the basic filter sections. These parametric equalizers offer three to five filter sections in one package. Each section is usually independently adjustable. Often, the frequency ranges of these sections are not the same, and there is considerable overlap between sections. For example, the first filter might be adjustable from 20 Hz to 2 kHz, and the second adjustable from 50 Hz to 5 kHz. By staggering the sections this way, the entire audio band may be covered without requiring excessive operating range from any one filter stage.

Most of the parametrics on the market divide the frequency adjustment ranges on each filter into two or three bands. The filter is made tunable within these bands via a multiturn potentiometer. To simplify tracking requirements, the range is generally limited to about 10:1. Selecting between bands (almost always in factors of 10) is done with a switch that changes capacitors in the filter. This approach provides fine resolution on frequency setting and allows approximate calibrations to be written on the equipment front panel.

Some parametrics provide switch-selected modes, such as constant-Q/reciprocal response or peaking/shelving. These features can significantly enhance the flexibility of the unit, eliminating the need to select between response types in advance of the purchase. Constant-Q cut capability is rarely needed except for sound system feedback reduction.

Flexibility and freedom from predefined frequencies and Qs make the parametric equalizer a powerful tool. This flexibility also makes the parametric equalizer a complicated tool. However, with experience, the operator can achieve precise equalization that can closely match the desired characteristics. Because the unit is really several simple filter sections in cascade, the effect of using two sections simultaneously is equal to the sum of their individual responses. This lack of interaction between controls makes the adjustment task easier. By using the sections one at a time, aberrations in system response can be removed in order of their significance. Complex response adjustments can be made by connecting two equalizers in series.

Graphic Equalizer. Graphic equalizers are so named because they contain a bank of filters on octave or fractional-octave center frequencies whose gain controls are arranged to create a graph of the resulting frequency response on the front panel of the unit. The row of linear sliders provides a simple operator interface, enabling instant recognition of the gain in each frequency band. Graphic equalizers are commonly available with frequency resolutions from one octave (9 or 10 sliders covering the audio band) to ⅓ octave (27 to 31 sliders). Graphic equalizers are almost always fixed-frequency, fixed-Q devices. These limitations are imposed primarily by cost and panel space considerations.

Unlike the filters used in a parametric equalizer, filters in a graphic equalizer are wired in parallel. As a result, the response with two sliders boosted or cut is not the same as the sum of the responses with each slider advanced individually. The Q of each filter is selected based on the spacing between center frequencies. There is, however, some latitude allowed. The actual Q, and some subtleties concerning the way in which the filters are connected, affects how well the filter responses combine. Graphic equalizers are generally reciprocal filter devices. The response obtained with any control setting may be undone with a complementary setting of the controls.

13.4.2 Hybrid Equalizer

Several hybrid approaches to equalization are available, as are products based on microcomputer technology. In the hybrid category, there are several commercial units that seek to provide the advantages of parametric and graphic equalizers in one package. One approach offers a graphic equalizer that has a fine frequency-adjustment control under each linear fader. This allows the user to trim the center frequency of the filter to exactly match the frequency of the desired response peak or dip. Another approach offers a graphic equalizer combined with several tunable notch filters. Still another approach helps the user visualize the resulting response of a parametric equalizer by configuring the boost/cut controls as linear faders arranged in a manner similar to a graphic unit. The positions of the boost/cut controls help convey the resulting response curve.

A number of programmable equalizers that use analog filters controlled by a microprocessor are also available. With a programmable equalizer, the user can set the desired frequency, Q, and boost or cut on a digital display. Accuracy and repeatability are assured. The resulting curve is displayed on a CRT or LCD display. By using a light pen or some other simple controls, the response may be adjusted to any desired shape within the capabilities of the system. The primary advantage that almost all programmable units offer is the capability of storing any number of equalizer settings in memory and recalling them when needed.

Digital audio technology can be expected to make significant changes in equalization and other forms of audio processing in the next few years. It has already revolutionized artificial reverberation, and several companies have demonstrated digital equalizers. The technology exists today to duplicate the functions of most commercial parametric or graphic equalizers in a digital device. Ultimately, the cost of digital systems will be competitive with that of top-quality analog-based equalizers.

13.5 AUDIO PROCESSING EQUIPMENT

From the earliest days of recorded sound, people have been trying to cram more signal into the recording and broadcast media. Because the dynamic range of the human voice or common musical instruments is much greater than that of conventional discs or tape, devices were developed to compress the dynamic range of signals, making the loud sounds softer and the soft sounds louder. Unfortunately, some audio processors actually create problems when dealing with the dynamic range of the human voice. These problems include amplifier overload, pickup of room noise, and excessive sibilance in speech.

13.5.1 Gain Control

Dynamic gain-control devices provide an output signal that ideally differs from the original signal only in level. The shape of the waveform theoretically remains the same, but its size (voltage) is made larger or smaller as necessary. System gain is an important characteristic for these types of devices. Therefore, their steady-state operation can be described by plotting input level versus output level on a graph. On a log-log scale (decibel output versus decibel input), the result would be a graph similar to that shown in Fig. 13.43, commonly referred to as a *transfer curve*. For a conventional amplifier, the graph is a straight line at a 45° angle. The gain of the amplifier determines where the line is positioned on the graph, but the slope is always the same.

Compressor. A compressor is a device that will increase the level of soft sounds and decrease the level of loud sounds in a somewhat predictable manner. An example of this action is graphed in Fig. 13.44. As the input signal amplitude increases, the output signal amplitude increases by a smaller amount. When the input signal amplitude decreases, the output signal amplitude decreases by a smaller amount. There is always a point at which the input level equals the output level, called the *unity-gain point*. The slope of the curve is called the *compression ratio*. A compressor whose output level increases by 1 dB for every 3 dB of input level increase is said to have a 3:1 compression ratio.

Compressors come in two types: *feedback* and *feedforward*. The classic block diagrams of each approach are shown in Fig. 13.45. The feedback-type compressor is the older and more common of the two. The output signal level is sensed and fed back to the gain-control element, which precedes it. As the input level is increased, the output level tries to increase. This is sensed by the level sensor circuit, which drives the gain-control element in an effort to reduce the amplitude of the output. Changing the gain after the level sensor changes the slope of the

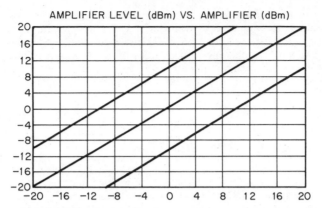

FIGURE 13.43 Basic amplifier gain transfer curve, showing output in decibels referred to 1 mW versus input in decibels referred to 1 mW (log plot). A conventional amplifier produces a line at a 45° angle, with gain determining the actual position on the graph. *(Source: Richard Cabot, "Limiters, Compressors and Expanders," Broadcast Engineering, August 1986. Reprinted with permission of Intertec Publishing Corp.)*

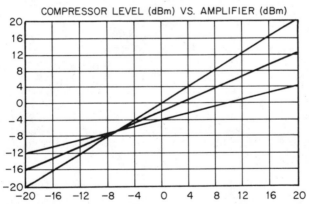

FIGURE 13.44 Typical compression graph, showing how the output level decreases or increases depending upon input level. *(Source: Richard Cabot, "Limiters, Compressors and Expanders," Broadcast Engineering, August 1986. Reprinted with permission of Intertec Publishing Corp.)*

compression characteristic. These circuits are easy to build and are self-correcting for errors in the gain element or level sensor. However, the approach guarantees that the output will overshoot its final value when the input level is suddenly increased.

The feedforward configuration senses the input level and generates the necessary control voltage for the gain element to make the output level change as desired. This avoids the overshoot problem, but places more stringent requirements on the accuracy of the level sensing and control circuitry.

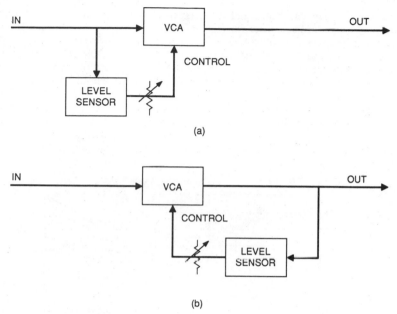

FIGURE 13.45 Simplified block diagrams of two common compressors. (*a*) Feedforward compressor. (*b*) Feedback compressor. (*Source: Richard Cabot, "Limiters, Compressors and Expanders,"* Broadcast Engineering, *August 1986. Reprinted with permission of Intertec Publishing Corp.*)

The graphs shown in Figs. 13.43 and 13.44 are all straight lines on linear decibel scales. This characteristic is obtained with level sensors that output a voltage proportional to the decibel signal level. Voltage-controlled amplifiers (VCAs) also exhibit similar characteristics, with the gain in decibels proportional to the voltage at the control input.

Broadcasters often use compressors to make the air signal sound louder. Many stations believe that, in the battle for listeners, the loudest signal will capture a larger audience. This theory has led to a proliferation of compressors optimized for broadcast use. These devices can be quite complex. Most compressors provide multiband operation. In this configuration, the compressor divides the frequency spectrum into several bands and processes each one separately. Although this approach produces a subjectively louder sound, the result is a frequency response that is a function of level.

Some broadcast audio processors are composed of both a compressor and a limiter. The compressor is used to reduce the dynamic range of the input signal, and the limiter prevents overmodulation of the station transmitter. Limiters intended for AM use sometimes treat the positive and negative signal peaks differently. The processors can allow slight overmodulation of the carrier in the positive direction, but not in the negative direction. This process keeps the modulated signal linear, eliminating the distortion that would occur if the carrier were allowed to disappear (the result of excessive negative modulation).

Limiter. Broadcasters are often faced with a signal that is generally fairly constant in level, but occasionally increases suddenly, causing the system to clip or to distort. To correct such signal fluctuation requires a limiter, a device that operates as a standard amplifier for signals below some input level, but becomes a compressor for signals above this level.

A transfer curve for a typical limiter is shown in Fig. 13.46. The level at which the limiter changes from unity gain to compression is called the *threshold* or *turnover point*. This point is usually variable, so that the threshold can be adjusted to match the requirements of the station and the program material. Above the threshold, the compression function is characterized by the slope of the transfer curve, as with a conventional compressor. The knee in the transfer curve may be sharp, as shown in Fig. 13.46, or it may be rounded. Some limiter manufacturers claim that the side effects from a rounded knee characteristic are less audible.

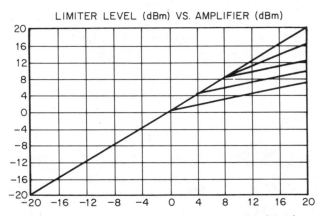

FIGURE 13.46 Transfer curve for a typical limiter, showing gain reduction of the output signal above the threshold or turnover point. (*Source: Richard Cabot, "Limiters, Compressors and Expanders,"* Broadcast Engineering, *August 1986. Reprinted with permission of Intertec Publishing Corp.*)

A compressor can be converted into a limiter by the addition of a diode before the gain-control element, as shown in Fig. 13.47. The dc voltage from the threshold pot is applied to the output side of the diode. This forces the signal level to exceed the threshold before compression can occur. As with compressors, limiters can be designed as either feedback or feedforward systems. The feedforward-type limiter requires predictable characteristics in the level sensor and voltage-controlled element. The feedback-type limiter does not require closely controlled elements if the exact compression slope is not a major concern. The limiting threshold is set by the diode bias voltage or its equivalent. The limiting function may be performed by a variety of devices, including an FET or a light-dependent resistor-LED combination.

Expander. Expanders are the functional inverse of compressors; they make soft signals softer and loud signals louder. The technique involved is graphed in Fig. 13.48. The slope of the lines is always greater than the 45° slope of an amplifier. If an expander produces an increase of 3 dB in output level for a 1-dB increase in input level, it is said to have an expansion ratio of 3:1. This will exactly cancel the dynamic range compression of a 3:1 compressor. An expander is shown in block

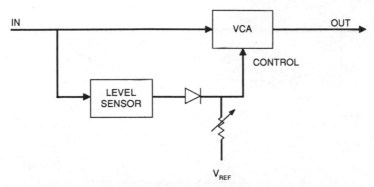

FIGURE 13.47 Basic block diagram of a limiter. A typical compressor may be essentially converted to a limiter through the addition of the diode shown. *(Source: Richard Cabot, "Limiters, Compressors and Expanders," Broadcast Engineering, August 1986. Reprinted with permission of Intertec Publishing Corp.)*

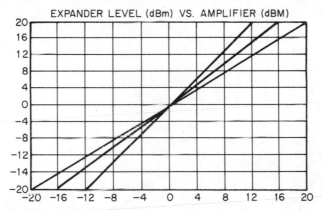

FIGURE 13.48 Typical transfer function for an expander circuit. *(Source: Richard Cabot, "Limiters, Compressors and Expanders," Broadcast Engineering, August 1986. Reprinted with permission of Intertec Publishing Corp.)*

diagram form in Fig. 13.49. The only change from a compressor is the addition of an inversion stage to make the gain increase with increased signal level.

Noise Gate. Broadcast, recording, and sound reinforcement pickups must sometimes rely on multiple-microphone setups, such as stage performances or panel discussions. Unfortunately, when there is no sound from the desired source, each microphone continues to pick up ambient noise. The noise gate offers a method of turning down the gain of a microphone or other audio source when the signal level drops below some preset value.

Noise gates are to expanders what limiters are to compressors. Above the threshold level, a noise gate operates as a normal amplifier. Below the threshold, the gain decreases with decreasing signal level, making soft sounds much softer. This effectively gates out or removes the noise, but does not affect the desired

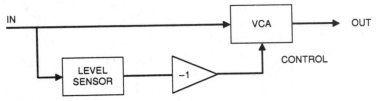

FIGURE 13.49 Simplified block diagram of an expander. *(Source: Richard Cabot, "Limiters, Compressors and Expanders," Broadcast Engineering, August 1986. Reprinted with permission of Intertec Publishing Corp.)*

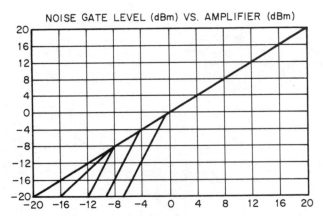

FIGURE 13.50 The transfer function for a typical noise gate. *(Source: Richard Cabot, "Limiters, Compressors and Expanders," Broadcast Engineering, August 1986. Reprinted with permission of Intertec Publishing Corp.)*

signal. This characteristic (graphed in Fig. 13.50) is similar to a limiter transfer curve that has been flipped diagonally. As with a limiter, there are two important parameters: the threshold level and the expansion ratio. With proper adjustment of the threshold level, the unit can discriminate between desired signal and unwanted background noise. If there is insufficient level difference between the two, erratic changes in gain will occur as the noise gate switches in and out of expansion.

The noise gate is shown in block diagram form in Fig. 13.51. It is similar to both an expander and a limiter. The inverter is used, as in an expander, to make gain increase with increasing signal level. The diode prevents the level sensor output from exceeding the desired threshold. When this occurs, the gain is clamped. Below the threshold, the unit functions as a conventional expander.

13.5.2 Dynamic Processing

So far, only the steady-state behavior of gain-control devices has been examined. When the signal amplitude changes with time, such signal analysis is not so easy. Audio signals are, by their nature, ac waveforms that go positive and negative many times per second. However, the signal amplitude must be controlled with-

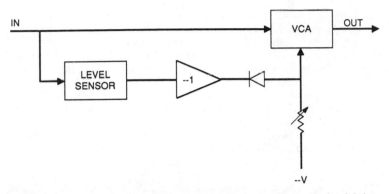

FIGURE 13.51 Simplified block diagram of a noise gate. *(Source: Richard Cabot, "Limiters, Compressors and Expanders," Broadcast Engineering, August 1986. Reprinted with permission of Intertec Publishing Corp.)*

out affecting the waveshape of these ac voltages. For example, if the signal amplitude is adjusted too quickly, the waveshape will be changed, causing audible distortion. If it is adjusted too slowly, the compressor or limiter will not be able to control the peaks.

Figure 13.52 illustrates the response time of a typical limiter. A tone burst

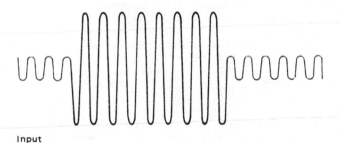

Input

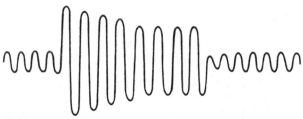

Output

FIGURE 13.52 Representative waveforms showing typical tone-burst response of a compressor. The waveforms illustrate the problems caused by inherent response delay of a compressor. *(Source: Richard Cabot, "Limiters, Compressors and Expanders," Broadcast Engineering, August 1986. Reprinted with permission of Intertec Publishing Corp.)*

changing 20 dB in amplitude is applied to the input. When the signal amplitude increases, the limiter requires a given time to respond, resulting in an overshoot at the output. As the limiter adjusts to the new gain level, the output amplitude decays to the desired value. When the signal amplitude drops, the output level also drops by the same amount. As the limiter readjusts to the new signal level, the output gradually increases.

Many audible problems are related to the time required to adjust the gain. These are the *pumping* and *breathing* sounds sometimes heard as medium-level background sound is modulated in amplitude by large-level sounds. If a limiter is designed to respond slowly to avoid these problems, it will not be able to prevent peak amplitudes from exceeding the desired level. With any design, there is a trade-off between audible side effects and incomplete processing.

Processor Features.   There are many features available on compressors and limiters that may be important for a particular application. Most professional compressors provide a visual indication of the gain or gain reduction taking place. Many units allow this display to be switched to also monitor the input or output signal levels. This visual indicator can be helpful when adjusting drive levels in a system.

Some compressors and limiters allow the control voltages in the level-sense path to be tied to other similar units for use in multichannel systems. If separate units are used on the two channels of a stereo sound mix or broadcast, tying together these points will prevent the image from shifting between the two channels because of unequal channel compression. A few units allow the level-sensing circuitry to be patched for special effects. The *sense* input is usually inserted into an EQ path for removing rumble or other noise that would disrupt the level-sensing action.

Compressors can sometimes be used as remote-controlled attenuators. Inputs for remote gain-control setting can be handy in special situations. For instance, a simple variable dc voltage could control the speaker or headset volume in a studio. This approach eliminates the problems associated with routing audio through remote volume controls.

Performance Specifications.   Standard audio distortion and signal-to-noise performance specifications are difficult to apply to dynamic range-modifying devices. Because the gain, as well as the selected ratios and threshold voltages, changes with input signal level, the performance measures also change. Noise generally becomes worse at high values of gain (low signal levels for expanders and compressors). Distortion will sometimes peak at intermediate values of gain and sometimes at the extremes of gain, depending on the type of gain-control element used. Specifications such as frequency response, common-mode rejection, and maximum input level typically are comparable to other types of signal-processing devices.

The precautions discussed in Chap. 10 (Sec. 10.4) regarding equipment interfacing apply especially to compressors, which can increase system hum and noise by many tens of decibels when the input signal is removed. Using such devices at different places in an audio system places differing constraints on residual noise and headroom. A careful study of the system gain structure is required before any particular device is specified.

It is difficult to quantify the specifications unique to limiters and compressors in a way that allows meaningful comparison of the audible performance of different units. Attack and release times are only two aspects of dynamic behavior.

Distortion performance of the limiter during attack will significantly alter the perceived distortion with actual program material. Some compressors and limiters have marginal headroom and hard clip on large inputs until the level sensor responds and reduces the gain. Other devices are designed with more headroom or a soft clip circuit, which greatly reduces the level of high-order distortion products during overdrive.

BIBLIOGRAPHY

Bartlett, Bruce, *Introduction to Professional Recording Techniques*. Howard W. Sams.

Beranek, Leo L. *Acoustics*. American Institute of Physics for the Acoustical Society of America.

Cheney, William C. "Bi- and Tri-amplification Sound Systems." *Broadcast Engineering*, January 1984.

Everest, F. Alton, *Successful Sound Operation*, Tab Books, Fender Application Manual, pp. 2224–2244, Fender Musical Instruments.

Mapp, Peter, *Audio System Design and Engineering*, Klark Teknik.

CHAPTER 14

LP AND CD RECORDING AND REPRODUCTION

K. Blair Benson
Television Technology Consultant, Norwalk, Connecticut

14.1 LONG-PLAYING PHONOGRAPH

14.1.1 Basic Elements

Acoustical or mechanical recording is the tree from which all recording technology grows. In the century since the filing on August 12, 1877, of Thomas Edison's basic patents, which envisioned the cylinder, disk, and mechanically encoded tape, the state of the art of recording and reproduction has addressed the multitude of process limitations and has driven the end product toward true fidelity. Though F. Langford Smith's 1952 evaluation of sound reproduction [1], that "it is manifestly impossible to reproduce at the two ears of the listener an exact equivalent of the sounds which he could hear in the concert hall," is technically true, analog disk recording and reproduction in the form of stereo and, in particular, discrete four-channel formats have come very close indeed. While significant re-

---

Portions of this chapter were adapted from K. Blair Benson (ed.), *Television Engineering Handbook*, McGraw-Hill, New York, 1986.

finements have advanced the art to near perfection, the basic elements of recording technology, present at the turn of the twentieth century, are either with us today or are periodically being rediscovered. For example:

1. *Sound-powered diaphragm:* In 1877, Edison's work in multiplex telegraphy spun off the technology of a sound-powered diaphragm lashed to an embossing or cutting tool capable of generating a trace in a moving substrate which could later be reproduced. The use of a diaphragm in sound transducers such as microphones and loudspeakers is still basic to audio technology today.

2. *Vacuum metallizing:* The use of vacuum metallizing was introduced by Edison in 1887 to prepare wax masters for electroforming [2]. This process was rediscovered in 1930 by Western Electric experimenters and, though not in general use for analog disks, is widely used today as a means of generating a reflecting surface in optical disks.

3. *Disk format:* The use of a disk format for recorded sound was promoted by Emile Berliner with his Gramophone (now gramophone) of 1887 [3]. Today the disk format is basic to audio recording and other high-density information uses.

4. *Electroforming:* The use of electroforming to generate tools for replicating recorded disks and cylinders has been basic to the industry from the beginning.

5. *Thermoplastic molding:* The use of thermoplastic materials and the process of molding for the making of copies have also been basic to the industry from the beginning.

6. *Vertical and lateral recording:* Cylinder records utilized vertical recording, and lateral recording began with disks. Today's analog stereo disk is a balance of the two techniques with two transducers acting at opposing 45° angles to the surface of the record.

For an excellent summary of early practices, Refs. 2 and 3 are recommended.

14.1.2 Comparison between Vertical and Lateral Recording

Vertical and lateral recording methods each had certain advantages and disadvantages which resulted in their close competition in the early days, but from now on this discussion will concentrate on the more modern technology of analog disk recording.

Today vertical recording really has no particular advantages, but it does exhibit several problems. Its chief disadvantage is that the playback stylus must resemble the recording stylus as closely as possible to reduce the geometric phenomenon known as *tracing distortion*. Records now are cut with a relatively sharp recording stylus and played with a somewhat rounded playback stylus. The greater the difference between the two styli, the greater the tracing distortion, which can be quite annoying.

Lateral recording has one very distinct advantage over vertical recording. It does not exhibit the nonlinear tracing distortion common with vertical modulation. Instead, it has a problem known as *diameter loss*, which is a loss of signal output with increasing frequency and decreasing groove diameter. This, too, is a result of the playback stylus's not exactly matching the recording stylus. There is also the phenomenon known as *pinch effect*, in which some of the desired recorded lateral motion is transformed into useless vertical motion.

Another advantage of lateral recording is that the groove can be a constant depth, which makes things easier for the disk-cutting engineer and for the manufacturing people. With vertical recording, the depth must be continuously varied, depending on the average loudness of the music, if the maximum recording time is to be realized. In both lateral and vertical recording, the groove pitch needs to be constantly varied to achieve maximum recording time.

Lateral recording, developed by Emile Berliner in 1887, superseded Edison's "hill-and-dale" vertical recording on cylinders in the marketplace, and the 78 r/min (actual speed of 78.26 r/min) disk recording remained dominant for half a century [4]. It was not until the 1950s that the 78 record was overtaken rapidly by the popularity of the LP and relegated to obsolescence. Consequently, the details of 78-disk characteristics are not covered in this book. Fortunately many of the recordings of lasting interest on the 78 format have been transferred from early disk and tape masters to LPs and more recently CDs. Readers interested in the technology are referred to the now-obsolete Electronic Industries Association (EIA) standard RS-211 for electrical and mechanical specifications.

14.1.3   Stereophonic (Two-Channel) Recording

Early Developments.   The battle between proponents of vertical and lateral recording was renewed with during the development of two-channel disk recording for stereo. In 1953 Emory Cook announced the first commercial system of recording two audio signals as separate tracks on one disk, called *binaural*—the term *stereophonic* was not in common use at that time [5]. Playing Cook's record on a monaural reproducer produced either the left or the right channel, not a mix of both.

John Mullin of Bing Crosby Enterprises developed a single-track system which utilized vertical and lateral recording for the left and right channels, respectively [6]. Stereo playback was achieved by the use of two pickup heads mounted at 90° to each other. However, as in the Cook system, a mix of the two channels was not possible on conventional monaural players which reproduced only the lateral channel.

45-45 System.   Monophonic compatibility was achieved with the *45-45 system*, invented by engineers for Westrex in 1957 [7]. Still the standard today, the system is so named because one side of the sound field is recorded with its axis of modulation rotated 45° to the left of vertical while the other side is recorded 45° to the right of vertical. (Refer to Fig. 14.1.) Notice that this places the two modulations at 90° to each other, and the system is orthogonal. Notice also that each side of the sound field has a component of its modulation lying laterally. There is also a vertical component of each. These components are smaller than the amplitude of the pure left or right modulations by a factor of 2, so there is some reduction in loudness. But when this record is played back on a monophonic player, equal parts of the left and right sides of the sound field are reproduced and the listener hears all the music—a compatible system.

The 45-45 system is the present world standard for two-channel analog disk recording; it will now be discussed in more detail. Assume a cross-sectional view of the groove, and assume that the disk is rotating toward the viewer. This corresponds to looking at the front of the pickup arm. Standard channel placement has the left-channel modulation recorded on the inner (left) groove wall and the right on the outer (right) wall.

For the system to be compatible with lateral monophonic playback, attention must be paid to the relative phase of the signals from the two sides of the sound

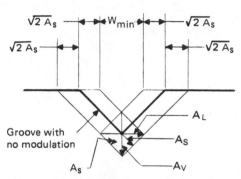

FIGURE 14.1 Cross-section of stereo groove. ["Disk Recording," *vol. 2*, J. Audio Eng. Soc., *379 (1981)*. In Audio Engineering Handbook, *K. Blair Benson, ed., McGraw-Hill, New York, 1988*]

field. Suppose that positive sound pressure on the left-microphone diaphragm eventually results in left-groove-wall modulation which causes the playback stylus to move toward the left (and down). If the same sound source is also acting on the right microphone and results in right-wall modulation which causes the stylus to move to the left (and up), the system is said to be *in phase*. The lateral direction is called the *sum* of the left and right signals, and the vertical direction is called the *difference*. From this example, it is clear why this is so. If the two microphones are placed very close together, the system acoustically and physically degenerates to a purely lateral monophonic one. There will be no difference in the sound fields picked up by the two microphones. The lateral components of left and right modulation will add in the same direction and reinforce each other, and the vertical components will cancel each other completely, resulting in a net purely lateral stylus motion.

No discussion of two-channel disk recording schemes would be complete without a mention of other, ill-fated propositions. One was similar to the Cook system with two independent grooves interleaved, or bifilar-wound, on the same side of the disk. The obvious problem with this system was that it, like the Cook system, reduced the playing time by half, and it required complex cutters and pickups. Another idea was to put one channel on each side of the disk. This would pose difficulties in manufacturing and in designing playback equipment. It would also halve the total playing time per disk. Another interesting scheme would prove to be providential: the carrier-disk system. This involved using the second audio channel to modulate a supersonic carrier frequency recorded in the same groove as the first channel. But at the time that this idea was conceived, technology was too primitive to permit successful recording or recovery of a signal of such a very high frequency from a disk. The system would later make a reappearance as the only discrete four-channel system put into commercial production with analog disks: CD-4.

The advantage of the 45-45 system over all two-groove systems is clear: It is the only system compatible with monophonic players, and that compatibility is reversible. If a monophonic (lateral) recording is played on a 45-45 stereo system, the same sound comes from both speakers in phase. The sound fields combine acoustically in the listening environment and create the effect of a single sound source located midway between the two reproducers. The other monogroove systems could have had this compatibility, too, if their designers had hit on the notion of electrically forming the sum-and-difference signals from the left and right sound sources and appropriately modulating their disks. This was done, in fact, with the later CD-4 system. However, the 45-45 system still has an advantage over a simple two-channel carrier-disk system in that it does not require expensive supersonic recording and reproducing equipment.

The chief disadvantage of the 45-45 system is the same one common to all systems that employ any form of vertical modulation: tracing distortion. It was a problem for Edison, and it continues to be a problem today. But technological advances have made possible reduced tracking forces, better transducers, and playback-stylus shapes that more closely approximate the recording stylus, so that tracing distortion has become much less of a problem than it once was.

14.1.4 Recording Characteristics

Preemphasis and Deemphasis. One of the primary objectives in recording technology is the maintenance of the lowest possible noise level. This has been achieved by a combination of stringent manufacturing quality control and the use of filters which tailor the frequency response to help mask the remaining defects. This is accomplished by using a technique known as *preemphasis* before recording and *deemphasis* upon playback. The most annoying defects heard in records are the result of spurious deflections of the playback stylus caused by disturbances such as dirt in the groove and roughness in the molded disk. The small size of these disturbances or defects corresponds to higher sound frequencies.

The function of preemphasis in disk recording is not the same as record and playback equalization in magnetic-tape recording, which are to compensate for head-inductance effects and head-to-tape gap and contact losses on playback. In addition, the recording characteristic is adjusted for the optimum level of tape magnetization, rather than to complement a playback deemphasis characteristic for reduction of playback signal and bias noise resulting from variations in the distribution of particles in the magnetic coating.

However, in analog disks, in addition to the playback noise from the record surface and foreign particles, the surface noise in recording is emphasized by the 6 dB per octave roll-off of high frequencies in the signal produced from the constant-velocity magnetic recording head. In early disk recording when magnetic playback heads were used almost universally for both recording and playback, their characteristics were complementary, resulting in flat overall response, but at the expense of boosting the spurious scratches and pops as well as the attenuated high frequencies of the recorded signal.

So the purpose of preemphasis in disk recording and of the complementary deemphasis in playback is to boost the amplitude of the recorded waveform at higher frequencies to improve the S/N ratio. A simple 6 dB per octave filter seems like a logical choice, since it would produce a constant-amplitude recording. But a glance at the response curve for the industry-standard emphasis filter reveals that it is not so simple (see Fig. 14.2). This emphasis, adopted as a standard by the Recording Industry Association of America and known as the *RIAA characteristic*, produces a shelved response which is essentially constant-amplitude below 50 Hz and above about 2 kHz and constant-velocity between these frequencies.

The reason for this curious filter shape is one of practicality and compromise. Prior to the adoption of the present standard in 1954, various "semistandards"

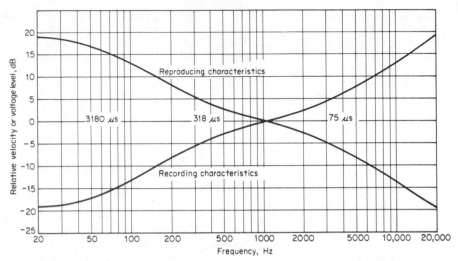

FIGURE 14.2 Characteristics for fine-groove disk records. *(Source: RIAA Dimensional Characteristics for 33¼ r/min Records. In* Audio Engineering Handbook, *K. Blair Benson, ed., McGraw-Hill, New York, 1988)*

were in use [8]. Some were intended for 78 r/min and some for LP; some were in-house agreements used by certain record companies; and there was general confusion. By the early 1950s, the National Association of Broadcasters (NAB) and the EIA had jointly proposed a deemphasis standard for disk playback. Essentially the same characteristic has been in use ever since, and it was formally adopted by the RIAA in 1967.

The low-frequency portion of this standard curve is a compromise among the various characteristics in use in the early 1950s. The high-frequency portion constitutes a "shelf" of about 9 dB above what would otherwise be a constant-amplitude extension of the low-frequency characteristic. This results in a corresponding amplitude reduction in the recorded waveform at high frequencies, which was needed to lessen the distortion inherent in the phonograph pickups of the day with their fairly high-mass, conically tipped stylus assemblies [9]. Those early pickups could not cope with the stylus velocities and accelerations at the high frequencies required by a constant-amplitude recording. Although greatly reduced, these same problems of tracking and tracing distortion are still troublesome.

Record Dimensions.   Current standards for 33⅓ r/min specify diameters of 7, 10, and 12 in. A detailed description of the physical dimensions can be found in EIA publication RS211-D [10].

14.1.5 Playback Equipment and Characteristics

Piezoelectric Pickups.   Crystal or ceramic transducers are also known as *piezoelectric pickups*. Piezoelectrics (PZTs, for piezoelectric transducers) were one of the first types developed for consumer applications and are still widely used in inexpensive record players. A PZT pickup is built with a piece of PZT material (usually a flat, thin, and narrow rectangular sliver) mounted so that one of its small ends is fixed to the body of the pickup. The opposite end is connected to the cantilever assembly so that the PZT material is twisted or bent by the motions of the stylus.

Stressing a PZT element in this way causes the element to develop an electric potential between its parallel broad faces. A PZT generator will produce a voltage proportional to its displacement from its mechanical equilibrium position. PZT pickups are therefore sensitive to the amplitude of the stylus motion and are known as *constant-amplitude transducers*. A PZT pickup is also a direct-current (dc) device. It will produce a constant unipolar (dc) voltage if its stylus is moved away from rest and held there. As with other types of constant-amplitude pickups, PZTs require a different deemphasis characteristic (playback equalization) in the preamplifier than do constant-velocity types.

The typical inexpensive PZT pickup has a rather large generator element which is also relatively stiff. This results in a pickup which produces high output voltages in the range of 0.1 to 1 V and with rather low compliance, which must then be used at fairly high tracking forces of about 3 to 10 g. This type of pickup, however, can then be used with a very simple low-gain amplifier with a very simple equalization circuit. The inexpensive PZT pickup which requires only an inexpensive amplifier has led to the popularity of this combination in economy record players.

PZT pickups are not necessarily inexpensive or of poor quality. Several manufacturers, notably Sonotone and Micro-Acoustics, have marketed premium PZT designs which compare favorably with competitive magnetic types. With the use of very small and compliant generator elements, low-mass moving systems, and quality styli, these pickups can perform as well as any other type. Owing to the smaller and more compliant generator, these pickups produce considerably less voltage than do the economy versions. Since the magnetic pickup has become the preferred type for high-quality applications, high-fidelity preamplifiers are usually equipped with proper equalization for only constant-velocity pickups. Premium PZT units therefore are usually designed with an electrical compensation network built into them or are supplied with such a network which can be connected between the pickup and the preamplifier. This network tailors the pickup's frequency response to look like that of a constant-velocity pickup.

Magnetic Pickups. The favored types for high-quality applications are *magnetic pickups*. There are three basic variations on the magnetic design, but they all result in producing the fundamental requirement of any magnetic generator: changing the magnetic lines of flux that pass through a coil of wire. Magnetic pickups are sensitive to the *change* of magnetic flux crossing the coil and not to the long-term intensity. Consequently, they respond to the velocity of stylus motion and are known as *constant-velocity generators*. They must be used with amplifiers employing a deemphasis characteristic (playback equalization) designed for this type of frequency response.

1. The most common type of magnetic pickup is the *moving-magnet pickup*. As the name implies, in this design the coil is held stationary, and the magnet structure itself is moved, which causes the lines of flux to vary in the coil. The magnet is usually mounted directly to the cantilever assembly. Some of the most popular models are manufactured by Stanton/Pickering, Shure Brothers, and Audio-Technica. A whole range of models, from the inexpensive to premium state-of-the-art designs, are available.

For some years, moving-magnet pickups were thought to be a moving high-mass design compared with the other magnetic types. However, today's high-intensity magnets (samarium cobalt) and high-permeability magnetic materials have allowed the moving-magnet pickup to employ an effective tip mass as low as that of any other design. Moving-magnet pickups also have the advantages that they are fairly inexpensive to build, the stylus assembly can be easily interchanged, and the generator itself can be very efficient and linear while also being quite small and of low mass. This allows the entire pickup to be lightweight and suitable for use in the best low-mass tone arms.

2. The second most popular type of magnetic pickup is the *variable-reluctance pickup*. In this design, neither the coil nor the magnet moves. Instead, some other permeable member of the magnetic circuit is connected to the cantilever. This moving member functions as a variable magnetic valve and causes the magnetic-flux intensity to vary at the coil.

Variable-reluctance pickups have essentially the same characteristics as the moving-magnet types, with interchangeable styli, linear generator systems, low moving masses, and light weight being the important features. They are also comparable in price and output voltage. Models are available from the inexpensive to premium-price ranges.

Developed by William S. Bachmann of the General Electric Company, this type of pickup was marketed by GE under the trademark Variable Reluctance for a number of years. GE has since discontinued the manufacture of magnetic phonograph pickups, and many other manufacturers have moved in with a wide range of variations on this basic design. Some of the most popular units are now made by Bang & Olufsen, which terms its design the *moving micro-cross*, and ADC, which calls its design the *induced-magnet principle*.

3. The third type of magnetic pickup is the *moving-coil pickup*. Just as the name implies, the magnet structure is held fixed in the cartridge body, and the coil is attached to the cantilever.

Other Types of Pickups.   In addition to the above, pickups based on principles of variable resistance and variable capacitance have found limited use.

Preamplifier Requirements.   The frequency responses of phonograph pickups can be grouped into two general categories. Pickups which are sensitive to the absolute displacement of the stylus regardless of frequency are said to have a *constant-amplitude frequency response*. Pickups which have an increasing output voltage with increasing frequency when the displacement is held constant are said to have a *constant-velocity response*.

Almost all phonograph pickups produce an output voltage which is too small to drive a loudspeaker or a headphone directly. The typical magnetic pickup produces about 5 mV at normal recording level at 1 kHz. This is also about 30 dB lower than the typical *line level* (about 220 mV) at which other consumer-type high-fidelity components such as tape decks or tuners operate. Therefore, it has become common practice to send the signal from the pickup through an additional amplifier, called a *preamplifier*, where it is boosted to a level comparable with that of other components. This line-level signal is the nominal voltage to properly drive most *power amplifiers*, which in turn drive headphones or loud-speakers.

The phonograph preamplifier also performs the important function of correcting the frequency response of the pickup to complement that used in cutting the record. The record is cut with a modified constant-amplitude characteristic. Since neither a constant-amplitude nor a constant-velocity pickup is properly suited to complement this characteristic, some response correction is necessary for either type.

14.1.6 Wow and Flutter

Audible *wow* is a condition in which the pitch of the music is slowly varying up and down. It is a form of frequency modulation and is usually caused by the record's being eccentric or warped. It can also be caused if the record is period-ically slipping on the turntable, which is sometimes the case with record changers employing heavy tracking forces. The modulating frequency is usually related to the rotational speed of the record, but in general it is a very low frequency in the range of about 0.1 to 4 Hz.

Flutter is a condition similar to wow, but with a higher unwanted modulating frequency, from about 4 to 30 Hz. Flutter in record players is usually the result of some defect in the turntable drive system. The motor shaft may be eccentric, or there may be a defect in the drive puck. In direct-drive players, the complicated motor itself may be exhibiting a problem with stability in its feedback speed control circuit.

References

1. R. Gelatt, *The Fabulous Phonograph*, Lippincott, Philadelphia, 1955, p. 18.

2. Ibid., p. 28.

3. Ibid., p. 192.

4. Ibid., p. 60.

5. E. Cook, "Binaural Disc Recording," *Disk Recording*, vol. 1, *J. Audio Eng. Soc.*, 179(1980).

6. J. T. Mullin, "Monogroove Stereophonic Disc Recording," *Disk Recording*, vol. 1, *J. Audio Eng. Soc.*, 182(1980).

7. J. G. Frayne and R. R. Davis, "Recent Developments in Stereo Disc Recording," *Disk Recording*, vol. 1, *J. Audio Eng. Soc.*, 185(1980).

8. H. E. Roys, "The RIAA Engineering Committee," *Disk Recording*, vol. 1, *J. Audio Eng. Soc.*, 466(1980).

9. "AES Standard Playback Curve," *Disk Recording*, vol. 1, *J. Audio Eng. Soc.*, 451(1980).

10. Recording Industry Association of America, Inc., *Standards for Analog Disc Records*, Bulletin E-1, 1977.

Bibliography

Carlson, R. E. "Resonance, Tracking, and Distortion: An Analysis of Phonograph Pickup Arms," *Disk Recording*, vol. 2, *J. Audio Eng. Soc.*, **259**(1981).

Cooper, D. H. "Compensation for Tracing and Tracking Error," *Disk Recording*, vol. 1, *J. Audio Eng. Soc.*, **39**(1980).

———. "Integrated Treatment of Tracing and Tracking Error," *Disk Recording*, vol. 1, *J. Audio Eng. Soc.*, **44**(1980).

———. "On Tracking and Tracing Error Measurements," *Disk Recording*, vol. 1, *J. Audio Eng. Soc.*, **61**(1980).

Hunt, F. V. "The Rational Design of Phonograph Pickups," *Disk Recording*, vol. 2, *J. Audio Eng. Soc.*, **186**(1981).

Marcus, E. J., and M. V. Marcus. "The Diamond as an Industrial Material, with Special Reference to Phono Styli," *Disk Recording*, vol. 1, *J. Audio Eng. Soc.*, **315**(1980).

14.2 COMPACT-DISK (CD) DIGITAL SYSTEM

14.2.1 Basic Specifications

Overview. The Sony Corporation in Tokyo, Japan, and Philips N. V. in Eindhoven, The Netherlands, announced in June 1980 the completion of development and standardization of a compact disk (CD) digital audio system. The CD 12.0-cm (4.7-in) diameter disk provides a nominal playing time of 60 min which can be expanded to 74.7 min by using pulse-code modulation (PCM) to convert analog audio signals to a 16-bit digital code. Stereo audio signals are sampled at a rate of 44.1 kHz.

Unlike the conventional phonograph, which relies on mechanical contact between the disk grooves and a stylus transducer, the CD playback transducer is not in contact with the disk. A laser beam emitted by a solid-state gallium arsenide diode focused on the disk surface is modulated by shallow depressions, called *pits*, and reflected to a photodiode for detection as an electric signal. The tracking function, provided mechanically by grooves in phonograph records, is accomplished in CDs by means of electromechanical servo systems which focus the laser spot on the disk and control its radial tracking of the spiral path of pits.

Disk Specifications. The diameter of the disk is 12.0 cm (4.7 in), and the thickness is 1.2 mm. The track pitch of 1.6 μm is about one-sixtieth that of the LP record. The disk rotation is clockwise, as viewed from the top. The readout is from the opposite side. The recording method is *constant linear velocity* (CLV) at a rate of 1.25 m/s. This technique was chosen to maximize the recording density. This requires a change in rotational speed from 500 r/min at the start near the center to 200 r/min at the maximum diameter of the program area. This is shown in Fig. 14.3.

14.2.2 Digital Format

The basic specifications, that is, the audio specifications, signal format, and disk specifications, for the compact-disk digital format are summarized in Table 14.1.

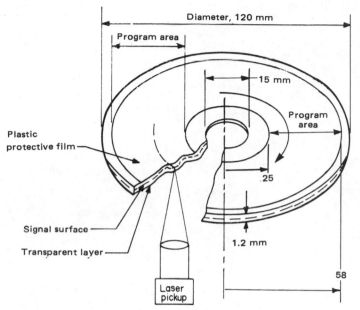

FIGURE 14.3 Compact-disk configuration and principal dimensions. *(Source: K. Blair Benson, ed.,* Audio Engineering Handbook, *McGraw-Hill, New York, 1988)*

Sampling Frequency. *Pulse-code modulation* is used to convert audio signals to digital signals. Stereo audio signals are sampled simultaneously at a rate of 44.1 kHz. This sampling frequency was chosen for the following reasons:

1. From the standpoint of filter design, a 10 percent margin with respect to the Nyquist frequency is needed. The frequency of 44 kHz is the minimum sampling frequency required to cover audible frequencies up to 20 kHz (20 kHz × 2 × 1.1 = 44 kHz).

2. The frequency of 44.1 kHz[1] is used in digital audio-tape recorders based on video-tape recorders (VTRs). Note that when a continuous signal in a limited bandwidth is sampled, it can be converted to a discrete signal with no degradation in sound quality.

Quantization. Quantization is the key factor in determining the sound quality of a digital system. A 16-bit linear quantization has been chosen to maintain the same quality as that of master tapes. Coding of 16 bits is also preferable from the point of view of digital-data application. The theoretical dynamic range of the system at maximum-amplitude input is about 97.8 dB, or substantially greater

---

1. There are 525 and 625 lines per frame in the NTSC and CCIR TV systems, respectively. Each line can record three words per channel including redundant data, and 490 and 588 lines per frame are available for data recording. Thus, 44.1 kHz comes from the following calculation:

3 × 490 × 30 Hz (NTSC bandwidth) = 3 × 588 × 25 Hz (CCIR) = 44.1 kHz

TABLE 14.1   Basic Specifications of the CD System

Recording method	
Signal detection	Optical
Linear recording density	43 kbit/in (1.2 m/s)
Area recording density	683 Mbit/in^2
Audio specifications	
Number of channels	2-channel stereo
Playing time	Approximately 60 min
Frequency response	20 ~ 20,000 Hz
Dynamic range	> 90 dB
Total harmonic distortion	< 0.01%
Channel separation	> 90 dB
Wow and flutter	Equal to crystal oscillator
Signal format	
Sampling frequency	44.1 kHz
Quantization	16-bit linear/channel
	2's complement
Preemphasis	No or $^{50}/_{15}$ μs
Modulation	EFM
Channel-bit rate	4.3218 Mbit/s
Error correction	CIRC
Transmission rate	2.034 Mbit/s
Redundancy	≈ 30%
Disk specifications	
Diameter of disk	120 mm
Thickness of disk	1.2 mm
Diameter of center hole	15 mm
Program area	50 ~ 116 mm
Scanning velocity	1.2–1.4 m/s, CLV
Revolution speed	500 ~ 200 r/min
Track pitch	1.6 μm
Pit size	0.11 × 0.5 × 0.9 ~ 3.2 (μm)

than that of conventional analog systems. This results from a lower noise level. To reduce quantization noise, preemphasis of a $^{15}/_{50}$-μs time constant can be used. Coding is 2' complement, so the positive peak level is 0111 1111 1111 1111, and the negative peak level is 1000 0000 0000 0000.

Error-Correction Code and Control.   An error-correction code is used in digital systems to insert digital redundancy in the data stream. The purpose is to protect the integrity of the system from spurious interference occurring after digital encoding. The CD system employs the optical noncontact readout method. Because the signal surface is protected by a plastic layer and the laser beam is focused on the signal surface, the disk surface itself is kept free from defects such as scratches. As a result, most of the errors which occur at and in the vicinity of the signal surface through the mastering and manufacturing process are random errors of several bits. Even though the CD system is resistant to fingerprints and scratches, defects exceeding the limit will naturally cause large burst errors. A typical bit error rate of a CD system is 10^{-5}, which means that a data error occurs 2×10^6 bits/s $\times 10^{-5}$ = 20 times per second. Such data errors, even though they may be 1-bit errors, cause unpleasant pulsive noise; so an error correction technique must be employed.

Unlike an error in computer data, an error in digital audio data (if the error can be detected) can be concealed. In fact, simple linear interpolation is sufficient in most cases. The error-correction code used in a CD system must satisfy the following criteria:

1. Powerful error-correction capability for random and burst errors
2. Reliable error detection in case of an uncorrectable error
3. Low redundancy

A number of error-correction codes have been developed in the last two decades. The *cross-interleave Reed-Solomon code* (CIRC) satisfies the CD criteria listed above.

14.2.3 Player Optical Requirements

The principle and some optical requirements for the readout of a digitally encoded signal from a disk by means of a scanning spot are described.

Basic Optics for Reading.   The basic optics for reading are shown in Fig. 14.4. This simple figure consists of a light source, a microscope objective lens to concentrate a spot onto the information layer of a disk, a beam splitter, and a pin diode as a photodetector which converts to electric current.

The optical principle of noncontact readout is based on diffraction theory. Although this phenomenon by means of a narrow slot is well known, an analogous situation occurs if a light beam impinges on a reflective signal surface with pitlike depressions. In the case of a flat surface (between pits), nearly all the light is reflected, whereas if a pit is present, the major part of the light is scattered and substantially less light is detected by the photodetector (see Fig. 14.5).

Laser Diode (LD).   The light source used in the CD system must satisfy the following conditions:

1. It must be small enough to be built into the optical pickup.
2. It uses coherent light in order to focus on an exceedingly small spot.
3. Enough light intensity for readout must be provided.

GaAlAs semiconductor laser diodes satisfy the above requirements. The specifications of such an LD are the following.

$$\begin{aligned}
\text{Wavelength} &= 0.78 \text{ to } 0.83 \ \mu\text{m} \\
\text{Light power} &= \text{about } 3 \text{ mW} \\
\text{Lateral mode} &= \text{fundamental} \\
\text{Transverse mode} &= \text{fundamental} \\
\text{Longitudinal mode} &= \text{multiple}
\end{aligned} \qquad (14.1)$$

When the light from the LD is returned from the reflective surface of the disk, it has an effect on the light-generating characteristics of the LD and generates large optical noise fluctuations. Thus, a *multiple longitudinal mode* is necessary to prevent the phenomenon.

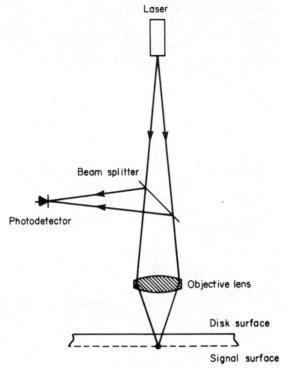

FIGURE 14.4 Basic optics for reading. *(Source: K. Blair Benson,* Audio Engineering Handbook, *McGraw-Hill, New York, 1988)*

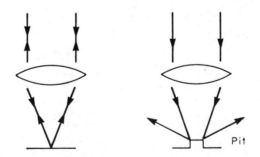

Scattering of light by phase object (pit)

FIGURE 14.5 Principle of noncontact readout. *(Source: K. Blair Benson, ed.,* Audio Engineering Handbook, *McGraw-Hill, New York, 1988)*

Lens.   The lens requirement can be specified by an optical parameter called the *numerical aperture* (NA). For a given index of refraction in a lens, the NA is proportional to the sine of half the angle of the cone of light projected by the lens, or

$$NA = n \sin 0 \qquad (14.2)$$

This is shown graphically in Fig. 14.5. Note that the first of a series of secondary responses of the projected beam, called an Airy disk, circles the focused spot. See Fig. 14.6.

A small value of NA is necessary for the lens system to focus on a small spot. Alternatively, a larger value will provide the desirable characteristics of greater depth of focus, greater allowance for skew (tilt), and greater allowance for variations in disk thickness. Taking all these factors into account, the value of NA for CD players must lie within the range of 0.45 to 0.50 at the wavelength of the laser diode.

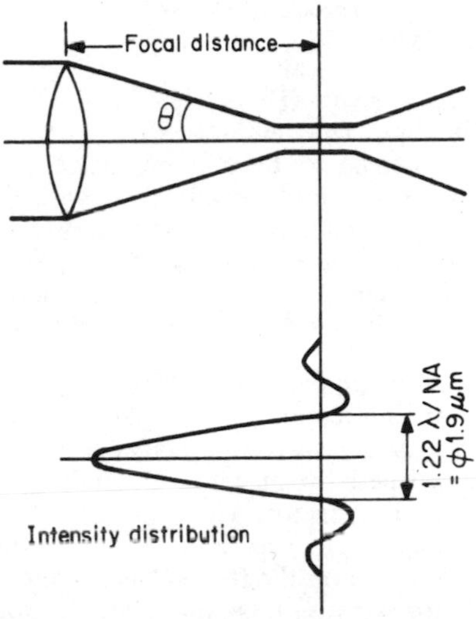

Modulation Transfer Function (MTF). The MTF is the amplitude-frequency characteristic of the optical system. In other words, it determines the smallest pits which can be detected in a disk. It includes the characteristics of the input channel, the disk and its reflective surface, and the output channel. The MTF of the CD optical system may be described as a form of low-pass filter with a gradual roll-off to a cutoff at the smallest pit that can be detected.

In numerical values, the CD optical system can detect pits as dense as 1154 per millimeter. The smallest pit of a

FIGURE 14.6   Numerical aperture of lens and Airy disk. (*Source: K. Blair Benson, ed., Audio Engineering Handbook, McGraw-Hill, New York, 1988)*

CD is about 0.87 μm at the linear track-reading velocity of 1.25 m/s. In terms of the temporal frequency, the cutoff is 1.44 MHz. These figures are for an ideal system and must be modified somewhat for actual operating conditions and available hardware.

14.2.4 Servo Systems

For tracking with a light beam, two position controls are necessary, one in the vertical and the other in the radial direction. These controls are called *focus-* and *radial-tracking controls*, respectively.

Generally, the servo system is composed of three subsystems, as shown in Fig. 14.7. The error of position is detected at the first block. The second block is the electronic compensation network, which is necessary for the stability of a closed-loop system. In the last stage, the electronic signal is converted to actual spot displacement by means of the electromechanical system.

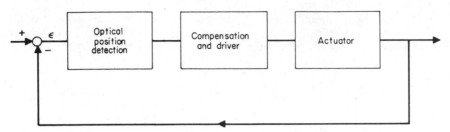

FIGURE 14.7   Block diagram of the servo system. *(Source: K. Blair Benson, ed.,* Audio Engineering Handbook, *McGraw-Hill, New York, 1988)*

Focus Servo.   This system keeps the laser beam focused on the reflective layer of the disk. The depth of focus must be great enough to permit minor variations in the distance of the rotating disk from the laser projection and receptor elements and in surface and thickness variations in the disk. The focus depth in a CD player nominally is ±2 μm. But the vertical deviation in distance for optimum focus can be as great as 0.5 mm, and the acceleration requirement for the focus adjustment is 10 m/s. Two methods for adjustment of focus, astigmatic and Foucault, and the actuator mechanism are described below.

1. *Astigmatic method:* One method to detect the light-beam position in the vertical direction is the astigmatic method (see Fig. 14.8). By using this method, it is necessary to modify the basic optics shown in Fig. 9.18. That is accomplished by placing a cylindrical lens between the beam splitter and the photodetector. The photodetector is divided into four segments. When the beam is focused on the disk surface within the focus depth, a circular spot is created on the four-segment detector surface. When the beam is focused before or after that point, elliptic spots are imaged on the detector. If an $(A + C)$ − $(B + D)$ operation is performed, the result is the focus-error signal.

2. *Foucault method:* There are differing forms of this method, one example of which is shown in Fig. 14.9. In this case a wedge is used instead of a cylindrical lens, and two two-segment detectors are employed. If the beam is in focus, the operation $(A + D)$ − $(B + D)$ is zero. If the disk and lens move closer, the image of the reflected light moves farther away. But if this distance increases, the resultant polarity of the signal becomes the opposite sign.

3. *Actuator:* The mechanism used in the vertical direction is the same as that employed in loudspeakers. For example, as in Fig. 14.10, an objective lens (or the complete pickup, if possible) can be attached to a voice coil, which moves up and down according to the electronic signal command from the focus-error detector through the phase-lead circuit.

Radial-Tracking Servo.   The laser spot should follow the center of the track to within about 0.1 mm without interfering with adjacent tracks merely 1.6 μm away. The variation in radial distance can be as great as 70 μm, and the acceleration requirement for tracking adjustment is 0.4 m/s. Two methods for adjustment of radial tracking and the actuator mechanism are described below.

1. *Twin-spot method:* When you are using this method for detection of the beam position in the radial direction, the basic optics must be modified. This is accomplished by placing a grating between the LD and the objective lens and

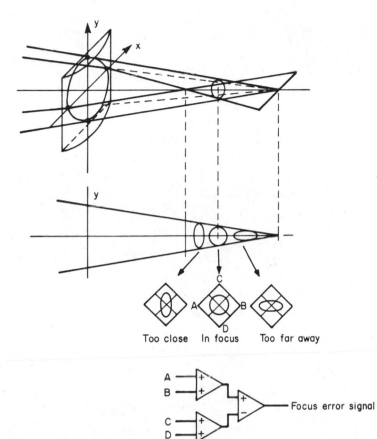

FIGURE 14.8 Astigmatic-focusing servo method. (*Source: K. Blair Benson, ed., Audio Engineering Handbook, McGraw-Hill, New York, 1988*)

two detectors in the detector plane. As shown in Fig. 14.11, two extra beams are projected, one on each side of the main beam. When the main beam is directly over the track, the edges of the secondary beams (E and F) barely encroach upon the track and the remainder of each beam is on the so-called mirror surface. In this case, the light output ($E - F$) is essentially zero. If all three beams should be misaligned slightly, the diffractions from E and F become different. Accordingly, the resultant output ($E - F$) becomes the radial-tracking-error signal.

2. *Push-pull method:* In contrast to the twin-spot method, in this method the radial error signal is obtained from one spot, as shown in Fig. 14.12a. When the position of the beam is exactly on the track (pit), there is an equal distribution in signal strength on the left and right sides of the detector. If this relationship of position should be changed slightly, however, the distribution of signal strength on the left and right sides becomes asymmetric. Accordingly, the resultant left half–right half becomes the radial-tracking-error signal.

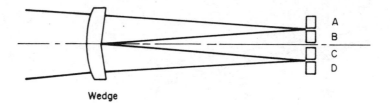

Wedge

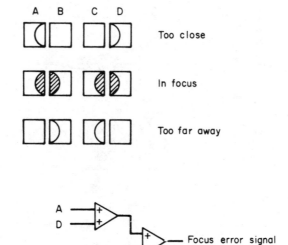

FIGURE 14.9 Foucault method for the focusing servo system. *(Source: K. Blair Benson, ed.,* Audio Engineering Handbook, *McGraw-Hill, New York, 1988)*

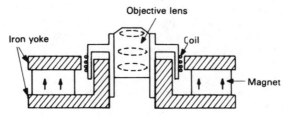

FIGURE 14.10 Actuator. *(Source: K. Blair Benson, ed.,* Audio Engineering Handbook, *McGraw-Hill, New York, 1988)*

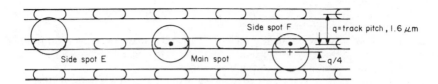

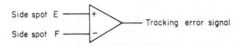

FIGURE 14.11 Twin-spot radial-tracking method. *(Source: K. Blair Benson, ed., Audio Engineering Handbook, McGraw-Hill, New York, 1988)*

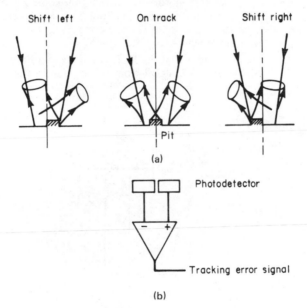

FIGURE 14.12 Push-pull method for radial tracking. (*a*) Beam-tracking position. *(Source: K. Blair Benson, ed., Audio Engineering Handbook, McGraw-Hill, New York, 1988)* (*b*) Block diagram of differential-amplifier system. *(Source: K. Blair Benson, ed., Audio Engineering Handbook, McGraw-Hill, New York, 1988)*

Although this method is extremely simple, there are some drawbacks: It is sensitive to variations in pit depth and pit shape, and when the lens is moved only to follow the track, a dc offset is produced in the radial-tracking-error signal.

3. *Actuator in radial direction:* There are two methods for moving the light beam in the radial direction. One uses the so-called two-axis device and slide move-

ment. The two-axis device is constructed by attaching additional coil assemblies to the voice coil of Fig. 14.12*b*. This method is a two-stage radical movement. A two-axis device mainly takes care of fine radial tracking and the slide drive of coarse tracking in the radial direction. This type of twin-spot actuator with astigmatic correction is a popular combination.

The other method is the so-called swing-arm pickup or segment meter. This pickup is mounted in a swing arm which describes an arc across the disk during playback. This type of actuator, a single-beam push-pull with Foucault focusing, is another popular combination.

Spindle Servo (Constant Linear Velocity). The specified CD linear speed typically is 1.25 m/s, but a speed of 1.2 to 1.4 m/s is permissible. But a CD player must rotate the disk at exactly the same speed as when the signal was recorded in mastering. Spindle servo control is accomplished in two sequential stages as follows (Fig. 14.13):

1. *Pull-in stage:* The disk spindle motor is controlled by some means so that it rotates within the capture range of the phase-lock loop (PLL) used for clock recovery. In most cases this is done by detecting $T_{\max}$ (the longest signal length) or $T_{\min}$ (the shortest signal length).
2. *Lock-in stage:* After confirming that the PLL is in *lock* condition, the spindle motor is locked to the reference signal from the crystal in the digital signal processor.

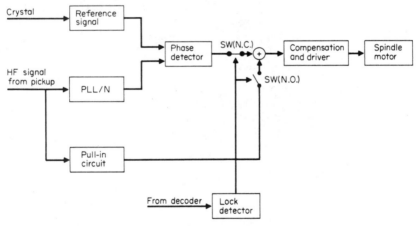

FIGURE 14.13 Spindle servo (CLV). *(Source: K. Blair Benson, ed.,* Audio Engineering Handbook, *McGraw-Hill, New York, 1988)*

14.2.5 Compact-Disk Player

This section describes the configuration of the CD player and signal processing from light modulated by pits on the disk to the output audio signal.

Functional Components. A block diagram of the CD player is shown in Fig. 14.14. The concentrated spot onto the information layer reads the signal, which

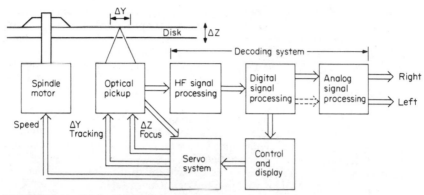

FIGURE 14.14 Configuration of the CD player. (*Source: K. Blair Benson, ed.*, Audio Engineering Handbook, *McGraw-Hill, New York, 1988*)

has been recorded on the disk in digitally encoded form. The readout signals are processed (added and/or subtracted) and separated into (1) servo status signals and (2) the audio program signal. The audio signal is processed in the decoding block into the conventional but highly precise audio signal waveforms for the right and left channels. Concurrently, the servo status signals drive the servo system, which maintains precise control of spindle speed and laser-beam tracking and focus. The control and display system, using a microprocessor, is a control center; it not only simplifies user operation but also provides a display of visual data (using subcoding channel Q information derived from the decoding block), which consists of brief notes about the musical selections as they are played.

High-Frequency Signal Processing.   After the compensation of frequency response, if necessary, we can obtain the so-called eye diagram, shown in Fig. 14.15. This is the result of processing by optical low-pass filter, expressed by MTF.

To convert to a two-level bit stream, it is necessary to take care of the pit distortion. By looking at Fig. 14.15 carefully, it can be understood that the center

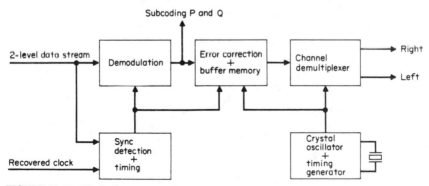

FIGURE 14.15 Block diagram of digital-signal processing. (*Source: K. Blair Benson, ed.*, Audio Engineering Handbook, *McGraw-Hill, New York, 1988*)

of the eye is not in the center of the amplitude. This is called *asymmetry*, a kind of pit distortion. It cannot be avoided when disks are produced in large quantities because of changes resulting from variations in mastering and stamping parameters as well as differences in the players used for playback. Accordingly, a form of feedback digitizer, using the fact that the dc component of the EFM signal is zero, is recommended. In addition, the clock for timing signals is regenerated with a PLL circuit locked to the channel-bit frequency (4.3218 MHz).

Digital Signal Processing.   Figure 14.16 is a block diagram of digital signal processing. The demodulation of EFM can be accomplished by using a conversion process. This provides the digital audio data and parities for error correction (CIRC). At the same time, the subcoding that directly follows the synchronization signal is demodulated and sent to the control and display block. The data and parities are then temporarily stored in a buffer memory (2K bytes) for the CIRC decoder circuit. The parity bits can be used here to correct errors or merely to detect them if they cannot be corrected. Although CIRC is one of the most powerful error-correcting codes, if more errors than a permissible maximum occur, they can only be detected and used to provide estimated data by linear interpolation between preceding and new data.

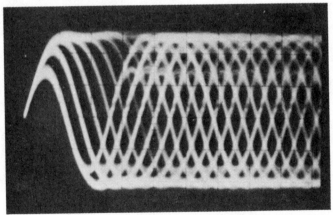

FIGURE 14.16 Eye diagram of the EFM signal. *(Source: K. Blair Benson, ed.,* Audio Engineering Handbook, *McGraw-Hill, New York, 1988)*

At the same time, the CIRC buffer memory operates as the deinterleaver of the CIRC and is used for time-base correction (TBC). If the data is written into the memory by means of the recovered clock signal with the PLL and then read out by means of the crystal clock after a certain amount of data has been stored, data can be arranged in accordance with a stable timing rate. In this way, the *wow* and *flutter* of the digital audio signal are reduced to a level equal to the stability of the crystal oscillator.

Analog Signal Processing.   The error-corrected and time-base-corrected digital data must be converted to the analog values. This is the role of the digital-to-analog converter (DAC), and the necessary conditions for the CD system are:

1. 16-bit resolution
2. Conversion speed of at least 15 μs
3. Low cost (monolithic integrated circuit)

For these requirements, several types of conversion methods have been developed:

1. R-2R ladder network
2. Dynamic element matching (DEM)
3. Integration method using a high-frequency clock

The popular R-2R ladder-type schematic diagram is shown in Fig. 14.17.

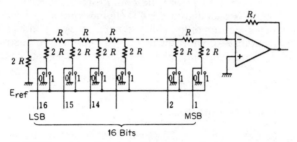

FIGURE 14.17 Type R-2R digital-to-analog converter. *(Source: K. Blair Benson, ed., Audio Engineering Handbook, McGraw-Hill, New York, 1988)*

At the last stage of analog signal processing is the low-pass filter to reduce the energy outside the band of the audible frequency range (20 Hz to 20 kHz). Instead of using only the analog filter, a combination of digital oversampling filter with a simple analog filter has recently become popular. A block diagram is shown in Fig. 14.18.

14.2.6 Compact-Disk-Based Systems

Read-Only Memory (ROM). The capacity of a CD of 540 Mbytes per side, compared to 362 kbytes for a 5¼-in magnetic disk, makes it an efficient and attractive storage medium for computer ROM data storage. In this application the data

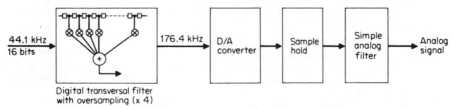

FIGURE 14.18 Digital-to-analog conversion using digital filtering. *(Source: K. Blair Benson, ed., Audio Engineering Handbook, McGraw-Hill, New York, 1988)*

track information is divided into addressable blocks of 2352 sequential bytes. The block size is based on a subcode frame equal to 98 frames of the CD format. A block consists of the following:

$$
\begin{aligned}
Sync\ field &= 12\ bytes \\
Header\ field &= 4\ bytes \\
\text{User data} &= 2048\ \text{bytes} \\
\text{Error-correcting code} &= 288\ \text{bytes}
\end{aligned}
\tag{14.3}
$$

In addition to data storage and retrieval, CD ROM is finding a rapidly expanding application in interactive video learning systems. The reader is referred to McGraw-Hill's *Guide to CD-ROM* for more information about these systems.

CD Video.   The principle is to record the composite video signal in its analog form. The audio signal is encoded in the CD digital format and recorded in the gaps in the frequency spectrum of the video signal. A playing time on the order of 10 min per side is suitable for short music videos and many industrial, advertising, and commercial applications. A block diagram of the signal system is shown in Fig. 14.19.

14.2.7 Compact-Disk Manufacture

Configuration of the CD.   Figure 14.20 shows a cross section of the CD. A laser beam is projected through a 1.2-mm-thick transparent substrate and focused on the surface of an aluminum reflective layer. The recorded digital signal, consisting of pits in the layer, is detected by modulation of the beam. The substrate protects the reflecting layer from dust or fingerprints which would introduce spurious noise disturbances in the signal. The replicated pits on the signal surface are about 0.1 μm deep, 0.5 μm wide, and several micrometers long. For protection, the reflective layer is coated with resin cured by ultraviolet light. The label is silkscreened onto the protective layer.

Pit Geometry and Optical Principles.   CD signal detection is based on the phenomenon of diffraction. The laser beam reflected from the disk surface is in opposite phase to that of the projected beam and therefore is canceled as shown at the left of Fig. 14.21. The depth of the pits relative to the wavelength of the laser beam has been chosen so that the beam reflected from the pits is in the same phase as the projected beam.

Therefore, the reflection from pits is reinforced by the resultant diffraction, as shown at the right of Fig. 14.21. In other words, a binary signal between beam-off and beam-on is generated by a pit as it passes under the projected laser beam.

The pit length and slope of the edges will vary with disk-processing operations, but are not critical.

Mastering-Process Flow Sequence.   Figure 14.22 shows the mastering-process flow sequence. During mastering 10 billion pits are made on the thin photosensitive layer. Mastering operations are carried out in a clean-air room to prevent deleterious contamination.

Preparation of Glass Master.   A highly polished and cleaned glass substrate is prepared in the mastering process. Flatness, smoothness, and lack of defects in the glass are important to produce good-quality disks. A positive photoresist about

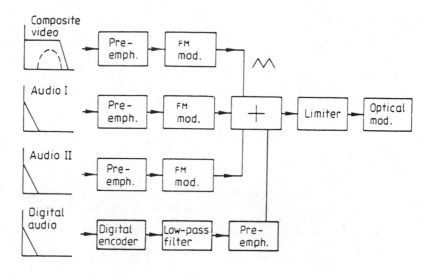

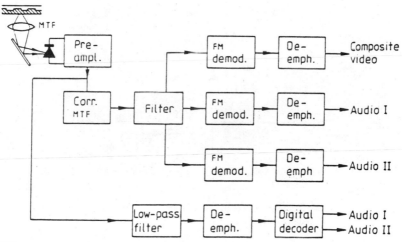

FIGURE 14.19 Block diagram of signal-processing and optical video-disk player (NTSC format). *(Source: K. Blair Benson, ed.,* Audio Engineering Handbook, *McGraw-Hill, New York, 1988)*

0.1 μm thick is coated on the glass surface by a spinning method. The positive photoresist is suitable for the production of optical disks because of its high resolution, smooth surface, and polarity of matrix. An ellipsometer is used to measure the resist thickness to determine the pit depth.

Laser Recording.   A photoresist on the glass is exposed by a laser spot that is intensely modulated by a CD signal. The recording laser should satisfy the following requirements:

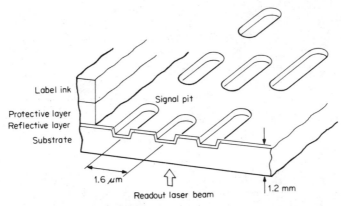

FIGURE 14.20 Cross-section of a replicated compact disk. The phase modulation of the laser beam by the reflective layer is shown in Fig. 14.18. *(Source: K. Blair Benson, ed.,* Audio Engineering Handbook, *McGraw-Hill, New York, 1988)*

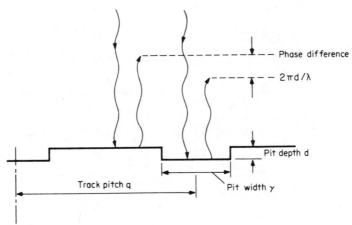

FIGURE 14.21 Phase difference of a reflected beam. *(Source: K. Blair Benson, ed.,* Audio Engineering Handbook, *McGraw-Hill, New York, 1988)*

1. Capability of producing a small spot (about 0.5 μm)
2. Sensitivity to photosensitive material
3. Continuous-wave emission from stable light source

The HeCd laser and Ar ion laser are suitable for recording.

A high-speed light modulator is also an important part of a laser recording system. An acousto-optic modulator (AOM) which uses ultrasonic waves and an electro-optic modulator (EOM) which makes use of the Pockels effect are employed to modulate laser light. Recording quality primarily depends on the performance of the *master code cutter*, the accuracy of track pitch, scanning velocity, and stability of exposure.

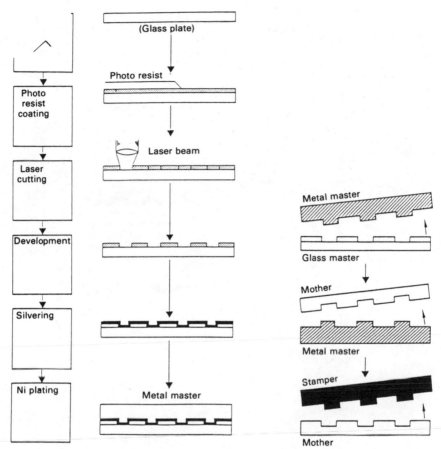

FIGURE 14.22   Mastering-process flow sequence. *(Source: K. Blair Benson, ed.,* Audio Engineering Handbook, *McGraw-Hill, New York, 1988)*

After the subsequent development process the exposed photoresist is washed away, leaving the signal pits.

Stamper Production.   Since the molding process needs a high temperature and high pressure, a glass master cannot be used. The information pits must be transferred to a nickel stamper.

As shown in Fig. 14.22, silvering metallizes on the signal surface. After silvering, nickel is electroplated to make a metal master about 300 μm thick. The process from a metal master to a stamper is similar to that of an analog long-playing disk. Several mothers and stampers can be obtained from one metal master by use of the matrix.

Replication-Process Flow Sequence.   Replication is the process whereby many CD replicas are produced from the stamper. The flow sequence is shown in Fig. 14.23.

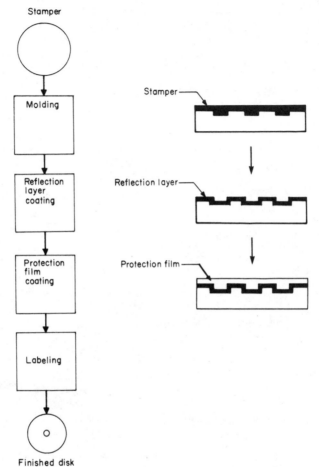

FIGURE 14.23 Replication-process flow sequence. *(Source: K. Blair Benson, ed.,* Audio Engineering Handbook, *McGraw-Hill, New York, 1988)*

Molding Method. There are about 9 Gbits of information on the signal surface of a CD stamper. The molding process is designed to make many good replicas with high productivity. Typical molding methods are outlined below.

In *injection molding* melted resin is injected at high pressure into a mold and solidified (see Fig. 14.24). Injection molding is widely used in CD production because of its high productivity.

Compression Molding. The stamper is attached to one side of the molding die, and both sides are heated by steam. Plastic cake is then inserted and molded by a hydraulic press (Fig. 14.25). LP disks are made by this method.

Photopolymerization Method. Figure 14.26 shows the photopolymerization process. A transparent plate is prepared, and ultraviolet-light-curing resin is inserted

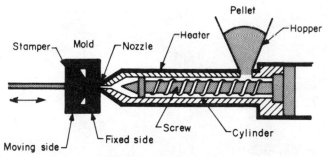

FIGURE 14.24 Injection mold in replication process. *(Source: K. Blair Benson, ed.,* Audio Engineering Handbook, *McGraw-Hill, New York, 1988)*

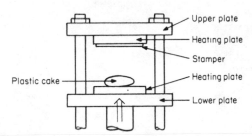

FIGURE 14.25 Compression mold in replication process. *(Source: K. Blair Benson, ed.,* Audio Engineering Handbook, *McGraw-Hill, New York, 1988)*

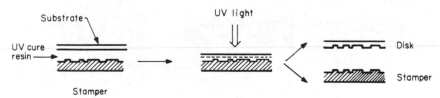

FIGURE 14.26 Photopolymerization process for replication. *(Source: K. Blair Benson, ed.,* Audio Engineering Handbook, *McGraw-Hill, New York, 1988)*

tween plate and stamper. From the transparent-plate side, an exposure to ultraviolet light is made, after which the resin is cured.

The most appropriate molding method is selected, for example, by conducting a study of the materials to be used, producibility, and facilities.

Coating. After molding, a layer of metal aluminum is evaporated onto the pit surface. Subsequently, a protective layer is spun onto the reflective layer. This protective layer consists of ultraviolet-light-curing resin about 10 μm thick.

The label is printed on the protective layer. The compact disk is now finished and ready for shipment after final inspection. Figure 14.27 shows the complete production process of the CD.

1. Mastering
Digital 2-channel recording is done on a 3/4″ U-matic tape, while subcode editing is performed to make the tape complete with all the data necessary for cutting.

2. Cutting
A glass master is made by laser-cutting data in real time on the photoresist layer of a glass plate.

3. Stamper Production
A metal master is produced from the glass master, a metal mother from the metal master, and production stampers from the metal mother.

4. Disk Injection Molding
Injection molding of polycarbonate resin results in volume production of disks from the same stamper.

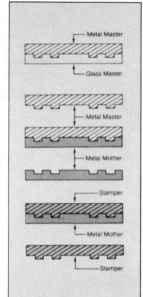

5. Reflective-Layer Coating
Aluminum is evaporated and deposited on the signal surface of each disk to enable data detection with a laser beam.

6. Protective-Layer Coating
Ultraviolet curing resin is applied over the aluminum coating and exposed to ultraviolet rays to form a transparent protective layer.

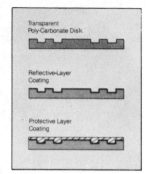

7. Inspection
The disks are quality-tested invidually, using a computerized automatic inspection system.

8. Label Printing/Packaging
Each disk is labeled and wrapped for shipment as the final product.

FIGURE 14.27 Compact-disk production process. *(Source: K. Blair Benson, ed.,* Audio Engineering Handbook, *McGraw-Hill, New York, 1988)*

Bibliography

Bouwhuis, G., and J. J. M. Braat. "Recording and Reading of Information on Optical Disks," in *Applied Optics and Optical Engineering*, vol. IX, Academic, New York, 1983, chap. 3.

————, ————, A. Pasman, G. van Rosmalen, and K. A. Schouhamer Immink. *Principles of Optical Disc Systems*, Adam Hilger Ltd., Bristol, England, 1985.

Driessen, L. M. H. E., and L. B. Vries. "Performance Calculations of the Compact Disc Error Correcting Code on a Memoryless Channel," International Conference on Video and Data Recording," University of Southampton (April 1982).

Immink, K. A., Schouhamer, A. H. Hoogendijk, and J. A. Kahlman. "Digital Audio Modulation in the PAL and NTSC Laser Vision Video Disc Coding Formats," presented at the 74th Convention of the Audio Engineering Society, New York (1983), preprint 1997.

Isailovic, J. *Videodisc and Optical Memory Systems*, Prentice-Hall, Englewood Cliffs, N.J., 1985.

Miyaoka, S. "Digital Audio Is Compact and Rugged," *IEEE Spectrum* (March 1984).

Odaka, K., T. Furuya, and A. Taki. "LSI's for Digital Signal Processing to Be Used in Compact Disc Digital Audio Players," presented at the 71st Convention of the Audio Engineering Society (March 1982), preprint 1860 (G-5).

———— and L. B. Vries. "CIRC: The Error Correcting Code for the Compact Disc Digital Audio System," presented at the Premier Audio Engineering Society Conference (June 1982).

Ogawa, H., and K. A. Schouhamer Immink. "EFM—The Modulation Method for the Compact Disc Digital Audio System," presented at the Premier Audio Engineering Society Conference (June 1982).

Philips Tech. Rev., **40**(6) (1982). This issue contains four articles about the CD system.

Sako, Y., and T. Suzuki. "CD-ROM System," Topical Meeting on Optical Data Storage, WCCI (October 1985).

Vries, L. B., et al. "The Compact Disc Digital Audio System: Modulation and Error Correction," presented at the 67th Convention of the Audio Engineering Society (October 1980), preprint 1674 (H-8).

CHAPTER 15
MAGNETIC-TAPE MANUFACTURE AND USE

E. Stanley Busby, Jr.
Ampex Corporation (retired), Redwood City, California

15.1 MAGNETIC-TAPE MATERIALS

There are four distinct components of magnetic tape:

1. A finely powdered magnetic material.
2. A binder, or glue, which surrounds the particles, holds them apart, and binds them to a plastic support.
3. A plastic support, usually polyethylene terephthalate, also called *polyester*. After coating and drying, if it is cut into strips, it is tape. If it is punched into a round torus, it is a floppy disk.

Portions of this chapter were adapted from K. Blair Benson, ed., *Audio Engineering Handbook*, McGraw-Hill, New York, 1988.

4. A conductive back coating is usually applied to tape. It is fairly rough and helps spill air from adjacent layers during fast winding, and inhibits the generation of static electricity.

The last three items above are covered in Sec. 15.2.

15.1.1   Iron Oxide

Having a coercivity of 300 to 360 Oe, gamma ferric oxide is by far the most widely used recording material. The first step in its preparation is the precipitation of seeds of geothite [alpha FeO(OH)], from scrap iron dissolved in sulfuric acid, or of lepidocrocite [gamma FeO(OH)], produced from ferrous chloride.

After further growth, the seeds are dehydrated to hematite (alpha Fe_2O_3) and then reduced to magnetite (Fe_3O_4). They are then reduced to maghemite (gamma Fe_2O_3), which not only is magnetic but also has the desired acicular (rod-shaped) form with an aspect ratio of 5-to-1 to 10-to-1. The length of the particles ranges from 0.2 to 1.0 μm.

15.1.2   Cobalt-Doped Iron Oxide

Having a coercivity of 500 to 1200 Oe, the preferred preparation causes cobalt ions to be adsorbed onto the surface of gamma ferric oxide as an epitaxial layer. The result is a magnetic particle that can sustain a higher level of magnetization and therefore can provide more output. This material and chromium dioxide are generally referred to as *high energy* or *high bias* or *Type II*, requiring a higher bias current and sustaining a slightly higher record level. There is little improvement in the noise contributed by the granularity of the particles.

15.1.3   Chromium Dioxide

Having coercivities of 450 to 650 Oe, this material offers saturation magnetization slightly higher than that of gamma ferric oxide. Chromium dioxide has high acicularity and lacks voids and dendrites. It has a low Curie temperature, making it a candidate for contact duplication of short-wavelength (digital) recordings.

Chromium dioxide is more abrasive than gamma ferric oxide. It is less stable chemically than iron oxide. At extremes of temperature and humidity it can degrade to nonmagnetic compounds of chromium. In some countries, chromium dioxide is considered a toxic substance, so disposal of used tapes can be a problem.

Tapes made with chromium dioxide or cobalt-doped iron oxide yield output levels 5 to 7 dB greater than tapes with gamma ferric oxide of the same coating thickness and require increased bias and erase currents.

15.1.4   Iron Particle

Tapes made from dispersions of finely powdered metallic iron particles are capable of a 10- to 12-dB greater signal output than gamma ferric oxide tapes. These tapes have a retentivity of 1000 to 1500 Oe. They can be difficult to bulk-erase.

The particle size is small, an advantage in digital recording at very high packing density, as in the R-DAT (*r*otary *d*igital *a*udio *t*ransport) digital audio format.

Several processes generate suitable metal particles. The reduction of iron oxide in hydrogen is one. Another is the reduction of ferrous salt solutions with borohydrides.

The small size of the particles presents two difficulties: the mixing time is increased to ensure good dispersion, and if dry particles of iron are exposed to the atmosphere, they are subject to rapid oxidation (explosion). Corrosion at elevated temperatures and humidity was once a problem, but it appears that solutions have been found. Iron-particle tape is also available in the popular Phillips audio cassette and in 0.5-in cassettes used in the D2 composite television recording format.

Recent digital video-tape recorder designs specify the exact type of magnetic material to be used for acceptable or optimum performance. This is especially important in the cases of the D1 component and D2 composite television formats. The cassette design is the same for both, but D1 requires high-bias Type II coating, whereas D2 specifies metal tape.

Operators should be cautioned that video tapes cannot be used for audio-signal recording because (1) the particle orientation is at the proper angle for a rotating head and (2) the formulation is selected for the short wavelengths of high-frequency video carriers, rather than low-frequency base-band audio signals. By the same token, audio tape cannot be used on video recorders because the binders cannot withstand the higher pressure per unit of surface area created by the impact of the video heads. More detail on the selection of tapes for different applications is given in for video in Sec. 7.2.21, and for audio in Sec. 7.2.22.

15.1.5 The Substrate

Virtually all flexible magnetic recording media are made by coating a plastic substrate, or base film, with a slurry of magnetic particles suspended in a volatile solvent. The plastic of choice is polyethylene terephthalate, also called polyester or PET.

PET has good tensile strength, excellent tear resistance, and good chemical stability, and its cost is reasonable. Its strength is improved by stretching in two dimensions after extrusion from the melt. In the case of very thin base films, of less than 0.5 mil, additional strength is needed in the direction of motion. This is provided by additional stretching in that direction. Films receiving the extra stretch are said to be *tensilized*.

Organic particles of very small size are added to PET to give the surface optimum roughness. If the surface is too smooth, the film is difficult to handle and there are adhesion problems. If the surface is too rough, the smoothness of the finished coating can be compromised. In audio direct recording, excessive surface roughness will decrease short-wavelength response and contribute to modulation noise (noise produced only in the presence of a recorded signal). If the recording surface is too smooth, the tape can "wring," or suffer molecular adhesion to smooth nonrotating tape-path elements. These include guides and heads.

Spools of base film range to 15,000 ft long, from 0.2 to 1.5 mils thick, and 12 to 60 in wide before slitting. Thin films and thin coatings are typical for video and digital recordings. Analog audio recordings tend to have thicker coatings and a moderately thick base. This is because analog audio recordings involve a wide range of wavelengths.

15.2 MAGNETIC-TAPE MANUFACTURE

In this section keep in mind that most of the processes employed in the manufacture of tape are trade secrets closely guarded by the several competing manufacturers and unlikely to be published. Patent protection is not perfect. Sometimes simple secrecy does a better job, especially when the product cannot be dissected so as to reveal the process which produced it.

15.2.1 The Production Process

Mixing. The first step in manufacture is the dispersion of the dry magnetic powder in a liquid mixture consisting of the following components:

1. Binders, to enhance adhesion of the coating to the base film.
2. Stabilizers, to prevent aging of the binders.
3. Dyes and conductive agents, to color the coating and inhibit the generation of static electricity during winding.
4. Plasticizers, to make the coating as flexible as the base film. The complete evaporation of plasticizers marks the end of the lifetime of a tape.
5. Lubricants, to reduce the tendency for stick-slip oscillation, also known as the *violin string effect*. Lubricants also reduce head wear.
6. Dispersants, which have high surface tension, to aid the dispersion of the magnetic particles and reduce their tendency to clump.
7. Abrasives, usually aluminum oxide, in very small quantities are added to the mix to automatically clean the heads, which tend to collect a thin layer of binder rubbed off the tape.

Dispersion takes place in a mill containing either steel balls or pebbles. Since the magnetic particles are equivalent to single-domain magnets, they tend to agglomerate. Milling defeats this tendency, especially if the viscosity of the mix is properly chosen. A uniform dispersion leads to minimum modulation noise.

Dry oxide is abrasive. One purpose of milling is to ensure that each magnetic particle is individually coated with the relatively soft binder and separated from other particles. Abrasiveness is also influenced by the surface treatment applied later in the process.

Coating. All known coating processes are continuous; i.e., if the process is stopped, the results are ruined. Figure 15.1 illustrates three coating methods; below five are discussed.

1. *Reverse-roll coating:* The total wet thickness is the distance between the surfaces of the rollers.
2. *Gravure coating:* The depth of the indentations in the roller controls the coat-

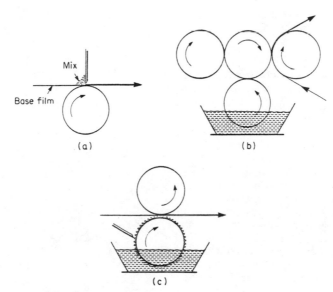

FIGURE 15.1 Coating methods. (*a*) Knife coating. (*b*) Reverse-roll coating. (*c*) Gravure coating. (*Source: K. Blair Benson, ed.,* Television Engineering Handbook, *McGraw-Hill, New York, 1986*)

ing thickness, and the viscosity of the mix controls the evenness of the coating.

3. *Knife-blade coating:* The coating material is applied by gravity, and a blade is used to scrape off the excess. Coating thickness is determined by the spacing between the blade and the base-film surface. This method is falling into disuse.

4. *Extrusion coating:* A carefully controlled amount of coating mixture is extruded onto a highly polished large-diameter roller. Later the resulting film is transferred to the base film by the roller.

5. *Spray coating:* Pneumatic or electrostatic forces are used to propel the mix onto the base film.

Particle Orientation. Before the coating hardens, while it is still viscous, it is passed through a strong magnetic field oriented in the direction of head-to-tape motion. Magnetic particles which are not oriented in the direction of tape-to-head motion tend to rotate so as to become oriented thus. The extent to which they succeed determines the squareness ratio, or the ratio of saturation flux to remanent flux. If all magnetic particles were perfectly formed rods of one magnetic domain each and all became physically aligned while passing through the aligning field, the squareness ratio would be 1.

Surfacing. The surface of the coating that contacts the recording and reproducing heads must be smoothed to an optimum roughness. The best roughness is just rougher than that at which molecular adhesion occurs at the heads, fixed guideposts, or other smooth surfaces which the tape contacts.

The methods which have been used include the following:

1. Brushing with a horsehair or nylon brush to dislodge protruding oxide particles from the surface.

2. Scraping with a sharp edge, either lifting out protruding crystals or fracturing them at the surface.

3. Abrasion by rubbing the coating over itself. In this case, protruding particles are used to break off other protruding particles.

4. Ironing particles, or pushing them into the surface of the binder by using a highly polished roller, results in a smooth surface with few broken oxide particles at the surface.

Back Coating. The noncontacting side of the tape is usually coated with a thin layer of conductive material which is relatively rough. This layer further suppresses the generation of static electricity, and its roughness helps the air which is trapped between layers of tape during rewind to escape before one turn of the reel is finished. If contact with the layer below is not accomplished within the time of 1 r, air, acting as a lubricant, will allow the tape to slide in respect to previous layers, especially in transports with reels in the horizontal plane. The tape edges should never touch the flanges of the reel. See Fig. 15.2.

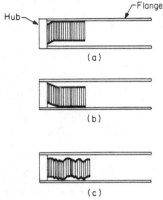

FIGURE 15.2 Poor tape packs. (*a* and *b*) Tape pack resulting from mechanical misalignment. (*c*) Tape pack resulting from excessive winding speed. (*Source: K. Blair Benson, ed., Audio Engineering Handbook, McGraw-Hill, New York, 1988*)

15.3 ENVIRONMENTAL AND OPERATIONAL PRACTICES[1]

15.3.1 The Operating Environment

A dust-free, smoke-free, humidity- and temperature-controlled environment is always preferable when you are working with tape. The ideal environment for tape is 35 to 45 percent relative humidity (RH) and 65 to 70°F. Figure 15.3 is a generalized illustration of the cost trade-offs.

Tape is made up of components which are hygroscopic and absorb or lose moisture, depending on the humidity. Tests indicate that tape handling on a recorder is dependent on *absolute* humidity rather than on *relative* humidity. Absolute humidity is the amount of moisture contained in a given amount of air. A psychrometric chart can be used to determine absolute humidity. Figure 15.4 is a plot of constant-moisture curves, with the "normal" room environment of 68°F

---

1. The following four sections incorporate whole paragraphs from "Long Term Storage of Videotape" and "Increasing the Life of Your Audio Tape," published in Refs. 1 and 2, respectively, by Jim Wheeler and iterated here with the permission of Jim Wheeler and the Ampex Corporation. A few editorial liberties were taken to allow for the stress of this chapter on *audio* tape.

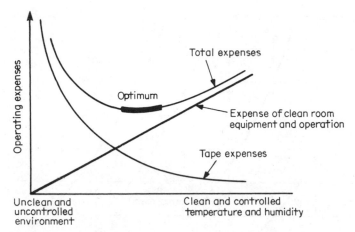

FIGURE 15.3 Generalized clean-environment expenses. *(Source:* SMPTE Journal, *June 1983, p. 651)*

and 40 percent RH assigned a value of 1 (0.006 lb of water for 1 lb of dry air). Doubling the moisture content will produce curve 2, tripling it will produce curve 3, etc. Experiments show that a good tape and recorder will operate properly up to about curve 5, provided that 100 percent RH *never* occurs. Any condensation of moisture on the tape surface will cause the tape to cling to the tape-guiding surfaces. Tape movement will then probably stop, because the water destroys the air film between the tape and the guides.

Tape should never be kept in a hot, wet environment very long because of the deteriorating effect of such an environment. Investigations into tape with polyester-urethane binder systems show that hydrolysis can occur in a hot, wet environment, causing the binder to deteriorate. The rate of deterioration depends on the type of binder system used. If the binder is allowed to deteriorate to the point at which all the ester molecules are consumed, the tape will be permanently degraded [3]. Before the point where total degradation occurs, the tape should be able to be restored to a condition suitable for copying.

The Bertram-Eshel report [4] has temperature-humidity curves that illustrate the regions in which the binder is consumed, is stable, and is reconstituted. Although the hydrolysis curves in the Bertram-Eshel report are not the same as the constant-moisture curves of Fig. 15.4, they are close enough to use only the one set of curves.

Moisture curve 2 of Fig. 15.4 approximates the upper tape storage limit of Fig. 40 in the Bertram-Eshel report [4] and provides a reasonable definition for the low end of a hot, wet environment. This applies to unrecorded tape as well as archival tape. If a tape has been subjected to a hot, wet environment, it should be placed in a cool, dry (low-moisture) environment for several days before use.

No matter the environment in which a tape is recorded, the tape should *always* be rewound in the environment in which it is to be stored.

15.3.2 Tape Handling

Winding Tape. Tape wound at low tension will be loosely packed. In storage, stresses in the pack will cause the layers to creep, and eventually a wavy pack

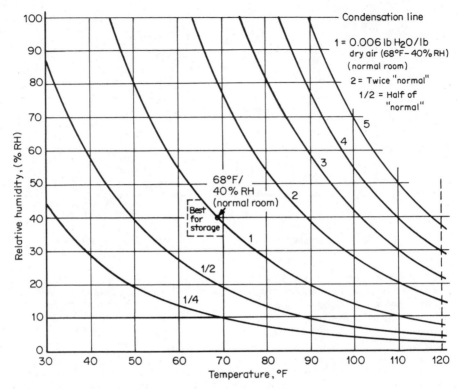

FIGURE 15.4 Constant-absolute-humidity curves (normalized). *(Source:* SMPTE Journal, *June 1983, p. 651)*

distortion will result, particularly if the tape is subjected to environmental changes. If a loosely packed reel is subsequently wound on a machine with fast acceleration and/or fast stop, the tape in the pack will slip, and permanent tape damage will probably result. A loose pack can be determined by holding the reel with one hand and pulling the tape end with the other hand. If the pack is loose, it will rotate. Tape wound at an excessively high tension, however, will create large internal pack pressures that can damage the oxide surface. It is difficult to define an ideal wind tension because of the many variables: outer pack radius, winding speed, flange windows, back coat friction, physical condition of the tape, and tension "profile" of the winding machine. Below we describe each variable briefly.

Pack Radius.   A small ratio of outer radius to inner radius is desired. Since the inner radius is 2.25 in on professional recorders, the outer radius is the variable. The moment of inertia of a reel of tape varies as the fourth power of its radius. This means that the larger the radius, the more likely it is that pack slippage will occur during acceleration or deceleration. In short, reels with a small diameter are preferred over reels with a large diameter.

Flange Windows. Windows allow trapped air to bleed out. Less air means better layer-to-layer stacking and therefore less pack creep.

Back Coat Friction. Higher back coat friction decreases the potential of pack creep because of better layer-to-layer stacking. Unfortunately, a very high back coat friction can create machine friction problems when the tape is being played.

15.3.3 Physical Condition of the Tape

If one or both edges of the tape have been damaged, the layers will not pack closely together when the tape is wound. Scratches on either the oxide or back coat or other surface irregularities will also tend to decrease the layer-to-layer stacking. The tape should be wound in the same environment in which it will be stored. However, it may be wound at a temperature a little higher than the storage temperature and/or at a humidity a little lower than the storage humidity.

Winding-Tension Profile. Constant-torque winding, with an increased tension near the outer radius of the reel, induces a preferred wound-in tension pattern. It is believed, however, that the tension pattern is not a critical parameter and that constant-tension winding is reasonable—at least for reels with up to 10.5-in diameter [2]. Figure 15.5 shows the nominal and upper and lower limits of tape tension for constant-tension winding.

The tape wind speed and the rate of reel deceleration at the end of the tape are very important. Typical wind speeds are between 300 and 500 in/s, and speeds this high will create trapped air. Deceleration time should be longer than 0.5 s for a 1-h reel of tape. The most desirable situation would be to wind archival tape on a machine with a constant-torque wind characteristic at about 100 in/s wind speed (or less) and with a reel deceleration time greater than 1 s.

15.4 TEMPORARY AND ARCHIVAL STORAGE

15.4.1 Tape Storage Room

The room used for tape storage should have the following characteristics:

- *Temperature and humidity control:* Humidity control is not a necessity if the tapes are sealed in plastic bags.
- *Physical protection: No window* that will allow the sun to shine directly upon the tape.
- *Low dust level*
- *Fire-resistant construction*
- *Located near the recorders*

15.4.2 Storage Environment

The "best for storage" region in Fig. 15.4 was determined by considering hydrolysis, tape-pack mechanics, and practical limitations. The *ideal* storage re-

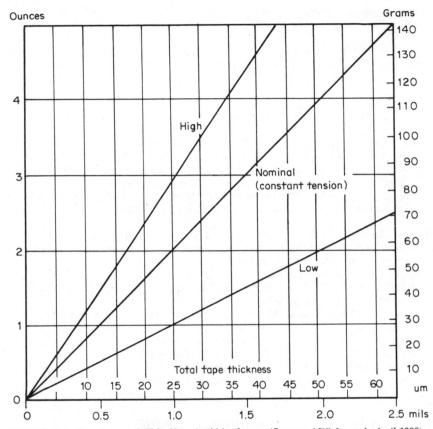

FIGURE 15.5   Tension per 0.25-in (6-mm) width of tape. *(Source:* AES Journal, *April 1988)*

gion would be at a lower temperature, with smaller variations in temperature and humidity. *If* the tape has been stored at a low temperature, the tape should sit for a few hours in the working-room environment before it is used. If a tape has been subjected to a hot, wet environment, it should be stored in a cool, dry place for a few days. It should then be wound under the same conditions in which it is to be stored. If the pack is loose, the winding should be done at a slow tape speed.

15.4.3   Guidelines for Storing a Reel of Tape

1. Only back-coated tape should be used.
2. The tape must be wound end to end before being put into storage. Stopping in the middle of the reel will create pack-tension distortions.
3. The tape should be wound onto a rigid hub. Tape should *never* be stored on a hub with a rubber sleeve because the soft material will compress and cause compaction of the tape layers.
4. The end of the tape degrades with usage, so prior to storage any damage or

greasy section should be cut off. This will provide better adhesion for the hold-down tab. Any residue from the adhesive tab should be removed prior to playback.

5. The end of the tape should be secured with an approved hold-down tab. Tabs that leave a sticky residue should never be used. The 3M hold-down tab tape 8125 is good for this purpose.

6. The tape should be stored in a quick-seal plastic bag and/or a sealed container. This will minimize the effect of fluctuations in the room environment and contact with air. Tape should *never* be sealed in a container or plastic bag while it is in a high-humidity environment because of the potential of future moisture condensation inside the container.

Print-Through. Most consumers are not even conscious of print-through, but the audiophile and the professional are well aware of it. Six major variables will affect print-through:

1. *Temperature:* High temperature is bad.
2. *Tape thickness:* Thin tape is bad.
3. *Time:* Print-through becomes worse with time.
4. *Rewind:* Rewinding the tape will reduce print-through.
5. *Magnetic field:* Presence of a field is bad.
6. *Wavelength:* Long-wavelength recordings (low frequencies) are worse.

To minimize print-through:

1. Use thick tape.
2. Store at a low temperature, but not below freezing.
3. Ship tape in an insulated container to minimize temperature variations. Ship master recordings in a steel container to avoid magnetic fields.
4. Rewind every year or two.
5. Store with tails out. This has two advantages: The tape must be rewound before playing, which will decrease print-through, and an analysis of print-through shows that the preprint signal is greater than the postprint signal (echo). So during replay after tails-out storage, the echo becomes the louder signal (which seems natural). This gives the impression of lower print-through. (This is only effective on tapes recorded in one direction.)
6. Do not copy a tape with a print-through problem. It will only make the problem permanent.

Maintenance of Stored Tape. Figure 15.6 shows the significance of storage temperature in determining the permissible rewind interval. Fluctuations in temperature will shorten the time between required rewinds. A radical change in the environment (such as an air-conditioner failure) is difficult to account for analytically [4].

Each reel should have a colored or numbered label that indicates when it was last rewound. For instance, a different color could be used for each rewind period, and the tapes can be grouped in periods of 6 months. This way, the rewind task can be done every 6 months. Three methods for establishing rewind time are listed here:

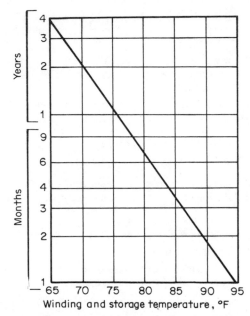

FIGURE 15.6 Estimated time between rewinds by using only creep-rate criteria. (*Note*: This curve is very general and should not be extrapolated below about 65°F.) (*Source:* SMPTE Journal, *June 1983, p. 653*)

1. The curve in Fig. 15.6 can be used to approximate the rewind interval. If the tape storage environment is reasonably constant, the curve can be used without modification. If there is any short-term environmental variation, the curve must be modified. For example, assume the tape was wound at 65°F. The curve shows that it should be rewound in 4 years. If the storage temperature increases to 85°F for one day, 250 days should be subtracted [4].

2. Stainless-steel foil can be placed in the tape pack to determine when the pack pressure is so low that pack creep can occur. Stainless-steel foil that is only 0.001 in thick, which can be obtained from a metals supply firm, should be used, or feeler-gage stock can be used. It should not have sharp edges, and the surface finish of each piece must be the same. Each strip should be about 0.5 in wide and equal to the tape width plus 1 in.

To insert the foil, wind the tape off its hub. Manually wind on about five turns of tape. Place the foil through the flange window so that it protrudes out both sides. Manually wind on a couple of turns of tape, and check that the foil is properly positioned. Wind the tape completely onto the reel. Attach a force gage (with a clamp) and determine how much force is required to move the foil. Write this number in a conspicuous place on the tape-storage container.

About twice a year, the tab should be checked to determine how much force is required to move it. When the force has decreased to about half of the original force, the reel should be rewound. The pack pressure can be allowed to decrease to about one-fourth the original force if the tape is *never* allowed on a fast-acceleration transport before it has been rewound on a lower-acceleration machine.

A single control reel can be used for a group of tapes stored within a month or two of each other. There must be a control reel for each category of tape: grouped by pack diameter and tape type. Or, rather than placing tabs in packs of different diameters, the tab can be placed in only the larger-diameter packs. Tabs in each tape type are necessary because of the difference in layer-to-layer friction between tape brands.

Another variable is the machine on which the tapes were wound. If all the tapes are wound on the same type of machine, then this is not a variable. If different types of machines are used, tabs are required for each type, because each one may have a different wind-tension profile.

3. Rewind tapes every 3 years. This should be done on a tape winder or on a transport adjusted for slow acceleration, deceleration, and wind speed. Each tape should be visually monitored during winding to determine if there is any damage.

If all three of the suggestions listed in this section are followed, and if the tape is normalized on a slow acceleration-deceleration machine before being used, the time between rewinds can be increased. A rewind interval of 10 years should be satisfactory.

15.4.4 Repair of Stored Tape

Exposure to High Temperature. Reel tapes can take up to 160°F (70°C) with no apparent problems, but cassettes can have a warpage problem above about 130°F (55°C). The inside of a delivery van can reach 160°F (70°C) on a very hot day. If a tape has been subjected to a high temperature, cool it (under 20°C if possible) for a couple of days. Then rewind it to relieve the stresses in the pack.

Water. If tape has been subjected to moisture, dry it before fungus grows. Place the tape in an oven at 120°F (50°C) for about a day. After it has cooled down to room temperature, shuttle the tape to the end and back on a tape recorder with few tape-contact parts. If the tape has been submerged in water, then repeat the heat-shuttle cycle a second time. To counteract the effects of the water, store the tape in a cold environment (not freezing) for a few days. After returning the tape to room temperature, shuttle it to the end and back to relieve stresses in the pack.

Friction. In the early days of tape making, some tape manufacturers used lubricants which eventually migrated out of the coating. In such a case, the actual recording is still okay; the problem is the high friction. (In extreme cases, the stick-slip phenomenon causes the tape to squeal loudly as it progresses through the tape path.) Such a high-friction tape can be "rejuvenated" for at least a few passes by applying Krytox to the oxide surface. This can be done by running the tape over a pad soaked in a solution of less than 1 percent Krytox and about 99 percent Freon TF.

REFERENCES

1. J. Wheeler. "Long-Term Storage of Videotape." *SMPTE J.*, June 1983.
2. J. Wheeler. "Increasing the Life of Your Audio Tape." *J. Audio Eng. Soc.*, **36** (no. 4) (April 1988).
3. E. Cuddihy. "Aging of Magnetic Tape." *IEEE Trans.*, **16** (no. 4) (July 1980).
4. N. Bertram and A. Eshel. "Recording Media Archival Attributes (Magnetic)." Ampex Corp. Report RR 80-01, November 30, 1979, USAF Contract No. F30602-78-C-0181, Pr N. 1-8-4008.

BIBLIOGRAPHY

Bertram, N., and E. Cuddihy. "Kinetics of the Humid Aging of Magnetic Recording Tape." *IEEE Trans. on Magnetics*, **18** (no. 5) (September 1982).

Bertram, N., M. Stafford, and D. Mills. "The Print Through Phenomenon and Its Practical Consequences." Preprint no. 1124E-1, 54th Audio Eng. Soc. Meeting, Los Angeles, May 1976.

Cuddihy, E. "Hygroscopic Properties of Magnetic Recording Tape." *IEEE Trans.*, **12** (March 1976), pp. 126–135.

Wheeler, J. "Videotape Storage, How to Make Your Videotapes Last for Decades—or Centuries." *American Cinematographer*, January 1982.

CHAPTER 16
TEST MEASUREMENT

Jerry Whitaker
Editorial Director, Broadcast Engineering *Magazine*
Carl A. Bentz
Intertec Publishing Corporation, Overland Park, Kansas

16.1 INTRODUCTION

Most measurements in the audio-video field involve characterizing fundamental parameters. These include signal level, phase, and frequency. Most other tests

---

Parts of Sections 16.1 through 16.8 are adapted from "Audio Tests and Measurements," by Dr. Richard Cabot, P.E., *Audio Engineering Handbook*, K. Blair Benson (ed.), McGraw-Hill, New York, 1988, Chapter 16. Parts of Section 16.8 are adapted from "Sync Generation and Distribution," by James Michener, *Television Engineering Handbook*, K. Blair Benson (ed.), McGraw-Hill, New York, 1986.

consist of measuring these fundamental parameters and displaying the results in combination by using some convenient format. For example, signal-to-noise ratio (SNR) consists of a pair of level measurements made under different conditions expressed as a logarithmic, or decibel (dB), ratio. When characterizing a device, it is common to view it as a box with input terminals and output terminals. In normal use a signal is applied to the input, and the signal, modified in some way, appears at the output. In the case of an equalizer, for example, the modification to the signal is an intentional change in the gain with frequency (frequency response). Often it is important to know or verify the details of this gain change. This is accomplished by measurement. Real-world behavior being what it is, audio and video devices will also modify other parameters of the input signal that should have been left alone. To quantify these unintentional changes to the signal, we again turn to measurements. Using the earlier example of an equalizer, changes to the amplitude versus frequency response of the signal inevitably bring changes in phase versus frequency. Some measurements are *one-port* tests, such as impedance or noise level, and are not concerned with input-output signals, only with one or the other.

16.2 AUDIO-MEASUREMENT OBJECTIVES

Measurements are made on audio equipment to check performance under specified conditions and to assess suitability for use in a particular application. The measurements may be used to verify specified system performance or as a way of comparing several pieces of equipment for use in a system. Measurements may also be used to identify components in need of adjustment or repair. Whatever the application, audio measurements are a key part of audio engineering.

Many parameters are important in audio devices and merit attention in the measurement process. Some common audio measurements are frequency response, gain or loss, harmonic distortion, intermodulation distortion, noise level, phase response, and transient response.

Measurement of level is fundamental to most audio specifications. Level can be measured either in absolute terms or in relative terms. Power output is an example of an absolute level measurement; it does not require any reference. SNR and gain or loss are examples of relative, or ratio, measurements; the result is expressed as a ratio of two measurements. Although it may not appear so at first, frequency response is also a relative measurement. It expresses the gain of the device under test as a function of frequency, with the midband gain as a reference.

Distortion measurements are a way of quantifying the amount of unwanted components added to a signal by a piece of equipment. The most common technique is total harmonic distortion (THD), but others are often used. Distortion measurements express the amount of unwanted signal components relative to the desired signal, usually as a percentage or decibel value. This is another example of multiple level measurements that are combined to give a new measurement figure.

16.3 AUDIO-LEVEL MEASUREMENTS

The simplest definition of a level measurement is the alternating-current (ac) amplitude at a particular place in the audio system. However, in contrast to direct current (dc) measurements, there are many ways of specifying ac voltage. The most common methods are average response, root-mean-square (rms) response, and peak response. Strictly speaking, the term *level* refers to a logarithmic, or decibel, measurement. However, common parlance employs the term for an ac amplitude measurement, and that convention will be followed here.

16.3.1 Root Mean Square

The rms technique measures the effective power of the ac signal. It specifies the value of the dc equivalent that would dissipate the same power if either were applied to a load resistor. This process is illustrated in Fig. 16.1 for voltage measurements. The input signal is squared, and the average value is found. This is equivalent to finding the average power. The square root of this value is taken to get the signal from a power value back to a voltage. For the case of a sine wave the rms value is 0.707 of its maximum value.

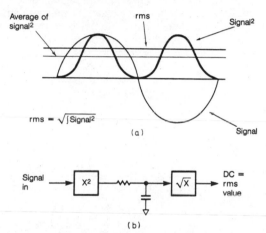

FIGURE 16.1 Root-mean-square (rms) voltage measurements. (*a*) The relationship of rms and average values. (*b*) The rms measurement circuit. *[Source: K. Blair Benson (ed.), Audio Engineering Handbook, McGraw-Hill, New York, 1988. Reprinted with permission.]*

Suppose that the signal is no longer a sine wave but rather a sine wave and several of its harmonics. If the rms amplitude of each harmonic is measured individually and added, the resulting value will be the same as an rms measurement on the signals together. Because rms voltages cannot be added directly, it is necessary to perform an rms addition. This is illustrated in Eq. 16.1. Each voltage is squared, and the squared values are added.

$$V_{\text{rms total}} = \sqrt{V_{\text{rms 1}}^2 + V_{\text{rms 2}}^2 + \cdots + V_{\text{rms } n}^2} \qquad (16.1)$$

The square root of the resulting value is taken, yielding the rms voltage of their combination. Note that the result is not dependent on the phase relationship of the signal and its harmonics. The rms value is determined completely by the amplitude of the components. This mathematical predictability is very powerful in practical applications of level measurement, enabling measurements made at different places in a system to be correlated. It is also important in correlating measurements with theoretical calculations.

16.3.2 Average-Response Measurements

Average-responding voltmeters were the most common in audio work until a few years ago, mainly because of their low cost. They measure ac voltage by rectifying it and filtering the resulting waveform to its average value as shown in Fig. 16.2. This results in a dc voltage that can be read on a standard dc voltmeter. As shown in the figure, the average value of a sine wave is 0.637 of its maximum amplitude. Average-responding meters are usually calibrated to read the same as an rms meter for the case of a single-sine-wave signal. This results in the measurement being scaled by a constant K of 0.707/0.637, or 1.11. Meters of this type are called *average-responding, rms calibrated*. For signals other than sine waves, the response will be different and hard to predict. If multiple sine waves are applied, the reading will depend on the phase shift between the components and will no longer match the rms measurement. A comparison of rms and average-response measurements is made in Fig. 16.3 for various waveforms. If the average readings are adjusted as described previously to make the average and rms values equal for a sine wave, all the numbers in the *average* column should be increased by 11.1 percent, while the *rms-average* numbers should be reduced by 11.1 percent.

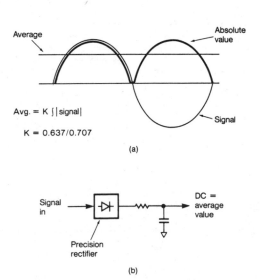

FIGURE 16.2 Average voltage measurements. (*a*) Illustration of average detection. (*b*) Average measurement circuit. *[Source: K. Blair Benson (ed.), Audio Engineering Handbook, McGraw-Hill, New York, 1988. Reprinted with permission.]*

Waveform		rms	Avg.	rms avg.	Crest factor
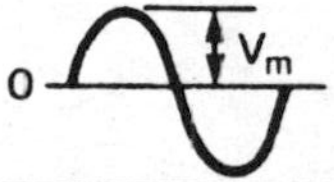 Sine wave		$\dfrac{V_m}{\sqrt{2}}$ 0.707 V_m	$\dfrac{2V_m}{\pi}$ 0.637V_m	$\dfrac{\pi}{2\sqrt{2}}$ = 1.111	$\sqrt{2}$ = 1.414
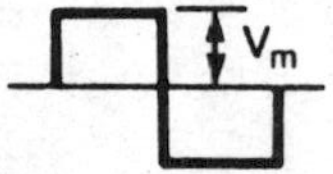 Symmetrical square wave or DC		V_m	V_m	1	1
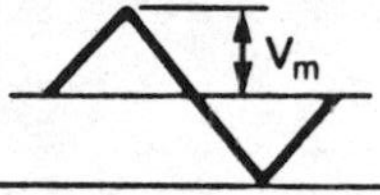 Triangular wave or sawtooth		$\dfrac{V_m}{\sqrt{3}}$	$\dfrac{V_m}{2}$	$\dfrac{2}{\sqrt{3}}$ = 1.155	$\sqrt{3}$ = 1.732

FIGURE 16.3 Comparison of rms and average voltage characteristics. (*Source: Courtesy of EDN, January 20, 1982.*)

16.5

16.3.3 Peak-Response Measurements

Peak-responding meters measure the maximum value that the ac signal reaches as a function of time. This is illustrated in Fig. 16.4. The signal is full-wave-rectified to find its absolute value and then passed through a diode to a storage capacitor. When the absolute value of the voltage rises above the value stored on the capacitor, the diode will conduct and increase the stored voltage. When the voltage decreases, the capacitor will maintain the old value. Some means for discharging the capacitor is required to allow measuring a new peak value. In a true peak detector, this is accomplished by a switch. Practical peak detectors usually include a large resistor to discharge the capacitor gradually after the user has had a chance to read the meter.

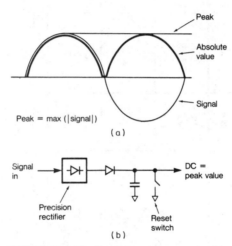

FIGURE 16.4 Peak voltage measurements. (a) Illustration of peak detection. (b) Peak measurement circuit. [*Source: K. Blair Benson (ed.), Audio Engineering Handbook, McGraw-Hill, New York, 1988. Reprinted with permission.*]

The ratio of the true peak to the rms value is called the *crest factor*. For any signal but an ideal square wave, the crest factor will be greater than 1 (see Fig. 16.5). As the measured signal becomes more peaked, the crest factor will increase.

By introducing a controlled charge and discharge time, a *quasi-peak* detector is achieved. The charge and discharge times may be picked, for example, to simulate the ear's sensitivity to impulsive peaks. International standards define these response times and set requirements for reading accuracy on pulses and sine-wave bursts of various durations. The gain of a quasi-peak detector is normally calibrated so that it reads the same as an rms detector for sine waves.

Another method of specifying signal amplitude is called *peak-equivalent sine*. It is the rms level of a sine wave having the same peak-to-peak amplitude as the signal under consideration. This is the peak value of the waveform scaled by the correction factor 1.414, corresponding to the peak-to-rms ratio of a sine wave. This is useful when specifying test levels of waveforms in distortion measurements. If the distortion of a device is measured as a function of amplitude, a point

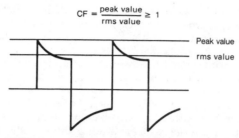

$$CF = \frac{\text{peak value}}{\text{rms value}} \geq 1$$

Peak value

rms value

FIGURE 16.5 Illustration of the crest factor in voltage measurements. *[Source: K. Blair Benson (ed.),* Audio Engineering Handbook, *McGraw-Hill, New York, 1988. Reprinted with permission.]*

will be reached where the output level cannot increase any further. At this point the peaks of the waveform will be clipped, and the distortion will rise rapidly with further increases in level. If another signal is used for distortion testing on the same device, it is desirable that the levels at which clipping is reached correspond. Signal generators are normally calibrated in this way to allow changing between waveforms without clipping or readjusting levels.

16.3.4 Types of Meters

Most meters of a few years ago were of the average-responding, rms-calibrated type. Newer meters measure the *true rms* value of the waveform by using special integrated circuits. This allows accurate measurement of voltage for all signals, not just sine waves. As mentioned earlier, rms measurements accurately reflect the heating power of the waveform in a resistor or a loudspeaker. This is critical to many measurements such as specification of power in amplifiers. However, many noise specifications were developed in the days of average-responding meters, and verifying these requires an average-responding unit. Therefore, good-quality audio test equipment allows selection between these two responses, giving compatibility with old and new techniques.

One obvious difference between various audio voltmeters is the type of display, analog or digital. Each type has its advantages. Analog meters are not easy to use for exact measurements because of their multiple scales and the need to interpolate numbers from the printed scale. In contrast, a digital meter gives a direct readout of the value to more digits of precision than could ever be obtained from an analog meter scale. This enables very precise measurement of gain and output power with little chance for error.

Digital meters, however, are suited only for measuring relatively stable signals. When trying to monitor audio program material to determine system operating levels under actual use, it becomes very difficult to extract a single number from the mass of flashing digits. The analog meter can handle this job with ease. Another application for analog meters is monitoring the results of an adjustment for a peak or a null. Some manufacturers have put both analog and digital displays on the same instrument to provide the best of both worlds.

Meter Bandwidth.   The bandwidth of a voltmeter can have a significant effect on the accuracy of the reading. For a meter with what is called a *single-pole rolloff* (i.e., one bandwidth-limiting component in the signal path), significant errors can occur in measurements. For such a meter with a specified bandwidth of 100 kHz, there will be a 10 percent error in the measurement of signal at 50 kHz. To obtain 1 percent accurate measurements (with other error sources in the meter ignored), the signal frequency must be less than 10 kHz.

Another problem with limited bandwidth measuring devices is illustrated in Fig. 16.6, which shows a distorted sine wave being measured by two meters with different bandwidths. The meter with the narrower bandwidth does not respond to all the harmonics and gives a lower reading. The severity of this effect varies with the frequency being measured and the bandwidth of the meter; it can be especially severe when measuring wideband noise. Most audio requirements are adequately served by a meter with a 500-kHz bandwidth. This allows reasonably accurate measurement of signals to about 100 kHz. Peak measurements are even more sensitive to bandwidth effects. Systems with restricted low-frequency bandwidth will produce *tilt* in a square wave, and bumps in the high-frequency response will produce an *overshoot*. The effect of either will be an increase in the peak reading.

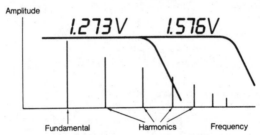

FIGURE 16.6   The effects of instrument bandwidth on voltage measurements. *[Source: K. Blair Benson (ed.), Audio Engineering Handbook, McGraw-Hill, New York, 1988. Reprinted with permission.]*

Meter Accuracy.   Accuracy is a measure of how well a meter quantifies a signal at a midband frequency, usually 1 kHz. This sets a basic limit on the performance of the meter in establishing the absolute amplitude of a signal. It is also important to look at the flatness specification to see how well this performance is maintained with changes in frequency. Flatness describes how well the measurements at any other frequency tracks those at 1 kHz. If a meter has an accuracy of 2 percent at 1 kHz and a flatness of 1 dB (10 percent) from 20 Hz to 20 kHz, the inaccuracy can be as wide as 12 percent at 20 kHz.

Meters often have a specification on accuracy that changes with voltage range, being most accurate only in the range in which the instrument was calibrated. A meter with 1 percent accuracy on the 2-V range and 1 percent accuracy per step would be 3 percent accurate on the 200-V scale. By using the flatness specification given earlier, the overall accuracy for a 100-V, 20-kHz sine wave is 14 percent. In many meters an additional accuracy derating is given for readings as a percentage of full scale, making readings at less than full scale less accurate.

However, the accuracy specification is not normally as important as the flatness. When performing frequency response or gain measurements, the results are relative and are not affected by the absolute voltage used. When measuring gain, however, the attenuator accuracy of the instrument is a direct error source. Similar comments apply to the accuracy and flatness specifications for signal generators. Most are specified in the same manner as voltmeters, with the inaccuracies adding in much the same way.

16.3.5 Decibel Measurements

Measurements in audio work are often expressed in decibels (dB). Audio signals span a wide range of level. The sound pressure of a rock-and-roll band is about 1 million times that of rustling leaves. This range is too wide to be accommodated on a linear scale. The decibel is a logarithmic unit that compresses this wide range down to a more easily handled range. Order-of-magnitude (factor-of-10) changes result in equal increments on a decibel scale. Furthermore, the human ear perceives changes in amplitude on a logarithmic basis, making measurements with the decibel scale reflect audibility more accurately.

A decibel may be defined as the logarithmic ratio of two power measurements or as the logarithmic ratio of two voltages. Equations 16.2 and 16.3 define the decibel for both power and voltage measurement.

$$dB = 20 \log \left(\frac{E_1}{E_2}\right) \qquad\qquad (16.2)$$

$$dB = 10 \log \left(\frac{P_1}{P_2}\right) \qquad\qquad (16.3)$$

There is no difference between decibel values from power measurements and decibel values from voltage measurements if the impedances are equal. In both equations the denominator variable is usually a stated reference. It is illustrated with an example in Fig. 16.7. Whether the decibel value is computed from the power-based equation or from the voltage-based equation, the same result is obtained.

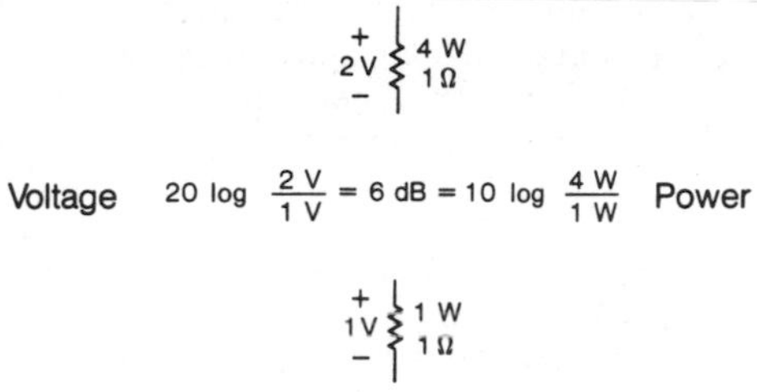

$$\begin{array}{c} + \\ 2\,V \\ - \end{array} \quad \begin{array}{l} 4\,W \\ 1\,\Omega \end{array}$$

Voltage $20 \log \dfrac{2\,V}{1\,V} = 6\ dB = 10 \log \dfrac{4\,W}{1\,W}$ Power

$$\begin{array}{c} + \\ 1\,V \\ - \end{array} \quad \begin{array}{l} 1\,W \\ 1\,\Omega \end{array}$$

FIGURE 16.7 Example of the equivalence of voltage and power decibels. *[Source: K. Blair Benson (ed.), Audio Engineering Handbook, McGraw-Hill, New York, 1988. Reprinted with permission.]*

A doubling of voltage will yield a value of 6.02 dB, while a doubling of power will yield 3.01 dB. This is true because doubling voltage results in a factor-of-4 increase in power. Table 16.1 shows the decibel values for some common voltage and power ratios.

Audio engineers often express the decibel value of a signal relative to some

TABLE 16.1   Common Decibel Values and Conversion Ratios

dB value	Voltage ratio	Power ratio
0	1	1
+1	1.122	1.259
+2	1.259	1.586
+3	1.412	1.995
+6	1.995	3.981
+10	3.162	10
+20	10	100
+40	100	10,000
−1	0.891	0.794
−2	0.794	0.631
−3	0.707	0.501
−6	0.501	0.251
−10	0.3163	0.1
−20	0.1	0.01
−40	0.01	0.0001

Source: K. Blair Benson (ed.), *Audio Engineering Handbook*, McGraw-Hill, New York, 1988.

standard reference instead of another signal. The reference for decibel measurements may be predefined as a power level, as in dBm (decibels above 1 mW), or it may be a voltage reference. When measuring dBm or any power-based decibel value, the reference impedance must be specified or understood. For example, 0 dBm (600 Ω) would be the correct way to specify level. Both 600 and 150 Ω are common reference impedances in audio work.

The decibel equations assume that the circuit being measured is terminated in the reference impedance used in the decibel calculation. However, most voltmeters are high-impedance devices and are calibrated in decibels relative to the voltage required to reach 1 mW in the reference impedance. This voltage is 0.775 V in the 600-Ω case. Termination of the line in 600 Ω is left to the user. If the line is not terminated, it is not correct to speak of a dBm measurement. The case of decibels in an unloaded line is referred to as dBu (or sometimes dBV) to denote that it is referenced to a 0.775-V level without regard to impedance.

Another common decibel reference in voltage measurements is 1 V. When using this reference, measurements are presented as dBV. Often it is desirable to specify levels in terms of a reference transmission level somewhere in the system under test. These measurements are designated dBr, where the reference point or level must be separately conveyed.

16.3.6   Crosstalk and Separation

Crosstalk is one application of decibel measurements using two measured quantities. *Crosstalk* is the leakage of signal from one audio channel into another. In the general case of crosstalk the channels are not necessarily related. For the special case of crosstalk between the two channels of a stereo system, the term *sep-*

aration is used. As a general rule, if the two channels under consideration carry two channels of the same audio program, the term "separation" is appropriate. If the two channels are unrelated, the term "crosstalk" is preferred.

Crosstalk or separation is defined as the difference in decibels between the interfering signal level in the source channel and the receiving channel. Crosstalk and separation specifications are usually expressed as a positive decibel value. Measuring crosstalk or separation consists of applying a sine wave to one channel of the device under test and measuring the level of the signal in the other channel. The amount of leakage is likely to depend on the frequency used, so measurements are normally made as a function of frequency and expressed in graphical form, as with an amplitude frequency response curve.

Take the case of a 10-V, 1-kHz sine wave in the left channel of a stereo system. If the right channel is not driven with signal, there should be no 1 kHz present at its output. However, if a level measurement on the right channel yields 10 mV of signal, the separation is defined to be 60 dB.

One possible problem with this procedure is that the 10-mV signal may not be 1 kHz leaking from the other channel but instead may represent the noise floor of the system under test. If this is true, the separation measurement is inaccurate. The solution is to use a bandpass filter tuned to the frequency of the test tone, thereby rejecting system noise and other interfering components. If the measurements are to be made as a function of frequency, the bandpass filter frequency must be slaved to the generator frequency. This is illustrated in Fig. 16.8. One channel is driven from a sine-wave generator at its normal operating level and terminated in its normal load impedance. The other channel input is terminated by the normal source impedance, and the output is terminated by the normal load impedance. The level at the output of the driven channel is measured with a voltmeter, and the output of the undriven channel is measured by a voltmeter with a bandpass filter centered on the test signal frequency. The level difference between these measurements, expressed in decibels, is the separation. When the two channels have different gains, it is common to correct the measurements for the gain difference so as to present the crosstalk referred to the channel inputs.

Crosstalk between two audio channels is sometimes nonlinear. The presence of a signal in one channel will sometimes yield tones of other frequencies in the receiving channel. This is especially true in transmission systems where cross-

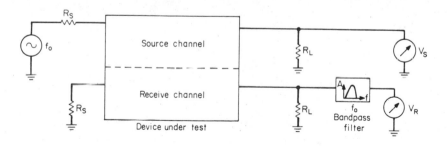

$$\text{Crosstalk (dB)} = 20 \log_{10} \left(\frac{V_R}{V_S} \right)$$

FIGURE 16.8 Procedure for making crosstalk measurements incorporating a bandpass filter. *[Source: K. Blair Benson (ed.),* Audio Engineering Handbook, *McGraw-Hill, New York, 1988. Reprinted with permission.]*

modulation can occur between carriers. These tones disappear when the source signal is removed, clearly indicating that they are caused by the suspected source. Measuring them can be tedious if sine waves are used because the frequency of the received interference may not be easily predictable. There may also be some test frequencies that cause the interference products to appear outside the channel bandwidth, hiding the effect being tested. Therefore, crosstalk tests are sometimes performed with a random noise source so that all possible interference frequencies are checked.

16.3.7 Noise Measurements

Noise measurements are simply specialized level measurements. It has long been recognized that the ear's sensitivity varies with frequency, especially at low lev-

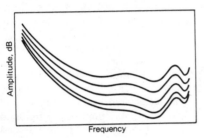

els. This effect was studied in detail by Fletcher and Munson and later by Robinson and Dadson. The Fletcher-Munson hearing-sensitivity curve for the threshold of hearing and above is given in Fig. 16.9. The ear is most sensitive in the region of 2 to 4 kHz, with rolloffs above and below these frequencies. This effect is responsible for the apparent loss of bass when the volume is reduced on a hi-fi system. To predict how loud something will sound, it is necessary to use a filter that duplicates this nonflat behavior electrically. The filter *weights* the signal level on the basis of frequency, thus earning the name *weighting filter*.

FIGURE 16.9 The Fletcher-Munson curves of hearing sensitivity versus frequency. *[Source: K. Blair Benson (ed.),* Audio Engineering Handbook, *McGraw-Hill, New York, 1988. Reprinted with permission.]*

Various efforts have been made to do this, resulting in several standards for noise measurement. Some of the common weighting filters are shown overlaid on the hearing threshold curve in Fig. 16.10.

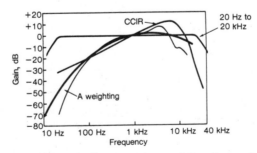

FIGURE 16.10 Response characteristics of several common weighting filters for audio measurements. *[Source: K. Blair Benson (ed.),* Audio Engineering Handbook, *McGraw-Hill, New York, 1988. Reprinted with permission.]*

The most common filter used in the United States for weighted noise measurements is the A-weighting curve. This filter is placed in front of a high-sensitivity voltmeter, as shown in Fig. 16.11. An average-responding meter is often used for A-weighted noise measurements, although rms meters are gaining acceptance for this application.

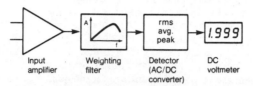

Input   Weighting   Detector   DC
amplifier   filter   (AC/DC)   voltmeter
                              converter)

FIGURE 16.11 The architecture of a weighted-voltmeter for audio applications. [*Source: K. Blair Benson,* Audio Engineering Handbook, *McGraw-Hill, New York, 1988. Reprinted with permission.*]

International Radio Consultive Committee (CCIR) Noise Measurements. European equipment is usually specified with a CCIR filter and a quasi-peak detector. The CCIR-filter response is shown with the A-weighting curve in Fig. 16.10. It is significantly more peaked than the A curve and has a sharper rolloff at high frequencies. The CCIR quasi-peak standard was developed to quantify the noise in telephone systems. The quasi-peak detector more accurately represents the ear's sensitivity to impulsive sounds that occur when telephone switching systems operate. When used with the CCIR filter curve, it is supposed to correlate better with the subjective level of the noise than A-weighted average-response measurements do.

Other Noise Measurements. Some audio equipment manufacturers specify noise with a 20-Hz to 20-kHz bandwidth filter and an rms-responding meter. This is done to specify noise over the audio band without regard to the ear's differing sensitivity with frequency. The International Electrotechnical Commission (IEC) defines the audio band as all frequencies between 22.4 Hz and 22.4 kHz. Measurements over such a bandwidth are referred to under IEC standards as *unweighted*.

About a dozen different weighting filter voltmeter combinations have been proposed for measuring noise. Each technique has its group of supporters. The important thing is that the measurement technique used be specified and be consistent from test to test. Meters designed for measuring noise and containing appropriate weighting filters are sometimes called *psophometers*. This term is more common in Europe than in the United States.

European telephone noise levels are often specified as dBmp, meaning that a dBm value has been measured in the system impedance by using a weighting filter corresponding to the International Telegraph and Telephone Consultative Committee (CCIT) psophometric curve. In the United States telephone noise measurements are normally made with a *C-message* weighting filter and an average-responding meter. The C-message curve is designed to integrate the amplitude versus frequency response of the ear with the response of a typical telephone handset, denoted dBrnc. Measurements made with an A-weighting filter are termed dBA. The term dBq refers to noise voltages measured with a quasi-peak detector and one of the filters specified in CCIR Recommendation 468-1. These may be either the CCIR weighting curve or the 22.4-Hz to 22.4-kHz audio band filter.

Noise Analysis. Power-line (mains) hum is low-frequency interference at the power-line frequency or its multiples. This interference is generally grouped with noise when making measurements. In countries using a 60-Hz primary frequency (North America and others), the components occur at 60 Hz, 120 Hz, 180 Hz, 240 Hz, etc. For 50-Hz countries (Europe and others), the corresponding frequencies are 50 Hz, 100 Hz, 150 Hz, 200 Hz, etc. The dominant components are usually fundamental, second, and third harmonic, which can be removed using a high-pass filter at approximately 400 Hz. With a three-pole or sharper filter, 1-dB accurate measurements may be made down to 1 kHz. By making unweighted noise measurements both with and without the high-pass filter, the effects of hum may be estimated. Subtracting the two decibel measurements results in a level ratio measurement called the *hum-to-hiss ratio*. Good-quality equipment will have hum-to-hiss ratios of less than 1 dB.

Even more information about the underlying sources of noise may be obtained from a special analysis of the system under test. Figure 16.12 shows a spectrum analysis of the output noise from a professional equalizer. The measuring equipment uses a sweeping one-third-octave filter and an rms detector. Note the presence of power line regulated signals, both 120- and 180-Hz components. The 120-Hz product is due to asymmetrical charging currents in the power supply, while the 180-Hz product is from transformer field leakage.

Another approach that is often useful in noise analysis is to view the noise on an oscilloscope and trigger the oscilloscope with an appropriate synchronization signal. The components of the noise related to the synchronizing signal will remain stationary on the screen, and the unrelated energy will produce fuzz on the trace. For example, to investigate line-related components, the oscilloscope should be triggered on the power line. If interference is suspected from a nearby television signal, the oscilloscope can be triggered on the television vertical and/or horizontal sync signals. The scope photograph in Fig. 16.13 is an example of the display when measuring power line noise with this technique.

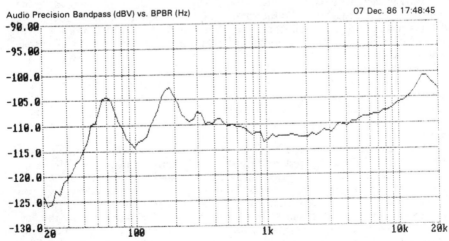

Audio Precision Bandpass (dBV) vs. BPBR (Hz)   07 Dec. 86 17:48:45

FIGURE 16.12 Typical spectrum of noise and hum measured with a sweeping one-third-octave filter. *[Source: K. Blair Benson (ed.),* Audio Engineering Handbook, *McGraw-Hill, New York, 1988. Reprinted with permission.]*

Noise may be expressed as an absolute level (usually in dBm or dBu) by simply measuring the weighted voltage (proper termination being assumed in the case of dBm) at the desired point in the system. However, this is often not very meaningful. A 1-mV noise voltage at the output of a power amplifier may be quite good, while 1 mV of noise at the output of a microphone would render it useless. Specifying the noise performance as the *signal-to-noise ratio* (SNR) is a better approach. SNR is a decibel measure of the noise level using the signal level measured at the same point as a reference. This makes measurements at different points in a system or in different systems directly comparable. A signal with a given SNR

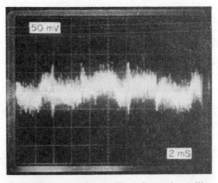

FIGURE 16.13 Power-line-triggered oscilloscope display of noise and hum. *[Source: K. Blair Benson (ed.),* Audio Engineering Handbook, *McGraw-Hill, New York, 1988. Reprinted with permission.]*

can be amplified with a perfect amplifier or attenuated with no change in the SNR. Any degradation in SNR at later points in the system is the result of limitations of the equipment that follows.

16.4 NOISE MEASUREMENT

When a signal is applied to the input of a device, the output will appear later. For sine-wave excitation, this delay between input and output may be expressed as a proportion of the sine-wave cycle, usually in degrees. One cycle is 360°, one-half cycle is 180°, and so on. This measurement is illustrated in Fig. 16.14. The phasemeter input signal no. 2 is delayed from, or is said to be *lagging*, input no. 1 by 45°.

Most audio test equipment checks phase directly by measuring the proportion of one signal cycle between zero crossings of the signals. This can be done with an edge-triggered set-reset flip-flop as shown in Fig. 16.14. The output of this flip-flop will be a signal that goes high during the time between zero crossings of the signals. By averaging the amplitude of this pulse over 1 cycle (i.e., measuring its *duty cycle*), a measurement of phase results.

Phase is typically measured and recorded as a function of frequency over the audio range. For most audio devices, phase and amplitude responses are closely coupled. Any change in amplitude that varies with frequency will produce a corresponding phase shift. A typical phase and amplitude versus frequency plot of a graphic equalizer is shown in Fig. 16.15.

A fixed time delay will introduce a phase shift that is a linear function of frequency. This time delay can introduce large values of phase shift at high frequencies that are of no significance in practical applications. The time delay will not distort the wave shape of complex signals and will not be audible in any way. There can be problems, however, with time delay when the delayed signal is used in conjunction with an undelayed signal. This would be the case if one channel of a stereo signal was delayed and the other was not.

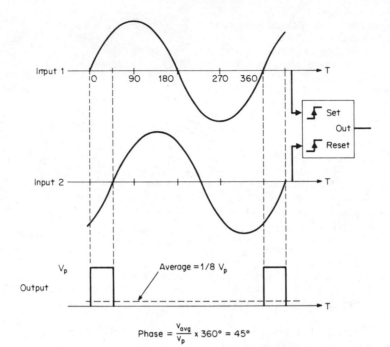

FIGURE 16.14 Illustration of the measurement of a phase shift between two signals. *[Source: K. Blair Benson (ed.),* Audio Engineering Handbook, *McGraw-Hill, New York, 1988. Reprinted with permission.]*

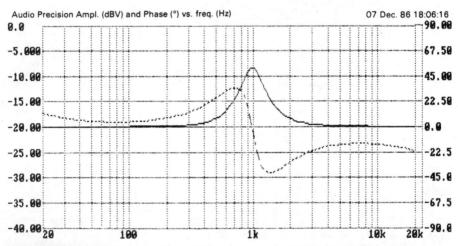

FIGURE 16.15 Typical phase and amplitude versus frequency plot of a graphic equalizer. The solid line represents amplitude (dBV), and the broken line represents phase (degrees). *[Source: K. Blair Benson (ed.),* Audio Engineering Handbook, *McGraw-Hill, New York, 1988. Reprinted with permission.]*

Relation to Frequency.   When dealing with complex signals, the meaning of phase can become unclear. Viewing the signal as the sum of its components according to Fourier theory, we find a different value of phase shift at each frequency. With a different phase value on each component, which one is to be used as the reference? If the signal is periodic and the waveshape is unchanged passing through the device under test, a phase value may still be defined. This may be done by using the shift of the zero crossings as a fraction of the waveform period. Indeed, most commercial phasemeters will display this value. However, if there is differential phase shift with frequency, the waveshape will be changed. It is then not possible to define any phase-shift value, and phase must be expressed as a function of frequency.

Group delay is another useful expression of the phase characteristics of an audio device. Group delay is the slope of the phase response. It expresses the relative delay of the spectral components of a complex waveform. If the group delay is flat, all components will arrive together. A peak or rise in the group delay indicates that those components will arrive later by the amount of the peak or rise. It is computed by taking the derivative of the phase response versus frequency. Mathematically:

$$\text{Group delay} = \frac{-(\text{phase at } f_2 - \text{phase at } f_1)}{f_2 - f_1} \tag{16.4}$$

This requires that phase be measured over a range of frequencies to give a curve that can be differentiated. It also requires that the phase measurements be performed at frequencies close enough together to provide a smooth and accurate derivative.

The most common application of phase measurement in a studio is aligning tape recorder heads. If a multitrack tape head is tilted relative to the direction of tape travel (an azimuth error), the signals in each channel will be slightly delayed. This delay results in a phase shift on sine waves. Because azimuth error results in a fixed time delay, the phase error will increase with increasing frequency. At a sufficiently high frequency, the phase error may reach or even exceed 360°. By measuring the phase shift at several frequencies, head misalignment becomes relatively easy to see and correct.

16.5 NONLINEAR AUDIO DISTORTION

Distortion is a measure of signal impurity. It is usually expressed as a percentage or decibel ratio of the undesired components to the desired components of a signal. Distortion of a device is measured by feeding it one or more sine waves of various amplitudes and frequencies. In simplistic terms, any frequencies at the output that were not present at the input are distortion. However, strictly speaking, components caused by power line interference or other spurious signals are not distortion. There are many methods of measuring distortion in common use: harmonic distortion and several types of intermodulation distortion.

16.5.1 Harmonic Distortion

The transfer characteristic of a typical amplifier is shown in Fig. 16.16. The transfer characteristic represents the output voltage at any point in the signal waveform for a given input voltage; ideally this is a straight line. The output waveform is the projection of the input sine wave on the device transfer characteristic. A change in the input produces a proportional change in the output. Because the actual transfer characteristic is nonlinear, a distorted version of the input waveshape appears at the output.

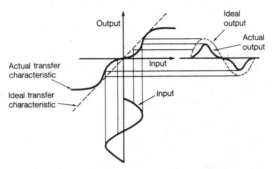

FIGURE 16.16 Illustration of total harmonic distortion (THD) measurement of an amplifier transfer characteristic. *[Source: K. Blair Benson (ed.),* Audio Engineering Handbook, *McGraw-Hill, New York, 1988. Reprinted with permission.]*

Harmonic distortion measurements excite the device under test with a sine wave and measure the spectrum of the output. Because of the nonlinearity of the transfer characteristic, the output is not sinusoidal. By using Fourier series, it can be shown that the output waveform consists of the original input sine wave plus sine waves at integer multiples (harmonics) of the input frequency. The spectrum of the distorted signal is shown in Fig. 16.17 for a 1-kHz input, and output signal consisting of 1 kHz, 2 kHz, 3 kHz, etc. The harmonic amplitudes are proportional to the amount of distortion in the device under test. The percentage of harmonic distortion is the rms sum of the harmonic amplitudes divided by the rms amplitude of the fundamental.

Harmonic distortion may also be measured with a spectrum analyzer. The procedure is illustrated in Fig. 16.17. The fundamental amplitude is adjusted to the 0-dB mark on the display. The amplitudes of the harmonics are then read and converted to linear scale. The rms sum of these values is taken, which represents the THD. This procedure is time-consuming and difficult for an unskilled operator. Even skilled operators have trouble in obtaining accuracies better than 2 dB in the final result because of equipment limitations and the problems inherent in reading numbers off a trace on the screen of an analyzer.

Notch-Filter Analyzer. A simpler approach to the measurement of harmonic distortion can be found in the notch-filter distortion analyzer. This device, commonly referred to as simply a *distortion analyzer*, removes the fundamental of the signal to be investigated and measures the remainder. A block diagram of such a unit is shown in Fig. 16.18. The fundamental is removed with

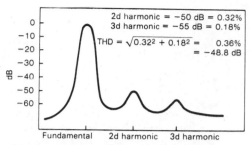

FIGURE 16.17 Example of reading THD from a spectrum analyzer. [*Source: K. Blair Benson (ed.), Audio Engineering Handbook, McGraw-Hill, New York, 1988. Reprinted with permission.*]

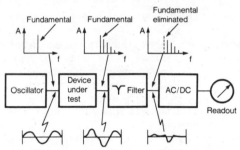

FIGURE 16.18 Simplified block diagram of a harmonic distortion analyzer. [*Source: K. Blair Benson (ed.), Audio Engineering Handbook, McGraw-Hill, New York, 1988. Reprinted with permission.*]

a notch filter, and the output is measured with an ac voltmeter. Because distortion is normally presented as a percentage of the fundamental signal, this level must be measured or set equal to a predetermined reference value. Additional circuitry (not shown) is required to set the level to the reference value for calibrated measurements. Some analyzers use a series of step attenuators and a variable control for setting the input level to the reference value. More sophisticated units eliminate the variable control by using an electronic gain control. Others employ a second ac-to-dc converter to measure the input level and compute the percentage by using an analog divider or a microprocessor. Completely automatic units also provide autoranging logic to set the attenuators and ranges. This provision significantly reduces the effort and skill required to make a measurement.

The correct method of representing percentage distortion is to express the level of the harmonics as a fraction of the fundamental level. However, commercial distortion analyzers use the total signal level as the reference voltage. For small amounts of distortion these two quantities are equivalent. At large values of distortion the total signal level will be greater than the fundamental level. This makes distortion measurements on these units lower than the actual value. The relationship between the measured distortion and the true distortion is given in Fig. 16.19. The errors are not significant until about 20 percent measured distortion.

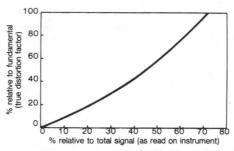

FIGURE 16.19 Conversion graph for indicated distortion and true distortion. [*Source: H. W. Tremaine, Audio Cyclopedia, Howard W. Sams, Indianapolis, 1975. In K. Blair Benson (ed.), Audio Engineering Handbook, McGraw-Hill, New York, 1988. Reprinted with permission.*]

The need to tune the notch filter to the correct frequency can make this work tedious if a manual instrument is used. Some manufacturers have circumvented this problem by including the measurement oscillator and analyzer in one package, placing the analyzer and oscillator frequency controls on the same knob or button. This eliminates the problem only when the signal source used for the test is the internal oscillator. If a tape recorder with a prerecorded test tape is being used, the generator frequency will not be the frequency coming from the tape and the user must resort to manual tuning. Similar problems occur when testing broadcast links or communications lines because the generator may be hundreds or even thousands of miles away. A better approach, used by a few manufacturers, is to measure the input frequency and tune the filter to the measured frequency. This eliminates any need to adjust the analyzer frequency.

Because of the notch-filter response, any signal other than the fundamental will influence the results, not just the harmonics. Some of these interfering signals are illustrated in Fig. 16.20. Any practical signal contains some hum and noise, and the distortion analyzer will include these in the reading. Because of these added components, the correct term for this measurement is total harmonic distortion and noise (THD + N). Although this fact does limit the reading of very low THD levels, it is not necessarily bad. Indeed, it can be argued that the ear hears all components present in the signal, not just the harmonics.

Additional filters are included on most distortion analyzers to reduce unwanted hum and noise as illustrated in Fig. 16.21. These usually consist of one or more high-pass filters (400 Hz is almost universal) and several low-pass filters. Common low-pass filter frequencies are 22.4, 30, and 80 kHz. Better equipment will include filters at all these frequencies to ease the trade-off between limiting bandwidth to reduce noise and the reduction in reading accuracy that results from removing desired components of the signal. When used in conjunction with a good differential input stage on the analyzer, these filters can solve most noise problems.

The use of a sine-wave test signal and a notch-type distortion analyzer has the distinct advantage of simplicity in both design and use. This simplicity has an additional benefit in ease of interpretation. The shape of the output waveform from a notch-type analyzer indicates the slope of the nonlinearity. Displaying the residual components on the vertical axis of an oscilloscope and the input signal on

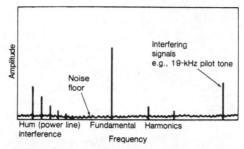

FIGURE 16.20 Example of interference sources in distortion and noise measurements. *[Source: K. Blair Benson (ed.),* Audio Engineering Handbook, *McGraw-Hill, New York, 1988. Reprinted with permission.]*

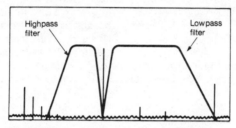

FIGURE 16.21 The use of filters to reduce interference in distortion measurements. *[Source: K. Blair Benson (ed.),* Audio Engineering Handbook, *McGraw-Hill, New York, 1988. Reprinted with permission.]*

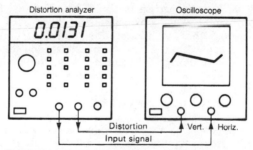

FIGURE 16.22 Transfer function monitoring using an oscilloscope and distortion analyzer. *[Source: K. Blair Benson (ed.),* Audio Engineering Handbook, *McGraw-Hill, New York, 1988. Reprinted with permission.]*

the horizontal axis gives a plot of the deviation of the transfer characteristic from a best-fit straight line. This technique is diagrammed in Fig. 16.22. The trace will be a horizontal line for a perfectly linear device. If the transfer characteristic curves upward on positive input voltages, the trace will bend upward at the right-hand side.

Examination of the distortion components in real time on an oscilloscope will show such things as oscillation on the peaks of the signal, crossover distortion, and clippings. This is a valuable tool in the design and development of audio circuits and one that no other distortion measurement method can fully match. Viewing the residual components in the frequency domain by using a spectrum analyzer also gives much information about the distortion mechanism inside the device under test.

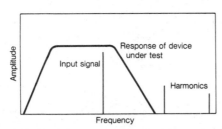

FIGURE 16.23 Illustration of problems that occur when measuring harmonic distortion in band-limited systems. *[Source: K. Blair Benson (ed.), Audio Engineering Handbook, McGraw-Hill, New York, 1988. Reprinted with permission.]*

Bandwidth limitations when measuring distortion at high frequencies are the major problem with THD testing, as illustrated in Fig. 16.23. Because the components being measured are harmonics of the input frequency, they may fall outside the passband of the device under test. A tape recorder with a cutoff frequency of 22 kHz (typical for a good machine) will allow measurement only up to the third harmonic of a 7-kHz input. THD measurements on a 20-kHz signal are impossible because all the distortion components are filtered out by the recorder.

16.5.2   Intermodulation Distortion

Many methods have been devised to measure the intermodulation (IM) of two or more signals passing through a device simultaneously. The most common of these is SMPTE IM, named after the Society of Motion Picture and Television Engineers, which first standardized its use. IM measurements according to the SMPTE method have been in use since the 1930s. The test signal is a low-frequency tone (usually 60 Hz) and a high-frequency tone (usually 7 kHz) mixed in a 4:1 amplitude ratio. Other amplitude ratios and frequencies are used occasionally. The signal is applied to the device under test, and the output signal is examined for modulation of the upper frequency by the low-frequency tone. The amount by which the low-frequency tone modulates the high-frequency tone indicates the degree of nonlinearity. As with harmonic distortion measurement, this test may be done with a spectrum analyzer or with a dedicated distortion analyzer.

The modulation components of the upper signal appear as sidebands spaced at multiples of the lower-frequency tone, as illustrated in Fig. 16.24. The rms amplitudes of the sidebands are summed and expressed as a percentage of the upper-frequency level.

The most direct way to measure SMPTE IM distortion is to measure each component with a spectrum analyzer and add their rms values. The spectrum analyzer approach has a drawback in that it is sensitive to frequency modulation of the carrier as well as amplitude modulation. Since Doppler effects cause frequency modulation, this approach cannot be used on loudspeakers. Similar problems result from the wow and flutter in tape recorders and disk recording equipment.

A distortion analyzer for SMPTE testing is quite straightforward. The signal to be analyzed is passed through a high-pass filter to remove the low-frequency tone, as shown in Fig. 16.25. The high-frequency tone is then demodulated as if it were an amplitude-modulated signal to obtain the sidebands. The sidebands pass through a low-pass filter to remove any remaining high-frequency energy. The resulting demodulated low-frequency signal will follow the envelope of the high-frequency tone. This low-frequency fluctuation is the distortion component and is displayed as a percentage of the amplitude of the high-frequency tone.

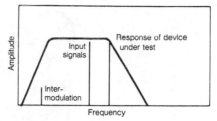

FIGURE 16.24 Measuring intermodulation distortion in band-limited systems. [*Source: K. Blair Benson (ed.),* Audio Engineering Handbook, *McGraw-Hill, New York, 1988. Reprinted with permission.*]

Because low-pass filtering sets the measurement bandwidth, noise has little effect on SMPTE IM measurements. The analyzer is tolerant of harmonics of the two input signals, allowing fairly simple oscillators to be used to generate the test signal. Indeed, early analyzers used a filtered version of the power line for the low-frequency tone: hence the 60-Hz low tone frequency in the SMPTE standard. It is important, however, that no harmonics of the low-frequency signal generator extend into the measurement range of the high-frequency tone. The analyzer will be unable to distinguish these from sidebands. After the first stage of filtering in the analyzer, there is little low-frequency energy left to create IM in the analyzer. This considerably simplifies the remaining circuitry.

Using the SMPTE IM Test.   Figure 16.26 shows the composite test signal applied to a device under test. The output waveform is a distorted version of the input because of amplifier nonlinearities. As the high-frequency tone is moved along the transfer characteristic by the low-frequency tone, its amplitude changes. This results in low-frequency amplitude modulation of the high-frequency tone. The IM test provides a good indication of problems such as crossover distortion and clipping. High-order nonlinearities create bumps in the transfer characteristic that produce large amounts of IM. SMPTE IM testing is also useful for exciting low-frequency thermal distortion. As the low-frequency signal moves around, exciting the thermal effects, the gain of the device changes, creating modulation distortion.

Testing the output inductance-capacitance (LC) stabilization networks in power amplifiers is another good application for IM measurements. Low-frequency signals may saturate the output inductor, causing it to become nonlinear. Because the frequency is low, very little voltage is dropped across the inductor, and there is little low-frequency harmonic distortion. The high-frequency tone will develop a signal across the inductor because of the rising impedance with frequency. When the low-frequency tone creates a nonlinear inductance, the high-frequency tone becomes distorted.

The inherent insensitivity of SMPTE IM to wow and flutter has fostered widespread use of the test in applications that involve audio recording. Much use is made of SMPTE IM in the disk-recording and film industries. When applied to disks, the test frequencies are usually 400 Hz and 4 kHz. This form of IM testing is especially sensitive to excessive polishing of the disk surface, even though harmonic distortion is not.

It is often claimed that because the distortion components in an SMPTE test

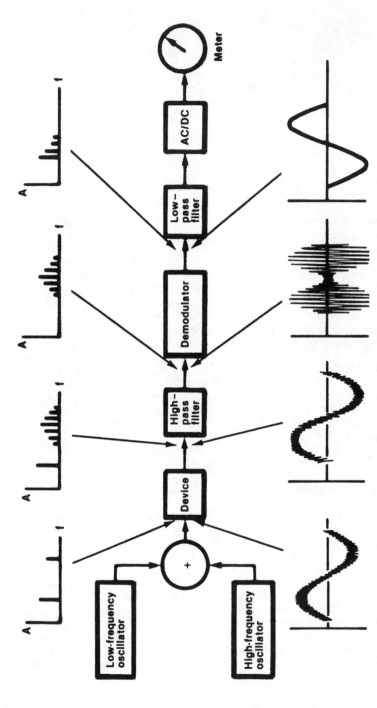

FIGURE 16.25 Simplified block diagram of an SMPTE intermodulation analyzer. *[Source: K. Blair Benson (ed.), Audio Engineering Handbook, McGraw-Hill, New York, 1988. Reprinted with permission.]*

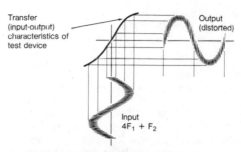

FIGURE 16.26 SMPTE intermodulation test of transfer characteristic. *[Source: K. Blair Benson (ed.),* Audio Engineering Handbook, McGraw-Hill, *New York, 1988. Reprinted with permission.]*

are not harmonically related to either input, they will be more objectionable to the ear. Musical instruments are rich in harmonics but contain few if any components that are not harmonic-related.

CCIT IM. Twin-tone intermodulation or CCIT difference frequency distortion is another method of measuring distortion by using two sine waves. The test signal consists of two closely spaced high-frequency tones as shown in Fig. 16.27. When the tones are passed through a nonlinear device, IM products are generated at frequencies related to the difference in frequency between the original tones. For the typical case of signals at 14 and 15 kHz, the IM components will be at 1, 2, 3 kHz, etc., and 13, 16, 12, 17 kHz, etc. Even-order or asymmetrical distortions produce low-frequency difference-frequency components. Odd-order or symmetrical nonlinearities produce components near the input signal frequencies. The most common application of this test measures only the even-order components because they may be measured with a multipole low-pass filter. Mea-

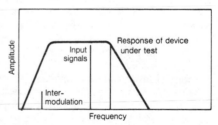

FIGURE 16.27 CCIT intermodulation test in band-limited systems. *[Source: K. Blair Benson (ed.),* Audio Engineering Handbook, *McGraw-Hill, New York, 1988. Reprinted with permission.]*

surement of the odd-order components requires a spectrum analyzer or selective voltmeter.

The CCIT test has several advantages over either harmonic or SMPTE IM testing. The signals and distortion components may almost always be arranged to be in the passband of the device under test. This method ceases to be useful below a few hundred hertz when the required selectivity in the spectrum analyzer or selective voltmeter becomes excessive. However, FFT-based devices can extend the practical lower limit substantially below this point.

The distortion products in the CCIT IM test are usually far removed from the input signal. This positions them outside the range of the auditory system's masking effects. If a test that measures what the ear might hear is desired, the CCIT test is the most likely candidate.

16.5.3 Distortion-Measurement Hardware

Distortion measurements are best performed using an rms-responding meter in the distortion analyzer. This is necessary to make the reading represent the true power in the distortion. Older instruments used average-responding meters, but these are rapidly becoming outdated. With most practical distortion measurements, the rms response will read about 2 dB higher than the average response. Some arguments can be made for measuring distortion with peak or quasi-peak detectors. The averaging time of the ear is between 50 and 250 ms, depending on test conditions and procedure. This has led to the definition of the quasi-peak meter for measuring telephone noise as described earlier.

The accuracy of most distortion analyzers is specified at 1 dB, but this can be misleading. Separate specifications are often put on the bandwidth and ranges, as explained previously (Sec. 16.3.4) for voltmeters. The residual distortion of the measurement system is a more important specification for distortion measurements. Manufacturers of distortion analyzers often specify the oscillator and the distortion analyzer separately. A system in which the oscillator and the analyzer are each specified at 0.002 percent THD can have a system residual distortion of 0.004 percent. If the noise of the analyzer and/or the oscillator is specified separately, this must be added to the residual specification to find the residual THD + N of the system. For example, an analyzer specified at 0.002 percent distortion and 20 μV input noise will have a 0.003 percent residual at 1-V input and 0.02 percent at 0.1-V input. These voltages are common when measuring mixing consoles and preamplifiers, resulting in a serious practical limitation with some distortion analyzers.

Many commercial units specify residual distortion at only one input voltage or at full scale of one range. Performance usually degrades by as much as 10 dB when the signal is at the bottom of an input range. This occurs because THD + N measurements are a ratio of the distortion components and noise to the signal level. At the full-scale input voltage, the signal to the notch filter is at maximum and the filter noise contribution is minimized. As the level drops, residual noise in the notch filter becomes a larger percentage of the reading.

16.5.4 Addition and Cancellation Hardware

Another often-overlooked problem is that of distortion addition and cancellation in the test equipment or the device under test. Consider the examples in Figs. 16.28 and 16.29. Suppose one device under test has a transfer characteristic similar to that diagrammed at the top of Fig. 16.28a and another has the characteristic diagrammed at the bottom. If the devices are cascaded, the resulting transfer-characteristic nonlinearity will be magnified as shown. The effect on sine waves in the time domain is illustrated in Fig. 16.28b. The distortion component generated by each nonlinearity is in phase and will sum to a component of twice the original magnitude. However, if the second device under test has a complementary transfer characteristic as shown in Fig. 16.29, quite a different result is obtained. When the devices are cascaded, the effects of the two curves cancel,

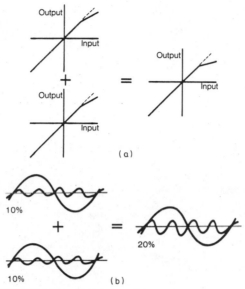

FIGURE 16.28 Illustration of the addition of distortion components. (*a*) Addition of transfer-function nonlinearities. (*b*) Addition of distortion components. *[Source: K. Blair Benson (ed.),* Audio Engineering Handbook, *McGraw-Hill, New York, 1988. Reprinted with permission.]*

yielding a straight line for the transfer characteristic (Fig. 16.29*a*) The corresponding distortion products are out of phase with each other, resulting in no measured distortion components in the final output (Fig. 16.29*b*).

This problem is common at low levels of distortion, especially between the test equipment and the device under test. For example, if the test equipment has a residual of 0.002 percent when connected to itself and readings of 0.001 percent are obtained from the circuit under test, cancellations are occurring. It is also possible for cancellations to occur in the test equipment itself, with the combined analyzer and signal generator giving readings lower than the sum of their individual residuals. If the distortion is the result of even-order (asymmetrical) nonlinearity, reversing the phase of the signal between the offending devices will change a cancellation to an addition. If the distortion is from an odd-order (symmetrical) nonlinearity, phase inversions will not affect the cancellation.

16.6 VIDEO-MEASUREMENT OBJECTIVES

The goal of every television broadcast or video production facility is to produce the best possible pictures for their viewers. Good pictures do not just happen. They are the result of a carefully planned maintenance program. Such a program involves correct setup of the equipment whenever a production is scheduled and a routine preventive-maintenance schedule for every piece of electronic equip-

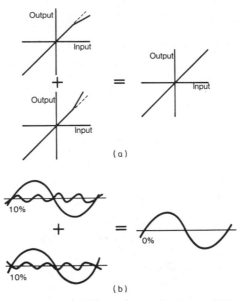

FIGURE 16.29 Illustration of the cancellation of distortion components. (*a*) Cancellation of transfer-characteristic nonlinearities. (*b*) Cancellation of distortion waveform. *[Source: K. Blair Benson (ed.), Audio Engineering Handbook, McGraw-Hill, New York, 1988. Reprinted with permission.]*

ment to keep the system operating at its peak performance. Unfortunately, no program avoids the inevitable failure, but careful observation during setup and preventive maintenance can detect problems while they are still minor. An integral part of the maintenance program is proper test equipment and an understanding of how and why to use it.

Setting up a maintenance shop for a television facility can be an expensive proposition. However, if the investment is going to be made for cameras, recorders, monitors, and the many other parts of a modern television system, acquiring the proper test equipment is much the same as an insurance policy on that investment. A thorough knowledge of the use of the test equipment only can produce a better yield on the investment.

16.7 ESSENTIAL TEST EQUIPMENT

In reality, test equipment requirements to perform most maintenance and adjustment procedures on video equipment are relatively limited. A technician, armed with the following list of test units, will be able to deal with most common video problems in the studio:

1. A set of camera test charts
2. A precision test signal generator

3. A wideband oscilloscope
4. A waveform monitor
5. A vectorscope
6. A spectrum analyzer

In the remainder of this section, we examine each item in terms of applications to video equipment adjustment and maintenance.

16.7.1 Camera Test Charts

The camera is the eye of the television system and is the only method we have to import the real environment into the world of recorded video. Although we may eventually change many characteristics of the pictures produced by the camera, we would like to begin as close to reality as possible. We want our camera to see the scene as we see it. A series of camera test charts provides references by which we can adjust the camera controls to achieve the best match between image and reality. Three charts of primary importance are the gray scale, registration, and resolution.

The most often used chart is the *10-step gray-scale* chart (Fig. 16.30). (For CCD cameras, an 11-step chart is preferred for gray-scale adjustments.) The chart contains two series of *paint chips* ranging from black to white. The coloring

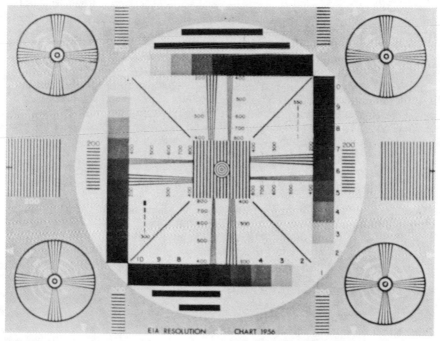

FIGURE 16.30 EIA resolution chart. *[Source: Electronics Industries Association. In K. Blair Benson (ed.),* Television Engineering Handbook, *McGraw-Hill, New York, 1986]*

and surface of the chips are such that they equal 10 levels of light reflectance. One series is black to white from left to right. The other series is in the opposite order. The chart is placed directly in front of the camera and should be lighted with the same amount of white light that is normally used on the set. The lens is adjusted to fill the screen with the chart. The signal produced from this chart is used to perform the following adjustments:

1. Peak-white level (master gain)
2. Black level (master pedestal)
3. Black balance (red and blue pedestals)
4. White balance (red and blue gain)
5. Gamma (a compensation for the nonlinearity of the cathode ray tube used in the receiver and monitor)

The first four adjustments should be performed each time the cameras are prepared for a production session. When the adjustments are correct, the result will appear to be an X shape on a waveform or oscilloscope screen. The picture on a color monitor should show only shades of gray without any red, green, or blue color tinting. The adjustment for incorrect gamma is necessary when the crossover point of the X is too high or too low, which causes impure coloring of the picture.

Another chart in the series is used for *registration* of the images from the three camera tubes. The pattern on the chart is a crosshatch of horizontal and vertical lines at precisely equal intervals with a circle superimposed over the lines. The circle in the picture produced by the camera should be round. If it is not, there are problems in horizontal or vertical sweep linearity or height and width. At any time registration is needed, the lines are used to register the blue and red images over the green image. Mode selector switches produce a combination of red-green and blue-green to be displayed on a picture monitor to make the registration adjustment easier.

The EIA resolution chart in Fig. 16.30 is useful also to check the frequency response of *resolution* of a camera. The numbers adjacent to each pattern of tapered horizontal or vertical lines indicate the line spacing in terms of the number of equivalent horizontal-scanning lines. The limiting resolution is where the individual lines of the wedge merge together.

16.7.2 Test-Signal Generators

The maintenance procedures for many video products depend upon a precision signal source to which measurements can be referred. The video test-signal generator is that reference. To be useful, the test signals must be locked to the station master sync generator. For that reason, many units suitable for master sync sources include circuits to produce the major test signals. If the choice is made to have a signal source separate from the master sync generator, the test-signal source should include gen-lock, allowing it to be used as part of the overall system, as well as a stand-alone instrument in the maintenance shop.

Color Bars. The most widely used test signal in television is the *color-bar pattern*. It is sufficiently important that a color-bar generator is included in the encoder section of nearly every television camera designed for broadcast use. Not only is the standard array of color bars (see Fig. 16.31) essential for adjustment of the camera encoder, a short period of the signal is commonly recorded at the beginning of a video tape, along with a standard audio tone level. Bars and tone are useful test signals for setup of the video recorder before the recording as well as for the eventual playback.

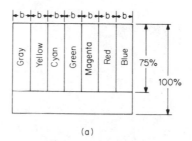

(a)

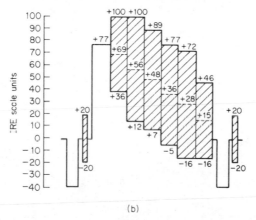

(b)

FIGURE 16.31 EIA RS189A color-bar displays. (*a*) Color-monitor display of gray and color bars. (*b*) Waveform display of reference gray and primary and complementary colors (upper part of picture display). *Note*: Dotted lines in *b* indicate luminance values. [*Source: Electronics Industries Association. In K. Blair Benson (ed.),* Television Engineering Handbook, *McGraw-Hill, New York, 1986]*

Several different display formats of color-bar signals can be used. A full-field bar signal displays the eight bars the full height of the screen. The more flexible split-field bar signal includes the color segments in the upper portion of the screen, while the lower section displays white and black along with color difference components, *I* and *Q* or *R-Y* and *B-Y*. Still another display format places a short segment of the bars in reverse order across the middle of the screen and may incorporate a pluge signal, a special waveform used to make adjustments to black levels more precisely. Ideally, the color-bar signal will at least provide references to set black-and-white levels as well as the chrominance level and color phase. The signal generator may provide 75 and 100 percent color-bar signals, which refers to a change in signal amplitude, not color saturation.

Multiburst.   The *multiburst pattern* (Fig. 16.32) includes bursts or packets of several specific frequencies across the screen. This pattern is designed for checking frequency response of equipment. Usually the highest frequency is 4.2 MHz, the greatest frequency that should theoretically pass through the television system. A second packet is centered on the color-subcarrier frequency of 3.579545 MHz. The remainder of the bursts are lower frequencies, usually down to about 500 kHz.

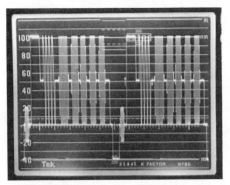

FIGURE 16.32   Waveform-monitor display of an electronically generated multiburst pattern used for checking the frequency response of an amplifier or a video transmission system. *(Source: Courtesy of Tektronix, Inc.)*

An alternative signal for checking frequency response is a video sweep signal. To form a video sweep, an oscillator is controlled by a ramp waveform that is closely related to the horizontal sweep signal driving camera and picture tubes. As the beam moves across the screen, the ramp increases, causing the frequency of the oscillator to increase from a nominal 500 kHz at the left side of the screen to 4.2 MHz or greater at the right side. Markers may be provided at specific frequencies to help in determining where response problems occur.

Stairstep.   The *stairstep signal* may include a choice for modulation by the color subcarrier or no modulation. Closely related is a modulated ramp signal. These signals are used in determining the extent of differential phase and differential gain errors in the television system. They will be discussed in greater detail later in this section.

Flat Field.   Flat-field signals are useful in checking color purity of monitor displays. The image that appears on the screen should be a solid color without any blemished areas. If the shade of the color—often red is preferred—varies, the placement of convergence magnets or deflection coils is wrong.

The flat field may be combined with an average picture level (APL) test, which causes the luminance value of the screen to change, typically from 10 to 90 percent. As the picture level changes, chroma phase and image size should remain fixed.

Convergence Patterns.   Convergence patterns are used to maintain the color registration of the monitor CRT. In delta-gun, delta-dot, dot-mask picture tubes, a special yoke is necessary on the neck of the picture tube to correct beam-landing, deflection errors on the three electron beams forming the picture. Some monitors may require static (magnetic) and dynamic (electromagnetic) convergence adjustments that add sinusoidal and ramp corrections to cause the three beams to pass through the same pin hole in the dot mask. The pattern most often used for these adjustments is a horizontal-vertical crosshatch with very small dots located in the center of the squares created by the lines.

Composite VITS Signal.   Figure 6.33 is a combination of several signals that are inserted onto one line at the top of the television picture, just prior to the start of the active video picture. The illustration includes a window, sine² pulse, and five-step staircase. Even though the signal is found on only 1 line of the 525-line picture, it can be used as a test for system degradation during regular programming.

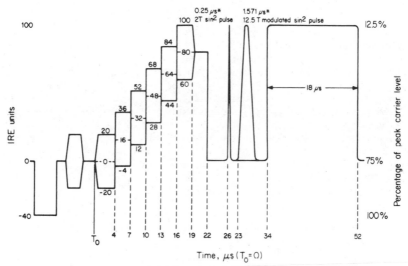

FIGURE 16.33   Composite vertical-interval test signal (VITS) inserted in field 1, line 18. The video level in IRE-scale units is on the left, and the radiated carrier-signal level is on the right. *Notes*: (1) Subcarrier of staircase is in phase with the color-synchronizing burst. (2) The nominal start of the active picture segment of the line (18) is T_0. *[Source: FCC. In K. Blair Benson (ed.),* Television Engineering Handbook, *McGraw-Hill, New York, 1986]*

A number of different test signals have been developed for advanced television systems. Most of the generators use digital techniques to count from the color-subcarrier frequency for pattern accuracy. Some provide a method for the user to program special-purpose tests.

16.7.3   Monitoring Instruments

Although there are now a number of computer-based, television signal monitors capable of measuring video, sync, chroma, and burst levels, as well as pulse widths and other timing factors, the best way to see what a signal is doing is to monitor it visually. Probably the most useful and flexible single test instrument for television is a precision, wideband, dual-channel *oscilloscope* with a dual time base and delayed sweeps. The unit must have a high enough vertical frequency response for the fast rise times of pulse and digital signals. The sweep speed range must go low enough to view fields and frames and high enough to resolve the color subcarrier (3.579545 MHz) and beyond. Once an operator is fully acquainted with the oscilloscope, it often serves to measure voltages, frequency, time intervals, and more. An instrument that provides on-screen alphanumeric readouts of the measured parameters is even more convenient as the universal test equipment.

The dual-channel capability allows two related signals to be monitored at the same time. By using the two channels to watch events at the input and output of a component, one can quickly determine if the component is operating correctly. Observation of the relationship between two signals is often the key to finding a circuit failure.

In some instances, the point of a waveform that we need to view is not conveniently visible because of the requirements of synchronizing the display to the signal. In such cases, a time base system with delayed sweep function can be used to sync the scope to a stable point on the waveform. The signal is routed through a precision delay line before it is displayed, making the point of interest visible.

Waveform and vector monitors are oscilloscopes especially adapted for the video environment and are typically used as monitoring equipment with products such as cameras and VTRs. The *waveform monitor*, like a traditional oscilloscope, operates in a voltage versus time mode. While an oscilloscope time base can be set over a wide range of intervals, the waveform monitor time base triggers automatically on sync pulses in the TV signal, producing line- and field-rate sweeps as well as multiple lines, multiple fields, and shorter time intervals. Filters, clamps, and other circuits process the video signal for specific monitoring needs.

Vectorscope. The *vectorscope* operates in an *X-Y* voltage versus voltage mode to display chrominance information. It decodes the signal in much the same way as a television receiver or a video monitor to extract color information and to display phase relationships.

The two instruments serve separate, distinct purposes as they sit side by side in a rack, monitoring the same signal. Some models combine the functions of both types of monitor in one chassis with a single CRT display. Others include a communications link between two separate instruments.

Beyond basic signal monitoring, these instruments provide a means to identify and analyze signal aberrations. If the signal is distorted, these instruments allow a technician to learn the extent of the problem and to locate the offending equipment. Because of the importance of these two instruments in the television system, more in-depth information on their use in measurements is provided in the following section.

16.8 BASIC WAVEFORM MEASUREMENTS

16.8.1 Test-Signal Parameters

Waveform monitors are used to evaluate the amplitude and timing of video signals and to show timing relationships between two or more signals. The familiar color-bar pattern is the only signal required for these basic tests. It is important to realize that all color bars are not created equal. Some generators offer a choice of 75 or 100 percent amplitude bars. Sync, burst, and setup amplitudes remain the same in the two color-bar signals, but the peak-to-peak amplitudes of high-frequency chrominance information and low-frequency luminance levels change. The saturation of color, a function of chrominance and luminance amplitudes, remains constant at 100 percent in both modes. The 75 percent bar signal has 75 percent amplitude with 100 percent saturation. In 100 percent bars, amplitude and saturation are both 100 percent.

Chrominance amplitudes in 100 percent bars exceed the maximum amplitude that should be transmitted. Therefore, 75 percent amplitude color bars, with no chrominance information exceeding 100 IRE, are the standard amplitude bars for NTSC. In the 75 percent mode, a choice of 100 IRE or 75 IRE white reference level may be offered. Figure 16.34 shows 75 percent amplitude bars with a 100 IRE white level. Either white level can be used to set levels, but operators must be aware of which signal has been selected. SMPTE bars have a white level of 75 IRE as well as a 100 IRE white flag.

To make precise evaluations of signals, the waveform monitor itself must be functioning properly. It should be periodically taken to a service center for calibration. The internal calibration signal for precise gain and sweep adjustments should be used regularly. The calibration signal is selected on the front panel, and, if necessary, the vertical gain calibration control is adjusted until the resulting square-wave signal is exactly 140 IRE units in amplitude. (Some instruments may require settings of 100 IRE units. Consult the manual.)

Many waveform monitors include a sweep calibrator. With the calibration signal enabled, you adjust the horizontal calibration control until the square wave crosses the graticule base line at the major division marks. Figure 16.35 shows a typical calibrator signal. The adjustments are generally done with a screwdriver. (Do not confuse the calibration adjustment controls with the variable-gain knobs.)

DC-Restoration. Before making measurements with the waveform monitor, it is wise to check the setting of the dc restorer. The dc restorer normally should be on, to stabilize the display from variations in average picture level (APL). If the

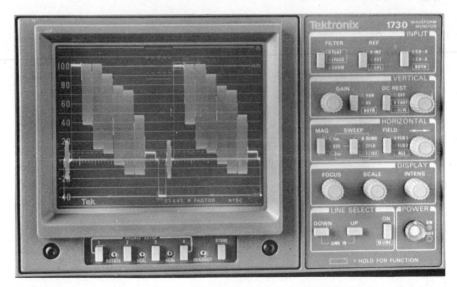

FIGURE 16.34 Waveform-monitor display of color-bar signal at line rate. *(Source: Courtesy of Tektronix, Inc.)*

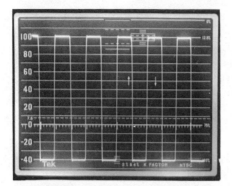

FIGURE 16.35 Waveform monitor display of typical calibration signal. *(Source: Courtesy of Tektronix, Inc.)*

instrument offers slow or fast dc restorer speeds, remember that the slow speed stabilizes the display at an average dc level but permits the observation of low-frequency abnormalities such as 60-Hz hum. The fast speed removes most of the hum.

The vertical response of waveform monitors depends upon filters that process the signal in order to display certain components. The flat response displays all components of the signal. A chroma response filter removes luminance and displays only chrominance. The low-pass filter removes chrominance, leaving only low-frequency luminance levels in the display. Some monitors include an IRE filter, designed to average out high-level, fine-detail peaks on a monochrome video signal, to aid the operator in setting brightness levels. The IRE response removes most, but not all, of the chrominance.

If the waveform monitor has a dual-filter mode, the operator can observe luminance levels and overall amplitudes at the same time. The instrument switches between the flat and low-pass filters. With a 2H sweep, the display on the left is low-pass filtered, while information to the right is unfiltered. The line-select mode is another useful feature for monitoring live signals.

16.8.2 Video Amplitudes

The overall amplitude of a video signal is an important parameter to monitor. Deviations from the nominal 1-V signal, expressed as IRE units or percentages, are referred to as an *insertion gain* or *loss*. Any equipment in the video path may change the gain. Insertion gain errors, whether they are too large or too small a signal amplitude, may eventually manifest themselves as signal distortions. It is, therefore, important for each piece of equipment to accurately transfer a 1-V signal at its input to a 1-V signal at the output. Insertion gain is measured at the output of every active device in the signal path.

To check overall amplitude, the displayed waveform is positioned vertically with the blanking (back porch) level overlaying the 0 IRE graticule line. The vertical scale of the graticule divides the standard 1-V video-signal display into 140 IRE units—100 IRE above the base line and 40 IRE below. The white level should extend just to the 100 or 75 IRE mark, depending on the color-bar source used. Sync extends to −40 IRE. Setup (the black part of the color bars) should be at 7.5 IRE. Setup and peak white are generally the only luminance levels checked with the color-bar signal. A linearity signal is used to check intermediate luminance levels, but linearity is usually not necessary for basic insertion gain evaluation.

Some chrominance checks are part of an insertion gain measurement procedure. The burst amplitude should be 40 IRE peak-to-peak, centered on the 0 IRE mark. Peaks of chrominance on the first and second 75 percent signal bars reach the 100 IRE line. A vectorscope is better suited to precisely evaluate chrominance amplitudes and will be discussed later in this section.

16.8.3 Sync-Pulse Levels and Timing

Duration and frequency of the sync pulses must be monitored. Horizontal-sync width should be watched closely. Most waveform monitors include 0.5- or 1-μs per division magnification (MAG) modes, which can be used to verify an H-sync width between 4.4 and 5.1 μs. The width is measured at the −4 IRE point. On waveform monitors with good MAG registration, sync appearing in the middle of the screen in the 2-line mode remains centered when the sweep is magnified.

It is a good idea to check the rise and fall times of sync and the widths of the front porch and entire blanking intervals. Examine burst, and verify that there are between 8 and 11 cycles of subcarrier.

Check the vertical intervals for correct format, and measure the timing of the equalizing pulses and vertical-sync pulses. The acceptable limits for these parameters are shown in the most recent FCC pulse-width specification, reproduced in Fig. 16.36.

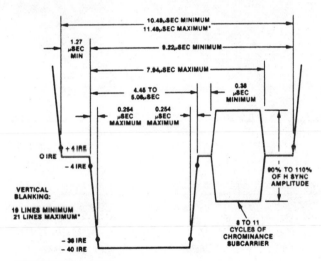

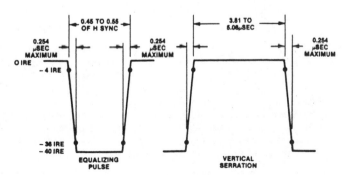

FIGURE 16.36 Sync-pulse widths and relative signal levels specified by the FCC as of March 14, 1985.

System Timing.   The sync pulses of all signals must be in the same phase at the point in the studio at which they are to be combined, for example, at the video switcher. If they are not properly timed, the viewer sees a horizontal shift when the program switches from one source to another. All signals may be locked to the house reference sync source, but timing errors arise as signals travel through different cable lengths. Therefore, the timing delay for each piece of equipment must be adjusted to bring all signals into coincidence at the switcher.

Table 16.2 shows pulse-timing specifications and tolerances for the various sync-signal components, and Table 16.3 shows frequency and intersync-signal relationships.

TABLE 16.2   Pulse Timing†

Pulse name	NTSC M ± tol, µs	PAL B ± tol, µs
Horizontal sync	4.7 ± 0.1	4.7 ± 0.1
Equalizing pulse	2.3 ± 0.1	2.35 ± 0.1
Vertical serration	4.7 ± 0.1	4.7 ± 0.1
Burst start	5.3	5.6 0 ± 0.1
Burst end	7.82	7.85
Front porch	1.5 + 0.7 − 0.1	1.55 ± 0.25
Back porch	9.2 + 1.1 − 0.1	10.5 ± 0.7

†All times measured with the leading edge of sync as the datum.
Source: K. Blair Benson (ed.), *Television Engineering Handbook*, McGraw-Hill, New York, 1986.

TABLE 16.3   Frequency and Intersync-Signal Relationships

Pulse name	NTSC system M	PAL system B
Subcarrier frequency	3,579,545 ± 10 Hz	4,433,618.75 Hz
Horizontal frequency	15,734.2657 Hz	15,625 Hz
Vertical frequency	59.94 Hz	50 Hz
Lines per frame	525	625
H to Sc relationship	$H = \dfrac{1 \times Sc}{455}$	$H = \dfrac{4 \times Sc}{1135 + (4/625)}$
Sc to H relationship	$Sc = 455 \times H/2$	$Sc = (283.75) \times H + 25$

Source: K. Blair Benson (ed.), *Television Engineering Handbook*, McGraw-Hill, New York, 1986.

Subcarrier-to-Horizontal Phasing.   The ScH phase relationship has been established for the NTSC system by the EIA TR4.4 subcommittee to revise RS170A. The EBU has also drafted a statement defining a preferred ScH phase for PAL recordings. The current definitions are given below:

1. *NTSC ScH phase definition:*

The extrapolation of the reference subcarrier burst should intersect the 50 percent point of the leading edge of sync at the zero crossing of subcarrier. Furthermore field one of the four field color sequence is identified in that on all even numbered lines, the extrapolated subcarrier will be observed rising at the leading edge of sync.

A tolerance at the time of writing has not been given for ScH phasing. Any error exceeding ±40° is considered as having an ambiguous ScH phase. Therefore, recordings should clearly have ScH phase errors of less than 40°. A timing tolerance of ±10° is a good long-term goal; however, ±30° is sufficient.

2. *PAL ScH phase definition:*

The CCIR standards for PAL specifically state that phase of subcarrier has no specific relationship to the horizontal-sync pulse. In 1979 the EBU drafted a statement first defining a preferred ScH phase relationship for PAL recordings. It states:

The subcarrier to line-sync (ScH) phase is defined as the phase of the Eu' component of color burst extrapolated to the half-amplitude point of the leading edge of the synchronizing pulse of Line 1 of Field 1. In the definition of Field 1 of the eight PAL fields, the EBU has adopted the value of zero degrees for this central value of the Sc-H phase.

The EBU gave a target tolerance of ±20° in the statement.

Sync-to-Subcarrier Phasing. As outlined above, in NTSC and PAL systems/four and eight fields, respectively, are required before the sequence for both sync and subcarrier repeats. This parameter of video is of no significance in television viewing; it is of great significance during video tape editing. A typical example of an NTSC edit where the field sequence recorded on the tape with both playback machines locked to the edit-room sync generator is illustrated below:

Edit

Video A   *Video B*

Field #   1 2 3 4 1 2 3 4 1 2 1 2 3 4 1 2 3 4 1 2 3 4 1 2 3

At the edit point during playback, since the tape machine must remain locked to the edit-room generator, and since the subcarrier must match the reference subcarrier in phase, the recording tape machine is forced to do one of two things: either shift the horizontal phase by one-half cycle of subcarrier timing (140 ns), which is the normal playback mode of operation, or unlock the servosystem and slide the tape one frame to realign the tape video to the same color frame as the edit room. This is operation of the machine in a color-frame playback mode. If the 140-ns shift occurs, and if the edit occurred at a time when there is a portion of the B video that is identical to the A video, a very noticeable shift will occur. Also, since the output processor of the tape machine is adding new sync and new blanking, sync appears to remain stationary, and the picture moves either right or left. The tape machine may be inserting a new horizontal blanking interval in the video, and when the picture shifts, a portion of active picture may be blanked, causing a widening of blanking and a narrowing of the active picture.

In cases where a program undergoes an extensive amount of editing and where color-frame editing is not followed, if the tape machine blanking width is set for the maximum standard (10.6 μs), the growth of blanking can be serious enough to exceed FCC tolerances. For example, after five edits there is a possibility of as much as a 1-μs increase in blanking and an equivalent horizontal picture shift.

The only solution to this problem is to ensure that all video on the tape has a consistent and stable sync-to-subcarrier phase relationship. For the sake of uniformity and interchangeability of tapes among machines, RS170A states the preferred sync-to-subcarrier relationship. If all video that can be recorded is timed and properly ScH-phased, and if the tape machine is referenced to a stable and properly phased source, the tape machine color-frame editor will provide contiguous color-frame timing.

A video processor ahead of the tape machine can alleviate color-frame editing problems. The processor at the output of a switcher, or at the input to a tape machine, can be operated with external sync and subcarrier to ensure consistent ScH-phased signals to the tape machines. If the processor is fed mistimed sources, it can still shift the picture and widen blanking before recording; however, the processor will prevent any picture shift during playback.

An operating practice is outlined in the EIA RS170A specifications which prevents blanking growth when improperly timed sources are passed through an externally referenced processor. The practice includes recommended blanking widths at various locations in a facility. It uses sources with narrow blanking and gates wide blanking in just before transmission to the consumer. Running narrow blanking internal to the facility ensures that there will be ample picture available to meet FCC specifications for over-the-air broadcast even if several non-color-frame edits, or passes of mistimed video through an externally referenced processor, occur.

PAL editing has the particular problem that color-frame editing of video can be achieved only when the two video sources are aligned at eight fields, or ⅛-s intervals. For exacting vocal editing and video effects, this restriction is prohibitive. If a color-frame edit would result in awkward audio track editing under such circumstances, a decision has to be made as to which is most objectionable, a verbal editing flaw or a shift in horizontal position. When this creative decision has been made, the repositioning of the horizontal sync should be accomplished at video between tape playback and recorded with an externally referenced processor. The new tape that is generated will again have a consistent ScH-phased sequential frame relationship.

ScH Phase Measurement. An early method of timing a zero-time point of a switcher input without regard to ScH phase was to align the subcarrier from each source to be precisely in time and then adjust horizontal so that they all match. To time with regard to ScH phase adds the constraint on horizontal timing that all sources must be exactly timed and also aligned to the proper ScH. This can be easily accomplished using an oscilloscope.

In measuring ScH phase, there are two common problems that can cause confusion. First is sync-to-subcarrier time-base error. Since subcarrier rarely jitters in phase, in effect this is a measure of the short-term jitter of the horizontal sync. Many devices in a system can inject instabilities in the sync, including any sync generator (whether a master, source, or slave), sync regenerative processors, and regenerative pulse distribution amplifiers. The second problem is normal transmission distortions. Group delay at low video frequencies is very common on long coaxial runs, and even though a modulated 20-T pulse may be acceptable, low-frequency group delay can alter the ScH phase considerably. Smear, a low-frequency response problem, can also cause video to run into the sync portion of the video waveform and cause picture-dependent sync time-base errors.

To compare signal-timing adjustments, connect a waveform monitor at the output of the switcher. Select the external reference with the house reference (probably blackburst) signal connected to the external reference input. First, sclect this reference signal on the switcher and display it on the waveform monitor in the 2H MAG sweep mode. Adjust the horizontal position control on the waveform monitor to place the 50 percent amplitude point of the leading edge of sync on one of the horizontal axis graticule marks (do not change this horizontal position knob setting until the system timing is finished). Now, switch through the video sources one by one, adjusting each source so that the leading edge of sync falls on the same graticule mark as the reference signal. Any one signal could be used as the reference with the other signals matched to it.

16.9 PROGRAM MONITORING

The procedures described thus far are performed with test signals before anything goes on-air. During editing or broadcasting, the waveform monitor is usually set in the 2H sweep mode. The operator should keep an eye on it in case signal levels need adjustment. If something goes drastically wrong with the picture, check the waveform monitor first. Are all signals present? Are the amplitudes correct? The waveform monitor will provide clues to the nature of the problem.

16.9.1 Basic Vectorscope Measurements

The vectorscope displays chrominance amplitudes, aids hue adjustments, and simplifies matching of burst phases of multiple signals. These functions require only the color-bar test signal.

To evaluate and adjust the chrominance in the TV signal, observe color bars on the vectorscope. The instrument should be in its calibrated gain position. Adjust the vectorscope phase control to place the burst vector at the 9 o'clock position and note the vector dot positions with respect to the six boxes marked on the graticule. If everything is well adjusted, each dot falls on the crosshairs of its corresponding box, as shown in Fig. 16.37.

Chrominance Amplitude. The chrominance amplitude of a video signal determines the intensity or brightness of color. If the amplitudes are correct, the color dots fall on the crosshairs in the corresponding graticule boxes. If vectors overshoot the boxes, chrominance amplitude is too large; if they undershoot, it is too small.

The boxes at each color location can be used to quantify the error. In the radial direction, the small boxes indicate a ±2.51 RE error from the standard amplitude. The large boxes indicate a ±20 percent error.

Hue Control. The hue control on television equipment changes the phase of the color burst with response to the rest of the signal. Even small hue errors are undesirable because very slight skin tone variations are noticeable. When the hue control is adjusted, the burst remains in the 9 o'clock position as the other dots rotate, hopefully falling into their boxes. The graticule boxes again show the extent of the error in a circumferential direction. The small boxes represent ±2.5° error, while the large boxes show ±10° error.

Burst Phase. A vectorscope can be used to ensure that signals from various sources have the same phase of burst. If burst signals vary in phase, color shifts

FIGURE 16.37   Vectorscope display of color-bar signal. *(Source: Courtesy of Tektronix, Inc.)*

occur during switching between sources. Adjusting the phase is accomplished in much the same way as sync timing with a waveform monitor. Connect the vectorscope at the output of the switcher, select external reference on the vectorscope, and switch up the reference signal. Use the vectorscope phase control to set the burst at 9 o'clock. (Do not move the phase control until burst phasing is completed.) Then, switch up each source, adjusting the phase control on the source until its burst vector is also at 9 o'clock.

Picture-Signal Monitoring.   Vectorscopes are primarily setup tools and are less useful than waveform monitors for watching live video. Burst can be distinguished, but the picture information is usually a blur. An exception to this is the use of a vectorscope to match tints in backgrounds. The functions discussed thus far are the common uses for vector and waveform monitors and should be routinely performed in most studios. There are additional uses, however, that require an instrument with a more advanced feature set, a more skilled operator or both. Some of these involve identifying problems, while others are precise measurements to quantify equipment malfunctions.

Other test signals, including a modulated staircase or multiburst, are required for many of these tests. It is important to take a good look at how these signals appear immediately after they come out of the generator. Knowing what the undistorted signal looks like simplifies the identification of distortions.

Line Selector.   Some waveform monitors and vectorscopes have line-select capability. They can display one or two lines out of the entire video frame of 525 lines. (In the normal display all the lines are overlaid on top of one another.) The principal use of the single-line feature is to monitor vertical interval test signals (VITSs). VITSs allow in-service testing of the transmission system. A full-field line selector drives a picture monitor output with an intensifying pulse. The pulse causes a single horizontal line on a picture monitor to be highlighted. This indicates where the line selector is within the frame to correlate the waveform monitor display with the picture.

16.9.2 Amplitude-Frequency Response: Indications of Distortion Problems

The television system should respond uniformly to signal components of different frequencies. The response is generally evaluated with a waveform monitor. Different signals are required to check the various parts of the frequency spectrum. If the signals are all faithfully reproduced on the waveform monitor screen after passing through the video system, it is safe to assume that there are no serious frequency-response problems.

1. At very low frequencies, look for externally introduced distortions, such as power-line hum or power-supply ripple and distortions resulting from inadequacies in the video equipment itself. Low-frequency distortions will probably appear on the TV screen as flickering or slowly varying brightness. Low-frequency interference can be seen on a waveform monitor when the dc restorer is set to the slow mode and a two-field sweep is selected. Sine-wave distortion from ac power-line hum is quite evident in Fig. 16.38.

2. A bouncing average picture-level (APL) signal can be used to detect distortion in the system itself. Vertical shifts in the blanking and sync levels indicate the possibility of low-frequency distortion.

3. Field-rate distortions appear as a difference in shading from the top to the bottom of the picture. A field-rate 60-Hz square wave is best for measuring field-rate distortion. Distortion occurs as a tilt in the waveform in two-field mode with the dc restorer off. If a 60-Hz square wave is not available, a window signal can also be used.

4. Line-rate distortions appear as streaking, shading, or poor picture stability. To detect errors of this type, look for tilt in the bar portion of a pulse-and-bar signal. The waveform monitor should be in the 1H or 2H mode with the fast dc restorer selected for the measurement.

5. The multiburst signal is used to test the high-frequency response of a system. The multiburst includes packets of discrete frequencies within the television passband, with the higher frequencies toward the right of each line. The highest frequency packet is at about 4.2 MHz, which is the upper frequency limit of the NTSC system. The next packet to the left is near the color-subcarrier frequency (3.58 MHz) for checking the chrominance transfer characteristics. Other packets are included at intervals down to 500 kHz. The most common distortion is high-frequency rolloff, seen on the waveform monitor as reduced-amplitude packets for the higher frequencies. (See Fig. 16.39.) The television picture exhibits loss of fine detail and color intensity when this type of distortion is present. High-frequency peaking, appearing on the waveform as higher amplitude packets at the higher frequencies, causes ghosting on the picture.

16.9.3 Nonlinear, Level-Dependent Distortion

Differential Gain. *Differential gain* (*dG*) distortion refers to a change in chrominance amplitude with changes in luminance level. The vividness of a col-

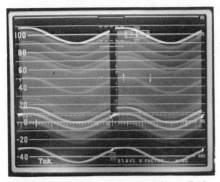

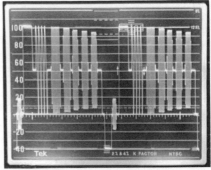

FIGURE 16.38 Waveform-monitor display showing additive 60-Hz hum. *(Source: Courtesy of Tektronix, Inc.)*

FIGURE 16.39 Waveform-monitor display of multiburst signal showing a fall-off in high-frequency response. *(Source: Courtesy of Tektronix, Inc.)*

ored object changes with changes in scene brightness. The modulated ramp or staircase is used to evaluate this impairment with the measurement taken on signals with different APL levels.

To measure differential gain with a vectorscope, set the vector to the 9 o'clock position and use the variable gain to bring it to the edge of the graticule circle. Differential-gain error appears as a lengthening of the vector dot in the radial direction. The dG scale at the left side of the graticule can be used to quantify the error. Figure 16.40 shows a dG error of 10 percent.

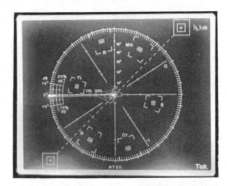

FIGURE 16.40 Vectorscope display of a 10 percent differential-gain error. *(Source: Courtesy of Tektronix, Inc.)*

Differential gain can be evaluated on a waveform monitor by using the chroma filter and examining the amplitude of the chrominance from a modulated staircase or ramp. With the waveform monitor in 1H sweep, use the variable gain to set the amplitude of the chrominance to 100 IRE. If the chrominance amplitude is not uniform across the line, there is dG error. With the gain normalized to 100 IRE, the error can be expressed as a percentage.

Finally, dG can be precisely evaluated with a swept display of demodulated video. This is similar to the single trace R-Y methods for differential phase. The B-Y signal is examined for tilt when the phase is set so that the B-Y signal is at its maximum amplitude. The tilt can be quantified against a vertical scale.

Differential Phase. A second nonlinear distortion is differential phase (d0), which occurs if a change in luminance level produces a change in the chrominance phase. If the distortion is severe, the hue of an object changes as its brightness changes. A modulated staircase or ramp is used to measure this. Either signal places chrominance of uniform amplitude and phase at different luminance levels. Figure 16.41 shows a 100 IRE modulated ramp. Because d0 may change with changes in APL, measurements at the center and at the two extremes of the APL range are necessary to fully evaluate system response.

To measure d0 with a vectorscope, increase the gain control until the vector dot is on the edge of the graticule circle. Use the phase shifter to set the vector to the 9 o'clock position. Phase error appears as circumferential elongation of the dot. The vectorscope graticule has a scale marked with degrees of d0 error. Figure 16.42 shows a d0 error of 5°.

More information can be obtained from a swept R-Y display, which is a feature of waveform monitor and vectorscope systems. If one or two lines of demodulated video from the vectorscope are displayed on a waveform monitor, differential phase appears as tilt across the line. In this mode, the phase control should be adjusted to place the demodulated video on the baseline, which is equivalent to phase to the 9 o'clock position of the vectorscope. Figure 16.43 shows a d0 error of 5° with the amount of tilt measured against a vertical scale.

This mode is useful in troubleshooting. By noting where along the line the tilt begins, it is possible to figure out at what dc level the problem starts to occur. In addition, field-rate sweeps enable the operator to look at d0 over the field.

A variation of the swept R-Y display may be available in some instruments for precise measurement of differential phase. Highly accurate measurements can be made with a vectorscope that has a precision phase shifter and a double-trace mode. This method involves nulling the lowest part of the waveform with the phase shifter, then using a separate calibrated phase con-

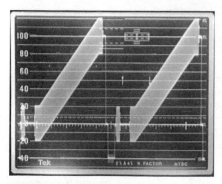

FIGURE 16.41 Waveform-monitor display of a ramp signal modulated with a high-frequency signal. *(Source: Courtesy of Tektronix, Inc.)*

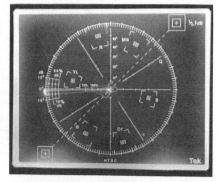

FIGURE 16.42 Vectorscope display of a 5 percent differential-phase error. *(Source: Courtesy of Tektronix, Inc.)*

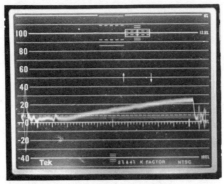

FIGURE 16.43 Waveform-monitor display showing, as a tilt on the vertical scale, a differential-phase error of 5°. *(Source: Courtesy of Tektronix, Inc.)*

trol to null the highest end of the waveform. A readout in tenths of a degree is possible.

K *Factor*. Lines and small boxes near the top of some waveform-monitor graticules are K-factor or quality-constant scales. The K-factor system is another means to quantify signal degradation. A series of subjective viewer reaction tests generated data that links relative degradation of the picture to a measured amount of all the distortions observed on a $sine^2$ pulse-bar signal. From the results of those tests, the K-factor established quality standards for TV signals from slight to severe picture degradation.

The K-factor markings line up horizontally with the pulse-bar portion of the FCC composite signal when the waveform monitor is set for a 1H sweep. The dashes and solid lines on the graticule represent ±2 and ±4 percent, K factor, respectively. A 5 percent K-factor distortion is said to be detectable by skilled observers, while 3 percent is not noticeable.

Incidental Phase Modulation (ICPM). Television receivers use a method known as *intercarrier sound* to reproduce audio information. Sound is recovered by beating the audio carrier against the video carrier, producing a 4.5-MHz IF signal, which is demodulated to produce the sound. From the interaction between audio and video portions of the signal, certain distortions in the video at the transmitter can produce audio buzz at the receiver. Distortions of this type are referred to as *incidental carrier phase modulation*, or *ICPM*.

Stereo audio for television has increased the importance of measuring this parameter at the transmitter because the buzz is more objectionable in stereo broadcasts. It is generally suggested that less than 3° of ICPM be present. ICPM is measured with a high-quality demodulator with a synchronous detector mode and an oscilloscope operated in a high-gain X-Y mode. Some waveform and vector monitors have such a mode as well. Video from the demodulator is fed to the Y input of the scope, and quadrature out is fed to the X input terminal. Low-pass filters make the display easier to resolve.

An unmodulated five-step staircase signal produces a polar display, which is shown in Fig. 16.44 on a special graticule developed for this purpose. Notice that

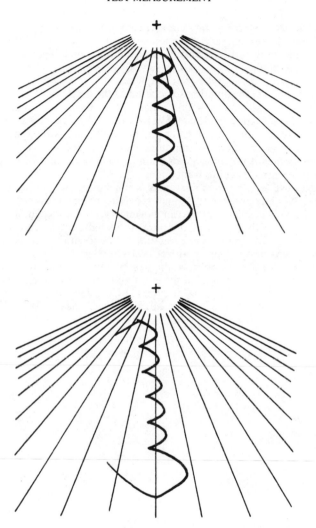

FIGURE 16.44 Waveform-monitor display, using the ICPM graticule of a five-level stairstep signal. (*a*) With no distortion. (*b*) With ICPM distortion of 5°.

the bumps all rest in a straight vertical line, if there is no ICPM in the system. Tilt indicates an error. The graticule is calibrated in degrees per radial division for differential-gain settings. Adjustment, but not measurement, can be performed without a graticule.

16.9.4 Spectrum Analyzers

While oscilloscope-type instruments display voltage levels referred to time, spectrum analyzers are special-purpose equipment used to indicate signal levels referred to frequency. The frequency components of the signal applied to the input of the analyzer instrument are detected and separated for display against a frequency-related time base. Spectrum analyzers are available in a variety of ranges with some models designed for use with video frequencies and others intended for use with transmitted-signal frequencies.

A spectrum analyzer is sometimes specified as the instrument to be used when setting modulation frequencies within videotape recorders. In such cases, one frequency is representative of peak whites. A second frequency is specified for blacks, while a third frequency is given for peak of sync. Although a frequency counter can be used to adjust the modulator of the video recorder, a spectrum analyzer with sufficient resolution can be used to monitor and display the complete range of the three specific frequencies.

The primary use for spectrum analyzers is the measurement and identification of RF signals. When connected to a small receiving antenna, the analyzer is used to measure the visual and aural carrier power levels. By expanding the sweep width of the display, the offset of the aural to visual carriers can be observed. The color subcarrier of the video information can be identified easily. By increasing the vertical sensitivity of the analyzer and adjusting the center frequency and sweep width, it is possible to observe the vestigial lower sideband of the RF signal and determine if any remnants of the color subcarrier are being transmitted by the lower sideband, and, if so, to what extent. FCC rules are quite specific on the maximum level (-48 dB below the visual carrier) that may exist.

CHAPTER 17

REFERENCE DATA AND TABLES

K. Blair Benson
Television Technology Consultant, Norwalk, Connecticut

17.1 REQUIREMENTS FOR STANDARDIZATION

17.1.1 Industrial Dependence on Standards

Broadly stated, industrial development is dependent to a large degree on the adoption of system and component standards to permit the exchange of products

---

Portions of this chapter were adapted from K. Blair Benson, ed., *Television Engineering Handbook*, McGraw-Hill, New York, 1986, and *Audio Engineering Handbook*, McGraw-Hill, New York, 1988.

and services. Thus, operational interchangeability is a prime consideration in the formulation of standards.

System Performance.   In a narrower view, a group of standards may be dedicated to a specific system to ensure that the system is designed, tested, and operated to meet, first, the user's requirements; second, the interface requirements of other interconnected systems; and third, the performance requirements of the overall system.

17.1.2   Scope of Standards Activities

Government Regulation.   In order to permit the end user, both professional and consumer, to acquire products that meet the performance requirements dictated by the intended application, test procedures and performance standards are necessary. These may be by government edict in the case of operations, such as broadcasting, or products for which either quality or safety for consumer protection may be a criterion.

Industry Regulation.   In addition to government regulation, there exists a large body of technical information consisting of recommended standards and practices that represent the consensus among those involved in a particular industry. These recommendations and industry-sponsored standards represent good engineering practice that ensures practical manufacturability within the current state of the art and, by consent of those skilled in the art, may serve as the basis for the generation of regulatory action by appropriate government agencies.

Development of International Standards Organizations.   The demand for a common language and for mutual agreement on technical items prompted the beginning of international work, which dates back to the establishment in 1875 of the International Bureau of Weights and Measures and in 1904 of the International Electrotechnical Commission (IEC). Much later, in 1926, the International Federation of National Standardizing Associations (ISA) was organized to cope with the growing international exchange of goods and services. Replaced in 1942 with the United Nations Standards Coordinating Committee (UNSCC), after 6 years it was reorganized as the International Standards Organization (ISO).

The ISO technical committees cover the fields of television and audio equipment and systems, broadcasting, recording, cinematography, and related consumer-equipment questions, in addition to maintaining liason with the standardizing committees of national professional organizations, for example, the American National Standards Institute (ANSI) in the United States.

17.1.3   Classification of Standards

Television radio standards may be classified in three broad categories as *spectrum allocation and radiation standards*, which relate primarily to the use of the available radio frequency spectrum; *signal generation and transmission standards*, which describe the specific signal parameters used to convey pictorial and aural information; and *equipment performance standards*, which relate to the apparatus used to generate the visual and aural signal components. These broad categories, in turn, may be divided into several operational and reference classifications: standards setting limits of performance, configuration standards, methods of measurement, procedures for operating equipment, and terminology. The most commonly recognized category is the *configuration standard*, which provides dimensional and operating limits for a device or a process. Such standards have provided interchangeable analog tape recordings, for example. Another category is the *method for measurement*, which is sometimes combined with a configuration standard and provides repeatable and reproduceable tests for a device or a process. Other categories include *standard definitions, procedures*, and *performance levels*. Such standards usually arise out of a need in the industry and are put into writing by several organizations not always in concert with one another.

17.2 STANDARDS-MAKING ORGANIZATIONS

17.2.1 Government Regulatory Standardization

In most countries, national governments sanction or organize standards-making organizations, usually for the purpose of issuing safety standards. Some countries, such as Germany and Japan, place most standards-making organizations in the governmental structure. Thus, Deutsche Industrie Normenausschus (DIN) and Japan Standards Association (JSA) standards have some mandatory aspects for nationals. Most governments also sponsor the voluntary standards organizations associated with the IEC.

All countries issue regulatory standards published by government agencies, although such regulations are often references to existing voluntary standards.

The United States is unique in that voluntary standards, including those for safety, are usually written by nongovernmental organizations.

Metrology. Metrology standards in the United States are written by the National Institute of Standards and Technology (NIST),[1] which maintains the physical materials on which the standards are based. Thus basic reference quantities for mass, length, time, charge, and some of their derivatives are maintained by the NIST. The NIST also has published derivative research reports in the field of acoustics, such as reports on the measurement of sound power, but has left the publication of standards based on the research to the voluntary standards organizations.

17.2.2 Quasi-governmental Standardization

Regional bodies that produce audio-related standards are found mainly in Europe. Typical are the European Broadcast Union (EBU) and the International Radio and Television Organization (OIRT). They are sponsored by trading groups such as the European Economic Community (Common Market) and COMECON (the Eastern European economic bloc).

The principal international regulatory bodies affecting audio are the Interna-

---

1. Formerly National Bureau of Standards (NBS). Renamed by Congress on August 23, 1988, in the Omnibus Trade and Competitiveness Act.

tional Radio Consultative Committee (CCIR), the Study Group on Television and Sound Transmission (CMTT), and the CCITT.

17.2.3 Voluntary Standardization

Commonly, when technologists refer to standards, they are thinking of voluntary consensus standards. Such standards carry the weight of regulation and of law only when adopted by regulatory agencies.

Voluntary consensus standards come about as a need is expressed by producers and users of products or processes. Such needs are usually expressed by trade organizations, producers, or users. In response, a standards-making organization will set up committees to discuss the need for the standard. If such a need can be shown, it will set up a writing or working group to attempt to develop the standard. The working group would consist of representatives of various producers and users. An effort is always made to get as wide a representation as possible so that no single interests are able to dominate the activities of the group. In the United States, in fact, such domination of the activities of the group can bring about serious legal problems.

The concept of consensus means that all parties involved in such working groups should agree substantially before a document is issued. This concept is carried through to the official standards-making bodies of the particular organizations. Such bodies are governed by careful rules and procedures that assure representative voting on the standards.

American National Standards Institute (ANSI).   In the United States, professional television and audio standards are published by ANSI under the direction of accredited standards committees administered by the Society of Motion Picture and Television Engineers (SMPTE) and the Audio Engineering Society (AES), discussed below. These committees review existing literature to fill the gaps in the American National Standards catalog, that is, to subject existing industry standards to the consensus process to see if they can be accepted by the representatives of producers, consumers, and general-interest groups that must vote on American National Standards. In addition, some existing American National Standards, as they come up for review, are examined in the light of current international standards.

Underwriters Laboratories (UL).   The particular case of voluntary consensus standards that take on the characteristics of regulation is found in the field of safety standards. In most countries a single government agency promulgates safety standards and also conducts investigations of the safety of products. In the United States this function is carried out by private organizations, in particular by the Underwriters Laboratories of the National Board of Fire Underwriters.

Underwriters Laboratories (UL) promulgates safety standards that are developed by committees consisting principally of manufacturers of the devices to be measured together with representatives of insurers. When a standard is of such general use that it may be desirable to obtain a wider consensus, UL will seek broader consensus by submitting it for approval through an ANSI procedure. In such cases other organizations, such as the S4 Committee, will become involved if the safety standard affects equipment or processes under their jurisdiction.

UL safety standards gain the force of law when they are adopted by government agencies such as cities promulgating fire codes, etc.

Electronic Industries Association (EIA). The largest trade organization that covers the television and audio fields is the EIA. It publishes a large catalog of standards, the most important to television and audio being those developed by its Parts Division (standards on components) and those by its Consumer Electronics Group (standards on consumer video and audio equipment).

The EIA is an accredited standards organization, so some of its standards are listed in a catalog published by ANSI.

Recording Industry Association of America (RIAA). A trade organization, the RIAA publishes configuration standards for phonograph records (see Chap. 14).

National Association of Broadcasters (NAB). The NAB publishes performance limits and configuration standards for transmitters and for video and audio equipment such as tape recorders used in broadcasting.

17.2.4 Professional Organization Standardization

Institute of Electrical and Electronic Engineers (IEEE). The IEEE issues standards both as a professional organization and as an accredited standards organization administrated by ANSI. The IEEE standards are concerned primarily with methods of measurements for video and audio.

Society of Motion Picture and Television Engineers (SMPTE). The first standards relating to motion-picture film and sound recording were published by SMPTE. SMPTE promulgates both standards that it publishes as an accredited standards organization coordinated by ANSI and also what it calls recommended practices, published as SMPTE documents.

Audio Engineering Society (AES). The primary international organization of users and producers of professional audio is the AES. The AES maintains a standards committee (AESSC), which supervises the work of several subcommittees and working groups covering the various fields of sound reinforcement, digital processing, etc. Because of the increasing interrelation of sound, film, and video, SMPTE and the AES maintain a close working relationship.

17.3 TELEVISION SERVICE

17.3.1 Signal Transmission Standards

Signal transmission standards describe the specific characteristics of the broadcast television signal radiated within the allocated spectrum. These standards may be summarized as follows:

1. Definitions of fundamental functions involved in producing the radiated signal format, to include relative carrier and subcarrier frequencies and tolerances as well as modulation-sideband spectrum and radio-frequency envelope parameters

2. Transmission standards describing the salient base-band signal values relating

visual psychophysical properties of luminance and chrominance values described in either the time or frequency domains

3. Synchronization and timing signal parameters, both absolute and relative

4. Specific test and monitoring signals and facilities

5. Relevant mathematical relationships describing the individual modulation signal components

The details of these signal transmission standards are contained, for U.S. monochrome and television broadcast services, in the FCC Rules and Regulations, Part 73, Subpart E, Secs. 73.676, 73.681, and 73.682.

Technical Standards Definitions.   Section 73.681 of the FCC Regulations provides basic definitions pertaining to transmission standards. Some of the more pertinent ones are listed below:

1. *Amplitude modulation (AM):* A system of modulation in which the envelope of the transmitted wave contains a component similar to the waveform of the base-band signal to be transmitted.

2. *Antenna height above average terrain:* The average of the antenna heights above the terrain from about 2 to 10 mi (3.2 to 16 km) from the antenna as determined for eight radial directions spaced at 45° intervals of azimuth is considered the antenna height. Where circular or elliptical polarization is employed, the average antenna height is based on the height of the radiation center of the antenna that produces the horizontal component of radiation.

3. *Antenna power gain:* The square of the ratio of the rms free space field intensity produced at 1 mi (1.6 km) in the horizontal plane, expressed in millivolts per meter for 1-kW antenna input power to 137.6 mV/m. The ratio is expressed in decibels (dB).

4. *Aspect ratio:* The ratio of picture width to picture height as transmitted. The standard is 4:3 for 525-line NTSC and 625-line PAL and SECAM systems.

5. *Chrominance:* The colorimetric difference between any color and a reference color of equal luminance, the reference color having a specific chromaticity.

6. *Effective radiated power:* The product of the antenna input power and the antenna power gain expressed in kilowatts and in decibels above 1 kW (dBk). The licensed effective radiated power is based on the average antenna power gain for each direction in the horizontal plane. Where circular or elliptical polarization is employed, the effective radiated power is applied separately to the vertical and horizontal components. For assignment purposes, only the effective radiated power for horizontal polarization is considered.

7. *Field:* A scan of the picture area once in a predetermined pattern.

8. *Frame:* Scanning all the picture area once. In the line-interlaced scanning pattern of 2/1, a frame consists of two interlaced fields.

9. *Frequency modulation (FM):* A system of modulation where the instantaneous radio frequency varies in proportion to the instantaneous amplitude of the modulating signal, and the instantaneous radio frequency is independent of the frequency of the modulating signal.

10. *Interlaced scanning:* A scanning pattern where successively scanned lines are spaced an integral number of line widths, and in which the adjacent lines are scanned during successive periods of the field rate.

11. *IRE standard scale:* A linear scale for measuring, in arbitrary IRE units, the relative amplitudes of the various components of a television signal as shown in Table 17.1.

TABLE 17.1 IRE Standard Scale

Level	IRE units	Modulation, %
Zero carrier	120	0
Reference white	100	12.5
Blanking	0	75
Sync peaks (max. carrier)	−40	100

Source: K. Blair Benson, ed., *Television Engineering Handbook*, McGraw-Hill, New York, 1986.

Figure 17.1 shows the IRE level units and modulation percentage for a standard SMPTE color-bar signal (SMPTE ECR 1-1978). In practice, as specified in EIA Standard RS-189-A, the gray, red, green, blue, yellow, cyan, and magenta bars are composed of 75 percent level red, green, and blue signals. The peak-white bar at the bottom of the field is composed of 100 percent red, green, and blue signals. This results in a transmitter modulation level no lower than 12.5 percent for the peak-white and yellow bars, thus avoiding differential gain and phase distortion that is inherent at near-zero carrier modulation.

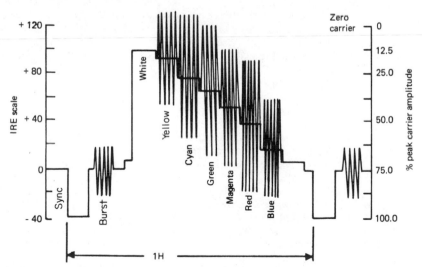

FIGURE 17.1   Standard SMPTE color-bar signal. *(Source: K. Blair Benson, ed.,* Television Engineering Handbook, *McGraw-Hill, New York, 1986)*

12. *Luminance:* Luminance flux emitted, reflected, or transmitted per unit solid angle per unit projected area of the source (the relative light intensity of a point in the scene).

13. *Negative transmission:* Modulation of the radio-frequency visual carrier in such a way as to cause an increase in the transmitted power with a decrease in light intensity.

14. *Polarization:* The direction of the electric field vector as radiated from the antenna.

15. *Scanning:* The process of analyzing successively, according to a predetermined method or pattern, the light values of the picture elements constituting the picture area.

16. *Vestigial sideband transmission:* A system of transmission wherein one of the modulation sidebands is partially attenuated at the transmitter and radiated only in part.

17.3.2 Channel Spectrum and Carrier Designations

The basic features of the technical transmission standards for television broadcasting in the United States as specified by the FCC are summarized and listed below:

1. The width of the television broadcast channel is 6 MHz.

2. The visual carrier frequency shall be nominally 1.25 MHz above the lower boundary of the channel with a tolerance of ± 1000 Hz.

3. The aural carrier frequency shall be 4.5 MHz ($\pm$ 1000 Hz) higher than the visual carrier frequency with a tolerance of ± 1000 Hz for the intercarrier spacing between the visual and aural carriers.

4. The visual amplitude characteristics of the sideband spectrum shall be in accordance with the vestigial sideband response as shown in Fig. 17.2 (Sec. 73.699, Fig. 5, of the FCC Rules and Regulations). However, the amplitude characteristics need not be totally vestigial in nature for stations operating on channels 14 to 83 and having a maximum peak visual power output of 1 kW or less.

17.3.3 Aural Carrier Modulation Standards

The FCC specifies frequency modulation (F3) of the aural carrier associated with the visual carrier for television broadcasting service. Some other systems in use by other administrations in the world employ amplitude modulation for audio, but most employ FM techniques.

FM has the advantage that (1) the service area for a given signal-to-noise ratio may be provided with a relatively low transmitted carrier power, and (2) when combined with a visual AM signal, it permits the use of *intercarrier* operation of the receiver, which may have economic advantages.

The maximum deviation for peak modulation is chosen as a compromise between signal-to-noise and bandwidth/stability constraints at the receiver. The U.S. system employs ± 25 kHz, while most of the systems in other parts of the world use ± 50 kHz. Both these values are less than the typical value of ± 75 kHz

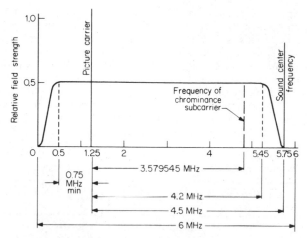

FIGURE 17.2 Idealized picture-transmission amplitude characteristic for all VHF and UHF channels with power outputs greater than 1 kW. *(Source: FCC. In* Television Engineering Handbook, *K. Blair Benson, ed., McGraw-Hill, New York, 1986)*

for FM broadcast service but are adequate to provide a service area approximately matching the service area of the associated visual signals.

As is customary in all FM systems involving speech and music, the energy relationships between low- and high-frequency sound energy and signal-to-noise characteristics make it appropriate to preemphasize the high-frequency components prior to modulation and provide a complementary deemphasis at the audio detector of the receiver. The preemphasis characteristic is stated in terms of impedance versus frequency that results from a combination of inductive reactance L and pure resistance R expressed as an L/R time constant in microseconds. The deemphasis is an RC time involving capacitive reactance and a pure resistance. The U.S. system specifies a time constant of 75 μs, while many other systems use 50 μs.

Multiplexing of the aural carrier may be permitted for purposes of transmitting telemetry and alerting signals from the transmitter site to the television station control point when used for remote control. The conditions for this special case are detailed in FCC Sec. 73.682, Subparagraph 23, and generally are aimed at providing the desired service without causing observable degradation of the associated visual or aural signals.

17.3.4 Channel Specifications and Carrier Frequencies in the United States

Broadcast Channels. The available radio-frequency spectrum space is assigned to two broad classes of television services that include standard broadcast and the related point-to-point transmissions. The point-to-point services include auxiliary functions such as network relays, studio-to-transmitter links, and both portable and mobile remote-pickup facilities. Certain ancillary activities such as broadcast translators and signal-booster operations, as well as instructional television, subscription television, satellite relay, and experimental television, are also included in the specific frequency assignments and channel specifications.[2]

The broadcast service is divided into two frequency ranges referred to as *very high frequency* (VHF) and the *ultrahigh frequency* (UHF) regions. Broadly speaking, the VHF region lies in a frequency range from about 40 MHz to slightly above 200 MHz, while the UHF region extends from about 470 MHz up to almost 1.0 GHz. Most of the point-to-point services are allocated in the 2- to 14-GHz region. The VHF range includes both frequency-modulation broadcasting and television broadcasting, while the UHF range contains only television channels, although there are gaps in both ranges that are assigned to other services. (See Table 17.2.)

TABLE 17.2 Designation of Television Channels in the United States

Channel no.	MHz	Channel no.	MHz	Channel no.	MHz	Channel no.	MHz
2	54–60	22	518–524	42	638–644	63	764–770
3	60–66	23	524–530	43	644–650	64	770–776
4	66–72	24	530–536	44	650–656	65	776–782
5‡	76–82	25	536–542	45	656–662	66	782–788
6†	82–88	26	542–548	46	662–668	67	788–794
7	174–180	27	548–554	47	668–674	68	794–800
8	180–186	28	554–560	48	674–680	69	800–806
9	186–192	29	560–566	49	680–686	70¶	806–812
10	192–198	20	566–572	50	686–692	71¶	812–818
11	198–204	31	572–578	51	692–696	72¶	818–824
12	204–210	32	578–584	52	698–704	73¶	824–830
13	210–216	33	584–590	53	704–710	74¶	830–836
14‡	470–476	34	590–596	54	710–176	75¶	836–842
15‡	476–482	35	596–602	55	716–722	76¶	842–848
16‡	482–488	36	602–608	56	722–728	77¶	848–854
17‡	488–494	37§	608–614	57	728–734	78¶	854–860
18‡	494–500	38	614–620	58	734–740	79¶	860–866
19‡	500–506	39	620–626	59	740–746	80¶	866–872
20‡	506–512	40	626–632	60	746–752	81¶	872–878
21	519–518	41	632–638	62	758–764	83¶	884–890

†In Alaska and Hawaii, channels 5 and 6 (76–82 MHz and 82–88 MHz) are allocated for nonbroadcast services.

‡Channels 14 through 20, previously assigned in 13 specific locations to land-mobile service, may not be reassigned to television broadcast (FCC Part 2, July 1981).

§Channel 37 (608–614 MHz) is reserved for radio astronomy.

¶Channels 70 through 83 are allocated for land-mobile service (FCC Part 2, July 1981), (FCC Parts 15, 73, 74 August 1982).

Source: K. Blair Benson, ed., *Television Engineering Handbook*, McGraw-Hill, New York, 1986.

Cable Channel. Frequency allocations for cable systems are shown in Table 17.3. The channel-numbering plan shown here has recently been developed by the EIA/NCTA Joint Committee on Receiver Compatibility and has been published by both organizations as an Engineering Standard.

---

2. FCC Rules and Regulations, Part 15, July 1981; Part 73, March 1980; Part 74, March 1980; Part 76, February 1981.

TABLE 17.3 Designation of Cable Television Channels in the United States

Pix carrier frequency, MHz			Band name	Channel designation	Historical reference‡
Std.	HRC	IRC			
55.25	54.00	55.25	↑	2	
61.25	60.00	61.25	⋮	3	
67.25	66.00	67.25		4	
†	72.00	73.25	Low VHF	1	4+, A-8
77.25	78.00	79.25	↓	5	A-7 (HRC, IRC)
83.25	84.00	85.25		6	A-6 (HRC, IRC)
91.25	90.00	91.25	↑	95	A-5
97.25	96.00	97.25		96	A-4
103.25	102.00	103.25		97	A-3
109.25	108.00	109.25		98	A-2
115.25	114.00	115.25		99	A-1
121.25	120.00	121.25	Midband	14	A
⋮	⋮	⋮		⋮	⋮
169.25	168.00	169.25	↓	22	I
175.25	174.00	175.25	↑	7	
⋮	⋮	⋮	High VHF	⋮	⋮
211.25	210.00	211.25	↓	13	
217.25	216.00	217.25	↑	23	J
⋮	⋮	⋮	Superband	⋮	⋮
295.25	294.00	295.25	↓	36	W
301.25	300.00	301.25	↑	37	AA
⋮	⋮	⋮		⋮	⋮
325.25	324.00	325.25		41	EE
⋮	⋮	⋮	Hyperband	⋮	⋮
397.25	396.00	397.25		53	QQ
⋮	⋮	⋮		⋮	⋮
463.25	462.00	463.25	↓	64	
469.25	468.00	469.25	↑	65	
⋮	⋮	⋮	Ultraband	⋮	⋮
493.25	492.00	493.25		69	
⋮	⋮	⋮		⋮	⋮
547.25	546.00	547.25		78	
⋮	⋮	⋮		⋮	⋮
†	⋮	⋮		⋮	⋮

Source: K. Blair Benson, ed., *Television Engineering Handbook*, McGraw-Hill, New York, 1986.

Standard frequencies refer to cable systems that transmit on the standard off-air frequencies for channels 2 to 6 and 7 to 13. Supplemental channels are in 6-MHz increments, counting down from channel 7 (175.25 MHz) to 91.25 MHz and upward from channel 13 (211.25 MHz).

The *harmonic-related carriers* (HRC) channeling plan refers to cable systems having a coherent head end and visual carriers located at multiples of 6 MHz starting at 54 MHz. All visual carriers in the system are phase-locked to a 6-MHz master oscillator. This ensures that all the carriers are harmonically related to 6 MHz and no matter what shift occurs to the master oscillator, all carriers maintain the same relative frequency separation. The second- and third-order intermodulation products resulting from any two carriers, therefore, always fall exactly on the visual carrier frequencies, and their undesirable effects on television pictures will be reduced or eliminated. When compared with the broadcast or standard plan, HRC channels are frequency displaced by -1.25 MHz on all standard and cable supplementary channels (midband, etc.) except channels 5 and 6, where the displacement is +0.75 MHz.

For the *interval-related carriers* (IRC) system, the cable channels are on frequencies starting at 55.25 MHz with increments of 6 MHz (6 N + 1.25 MHz). These channels are the same as standard frequencies except for channels between 67.25 and 91.25 MHz.

Specifications for Cable-Compatible Receivers.   EIA TV Systems Bulletin No. 2, *Cable Compatible Television Receiver and Cable System Technical Standards* (March 1975), specified desirable and standard characteristics for a cable-compatible receiver. Many advances in the cable industry since that time have made the EIA document obsolete. These issues have been updated by the EIA/NCTA Joint Engineering Committee.

By Canadian regulation, two types of receivers exist: the standard receiver and the cable compatible receiver, the latter having additional requirements as listed below.

Minimum selection of 18 preset channels, with all others in the VHF, mid-, and superband being accessible by mechanical means.

Fine tuning of AFC capability of ±0.55 MHz on VHF channels and -1.31 MHz on mid- and superband channels.

Noise figure not to exceed 10 dB, unless double conversion is utilized; noise figure can then be up to 13 dB.

No noticeable co-channel interference when receiver is in 100 mV/m radiated field from a broadcast station with receiver 75-Ω cable input adjusted to 1 mV.

No receiver overload for signal input up to 5 mV.

17.3.5   Grades of Service

In the process of authorizing the operation of television stations, two field-strength contours are considered. Specified as *grade A* and *grade B*, these indicate the approximate extent of coverage over average terrain in the absence of interference from other television stations. On the other hand, the true coverage may vary greatly because the actual terrain over a specific path may be considerably different from the average terrain on which field-strength charts are based. The required field strengths in decibels above 1 μV/m (dBμ) for the grade A and grade B service as well as the local community minimum values are shown in Table 17.4.

TABLE 17.4 Grades of Television Service, FCC Regulations†

Channel designations	Frequency band, Hz	Grade A service	Grade B service	Local community minimum
2–6 (low VHF)	54–88	68 dBμ; 2510 μV/m	47 dBμ; 224 μV/m	74 dBμ; 5010 μV/m
7–13 (high VHF)	174–216	71 dBμ; 3550 μV/m	56 dBμ; 631 μV/m	77 dBμ; 7080 μV/m
14–83 (UHF)	470–890	74 dBμ; 5010 μV/m	64 dBμ; 1580 μV/m	80 dBμ; 10,000 μV/m

†dBμ = decibels above 1 μV/m.
Source: K. Blair Benson, ed., *Television Engineering Handbook*, McGraw-Hill, New York, 1986.

17.3.6 Test and Monitoring Signal Standards

The FCC Rules and Regulations specify certain signals that may be used for modulating the transmitter for test and monitoring purposes. These signals may coexist with the broadcast of normal picture information and are permitted to be inserted in the interval of vertical blanking beginning with line 17 and continuing through line 21 of each field. (See Table 17.5.)

TABLE 17.5 Vertical Interval Test Signals and Line Number Allocations

	Line number	Signal format
(1)	17 (Field 1)	Multiburst test signal
(2)	17 (Field 2)	Color-bar test signal
(3)	18 (Field 2)	Composite radiated signal (*a*) Modulated stairstep, (*b*) 2T pulse,† (*c*) 12.5T pulse, (*d*) White bar
(4)	19 (each field)	Devoted exclusively to the vertical-interval reference (VIR) signal (FCC Sec. 73.699, Fig. 16)
(5)	21 (Field 1), one-half of 21 (Field 2)	Program-related data signal that is related to the aural channel information (FCC Sec. 73.699, Fig. 17a)‡
(6)	21 (every eighth frame)	Pulse for adaptive multipath equilizer decoder (FCC Sec. 73.699, Fig. 17b)‡
(7)	21 (Field 2)	A decoder test signal representing alphanumeric characters unrelated to program material; a framing code may be inserted during the first half of line 21 (FCC Sec. 73.699, Fig. 17c)‡

†T pulse = sin² pulse with a half-amplitude duration of 125 ns (½f_c).
‡Items (5) to (7) may be deleted and replaced by new information relating to *Teletext*.
Source: K. Blair Benson, ed., *Television Engineering Handbook*, McGraw-Hill, New York, 1986.

Test signals may include signals used to supply reference modulation levels so that light intensity variations will be faithfully transmitted, and certain signals designed to check the performance of the overall transmission system or its components.

The modulation by these signals shall be confined between reference white and blanking level except in certain cases relating to chrominance subcarrier excursions. In no case shall the signals extend beyond the peak of sync, or to zero carrier level. The use of these test and cue signals shall not degrade or impair the normal program material being transmitted nor produce emission outside the normal assigned channel specifications, and no test signals are permitted to extend into horizontal blanking period.

17.3.7 Worldwide Television Transmission Standards

Color-television broadcasting service, compatible with existing 525-line, 60-field monochrome transmissions, was approved on December 17, 1953, by the FCC, with regular broadservice authorized to start on January 23, 1954. The system adopted was based on the recommendations of the National Television Systems Committee (NTSC), an industry-supported group composed of broadcasters, manufacturers, and research laboratories.

In Europe and other portions of the world, color television was delayed several years until agreement was reached on a system compatible with their 625-line, 50-field monochrome standard. Rather than following the lead of America in adopting a system for simultaneous transmission of all the color information on a subcarrier interleaved with the monochrome signal, in the interest of reducing the possibility of color hue errors inherent in transmission of the simultaneous NTSC color subcarrier, Europe gave prime consideration to an alternate-line sequential transmission of the two color signals. A one-line delay is used in receivers to store each color-signal component for a repeat on the next line, concurrent with the other color component. It is to be noted that this approach halves the vertical resolution of the color signal.

SECAM.   One such system proposed by Henri de France, of CSF in Paris, was SECAM (*sequential couleur avec memoire*, or sequential color with memory). SECAM was officially adopted by France in 1967 and shortly thereafter by the USSR.

PAL.   The one-line-delay technique led to the development of the phase-alternation line (PAL) system by Walter Bruch of Telefunken in Germany. PAL reverses the phase of one of the color subcarrier components, line by line. Thus, any color-signal errors resulting from distortion in transmission are averaged.

The 625-line PAL system was selected by most of Europe in 1967 and later in many other countries throughout the world. However, although the basic PAL standards are common, differences exist among various countries in video bandwidths and RF carrier specifications.

PAL-M.   In addition, several countries have chosen PAL-M, a 525-line, 60-field version of PAL. M is the CCIR designation for 525-line systems (N identifies 625-line systems). Table 17.6 lists the video and synchronizing-signal characteristics of the systems agreed upon by the CCIR. Table 17.7 lists the systems used throughout the world.

TABLE 17.6 Basic Characteristics of Video and Synchronizing Signals

Characteristic	CCIR system identification										
	A	M	N	C	B, G	H	I	D, K	K1	L	E
Number of lines per frame	405	525	625	625	625	625	625	625	625	625	819
Number of fields per second	50	60 (59.94)	50	50	50	50	50	50	50	50	50
Line frequency f_H, Hz, and tolerances	10,125	15,750 15,734 (±0.0003%)	15,625 ±0.15%	15,625 ±0.02%	15,625 ±0.02% (±0.0001%)	15,625 ±0.02% (±0.0001%)	15,625 (±0.0001%)	15,625 ±0.02% (±0.0001%)	15,625 ±0.02% (±0.0001)	15,625 ±0.02% (±0.0001)	20,475
Interlace ratio	2/1	2/1	2/1	2/1	2/1	2/1	2/1	2/1	2/1	2/1	2/1
Aspect ratio	4/3	4/3	4/3	4/3	4/3	4/3	4/3	4/3	4/3	4/3	4/3
Blanking level, IRE units	0	0	0	0	0	0	0	0	0	0	
Peak-white level	100	100	100	100	100	100	100	100	100	100	100
Sync-pulse level	−43	−40	−40	−43	−43	−43	−43	−43	−43	−43	−43
Picture-black level to blanking level (setup)	0	7.5 ±2.5	7.5 ±2.5	0	0	0	0	0–7	0 color 0–7 mono	0 color 0–7 mono	0–5
Nominal video bandwidth, MHz	3	4.2	4.2	5	5	5	5.5	6	6	6	10
Assumed display gamma	2.8	2.2	2.2	2.8	2.8	2.8	2.8	2.8	2.8	2.8	2.8

†*Notes:* (1) Systems A, C, and E are not recommended by CCIR for adoption by countries setting up a new television service. (2) Values of horizontal line rate tolerances in parentheses are for color television. (3) In the systems using an assumed display gamma of 2.8, an overall system gamma of 1.2 is assumed. All other systems assumed an overall transfer function of unity.

Source: K. Blair Benson, ed., *Television Engineering Handbook*, McGraw-Hill, New York, 1986.

TABLE 17.7   Television Systems† Used throughout the World‡

Country	Band I/III (VHF)	Band IV/V (UHF)	Country	Band I/III (VHF)	Band IV/V (UHF)
Algeria	B/PAL	G§, H§/PAL	Jamaica	M	
Angola	I/PAL§		Japan	M/NTSC	M/NTSC
Argentina	N/PAL	N/PAL	Jordan	B/PAL	
Australia	B/PAL		Kenya	B/PAL	
Austria	B/PAL	G/PAL	Korea (South)	M/NTSC	M/NTSC
Bahrain	B/PAL		Kuwait	B/PAL	
Bangladesh	B/PAL		Lebanon	B/SECAM	
Barbados	N/NTSC		Liberia	B/PAL	
Belgium	B/PAL	H/PAL	Libya	B/SECAM	
Bermuda	M/NTSC		Luxembourg	C/SECAM	G/PAL
Bolivia	N/NTSC				L/SECAM
Brazil	M/PAL	M/PAL			
Brunei	B/PAL		Madeira	B/PAL	
Bulgaria	D/SECAM	K/SECAM	Malagasy	K1/	
Canada	M/NTSC	M/NTSC		SECAM	
Canary Islands	B/PAL		Malaysia (Fed. of)	B/PAL	
Chile	M/NTSC	M/NTSC			
China (People's Rep.)	D/PAL	K/PAL	Malta	B/PAL	H/PAL
			Martinique	K1/	
Colombia	M/NTSC	M/NTSC		SECAM	
Costa Rica	M/NTSC	M/NTSC	Mauritius	B/SECAM	
Cuba	M	M	Mexico	M/NTSC	M/NTSC
Cyprus	B/PAL	G,H§/PAL	Monaco	E/SECAM	G/PAL
Czechoslovakia	D/SECAM	K/SECAM			L/SECAM
Denmark	B/PAL		Morocco	B/SECAM	
Djibouti (Rep.)	K1/		Netherlands	B/PAL	G/PAL
	SECAM		Netherlands Antilles	M/NTSC	M/NTSC
Dominican Republic	M/NTSC	M/NTSC			
			New Caledonia	K1/	
Ecuador	M/NTSC	M/NTSC		SECAM	
Egypt (Arab Rep.)	B/SECAM	G§, H§/ SECAM	New Zealand	B/PAL	
			Nicaragua	M/NTSC	M/NTSC
El Salvador	M/NTSC	M/NTSC	Niger	K1/	
				SECAM	
Equatorial Guinea	B/PAL§		Nigeria	B/PAL	
			Norway	B/PAL	G/PAL
Ethiopia	B/PAL§		Oman (Sultanate of)	B/PAL	G/PAL
Finland	B/PAL	G/PAL			
France	E	L/SECAM	Pakistan	B/PAL	
Gabon	K1/SECAM				
Germany (East)	B/SECAM	G/SECAM	Panama	M/NTSC	M/NTSC
Germany (West)	B/PAL	G/PAL	Paraguay	N/PAL	
Ghana	B/PAL§		Peru	M/NTSC	M/NTSC
Gibraltar	B/PAL		Philippines	M/NTSC	M/NTSC
Greece	B/SECAM		Poland	D/SECAM	K/SECAM
Guadeloupe	K1/SECAM		Portugal	B/PAL	G/PAL
Guatemala	M/NTSC	M/NTSC	Qatar	B/PAL	
Haiti	M/NTSC	M/NTSC	Reunion	K1/SECAM	
Honduras	M/NTSC	M/NTSC	Sabah/Sarawak	B/PAL	
Hong Kong	B/PAL	I/PAL	Saudi Arabia	B/SECAM	G/SECAM
Hungary	D/SECAM	K/SECAM	Sierra Leone	B/PAL	
Iceland	B/PAL		Singapore	B/PAL	
India	B		South Africa	I/PAL	I/PAL
Indonesia	B/PAL				
Iran	B/SECAM		Spain	B/PAL	G/PAL
Iraq	B/SECAM				
Ireland	A	I/PAL	St. Kitts	M/NTSC	M/NTSC
Israel	B/PAL	G/PAL	Sudan	B/PAL§	
Italy	B/PAL	G/PAL			
Ivory Coast	K1/		Surinam	M/NTSC	M/NTSC
	SECAM		Swaziland	B/PAL	G/SECAM

TABLE 17.7 *(Continued)*

Country	Band I/III (VHF)	Band IV/V (UHF)	Country	Band I/III (VHF)	Band IV/V (UHF)
Sweden	B/PAL	G/PAL	United Arab Emirates	B/PAL	G/PAL
Switzerland	B/PAL	G/PAL	United Kingdom	A	I/PAL
Syrian Arab Rep.	B/SECAM		United States	M/NTSC	M/NTSC
Taiwan	M/NTSC	M/NTSC	Uruguay	N/PAL	N/PAL
Tanzania	B/PAL	B/PAL	USSR	D/SECAM	K/SECAM
(Zanzibar)			Venezuela	M/NTSC	M/NTSC
Thailand	B/PAL	M/PAL			
Togo	K1/SECAM		Yemen Arab Rep.	B/PAL	
Trinidad and	M/NTSC	M/NTSC	Yugoslavia	B/PAL	H/PAL
Tobago			Zaire	K1/SECAM	
Tunisia	B/SECAM		Zambia	B/PAL	
Uganda	B/PAL		Zimbabwe	B	

‡CCIR letter designations: B, D, I M, N, etc.
†As of February 1982.
§Planned.
Source: K. Blair Benson, ed., *Television Engineering Handbook*, McGraw-Hill, New York, 1986.

17.4 AUDIO SYSTEMS AND STANDARDS

17.4.1 Sound Levels

The ear responds to a very wide range of sound-pressure amplitudes. From the smallest audible sound to a level near the threshold of pain is a range of 1 million in sound pressure 10^6. Dealing with such large numbers is impractical; thus, a logarithmic scale is used. This is based on the *bel*, which represents a ratio of 10:1 in sound intensity or sound power (the power can be acoustical or electrical). More commonly, the decibel (dB), one-tenth of a bel, is used. A difference of 10 dB therefore corresponds to a factor-of-10 difference in sound intensity or sound power. Table 17.8 is a list of the relationships between decibels and several sound-power and sound-pressure ratios. The footnote describes the means for interpolation between any two adjacent values. In order to relate to the real world, Table 17.9 lists typical sound-pressure levels of some common sounds.

17.4.2 Acoustic Characteristics of Construction Materials

As an aid in the selection of the appropriate acoustic materials for either reduction in background noise or for live reverberation effects, Table 17.10 provides a tabulation of the sound-absorption characteristics of commonly used building and data to Table 17.10 for sound isolation, and an industry-developed rating figure (STC) that takes into account the variations across the range of audible frequencies.

17.4.3 Acoustic Design Requirements

The recommended sound isolation figures for a variety of workplaces and sound-studio facilities are listed in Table 17.12. Reverberation recommendations are provided in Fig. 17.3.

TABLE 17.8 Various Power and Amplitude Ratios and Their Decibel Equivalents†

Sound or electrical power ratio	Decibels	Sound pressure, voltage, or current ratio	Decibels
1	0	1	0
2	3.0	2	6.0
3	4.8	3	9.5
4	6.0	4	12.0
5	7.0	5	14.0
6	7.8	6	15.6
7	8.5	7	16.9
8	9.0	8	18.1
9	9.5	9	19.1
10	10.0	10	20.0
100	20.0	100	40.0
1,000	30.0	1,000	60.0
10,000	40.0	10,000	80.0
100,000	50.0	100,000	100.0
1,000,000	60.0	1,000,000	120.0

†Other values can be estimated by using this table and the following rules:

Power ratios that are multiples of 10 are converted into their decibel equivalents by multiplying the appropriate exponent by 10. For example, a power ratio of 1000 is 10^3, and this translates into 3 × 10 = 30 dB. Since power is proportional to the square of amplitude, the exponent of 10 must be doubled to arrive at the decibel equivalent of an amplitude ratio.

Intermediate values can be estimated by combining values in this table by means of the rule that the multiplication of power or amplitude ratios is equivalent to adding level differences in decibels. For example, increasing a sound level by 27 dB requires increasing the power by a ratio of 500 (20 dB is a ratio of 100, and 7 dB is a ratio of 5; the product of the ratios is 500). The corresponding increase in sound pressure or electrical signal amplitude is a factor of just over 20 (20 dB is a ratio of 10, and 7 dB falls between 6.0 and 9.5 and is therefore a ratio of something in excess of 2); the calculated value is 22.4). Reversing the process, if the output from a power amplifier is increased from 40 to 800 W, a ratio of 20, the sound pressure level would be expected to increase by 13 dB (a power ratio of 10 is 10 dB, a ratio of 2 is 3 dB, and the sum is 13 dB). The corresponding voltage increase measured at the output of the amplifier would be a factor of between 4 and 5 (by calculation, 4.5)..

Source: K. Blair Benson, ed., *Audio Engineering Handbook*, McGraw-Hill, New York, 1988.

17.4.4   Magnetic Audio Tape-Recording Practices

Tables 17.13 and 17.14 list industry-recommended usage of multichannel magnetic-tape recording tracks.

Table 17.15 lists the general operational parameters and applications of the SMPTE magnetic-tape editing time code as used throughout the world.

17.4.5   Symbols and Measurements

Tables 17.16 and 17.17 provide references on symbols and units of measurement.

TABLE 17.9 Typical Sound Pressure Levels and Intensities for Various Sound Sources[†]

Sound source	Sound pressure level, dB	Intensity, W/m^2	Listener reaction
Jet engine at 10 m	160	10^3 }	Immediate damage
	150		
	140	}	Painful feeling
	130		
SST takeoff at 500 m	120	1	Discomfort
Amplified rock music	110		
Chain saw at 1 m	100		fff
Power mower at 1.5 m	90	10^{-3}	
75-piece orchestra at 7 m	80		f
City traffic at 15 m	70		
Normal speech at 1 m	60	10^{-6}	p
Suburban residence	50		
Library	40		ppp
Empty auditorium	30	10^{-9}	
Recording studio	20		
Breathing	10	10^{-12}	Inaudible
	0‡		

†The relationships illustrated in this table are necessarily approximate because the conditions of measurement are not defined. Typical levels should, however, be within about 10 dB of the stated values.

‡0-dB sound pressure level (SPL) represents a reference sound pressure of 0.0002 µbar, or 0.00002 N/m^2.

Source: K. Blair Benson, ed., *Audio Engineering Handbook*, McGraw-Hill, New York, 1988.

TABLE 17.10 Typical Absorbtion Coefficients

Material	Sound absorption coefficient						NRC number
	125 Hz	250 Hz	500 Hz	1000 Hz	2000 Hz	4000 Hz	
Concrete masonry units, painted	0.08	0.05	0.05	0.07	0.08	0.08	0.06
Gypsum wallboard, ½ in thick, studs spaced 24 in on center	0.27	0.10	0.05	0.04	0.07	0.08	0.07
Typical window glass	0.30	0.22	0.17	0.13	0.07	0.03	0.15
Plaster on lath	0.15	0.10	0.06	0.05	0.05	0.03	0.07
Light fabric, flat against concrete wall	0.08	0.06	0.10	0.16	0.25	0.32	0.14
Thick drapery, draped to half area	0.15	0.36	0.55	0.70	0.73	0.75	0.59
Linoleum on concrete	0.02	0.03	0.03	0.03	0.03	0.03	0.03
Typical wood floor	0.15	0.12	0.10	0.06	0.06	0.06	0.09
Thin carpet on concrete	0.03	0.06	0.10	0.20	0.43	0.63	0.20
Thick carpet with underpadding	0.08	0.28	0.38	0.40	0.48	0.70	0.39
Typical ½-in-thick mineral-fiber acoustic ceiling tile	0.45	0.50	0.53	0.69	0.85	0.93	0.64
Typical ¾-in-thick glass-fiber acoustic ceiling tile	0.44	0.65	0.90	0.92	0.94	0.97	0.85

Source: K. Blair Benson, ed., *Audio Engineering Handbook*, McGraw-Hill, New York, 1988.

TABLE 17.11 Typical Sound Transmission Class (STC) Rating and Transmission Loss, in Decibels, for Various Building Materials

Material	Transmission loss, dB						STC rating
	125 Hz	250 Hz	500 Hz	1000 Hz	2000 Hz	4000 Hz	
Gypsum wallboard, ½ in thick	14	20	24	30	30	27	27
Two layers in gypsum wallboard, both ½ in thick	19	26	30	30	29	36	31
Flat concrete panel, medium weight, 6 in thick	37	43	51	59	67	73	55
One layer of ½-in-thick gypsum wallboard on each side of 2- by 4-in wood studs (16 in off center) with 2-in-thick glass-fiber batt in the cavity	20	28	33	43	43	40	38
Same as above, but with two layers of ½-in-thick gypsum wallboard on each side	24	37	44	49	50	50	46
Same as above, but with staggered studs	34	43	49	54	54	52	51
Same as above, but with double row of 2- by 4-in studs spaced 1 in apart on separate plates, using type X (fire-rated) gypsum wallboard, and two layers of 3-in glass-fiber batt in the cavity	45	54	63	66	66	64	63
Same as above, but with bracing across cavity at third points of studs	40	45	56	62	57	60	57
4-in face brick, mortared together	31	33	39	47	55	61	45
Two layers of mortared 4-in face brick separated by 2-in air space, with metal ties	36	36	46	54	61	66	50
Same as above, but air space filled with concrete grout	41	47	56	62	66	70	59
6-in-thick three-cell dense concrete masonry units, mortared together	36	38	42	49	53	60	48
2-in-thick hollow-core door, ungasketed	13	19	23	18	17	21	19
2-in-thick solid wood door with airtight gasketing and drop seal	29	31	31	31	39	43	35
Typical window glass, ⅛ in thick, single plate	15	23	26	30	32	30	29
Typical thermal glazing window (³⁄₁₆-in glass, ⅛-in air space, ³⁄₁₆-in glass)	22	21	29	34	30	32	30
½-in-thick laminated glass	34	35	36	37	40	51	39
Composite window (½-in laminated glass, 5-in air space, ¼-in glass)	33	54	60	57	55	63	55
Typical ½-in-thick mineral-fiber acoustic ceiling tile	6	10	12	16	21	21	17

Source: K. Blair Benson, ed., *Audio Engineering Handbook*, McGraw-Hill, New York, 1988.

TABLE 17.12 Survey of Sound Insulation between Studios, Control Rooms, and Dwellings, in Decibels

	Studio	Control-room sound	Control-room TV	Announcer's booth	Camera check	Editing room	Corridor	Office	Plant, workshop	Equipment room	Projection room	Disk jockey	Weather studio	Telephone studio	Concert hall
Studio	72/72	68/64	62/58	62/58	57/52	68/64	68/64	68/64	72/72	68/64	68/64	52/45	68/64	57/52	72/72
Control-room sound	68/64	68/64	62/58	62/58	57/45	68/64	52/45	68/64	68/64	62/58	62/58	52/45	62/58	57/52	68/64
Control-room TV	62/62	57/57	57/45	62/58	57/45	57/45	52/45	68/64	62/58	52/45	58/52	52/45	52/45	52/45	68/64
Announcer's booth	62/58	62/58	62/58	68/64	62/58	62/58	68/64	68/64	72/72	68/64	68/64	52/45	68/64	62/57	68/64
Camera check	57/52	57/45	57/45	62/58	68/64	62/58	52/45	45/37	62/58	58/52	52/45	52/45	52/45	57/52	62/58
Editing room	68/64	68/64	57/45	62/58	62/58	68/64	52/45	45/37	62/58	58/52	52/45	52/45	52/45	57/52	68/58
Corridor	68/64	52/45	52/45	68/64	52/45	52/45	: / :	45/37	45/37	45/37	45/37	52/45	52/45	52/45	68/64
Office	68/64	68/64	68/64	68/64	45/37	45/37	45/37	: / :	62/62	52/45	52/45	52/45	52/45	52/45	68/64
Plant, workshop	72/72	68/64	62/58	72/72	62/58	62/58	45/37	62/62	: / :	: / :	: / :	62/62	62/62	57/57	72/72
Equipment room	68/64	62/58	52/45	68/64	58/52	58/52	45/37	52/45	62/58	58/52	52/45	52/45	52/45	52/45	68/64
Projection room	68/64	62/58	58/52	68/64	52/45	52/45	45/37	52/45	62/58	52/45	: / :	52/45	52/45	52/45	68/64
Disk jockey	52/45	52/45	52/45	52/45	52/45	52/45	52/45	52/45	62/62	52/45	52/45	: / :	62/58	58/57	62/58
Weather studio	68/64	62/58	52/45	68/64	52/45	52/45	52/45	52/45	62/62	52/45	52/45	62/58	: / :	45/52	62/58
Telephone studio	57/52	57/52	52/45	62/57	57/52	57/52	52/45	52/45	57/57	52/45	52/45	58/57	45/52	: / :	62/58
Concert hall	72/72	68/64	68/64	68/64	62/58	62/58	68/64	68/68	72/72	68/64	68/64	62/58	62/58	62/58	72/72

Source: K. Blair Benson, ed., Audio Engineering Handbook, McGraw-Hill, New York, 1988.

17.22

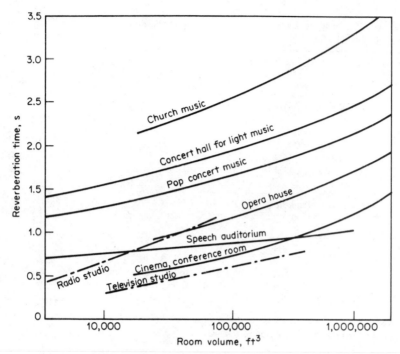

FIGURE 17.3 Typical reververation times for different sizes of rooms and auditoriums according to usage. *(Source: K. Blair Benson, ed.,* Audio Engineering Handbook, *McGraw-Hill, New York, 1988)*

TABLE 17.13 Magnetic-Film Track Dimensions, Numbers, and Locations for Two, Three, Four, and Six Tracks

A. Dimensions for Two Magnetic Sound Records

Dimensions	Inches	Millimeters
A	0.200 + 0.004 − 0	5.0 + 0.1 − 0
A_1	0.150 + 0.004 − 0	3.8 + 0.1 − 0
B	0.339 ± 0.002	5.6 ± 0.06
C	0.725 ± 0.002	18.4 ± 0.06
H ref.	1.337	34.97

B. Dimensions for Three Magnetic Sound Records

Dimensions	Inches	Millimeters
A	0.200 + 0.004 − 0	5.0 + 00.1 − 0
B	0.119 ± 0.002	8.6 ± 0.05
C	0.350 ± 0.002	8.9 ± 0.05
D	0.700 ± 0.002	17.8 ± 0.05
H ref.	1.377	34.97

C. Dimensions for Four Magnetic Sound Records

Dimensions	Inches	Millimeters
A	0.150 + 0.004 − 0	3.8 + 0.1 − 0
B	0.314 ± 0.002	7.9 ± 0.05
C	0.250 ± 0.002	6.4 ± 0.05
D	0.500 ± 0.002	12.8 ± 0.05
E	0.750 ± 0.002	19.2 ± 0.05
H ref.	1.377	34.87

D. Dimensions for Six Magnetic Sound Records

Dimensions	Inches	Millimeters
A	0.100 ± 0.002	2.40 ± 0.10
B	0.289 ± 0.002	7.34 ± 0.05
C	0.150 ± 0.002	4.06 ± 0.05
D	0.320 ± 0.002	8.12 ± 0.05
E	0.480 ± 0.002	12.18 ± 0.05
F	0.540 ± 0.002	15.24 ± 0.05
G	0.500 ± 0.002	20.30 ± 0.05
H ref.	1.377	34.97

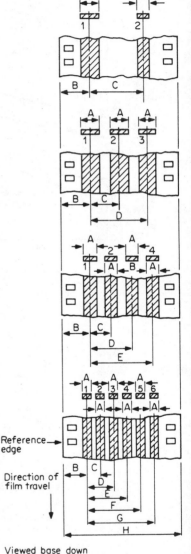

Reference edge

Direction of film travel

Viewed base down
(magnetic surface facing upward)

Note: The metric values listed in the tables are not exact conversions and deviate from accepted conversion practices. They are based upon the practice of those countries using the metric system. Head assemblies made to either system of dimensions will, for all practical purposes, be interchangeable.

Source: K. Blair Benson, ed., *Audio Engineering Handbook*, McGraw-Hill, New York, 1988.

TABLE 17.14 SMPTE Recommended Practice for Audio Channel Assignments of Multichannel Submasters Used in Preparation for Two-Track Masters for Transfer to Video

---

SMPTE RECOMMENDED PRACTICE   RP147

Audio Channel Assignments of Multi-Channel Sub-Masters used in Preparation for Two-Track Masters for Transfer to Video

1. Scope

This Recommended Practice specifies the audio channel assignments of multi-channel sub-master recordings for transfer to two-channel stereo audio tracks on video tape and video cassettes.

2. Format

2.1 Non-sprocketed tape. The recording may be 16 or 24 channel on 1 or 2″ audio tape.

2.2 35-mm magnetic film. The recording format may be on 3-track, 4-track, two 4-tracks interlocked, or 6-track.

3. Channel assignments

3.1 Using non-sprocketed tape or two 35-mm magnetic films interlocked, in the 4 track format (producing 8 channels):

Track	Program
1	Left Dialog
2	Right Dialog
3	Left Music
4	Right Music
5	Left Effects
6	Right Effects
7	Left Laugh or Sweetener
8	Right Laugh or Sweetener

3.2 Using 35-mm magnetic recording film, in the 3-track format, or the single 4-track format:

Track	Program
1	Left Music & Effects
2	Center Dialog
3	Right Music & Effects

3.3 Using 35-mm magnetic film, in the 6-track format:

Track	Program
1	Left Dialog
2	Right Dialog
3	Left Music
4	Right Music
5	Left Effects
6	Right Effects

3.4 For non-sprocketed formats the second to last track should be reserved for sync pulse and the last track should be reserved for SMPTE time code, using line-referenced non-drop frame. (However, if the video system is locked to a crystal or drop-frame code, then the same time base must be used for the audio reference.) The hour digits may be assigned to reel numbers and the minutes, seconds, and frames should be set to zero on the START frame of the SMPTE frame leader. (See SMPTE/RP136 Time and control codes for 24, 25 or 30 frame-per-second Motion Picture systems.)

Source K. Blair Benson, ed., *Audio Engineering Handbook*, McGraw-Hill, New York, 1988.

TABLE 17.15 SMPTE Time and Control Code

Frame rate, frames per second	Special frame sequence	Countries of widest application	Prime use
24	No	All, for "pure" film production	Most film production and specialized video generally for release on film
25	No	Those using PAL and SECAM television systems, including Europe	Both film and video for television production
29.97	No	Those using NTSC television systems, including United States	Video production with continuously ascending code numbers but without correction for consequent 3.6 s/h error from real time
29.97 (Drop)	Yes		Video production corrected to real (clock) time; frames "dropped" to accommodate
30	No	All	Most used in audio postproduction of film

Source: K. Blair Benson, ed., *Audio Engineering Handbook*, McGraw-Hill, New York, 1988.

TABLE 17.16 Symbols for Units of Measurement

Symbol	Unit	Quantity
A	ampere	Electric current
cd	candela	Luminous intensity
F	farad	Capacitance
H	henry	Inductance
Hz	hertz	Frequency per second
J	joule	Work by application of a force, or thermal energy produced by current in a conductor
K	kelvin	Metric temperature relative to absolute zero ($-273.16°C$)
lm	lumen	Luminance flux
lx	lux	Illuminance (lm/m)
m	meter	Length (1.094 yd)
Ω	ohm	Electric resistance or impedance
rad	radian	Plane angle between two radii of a circle
s	second	Time
T	tesla	Magnetic-induction flux
V	volt	Electric potential
W	watt	Power

TABLE 17.17 Prefixes for Decimal Multiples or Submultiples of SI Units

Multiple	SI prefix	Symbol	Multiple	SI prefix	Symbol
10^{18}	exa	E	10^{-1}	deci	d
10^{15}	peta	P	10^{-2}	centi	c
10^{12}	tera	T	10^{-3}	milli	m
10^{9}	giga	G	10^{-6}	micro	μ
10^{6}	mega	M	10^{-9}	nano	n
10^{3}	kilo	k	10^{-12}	pico	p
10^{2}	hecto	h	10^{-15}	femto	f
10	deka	da	10^{-18}	atto	a

INDEX

AC (*see* Alternating current)
ACC (automatic chroma control), **9.46**
A/D (analog-to-digital converter), **1.16, 1.17**
(*See also* Analog-to-digital conversion)
ADC (analog-to-digital converter), **1.16, 1.17**
(*See also* Analog-to-digital conversion)
AES (*see* Audio Engineering Society)
AESSC (*see* Audio Engineering Society)
AFC (*see* Automatic frequency control)
AFT (automatic fine tuning), **9.33**
Afterglow, **7.33**
(*See also* Cameras, telecine photoconductive)
AGC (*see* Automatic gain control)
Alpha, transistor characteristic, **2.13**
Alternating current:
amplitude measurement of, **16.2, 16.3**
root-mean-square, **16.3, 16.4**
definition of, **1.3**
AM (*see* Amplitude modulation)
American National Standards Institute
(ANSI), **2.2, 17.2, 17.4**
Ampex, video-tape development, **7.2**
Amplifiers, radio RF, **12.14 to 12.16**
gain control of, **12.14 to 12.16**
Amplitude-frequency measurements, **1.14**
(*See also* Test signals, video)
Amplitude linearity (*see* Linearity)
Amplitude modulation (AM), **1.9, 1.10,
11.1, 11.2**
definition of, FCC, **17.6**
(*See also* Radio)
Analog, definition of, **1.11**
Analog-to-digital conversion, **1.16, 1.17,
5.12, 5.13**
Analog-signal transmission, **1.11, 1.12**
AND, digital logic function, **5.2**
ANSI (*see* American National Standards
Institute)
Antennas:
AM-radio reception, **12.9, 12.10**
automobiles, **12.9**
coupling network, **12.9, 12.11**
AM-radio transmission, **11.17, 11.18**
audio-response, reception, **11.22, 11.23**

Antennas, AM-radio transmission (*Cont.*):
directional, **11.18, 11.19**
reference tower, directional system,
11.19
tuning unit (ATU), **11.18**
FM-radio reception, **12.9, 12.10**
coupling network, **12.9**
filter technology, **12.11**
FM-radio transmission, **11.19**
audio-response, received-signal, **11.22**
beam tilt, **11.20**
gain, **11.20**
multipath, **11.22**
panel configuration, wide-band, **11.21**
polarization, circular (CP), **11.20, 11.22**
television reception, **9.2, 9.3**
arrays, multielement, **9.7, 9.8**
bandwidth of, **9.4**
dipole, **9.5**
figure of merit, **9.3**
gain of, **9.3, 9.4**
input impedance of, **9.4**
monopole, quarter-wave, **9.7**
radiation resistance, **9.4**
television transmission, **8.6**
butterfly, VHF, **8.25**
candelabra, **8.22**
circular polarized, **8.24, 8.25**
helix, **8.24**
pylon, **8.24**
resonant, **8.23**
satellite, **8.29**
side-mounted, UHF, **8.25**
slot, **8.24**
tower-top, **8.21 to 8.23**
traveling-wave, **8.23**
turnstile, VHF, **8.23**
two-channel, **8.22, 8.23**
zigzag, **8.21, 8.24**
Antimony trisulphide, photoconductor
target, **3.18, 3.20**
spectral response of, **3.20**
APC (*see* Automatic phase control)
Aperture correction, television camera, **6.8**
horizontal, **6.8**

1

ABOUT THE AUTHORS

K. Blair Benson is one of the industry's foremost television and audio technology consultants. Over a career spanning nearly 50 years with companies including General Electric, Goldmark Communications, CBS, and Video Corporation of America, Mr. Benson has been responsible for many of the technical systems and innovations now used in modern television. He received an Emmy for CBS's participation with Ampex in the development and introduction of the first broadcast video-tape recorder. As chairman of an all-industry committee he initiated techniques to improve the uniformity and fidelity of color television transmission, including the well-known VIR signal. Mr. Benson is the editor of McGraw-Hill's *Television Engineering Handbook* and *Audio Engineering Handbook*.

Jerry Whitaker is Editorial Director of *Broadcast Engineering* and *Video Systems* magazines. He is a Fellow of the Society of Broadcast Engineers, as well as a member of the SMPTE, AES, ITA, and IEEE. He has written and lectured extensively on the topic of RF transmission systems and is a former chief engineer at two radio stations.